Lorenzo Magnani and Ping Li (Eds.)

Model-Based Reasoning in Science, Technology, and Medicine

Studies in Computational Intelligence, Volume 64

Editor-in-chief
Prof. Janusz Kacprzyk
Systems Research Institute
Polish Academy of Sciences
ul. Newelska 6
01-447 Warsaw
Poland
E-mail: kacprzyk@ibspan.waw.pl

Further volumes of this series
can be found on our homepage:
springer.com

Vol. 42. Bernd J. Krämer, Wolfgang A. Halang (Eds.)
Contributions to Ubiquitous Computing, 2007
ISBN 978-3-540-44909-6

Vol. 43. Fabrice Guillet, Howard J. Hamilton (Eds.)
Quality Measures in Data Mining, 2007
ISBN 978-3-540-44911-9

Vol. 44. Nadia Nedjah, Luiza de Macedo
Mourelle, Mario Neto Borges,
Nival Nunes de Almeida (Eds.)
Intelligent Educational Machines, 2007
ISBN 978-3-540-44920-1

Vol. 45. Vladimir G. Ivancevic, Tijana T. Ivancevic
Neuro-Fuzzy Associative Machinery for Comprehensive Brain and Cognition Modeling, 2007
ISBN 978-3-540-47463-0

Vol. 46. Valentina Zharkova, Lakhmi C. Jain
Artificial Intelligence in Recognition and Classification of Astrophysical and Medical Images, 2007
ISBN 978-3-540-47511-8

Vol. 47. S. Sumathi, S. Esakkirajan
Fundamentals of Relational Database Management Systems, 2007
ISBN 978-3-540-48397-7

Vol. 48. H. Yoshida (Ed.)
Advanced Computational Intelligence Paradigms in Healthcare, 2007
ISBN 978-3-540-47523-1

Vol. 49. Keshav P. Dahal, Kay Chen Tan, Peter I. Cowling (Eds.)
Evolutionary Scheduling, 2007
ISBN 978-3-540-48582-7

Vol. 50. Nadia Nedjah, Leandro dos Santos Coelho,
Luiza de Macedo Mourelle (Eds.)
Mobile Robots: The Evolutionary Approach, 2007
ISBN 978-3-540-49719-6

Vol. 51. Shengxiang Yang, Yew Soon Ong, Yaochu Jin
Honda (Eds.)
Evolutionary Computation in Dynamic and Uncertain Environment, 2007
ISBN 978-3-540-49772-1

Vol. 52. Abraham Kandel, Horst Bunke, Mark Last (Eds.)
Applied Graph Theory in Computer Vision and Pattern Recognition, 2007
ISBN 978-3-540-68019-2

Vol. 53. Huajin Tang, Kay Chen Tan, Zhang Yi
Neural Networks: Computational Models and Applications, 2007
ISBN 978-3-540-69225-6

Vol. 54. Fernando G. Lobo, Cláudio F. Lima
and Zbigniew Michalewicz (Eds.)
Parameter Setting in Evolutionary Algorithms, 2007
ISBN 978-3-540-69431-1

Vol. 55. Xianyi Zeng, Yi Li, Da Ruan and Ludovic Koehl (Eds.)
Computational Textile, 2007
ISBN 978-3-540-70656-4

Vol. 56. Akira Namatame, Satoshi Kurihara and
Hideyuki Nakashima (Eds.)
Emergent Intelligence of Networked Agents, 2007
ISBN 978-3-540-71073-8

Vol. 57. Nadia Nedjah, Ajith Abraham and Luiza de
Macedo Mourella (Eds.)
Computational Intelligence in Information Assurance and Security, 2007
ISBN 978-3-540-71077-6

Vol. 58. Jeng-Shyang Pan, Hsiang-Cheh Huang, Lakhmi
C. Jain and Wai-Chi Fang (Eds.)
Intelligent Multimedia Data Hiding, 2007
ISBN 978-3-540-71168-1

Vol. 59. Andrzej P. Wierzbicki and Yoshiteru
Nakamori (Eds.)
Creative Environments, 2007
ISBN 978-3-540-71466-8

Vol. 60. Vladimir G. Ivancevic and Tijana T. Ivacevic
Computational Mind: A Complex Dynamics Perspective, 2007
ISBN 978-3-540-71465-1

Vol. 61. Jacques Teller, John R. Lee and Catherine
Roussey (Eds.)
Ontologies for Urban Development, 2007
ISBN 978-3-540-71975-5

Vol. 62. Lakshmi C. Jain, Raymond A. Tedman
and Debra K. Tedman (Eds.)
Evolution of Teaching and Learning Paradigms in Intelligent Environment, 2007
ISBN 978-3-540-71973-1

Vol. 63. Wlodzislaw Duch and Jacek Mańdziuk (Eds.)
Challenges for Computational Intelligence, 2007
ISBN 978-3-540-71983-0

Vol. 64. Lorenzo Magnani and Ping Li (Eds.)
Model-Based Reasoning in Science, Technology, and Medicine, 2007
ISBN 978-3-540-71985-4

Lorenzo Magnani
Ping Li
(Eds.)

Model-Based Reasoning in Science, Technology, and Medicine

With 77 Figures, 10 in Color and 7 Tables

Springer

Lorenzo Magnani
Director, Computational Philosophy
Laboratory
Department of Philosophy
University of Pavia
Pizza Botta 6
27100 Pavia
Italy
E-mail: lmagnani@unipv.it

Visiting Professor
Department of Philosophy
Sun Yat-sen University
Guangzhou 510275
P.R. China

Ping Li
Department of Philosophy
Sun Yat-sen University
Guangzhou 510275
P.R. China
E-mail: hsslip@sysu.edu.cn

Library of Congress Control Number: 2007925684

ISSN print edition: 1860-949X
ISSN electronic edition: 1860-9503
ISBN 978-3-540-71985-4 Springer Berlin Heidelberg New York

This work is subject to copyright. All rights are reserved, whether the whole or part of the material is concerned, specifically the rights of translation, reprinting, reuse of illustrations, recitation, broadcasting, reproduction on microfilm or in any other way, and storage in data banks. Duplication of this publication or parts thereof is permitted only under the provisions of the German Copyright Law of September 9, 1965, in its current version, and permission for use must always be obtained from Springer-Verlag. Violations are liable to prosecution under the German Copyright Law.

Springer is a part of Springer Science+Business Media
springer.com
© Springer-Verlag Berlin Heidelberg 2007

The use of general descriptive names, registered names, trademarks, etc. in this publication does not imply, even in the absence of a specific statement, that such names are exempt from the relevant protective laws and regulations and therefore free for general use.

Cover design: deblik, Berlin
Typesetting by the editors using a Springer LaTeX macro package
Printed on acid-free paper   SPIN: 11904007   89/SPi   5 4 3 2 1 0

MBR CHINA 2006
Model-Based Reasoning in Science and Medicine

Guangzhou (Canton), China, July 3-5, 2006
Chairs: Ping Li and Lorenzo Magnani

in collaboration with
Department of Philosophy, Sun Yat-Sen University (China)
&
Department of Philosophy, University of Pavia (Italy)

When a chicken first emerges from the shell, it does not try fifty random ways of appeasing its hunger, but within five minutes is picking up food, choosing as it picks, and picking what it aims to pick. That is not reasoning, because it is not done deliberately; but in every respect but that, it is just like abductive inference.

Charles Sanders Peirce

Preface

The volume is based on the papers that were presented at the international conference *Model-Based Reasoning in Science and Medicine* (MBR'06 China), held at Sun Yat-sen University, Guangzhou, P.R. China in July 2006. The previous volume *Model-Based Reasoning in Scientific Discovery*, edited by L. Magnani, N.J. Nersessian, and P. Thagard (Kluwer Academic/Plenum Publishers, New York, 1999; Chinese edition, China Science and Technology Press, Beijing, 2000), was based on the papers presented at the first "model-based reasoning" international conference, held at the University of Pavia, Pavia, Italy in December 1998. Other two volumes were based on the papers presented at the second "model-based reasoning" international conference, held at the same place in May 2001: *Model-Based Reasoning. Scientific Discovery, Technological Innovation, Values*, edited by L. Magnani and N.J. Nersessian (Kluwer Academic/Plenum Publishers, New York, 2002) and *Logical and Computational Aspects of Model-Based Reasoning*, edited by L. Magnani, N.J. Nersessian, and C. Pizzi (Kluwer Academic, Dordrecht, 2002). Another volume *Model-Based Reasoning in Science and Engineering*, edited by L. Magnani (College Publications, London, 2006), was based on the papers presented at the third "model-based reasoning" international conference, held at the same place in December 2004.

The presentations given at the conference explored how scientific thinking uses models and explanatory reasoning to produce creative changes in theories and concepts. Some addressed the problem of model-based reasoning in technology, and stressed the issue of technological innovation and medical reasoning. Moreover, there still are some presentations in Chinese given at the conference that deal with problem-solving in science and ordinary reasoning, which were published in the volume *Philosophical Investigations from a Perspective of Cognition*, edited by L. Magnani and P. Li (Guangdong People's Publishing House, Guangzhou, 2006).

The study of diagnostic, visual, spatial, analogical, and temporal reasoning has demonstrated that there are many ways of performing intelligent and creative reasoning that cannot be described with the help only of traditional notions of reasoning such as classical logic. Understanding the contribution of modeling practices to discovery and conceptual change in science requires expanding scientific reasoning to include complex forms of creative reasoning that are not always successful and can lead to incorrect solutions. The study of these heuristic ways of reasoning is situated at the crossroads of philosophy, artificial intelligence, cognitive psychology, and logic; that is, at the heart of cognitive science. There are several key ingredients common to the various forms of model-based reasoning. The term "model" comprises both internal and external representations. The models are intended as interpretations of target physical systems, processes, phenomena, or situations. The models are retrieved or constructed on the basis of potentially satisfying salient constraints of the target domain. Moreover, in the modeling process, various forms of abstraction are used. Evaluation and adaptation take place in light of structural, causal, and/or functional constraints. Model simulation can be used to produce new states and enable evaluation of behaviors and other factors.

The various contributions of the book are written by interdisciplinary researchers who are active in the area of creative reasoning in science and technology: the most recent results and achievements about the topics above are illustrated in detail in the papers.

The editors express their appreciation to the members of the Scientific Committee for their suggestions and assistance: – Thomas Addis, Department of Computer Science and Software Engineering, University of Portsmouth, UK – Atocha Aliseda, Instituto de Investigaciones Filosoficas Universidad, Nacional Autonoma de Mexico (UNAM), Mexico City, MEXICO – Diderik Batens, Center for Logic and Philosophy of Science, Universiteit Gent, BELGIUM – David Brown, Institute of Industrial Research University of Portsmouth, Portsmouth, UK – Walter Carnielli, Centre for Logic, Epistemology and the History of Science, State University of Campinas, UNICAMP, Campinas, SP, BRAZIL – Xiang Chen, Department of Philosophy, California Lutheran University, CA, USA – Roberto Cordeschi, Department of Communication Sciences, University of Salerno, Salerno, ITALY – Dov Gabbay, Department of Computer Science, King's College, London, UK – Shenchun Gao, College of Philosophy and Sociology, Jilin University, Changchun, CHINA – Michael E. Gorman, School of Engineering and Applied Science, University of Virginia, Charlottesville, VA, USA – David Gooding, Science Studies Centre, Department of Psychology, University of Bath, Bath, UK – Sundari Krishnamurthy, Stella Maris College (Autonomous), University of Madras, INDIA – Michael Leyton, DIMACS, Busch Campus, Rutgers University, New Brunswick, NJ, USA – Ping Li, Department of Philosophy, Sun Yat-sen University, Guangzhou, CHINA – Xingmin Li, Graduate School, Chinese Academy of Sciences, Beijing, CHINA – Xiaoli Liu, Department of Philosophy, Beijing Normal University, Beijing, CHINA – Honghai Liu, Department of Computing Science, University of Aberdeen, Aberdeen, UK – Shangmin Luan, Institute of Software, Chinese Academy of Sciences,

Beijing, CHINA – Lorenzo Magnani, Department of Philosophy, University of Pavia, Pavia, ITALY – Joke Meheus, Center for Logic and Philosophy of Science, Universiteit Gent, BELGIUM – Claudio Pizzi, Department of Philosophy and Social Sciences, University of Siena, Siena, ITALY – Colin Schmidt, Le Mans University, Laval, FRANCE – Paul Thagard, Department of Philosophy, University of Waterloo, CANADA – Barbara Tversky, Stanford University, Stanford, CA, USA – Riccardo Viale, Lascomes (Laboratory of Cognitive, Methodological and Socio-Economic Sciences), Fondazione Rosselli, Torino, ITALY – John Woods, Department of Philosophy, University of British Columbia, Vancouver, CANADA, and Department of Computer Science, King's College, London, UK – Zhilin Zhang, Department of Philosophy, Sun Yat-sen University, Guangzhou, CHINA – Changle Zhou, Department of Computer Science, Xiamen University, Xiamen, CHINA.

Special thanks to R. Dossena and E. Bardone for their contribution in the preparation of this volume. The conference MBR06_China, and thus indirectly this book, was made possible through the generous financial support of the MIUR (Italian Ministry of the University), University of Pavia, Fondazione CARIPLO, and Sun Yat-sen University. Their support is gratefully acknowledged. The preparation of the volume would not have been possible without the contribution of resources and facilities of the Computational Philosophy Laboratory and of the Department of Philosophy, University of Pavia.

Several papers concerning model-based reasoning deriving from the previous conferences MBR98 and MBR01 can be found in Special Issues of Journals: in *Philosophica*: Abduction and Scientific Discovery, 61(1), 1998, and Analogy and Mental Modeling in Scientific Discovery, 61(2) 1998; in *Foundations of Science*: Model-Based Reasoning in Science: Learning and Discovery, 5(2) 2000, all edited by L. Magnani, N.J. Nersessian, and P. Thagard; in *Foundations of Science*: Abductive Reasoning in Science, 9, 2004, and Model-Based Reasoning: Visual, Analogical, Simulative, 10, 2005; in *Mind and Society*: Scientific Discovery: Model-Based Reasoning, 5(3), 2002, and Commonsense and Scientific Reasoning, 4(2), 2001, all edited by L. Magnani and N.J. Nersessian. Finally, other related philosophical, epistemological, and cognitive oriented papers deriving from the presentations given at the conference MBR04 have been published in a Special Issue of *Logic Journal of the IGPS*: Abduction, Practical Reasoning, and Creative Inferences in Science 14(1) (2006) and will be published in two Special Issues of *Foundations of Science*: Tracking Irrational Sets: Science, Technology, Ethics, and Model-Based Reasoning in Science and Engineering, all edited by L. Magnani.

Lorenzo Magnani
University of Pavia, Pavia, Italy and Sun Yat-sen University, Guangzhou, P.R. China
Ping Li
Sun Yat-sen University, Guangzhou, P.R. China

Pavia, Italy, February 2007

Contents

Part I Abduction, Problem Solving, and Practical Reasoning

Animal Abduction
From Mindless Organisms to Artifactual Mediators
Lorenzo Magnani 3

Communicative Gestures Facilitate Problem Solving for Both Communicators and Recipients
Sandra C. Lozano, Barbara Tversky 39

The Concept of Fallacy is Empty
A Resource-Bound Approach to Error
John Woods 69

Abductive Reasoning, Information, and Mechanical Systems
Maria Eunice Quilici Gonzalez, Mariana Claudia Broens and Fabricio Loffredo D'Ottaviano 91

Automated Abduction in Scientific Discovery
Oliver Ray 103

Abduction, Medical Semeiotics and Semioethics
Individual and Social Symptomatology from a Semiotic Perspective
Susan Petrilli 117

Abduction and Modeling in Biosemiotics and Sociosemiotics
Augusto Ponzio 131

Reason out Emergence from Cellular Automata Modeling
Leilei Qi, Huaxia Zhang 147

Belief Ascription and *De Re* Communication
Yuan Ren 161

Multiagent-Based Simulation in Biology
A Critical Analysis
Francesco Amigoni, Viola Schiaffonati 179

Mathematics through Diagrams: Microscopes in Non-Standard and Smooth Analysis
Riccardo Dossena, Lorenzo Magnani 193

Part II Models, Mental Models, Representations, and Medical Reasoning

---

Cognition, Environment and the Collapse of Civilizations
Michael E. Gorman 217

Cognitive Aspects of Tacit Knowledge and Cultural Diversity
Riccardo Viale, Andrea Pozzali 229

The Functional-Analogical Explanation in Chinese Science and Technology
A Case Study of the Theory of Yin-Yang and Five Elements
Huaxia Zhang, Zhilin Zhang 245

Model-Based Reasoning and Diagnosis in Traditional Chinese Medicine (TCM)
Zhikang Wang 261

Model-Based Reasoning in Cognitive Science
Yi-dong Wei 273

An Examination of Model-Based Reasoning in Science and Medicine in India
Sundari Krishnamurthy 293

Ontology, Artefacts, and Models of Reasoning
Pasi Pohjola 315

The Wondering Angels of the Fractal Art
Viorel Guliciuc 333

Part III Logical and Computational Aspects of Model-Based Reasoning

---

Polynomizing: Logic Inference in Polynomial Format and the Legacy of Boole
Walter Carnielli 349

Abductive Inference and Iterated Conditionals
Claudio Pizzi 365

Peircean Pragmatic Truth and da Costa's Quasi-Truth
Itala M. Loffredo D'Ottaviano, Carlos Hifume 383

**Sliding Mode Motion Control Strategies
for Rigid Robot Manipulators**
Antonella Ferrara, Lorenza Magnani 399

Model-Based Chemical Compound Formulation
Stefania Bandini, Alessandro Mosca and Matteo Palmonari 413

**Model-Based Reasoning for Self-Repair
of Autonomous Mobile Robots**
*Michael Hofbaur, Johannes Köb, Gerald Steinbauer
and Franz Wotawa* 431

**Application of Bayesian Inference
to Automatic Semantic Annotation of Videos**
Fangshi Wang, De Xu, Hongli Xu, Wei Lu and Weixin Wu 447

An Algebraic Approach to Model-Based Diagnosis
Shangmin Luan, Lorenzo Magnani and Guozhong Dai 467

**CYBERNARD: A Computational Reconstruction
of Claude Bernard's Scientific Discoveries**
Jean-Gabriel Ganascia, Claude Debru 497

**Do Computational Models of Reading Need
a Bit of Semantics?**
Remo Job, Claudio Mulatti 511

Part I

Abduction, Problem Solving, and Practical Reasoning

Animal Abduction
From Mindless Organisms to Artifactual Mediators

Lorenzo Magnani

Department of Philosophy and Computational Philosophy Laboratory
University of Pavia, Pavia, Italy and Department of Philosophy, Sun Yat-sen
University, Guangzhou, P.R. China
lmagnani@unipv.it

Summary. Many animals – traditionally considered "mindless" organisms – make up a series of signs and are engaged in making, manifesting or reacting to a series of signs: through this semiotic activity – which is fundamentally *model-based* – they are at the same time engaged in "being cognitive agents" and therefore in thinking intelligently. An important effect of this semiotic activity is a continuous process of "hypothesis generation" that can be seen at the level of both instinctual behavior, as a kind of "wired" cognition, and representation-oriented behavior, where nonlinguistic pseudothoughts drive a plastic model-based cognitive role. This activity is at the root of a variety of *abductive* performances, which are also analyzed in the light of the concept of affordance. Another important character of the model-based cognitive activity above is the externalization of artifacts that play the role of mediators in animal languageless reflexive thinking. The interplay between internal and external representation exhibits a new cognitive perspective on the mechanisms underlying the semiotic emergence of abductive processes in important areas of model-based thinking of mindless organisms. To illustrate this process I will take advantage of the case of affect attunement which exhibits an impressive case of model-based communication. A considerable part of abductive cognition occurs through an activity consisting in a kind of reification in the external environment and a subsequent re-projection and reinterpretation through new configurations of neural networks and of their chemical processes. Analysis of the central problems of abduction and hypothesis generation helps to address the problems of other related topics in model-based reasoning, like pseudological and reflexive thinking, the role of pseudoexplanatory guesses in plastic cognition, the role of reification and beliefs, the problem of the relationship between abduction and perception, and of rationality and instincts.

1 Mindless Organisms and Cognition

Philosophy has for a long time disregarded the ways of thinking and knowing of animals, traditionally considered "mindless" organisms. Peircean insight regarding the role of abduction in animals was a good starting point, but only more recent results in the fields of cognitive science and ethology about animals, and of developmental psychology and cognitive archeology about humans and infants, have provided the actual intellectual awareness of the importance of the comparative studies.

Philosophy has anthropocentrically condemned itself to partial results when reflecting upon human cognition because it lacked in appreciation of the more "animal-like" aspects of thinking and feeling, which are certainly in operation and are greatly important in human behavior. Also in ethical inquiry a better understanding of animal cognition could in turn increase knowledge about some hidden aspects of human behavior, which I think still evade any ethical account and awareness.

In the recent [1] I maintain that people have to learn to be "respected as things", sometimes, are. Various kinds of "things", and among them work of arts, institutions, symbols, and of course animals, are now endowed with intrinsic moral worth. Animals are certainly morally respected in many ways in our technological societies, but certain knowledge about them has been disregarded. It is still difficult to acknowledge respect for their cognitive skills and endowments. Would our having more knowledge about animals happen to coincide with having more knowledge about humans and infants, and be linked to the suppression of constitutive "anthropomorphism" in treating and studying them that we have inherited through tradition? Consequently, would not novel and unexpected achievements in this field be a fresh chance to grant new "values" to humans and discover new knowledge regarding their cognitive features? [2] Darwin has already noted that studying cognitive capacities in humans and non-humans animals "[...] possesses, also, some independent interest, as an attempt to see how far the study of the lower animals throws light on one of the highest psychical faculties of man" – the moral sense [3].

Among scientists it is of course Darwin [4] who first clearly captured the idea of an "inner life" (the "world of perception" included) in some humble earthworms [5]. A kind of mental life can be hypothesized in many organisms: Darwin wanted "to learn how far the worms acted consciously and how much mental power they displayed" [4, p. 3]. He found levels of "mind" where it was not presumed to exist. It can be said that this new idea, which bridges the gap between humans and other animals, in some sense furnishes a scientific support to that metaphysical synechism claimed by Peirce contending that matter and mind are intertwined and in some sense indistinguishable[1].

1.1 Worm Intelligence, Abductive Chickens, Instincts

Let us consider the behavior of very simple creatures. Earthworms plug the opening of their burrow with leaves and petioles: Darwin recognized that

---

[1] The recent discovery of the cognitive roles (basically in the case of learning and memory) played by spinal cord further supports this conviction that mind is extended and distributed and that it can also be – so to say – "brainless" [6].

behavior as being too regular to be random and at the same time too variable to be merely instinctive. He concluded that, even if the worms were innately inclined to construct protective basket structures, they also had a capacity to "judge" based on their tactile sense and showed "some degree of intelligence" [4, p. 91]. Instinct alone would not explain how worms actually handle leaves to be put into the burrow. This behavior seemed more similar to their "having acquired the habit" [4, p. 68]. Crist says: "Darwin realized that 'worm intelligence' would be an oxymoron for skeptics and even from a commonsense viewpoint 'This will strike everyone as very improbable' he wrote [4, p. 98]. ...] He noted that little is known about the nervous system of 'lower animals', implying they might possess more cognitive potential than generally assumed" [5, p. 5].

It is important to note that Darwin also paid great attention to those external structures built by worms and engineered for utility, comfort, and security. I will describe later on in this article the cognitive role of artifacts in both human and non-human animals: artifacts can be illustrated as *cognitive mediators* [7] which are the building blocks that bring into existence what it is now called a "cognitive niche"[2]. Darwin maintains that "We thus see that burrows are not mere excavations, but may rather be compared with tunnels lined with cement" [4, p. 112]. Like humans, worms build external artifacts endowed with precise roles and functions, which strongly affect their lives in various ways, and of course their opportunity to "know" the environment.

I have said their behavior cannot be accounted for in merely instinctual terms. Indeed, the "variability" of their behavior is for example illustrated by the precautionary capacity of worms to exploit pine needles by bending over pointed ends: "Had this not effectually been done, the sharp points could have prevented the retreat of the worms into their burrows; and these structures would have resembled traps armed with converging points of wire rendering the ingress of an animal easy and its egress difficult or impossible" [4, p. 112]. Cognitive *plasticity* is clearly demonstrated by the fact that Darwin detected that pine was not a native tree! If we cannot say that worms are aware like we are (consciousness is unlikely even among vertebrates), certainly we can acknowledge in this case a form of material, interactive, and embodied, manifestation of awareness in the world.

Recent research has also demonstrated the existence of developmental plasticity in plants [11]. For example developing tissues and organs "inform" the plant about their states and respond according to the signals and substrates they receive. The plant adjusts structurally and physiologically to its own development and to the habitat it happens to be in (for example a plasticity of organs in the relations between neighboring plants can be developed) [12, 13].

In this article I am interested in improving knowledge on abduction and model-based thinking. By way of introduction let me quote the interesting

---

[2] A concept introduced by Tooby and DeVore [8] and later on reused by Pinker [9, 10].

Peircean passage about hypothesis selection and chickens, which touches on both ideas, showing a kind of completely language-free, model-based abduction:

> How was it that man was ever led to entertain that true theory? You cannot say that it happened by chance, because the possible theories, if not strictly innumerable, at any rate exceed a trillion – or the third power of a million; and therefore the chances are too overwhelmingly against the single true theory in the twenty or thirty thousand years during which man has been a thinking animal, ever having come into any man's head. Besides, you cannot seriously think that every little chicken, that is hatched, has to rummage through all possible theories until it lights upon the good idea of picking up something and eating it. On the contrary, you think the chicken has an innate idea of doing this; that is to say, that it can think of this, but has no faculty of thinking anything else. The chicken you say pecks by instinct. But if you are going to think every poor chicken endowed with an innate tendency toward a positive truth, why should you think that to man alone this gift is denied? [14, 5.591]

and again, even more clearly, in another related passage

> When a chicken first emerges from the shell, it does not try fifty random ways of appeasing its hunger, but within five minutes is picking up food, choosing as it picks, and picking what it aims to pick. That is not reasoning, because it is not done deliberately; but in every respect but that, it is just like abductive inference[3].

From this Peircean perspective hypothesis generation is a largely instinctual and *nonlinguistic* endowment of human beings and, of course, also of animals. It is clear that for Peirce abduction is rooted in the instinct and that many basically instinctual-rooted cognitive performances, like emotions, provide examples of abduction available to both human and non-human animals. Also cognitive archeology [16, 17] acknowledges that it was not language that made cognition possible: rather it rendered possible the integration in social environments of preexistent, separated, domain-specific modules in prelinguistic hominids, like complex motor skills learnt by imitation or created independently for the first time [18]. This integration made the emergence of tool making possible through the process of "disembodiment of mind" that I recently illustrated in [19]. Integration also seeks out established policies, rituals, and complicated forms of social cognition, which are related to the other forms of prevalently nonlinguistic cognitive behaviors.

---

[3] Cf. the article "The proper treatment of hypotheses: a preliminary chapter, toward and examination of Hume's argument against miracles, in its logic and in its history" (1901) [15, p. 692].

1 2 Nonlinguistic Representational States

It can be hypothesized that some language-free, more or less stable, *representational* states that are merely model-based[4] are present in animals, early hominids, and human infants. Of course tropistic and classically conditioned schemes can be accounted for without reference to these kind of model-based "representations", because in these cases the response is invariant once the creature in question has registered the relevant stimuli.

The problem of attributing to those beings strictly nonlinguistic model-based inner "thoughts", beliefs, and desires, and thus suitable ways of representing the world, and of comparing them to language-oriented mixed (both model-based and sentential) representations, typical of modern adult humans, appears to be fundamental to comprehending the status of animal presumptive abductive performances.

Of course this issue recalls the traditional epistemological Kuhnian question of the incommensurability of meaning [21]. In this case it refers to the possibility of comparing cognitive attitudes in different biological species, which express potentially incomparable meanings. Such problems already arose when dealing with the interpretation of primitive culture. If we admit, together with some ethologists, animal behaviorists, and developmental psychologists, that in nonlinguistic organisms there are some intermediate representations, it is still difficult to make an analogy with those found in adult humans. The anthropologists who carried out the first structured research on human primitive cultures and languages already stressed this point, because it is difficult to circumstantiate thoughts that can hold in beings but only manifest themselves in superficial and external conducts (cf. Quine [22]).

A similar puzzling incommensurability already arises when we deal with the different sensorial modalities of certain species and their ways of being and of feeling to be in the world. We cannot put ourselves in the living situation of a dolphin, which lives and feels by using echolocations, or of our cat, which "sees" differently, and it is difficult to put forward scientific hypotheses on these features using human-biased language, perceptive capacities, and cognitive representations. The problem of the existence of "representation states" is deeply epistemological: the analogous situation in science concerns for example the status of the so-called theoretical terms, like quarks or electrons, which are not directly observable but still "real", reliable, and consistent when meaningfully legitimated/justified by their epistemological unavoidability in suitable scientific research programs [23].

I have already said that commitment to research on animal cognition is rare in human beings. Unfortunately, even when interested in animal

---

[4] They do not have to be taken like for example visual and spatial imagery or other internal model-based states typical of modern adult humans, but more like action-related representations and thus intrinsically intertwined with perception and kinesthetic abilities. Saidel [20] interestingly studies the role of these kinds of representations in rats.

cognition, human adult researchers, victims of an uncontrolled, "biocentric" anthropomorphic attitude, always risk attributing to animals (and of course infants) our own concepts and thus misunderstanding their specific cognitive skills [24].

2 Animal Abduction

2.1 "Wired Cognition" and Pseudothoughts

Nature writes programs for cognitive behavior in many ways. In certain cases these programs draw on cognitive functions and sometimes they do not. In the latter case the fact that we describe the behavioral effect as "cognitive" is just a metaphor. This is a case of *instinctual* behavior, which we should more properly name "wired cognition".

Peirce spoke – already over a century ago – of a wide semiotic perspective, which taught us that a human internal representational medium is not necessarily structured like a language. In this article I plan to develop and broaden this perspective. Of course this conviction strongly diverges from that maintained by the intellectual traditions which resort to the insight provided by the modern Fregean logical perspective, in which thoughts are just considered the "senses of sentences". Recent views on cognition are still influenced by this narrow logical perspective, and further stress the importance of an isomorphism between thoughts and language sentences (cf. for example Fodor's theory [25]).

Bermúdez clearly explains how this perspective also affected the so-called *minimalist view* on animal cognition (also called *deflationary view*) [18, p. 27]. We can describe nonlinguistic creatures as thinkers and capable of goal-directed actions, but we need to avoid assigning to them the type of thinking common to linguistic creatures, for example in terms of belief-desire psychology: "Nonlinguistic thinking does not involve propositional attitudes – and, a fortiori, psychological explanation at the nonlinguistic level is not a variant of belief-desire psychology" (*ibid.*). Belief-desire framework should only be related to linguistic creatures. Instead, the problem for the researcher on animal cognition would be to detect how a kind of what we can call "general belief" is formed, rather than concentrating on its content, as we would in the light of human linguistic tools.

Many forms of thinking, such as imagistic, empathetic, trial and error, and analogical reasoning, and cognitive activities performed through complex bodily skills, appear to be basically model-based and manipulative. They are usually described in terms of living beings that adjust themselves to the environment rather than in terms of beings that acquire information from the environment. In this sense these kinds of thinking would produce responses that do not seem to involve sentential aspects but rather merely "non-inferential" ways of cognition. If we adopt the semiotic perspective above, which does not reduce the term "inference" to its sentential level, but which includes the whole arena of sign activity – in the light of Peircean tradition – these kinds of thinking promptly appear full, inferential forms of thought. Let me recall that Peirce stated that all thinking is in signs, and signs can be icons, indices, or symbols, and, moreover, all *inference* is a form of sign activity, where the word sign includes "feeling, image, conception, and other representation" [14, 5.283].

From this perspective human and the most part of non-human animals possess what I have called *semiotic brains* [26], which make up a series of signs and which are engaged in making or manifesting or reacting to a series of signs: through this semiotic activity they are at the same time occasionally engaged in "being cognitive agents" (like in the case of human beings) or at least in thinking intelligently. For example, spatial imaging and analogies based on perceiving similarities – fundamentally context-dependent and circumstantiated – are ways of thinking in which the "sign activity" is of a nonlinguistic sort, and it is founded on various kinds of implicit naïve physical, biological, psychological, social, etc., forms of intelligibility. In scientific experimentation on prelinguistic infants a common result is the detection of completely language-free working ontologies, which only later on, during cognitive development, will become intertwined with the effect of language and other "symbolic" ways of thinking.

With the aim of describing the kinds of representations which would be at work in these nonlinguistic cognitive processes Dummett [27] proposes the term *protothought*. I would prefer to use the term *pseudothought*, to minimize the hierarchical effect that – ethnocentrically – already affected some aspects of the seminal work on primitives of an author like Lévi-Bruhl [28]. An example of the function of model-based pseudothoughts can be hypothesized in the perception of space in the case of both human and non-human animals. The perceived space is not necessarily three-dimensional and merely involves the apprehension of movement changes, and the rough properties of material objects. Dummett illustrates the case of the car driver and of the canoeist:

> A car driver or canoeist may have to estimate the speed and direction of oncoming cars and boats and their probable trajectory, consider what avoiding action to take, and so on: it is natural to say that he is highly concentrated in thought. But the vehicle of such thoughts is certainly not language: it would be said, I think, to consist in visual imagination superimposed on the visual perceived scene. It is not just that these thoughts are not in fact framed in words: it is that they do not have the structure of verbally expressed thoughts. But they deserve the name of "protothoughts" because while it would be ponderous to speak of truth or falsity in application to them, they are intrinsically connected with the possibility of their being mistaken: judgment, in a non-technical sense, is just what the driver and the canoeist need to exercise. [27, p. 122]

2.2 Plastic Cognition in Organisms' Pseudoexplanatory Guesses

To better understand what the study of nonlinguistic creatures teaches us about model-based and manipulative abduction (and go beyond Peirce's insights on chickens' "wired" abductive abilities), it is necessary to acknowledge the fact that it is difficult to attribute many of their thinking performances to innate releasing processes, trial and error or to a mere reinforcement learning, which do not involve complicated and more stable internal representations.

Fleeting and evanescent (not merely reflex-based) pseudorepresentations are needed to account for many animal "communication" performances even at the level of the calls of "the humble and much-maligned chicken", like Evans says:

> We conclude that chicken calls produce effects by evoking representations of a class of eliciting events [food, predators, and presence of the appropriate receiver]. This finding should contribute to resolution of the debate about the meaning of referential signals. We can now confidently reject reflexive models, those that postulate only behavioral referents, and those that view referential signals as imperative. The humble and much maligned chicken thus has a remarkably sophisticated system. Its calls denote at least three classes of external objects. They are not involuntary exclamations, but are produced under particular social circumstances. [29, p. 321]

In sum, in nonlinguistics animals, a higher degree of *abductive* abilities has to be acknowledged: chicken form separate representations faced with different events and they are affected by prior experience (of food, for example). They are mainly due to internally developed plastic capacities to react to the environment, and can be thought of as the fruit of learning. In general this plasticity is often accompanied by the suitable reification of external artificial "pseudorepresentations" (for example landmarks, alarm calls, urine-marks and roars, etc.) which artificially modify the environment, and/or by the referral to externalities already endowed with delegated cognitive values, made by the animals themselves or provided by humans.

The following is an example of not merely reflex-based cognition and it is fruit of plasticity: a mouse in a research lab perceives not simply the lever but the fact that the action on it affords the chance of having food; the mouse "desires" the goal (food) and consequently acts in the appropriate way. This is not the fruit of innate and instinctual mechanisms, merely a trial and error routine, or brute reinforcement learning able to provide the correct (and direct) abductive appraisal of the given environmental situation. Instead it can be better described as the fruit of learnt and flexible thinking devices, which are not merely fixed and stimulus driven but also involve "thought". "Pseudothought" – I have already said – is a better term to use, resorting to the formation of internal structured representations and various – possibly new – links between them. The mouse also takes advantage in its environment of an external device, the lever, which the humans have endowed with a fundamental predominant cognitive value, which can afford the animal: the mouse is able to cognitively pick up this externality, and to embody it in internal, useful representations.

Another example of plastic cognition comes from the animal activity of reshaping the environment through its mapping by means of seed caches:

> Consider, for example, a bird returning to a stored cache of seeds. It is known from both ethological studies and laboratory experiments that species such as chickadees and marsh tits are capable of hiding extraordinary number of seeds in a range of different hiding places and then retrieving them after considerable periods of time have elapsed. ([30], quoted in [18, p. 48])

It is also likely to hypothesize that this behavior is governed by the combination of a motivational state (a general desire for food) and a memory of the particular location, and how to get to it[5]. The possibility of performing such behavior is based on structured internal pseudorepresentations originating from the previous interplay between internal and external signs suitably picked up from the environment in a step-by-step procedure.

To summarize, in these cases we are no longer observing the simple situation of the Peircean, picking chicken, which "[...] has an innate idea of doing this; that is to say, that it can think of this, but has no faculty of thinking anything else". This "cognitive" behavior is the one already described by the minimalist contention that there is no need to specify any kind of internal content. It is minimally – here and now and immediately related to action – goal-directed, mechanistic, and not "psychological" in any sense, even in a metaphorical one, as we use the term in the case of animals [18, p. 49].

On the contrary, the birds in the example above have at their disposal flexible ways of reacting to events and evidence, which are explainable only in terms of a kind of *thinking* "something else", to use the Peircean words, beyond mere mechanistic pre-wired responses. They can choose between alternative behaviors founding their choice on the basis of evidence available to be picked up. The activity is "abductive" in itself: it can be *selective*, when the pseudoexplanatory guess, on which the subsequent action is based, is selected among those already internally available, but it can also be *creative*, because the animal can form and excogitate for the first time a particular pseudo-explanation of the situation at hand and then creatively act on the basis of it. The tamarins quickly learn to select the best hypothesis about the tool – taking into account the different tools on offer – that has to be used to obtain the most food in "varied" situations. To avoid "psychological" descriptions, animal abductive cognitive reaction at this level can be seen as an emergent property of the whole organism, and not, in an anthropocentric way, as a

---

[5] Of course the use of concepts like "desire", deriving from the "folk-psychology" lexicon, has to be considered merely metaphorical.

small set of specialized skills like we usually see them in the case of humans. By the way, if we adopt this perspective it is also easier to think that some organisms can learn and memorize even without the brain[6].

As I will illustrate in subsection 2.4, animals occupy different environmental niches that "directly" afford their possibility to act, like Gibson's original theory teaches, but this is only one of the ways the organism exploits its surroundings to be suitably attuned to the environment. When behaviors are more complicated other factors are at stake. For example, animals can act on a goal that they cannot perceive – the predator that waits for the prey for example – so the organism's appraisal of the situation includes factors that cannot be immediately perceived,

Well-known dishabituation experiments have shown how infants use model-based high-level physical principles to relate to the environment. They look longer at the facts that they find surprising, showing what expectations they have; animals like dolphins respond to structured complex gestural signs in ways that can hardly be accounted for in terms of the Gibsonian original notion of immediate affordance. A similar situation can be seen in the case of monkeys that perform complicated technical manipulations of objects, and in birds that build artifacts to house beings that have not yet been born. The problem here is that organisms can dynamically abductively "extract" or "create" – and further stabilize – affordances not previously available, taking advantage not only of their instinctual capacities but also of the plastic cognitive ones (cf. below subsection 2.4)[7].

2.3 Artifacts and Classical and Instrumental Conditioning

Other evidence supports the assumption about the relevance of nonlinguistic model-based thinking beyond the mere reflex-based level. The birth of what is called material culture in hominids, I will quote in the following subsection, and the use of artifacts as external cognitive mediators in animals, reflect a kind of *instrumental* thought that cannot be expressed in terms of the minimalist conception. The instrumental properties are framed by exploiting artificially made material cognitive tools that *mediate* and so enhance perception, body kinesthetic skills, and a full-range of new cognitive opportunities. Through artifacts more courses of action can be selected, where – so to say – "sensitivity" to the consequences is higher. In this case actions cannot be accounted for solely in terms of the mere perceptual level[8].

---

[6] It is interesting to note that recent neurobiological research has shown that neural systems within the spinal cord in rats are quite a bit smarter than most researchers have assumed, they can, for example, learn from experience [6]. Cf. also footnote 1 above.

[7] On the creation/extraction of new affordances through both evolutionary changes and construction of new knowledge and artifacts cf. [31].

[8] This sensitivity is already present in birds like ravens [32].

The difference has to be acknowledged between sensitivity to consequences, which is merely due to innate mechanisms and/or classical conditioning (where behavior is simply modified in an adaptive way on the basis of failures and successes), and the more sophisticated sensitivity performed through some doxastic/representational intermediate states:

> In classical conditioning, a neutral stimulus (e.g., the sound of a bell) is followed by an unconditioned stimulus (e.g., the presentation of food) that elicits a reaction (e.g., salivation). The outcome of classical conditioning is that the conditioned response (the salivation) comes to be given to the conditioned stimulus (the sound of the bell) in the absence of the unconditioned stimulus. In instrumental operant conditioning the presentation of the reinforcing stimulus is contingent on the animal making a particular behavioral response (such as a pecking lever). If the behavioral response does not occur, the reinforcing stimulus is withheld. Classical conditioning behavior is not outcome-sensitive in any interesting sense, since it is not the behavior that is reinforced. [18, p. 167]

It is evident that instrumental conditioning is also important in (and intertwined with) tool and artifact construction where for example the ability to *plan* ahead (modifying plans and reacting to contingencies, such as unexpected flaws in the material and miss-hits) is central.

2.4 Affordances and Abduction

Gibson's eco-cognitive concept of "affordance" [33] and Brunswik's interplay between proximal and distal environment [34] also deal with the problem of the so-called model-based pseudothoughts, which concern any kind of thinking far from the cognitive features granted by human language[9]. These kinds of cognitive tools typical of infants and of many animals (and still operating in human adults in various forms of more or less unexpressed thinking) are hypothesized to express the organic beings' implicit skills to act in a perceived environment as a distal environment ordered in terms of the possibilities to *afford* the action in response to local changes.

Different actions will be suitable to different ways of apprehending aspects of the external world. The objectification of the world made possible by language and other highly abstract organizing cognitive techniques (like mathematics) is not needed. An affordance is a resource or chance that the environment presents to the "specific" organism, such as the availability of water or of finding recovery and concealment. Of course the same part of the environment offers different affordances to different organisms. The concept can be also extended to artificial environments built by humans, my cat affords

---

[9] A detailed illustration of the relationships between affordances and abduction is given in [31].

her actions in the kitchen of my house differently than me, for example I do not find affordable to easily jump through the window or on the table! I simply cannot imagine the number of things that my cat Sheena is possibly "aware" of (and her way of being aware) in a precise moment, such as the taste of the last mouse she caught and the type of memory she has of her last encounter with a lizard[10]: "Only a small part of the network within which mouseness is nested for us extends into the cat's world" [37, p. 203].

It can be hypothesized that in many organisms the perceptual world is the only possible model of itself and in this case they can be accounted for in terms of a merely reflex-based notions: no other internal more or less stable representations are available. In the case of affordance sensitive organisms described above the coupling with the environment is more flexible because it is important in coupling with the niche to determine what environmental dynamics are currently the most relevant, among the several ones that afford and that are available. An individual that is looking for its prey and at the same time for a mate (which both immediately afford it without any ambiguity) is contemporarily in front of two different affordances and has to *abductively* select the most suitable one weighting them. Both affordances and the more or less plastic processes of their selection in specific situations can be stabilized, but both can also be modified, increased, and substituted with new ones. In animals, still at the higher level on not-merely reflex-based cognitive abilities, no representational internal states need be hypothesized [38].

The etheromorphism of affordances is also important: bats use echolocation, and have a kind of sensory capacity that exceeds that of any man-made systems; dolphins can for example detect, dig out, and feed on fish and small eels buried up to 45 cm beneath the sandy seabed and are able to detect the size, structure, shape, and material composition of distant objects. They can also discriminate among aluminum, copper, and brass circular targets, and among circles, squares, and triangular targets covered with neoprene [39]. These amazing cognitive performances in dolphins are processed through complex computations that transform one dimensional waves (and multiple echoes), arriving at each of their two ears, into representations of objects and their features in the organism's niche. The process is "multimodal" because dolphins also interface with their world using visual and other auditory signals, vocal and behavioral mimicry, and representational capabilities. It even seems that significant degrees of self-awareness are at work, unique to nonhuman

---

[10] The point of view of Gibson has been taken into account by several people in the computational community, for example by Brooks in robotics [35]. "Vision is not delivering a high level representation of the world, instead it cooperates with motor controls enabling survival behavior in the environment. [...] While it is very sensible that the main goal of vision in humans is to contribute to moving and acting with objects in the word, it is highly improbable that a set of actions can be identified as the output of vision. Otherwise, vision must include all sort of computations contributing to the acting behavior in that set: it is like saying that vision should cover more or less the whole brain activity" [36, pp. 369–370].

animals [40]. It is easy to imagine that we can afford the world in a similar way only by hybridizing ourselves using artificial instruments and tools like sonar: the fruit of modern scientific knowledge.

It is important to note that recent research based on Schrödinger's focusing on energy, matter and thermodynamic imbalances provided by the environment, draws the attention to the fact that all organisms, including bacteria, are able to perform elementary *cognitive functions* because they "sense" the environment and process internal information for "thriving on latent information embedded in the complexity of their environment" (Ben Jacob, Shapira, and Tauber [41, p. 496]). Indeed Schrödinger maintained that life requires the consumption of negative entropy, i.e. the use of thermodynamic imbalances in the environment. As a member of a complex superorganism – the colony, a multi-cellular community – each bacterium possesses the ability to sense and communicate with the other units comprising the collective and performs its work within a distribution task so, bacterial communication entails collective sensing and cooperativity through interpretation of chemical messages, distinction between internal and external information, and a sort of self vs. non-self distinction (peers and cheaters are both active).

In this perspective, "biotic machines" are *meaning*-based forms of intelligence to be contrasted with the *information*-based forms of artificial intelligence: biotic machines generate new information, assigning contextual meaning to gathered information: self-organizing organisms like bacteria are afforded – through a real cognitive act – and by "relevant" information that they subsequently couple with the regulating, restructuring, and *plastic* activity of the contextual information (intrinsic meaning) already internally stored, which reflects the intra-cellular state of the cells. Of course the "meaning production" involved in the processes above refers to structural aspects of communication that cannot be related to the specific sentential and model-based cognitive skills of humans, primates, and other simpler animals, but still shares basic functions with these like sensing, information processing, and collective abductive contextual production of meaning. As stressed by Ben Jacob, Shapira, and Tauber

> In short, bacteria continuously sense their milieu and store the relevant information and thus exhibit "cognition" by their ability to process information and responding accordingly. From those fundamental sensing faculties, bacterial information processing has evolved communication capabilities that allow the creation of cooperative structures among individuals to form super-organisms [41, p. 504].

Organisms need to become *attuned* to the relevant features offered in their environment and many of the cognitive tools built to reach this target are the result of evolution. The wired and embodied perceptual capacities and imagistic, empathetic, trial and error, and analogical devices I have described above already fulfill this task. These capabilities can be seen as devices adopted by organisms that provide them with potential "abductive" powers: they can provide an overall appraisal of the situation at hand and thus orient action. They can be seen as providing abductive "pseudoexplanations" of what is occurring "over there", as it emerges through that material contact with the environment grounded in perceptual interplay. It is through this embodied process that affordances can arise both in wild and artificially modified niches. Peirce had already contended more than one hundred years ago that abduction even takes place when a new born chick picks up the right sort of corn. This is an example, so to say, of spontaneous abduction – analogous to the case of some unconscious/embodied abductive processes in humans.

The original Gibsonian notion of affordance deals with those situations in which the signs and clues the organisms can detect, prompt, or suggest a certain action rather than others. They are immediate, already available, and belong to the normality of the adaptation of an organism to a given ecological niche. Nevertheless, if we acknowledge that environments and organisms evolve and change, and so both their instinctual and cognitive plastic endowments, we may argue that affordances can be related to the variable (degree of) "abductivity" of a configuration of signs: a chair affords sitting in the sense that the action of sitting is a result of a sign activity in which we perceive some physical properties (flatness, rigidity, etc.), and therefore we can ordinarily "infer" (in Peircean sense) that a possible way to cope with a chair is sitting on it. So to say, in most cases it is a spontaneous abduction to find affordances because this chance is already present in the perceptual and cognitive endowments of human and non-human animals.

I maintain that describing affordances that way may clarify some puzzling themes proposed by Gibson, especially the claim concerning the fact that organisms directly perceive affordances and that the value and meaning of a thing is clear on first glance. As I have just said, organisms have at their disposal a standard endowment of affordances (for instance through their wired sensory systems), but at the same time they can plastically extend and modify the range of what can afford them through the appropriate cognitive abductive skills (more or less sophisticated). As maintained by several authors [7, 42–44], what we see is the result of an embodied cognitive abductive process. For example, people are adept at imposing order on various, even ambiguous, stimuli [7, p. 107]. Roughly speaking, we may say that what organisms *see* (or *feel* with other senses) is what their visual (or other senses') apparatus can, so to say, "explain". It is worth noting that this process happens almost simultaneously without any further mediation. Perceiving affordances has something in common with it. Visual perception is indeed a more automatic and "instinctual" activity, that Peirce claimed to be essentially abductive. Indeed he considers inferential any cognitive activity whatever, not only conscious abstract thought: he also includes perceptual knowledge and subconscious cognitive activity. For instance he says that in subconscious mental activities visual representations play an immediate role [45].

We also have to remember that environments evolve and change and so the perceptive capacities especially when enriched through new or higher-level cognitive skills, which go beyond the ones granted by the merely instinctual levels. This dynamics explains the fact that if affordances are usually stabilized this does not mean they cannot be modified and changed and that new ones can be formed.

It is worth noting that the history of the construction of artifacts and various tools can be viewed as a continuous process of building and crafting new affordances upon pre-existing ones or even from scratch. From cave art to modern computers, there has been a co-evolution between humans and the environment they live in. Indeed, what a computer can afford embraces an amazing variety of opportunities and chances comparing with the ones exhibited by other tools and devices. More precisely, a computer as a Practical Universal Turing Machine [46] can mimetically reproduce even some of the most complex operations that the human brain-mind systems carry out (cf. Magnani [19]).

The hypothetical status of affordances reminds us that it is not necessarily the case that just any organisms can detect it. Affordances are a mere potentiality for organisms. First of all perceiving affordances results from an abductive activity in which we infer possible ways to cope with an object from the signs and cues available to us. Some of them are stable and in some cases they are neurally wired in the perceptual system. This is especially true when dealing with affordances that have a high cognitive valence. Perceiving the affordances of a chair is indeed not neurally wired but strongly rooted and stabilized in our cultural evolution. The differences that we can appreciate are mostly *inter-species* – so to speak. A chair affords a child as well as an adult. But this is not the case of a cat. The body of a cat – actually, the cat can sit down on a chair, but also it can sleep on it – has been shaped by evolution quite differently from us.

In higher-level cognitive performances there is something different, since *intra-species* differences seem to be strongly involved. For instance, only a person that has been taught about geometry can infer the affordances "inside" the new manipulated construction built on a geometrical depicted diagram in front of him/her. He/she has to be an "expert". First of all, artificial affordances are intimately connected to culture and the social settings in which they arise and the suitable availability of knowledge of the individual(s) in question. Secondly, affordances deal with learning. There are some affordances like those of an Euclidean triangle that cannot be perceived without a learning process (for instance a course of geometry): people must be somehow *trained* in order to perceive them. Of course acknowledging this last fact places much more emphasis upon the dynamic and also evolutionary character of affordances. The abductive process at play in these cases is very complicated and requires higher level education in cognitive information and skills.

I have already noted that an artificially modified niche (at both levels of biotic and abiotic sources) can be also called "cognitive niche". Recently it has been contended that cognitive niche construction is an evolutionary process in its own right rather than a mere product of natural selection. Through cognitive niche construction organisms not only influence the nature of their world, but also in part determine the selection pressure to which they and their descendants are exposed (and of course the selection pressures to which other species are subjected).

This form of feedback in evolution has been rarely considered in the traditional evolutionary analyses [47]. On this basis a co-evolution between niche construction and brain development and its cognitive capabilities can be clearly hypothesized, a perspective further supported by some speculative hypotheses given by cognitive scientists and paleoanthropologists (for example [16, 17, 48][11]. These authors first of all maintain that the birth of material culture itself was not just the product of a massive "cognitive" chance but also cause of it. In the same light the "social brain hypothesis" (also called "Machiavellian intelligence hypothesis" [49–51]), holds that the relatively large brains of human beings and other primates reflect the computational demands of complex social systems and not only the need of processing information of ecological relevance.

3 Perception as Abduction

3.1 Reifications and Beliefs

Some examples testify how animals are able to form a kind of "concept". These activities are surely at the basis of many possibilities to reify the world. Honey bees are able to learn/form something equivalent to the human concepts of "same" and "different"; pigeons, learn/form such concepts as tree, fish, or human [52, 53]. Sea lions abduce among already formed equivalence classes: a pup's recognition of its mother "[...] depends on the association of many sensory cues with the common reinforcing elements of warmth, contact, and nourishment, while a female recognition of her sisters may depend on their mutual association with the mother" [54].

Something more complicated than classical conditioning is at play when some animals are able to *reify* various aspects of the world using a kind of analogical reasoning. In this way they are able to detect similarities in a certain circumstance, which will be properly applied in a second following situation. Of course this capacity promotes the possibility to form a more contextual independent view about the objects perceived, for example it happens when

---

[11] I have treated this problem connecting it to some of Turing's insights on the passage from "unorganized" to "organized" brains in a recent article on the role of mimetic and creative representations in human cognition [19].

recognizing similarities in objects that afford food. The mechanism is analogous to the one hypothesized by philosophers and cognitive scientists when explaining concept formation in humans, a process that of course in this case greatly takes advantage of the resources provided by language. This way of thinking also provides the chance of grasping important regularities and the related power to re-identify objects and to predict what has to be expected in certain out-coming situations [54, p. 58]. It is a form of abduction by analogy, which forms something like general hypotheses from specific past event features that can be further applied to new ones.

Bermúdez [18, chapter four] maintains that the process of ascribing thoughts to animals is a form of what Ramsey called "success semantic" [55]. When for example we are confronted by the evidence that a chicken abstains "[...] from eating such caterpillars on account of unpleasant experience" a pseudobelief that something is poisonous can be hypothesized and equated to this event. "Thus any set of actions for whose utility P is a necessary and sufficient condition might be called a belief that P, and so would be true if P, i.e., if they are useful" [18, p. 65]. Success semantics adopts a "thought/truth" condition for belief, respecting the idea that thoughts can be true or false because they represent states of affairs as holding: thought is truth-evaluable. Utility condition of a belief is a state of affairs that when holding leads to the satisfaction of desires with which that belief is combined. The satisfaction condition is equally that state of affairs that "[...] extinguishes in the right sort of way the behavior to which the desire has given risen. [...] The utility condition of a belief in a particular situation is completely open to the third-person perspective of the ethologist or developmental psychologist [...] and provides a clear way of capturing how an adaptive creature is in tune with its environment without making implausibile claims at the level of the vehicle of representation" (pp. 65 and 68).

Hence, in success semantics the role of reinforcement through satisfaction is still relevant but it does not impede the fact that also internal representations can be hypothesized, especially when we are dealing with non-basic appetites. Indeed, following Bermúdez, we can say that in some cases representational states are at stake and are directly related to evolutionary pressures: "[...] the attunement of a creature to its environment niche is a direct function of the fact that various elements of the subpersonal representational system have evolved to track certain features of the distal environment" [18, p. 69], like in the case of so-called "teleosemantics" [56]. In other cases intelligent skills arise where it is difficult to hypostatize representational contents in situations where evolutionary notions do not play any role: here "Attunement to the environment arises at the level of organism, rather than at the level of subpersonal representational vehicles. That is to say, an organism can be attuned to the environment in a way that will allow it to operate efficiently and successfully, even if there has not been selective pressure for sensitivity at the subpersonal level to the relevant features of the distal environment" [18, p. 69].

3.2 Perception as Abduction

Bermúdez says: "A body is a bundle of properties. But a body is a thing that has certain properties. The simple clustering of collocated features can be immediately perceived, but to get genuine reification there needs to be an understanding (which may or may not be purely perceptual) of a form of coinstantiation stronger that mere spatio-temporal coinstantiation" [18, p. 73]. *Reification* that is behind coinstantiation is not necessarily a matter of the effect generated by the poietic activity of linguistic devices (names for example). Objects over there in the environment, grasped through perception, obey certain principles and behave in certain standard ways that can be reflected and ordered in creatures' brains. To perceive a body is to perceive a cluster of semiotic features that are graspable through different sensory modalities, "but" this process is far beyond the mere activity of parsing the perceptual array. This array has to be put in resonance – to be matched – with already formed suitable configurations of neural networks (endowed with their electrical and chemical processes), which combine the various semiotic aspects arrived at through senses.

These configurations are able for instance to maintain constant some aspects of the environment, like the edges of some standard forms, that also have to be kept constant with respect to kinesthetic aspects related to the motor capabilities of the organism in question. For example these neural configurations compensate variation of size and shape of a distal object with respect to an organism's movements. It is in this sense that we can say, by using a Kantian lexicon, that these neural configurations "construct" the world of the chaotic multiplicity gathered at the level of phenomena. The process is of course very different in different organisms – for example some creatures are not able to retain the size of an object through rotation – but still create a permanent cluster of other appropriate intertwined features[12].

Perception is strongly tied up with reification. Through an interdisciplinary approach and suitable experimentation some cognitive scientists (cf. for example Raftopoulos [60, 61]) have recently acknowledged the fact that in humans perception (at least in the visual case) is not strictly modular, like Fodor [62] argued, that is, it is not encapsulated, hard-wired, domain-specific, and mandatory. Neither is it wholly abductively "penetrable" by higher cognitive states (like desires, beliefs, expectations, etc.), by means of top-down pathways in the brain and by changes in its wiring through perceptual learning, as stressed by Churchland [63]. It is important to consider the three following levels: visual sensation (bodily processes that lead to the formation of retinal image which are still useless – so to say – from the high-level cognitive perspective), perception (sensation transformed along the visual neural pathways in a structured representation), and observation, which consists in all subsequent visual processes that fall within model-based/propositional cognition.

---

[12] On neural correlates of allocentric space in mammals cf. [57–59].

These processes "[...] include both post-sensory/semantic interface at which the object recognition units intervene as well as purely semantic processes that lead to the identification of the array – high level vision" [60, p. 189].

On the basis of this distinction it seems plausible – like Fodor contends – to think there is a substantial amount of information in perception which is theory-neutral. However, also a certain degree of theory-ladenness is justifiable, which can be seen at work for instance in the case of so-called "perceptual learning". However, this fact does not jeopardize the assumption concerning the basic cognitive impenetrability of perception: in sum, perception is informationally "semi-encapsulated", and also semi-hardwired, but, despite its bottom-down character, it is not insulated from "knowledge". For example, it results from experimentation that illusion is a product of learning from experience, but this does not regard penetrability of perception because these experience-driven changes do not affect a basic core of perception[13].

Higher cognitive states affect the product of visual modules only after the visual modules "[...] have produced their product, by selecting, acting like filters, which output will be accepted for further processing" [61, p. 434], for instance by selecting through attention, imagery, and semantic processing, which aspects of the retinal input are relevant, activating the appropriate neurons. I contend these processes are essentially *abductive*, as is also clearly stressed by Shanahan [65], who provides an account of robotic perception from the perspective of a sensory fusion in a unified framework: he describes problems and processes like the incompleteness and uncertainty of basic sensations, top-down information flow and top-down expectation, active perception and attention.

It is in this sense that a certain amount of *plasticity* in vision does not imply the penetrability of perception. As I have already noted, this result does not have to be considered equivalent to the claim that perception is not theory-laden. It has to be acknowledged that even basic perceptual computations obey high-level constraints acting at the brain level, which incorporate implicit and more or less model-based assumptions about the world, coordinated with motor systems. At this level, they lack a semantic content, so as they are not learnt, because they are shared by all, and fundamentally hard-wired.

High order physical principles are also important in reification: I have already cited the experiments on dishabituation in nonlinguistic infants and animals, which have shown that sensitivity to some physical principles starts at birth, and so before the acquisition of language both in phylogenetic and ontogenetic terms [18, pp. 78–79]. In these results it is particularly interesting to see how nonlinguistic beings are able to detect that objects continue to exist even if not perceived, thus clearly showing a kind of reification at work in the perception of an organized world.

---

[13] Evidence on the theory-ladenness of visual perception derived from case-studies in the history of science is illustrated in Brewer and Lambert [64].

In the various nonlinguistic organisms different sets of spatial and physical principles give rise to different ontologies (normally shared with the conspecifics at a suitable stage of development). The problem is to recognize how they are structured, but also how they "evolve". Of course different properties – constant and regular in an appropriate lapse of time – will be salient for an individual at different times, or for different individuals at a given time. This way of apprehending is basically explanatory and thus still abductive (selective or creative) in itself and of course related to the doxastic states I introduced above. Consequently, the "intelligent" organism exhibits a suitable level of flexibility in responding. To make an example, when a mouse is in a maze where the spatial location of food is constant, it is in a condition to choose different paths (through a combination of heuristics and of suitable representations), which can permit it to reach and take the food[14]. This means that in mouse spatial cognition, various forms of prediction/anticipation are at play.

4 Is Instinct Rational? Are Animals Intelligent?

4.1 Rationality of Instincts

Instincts are usually considered irrational or at least a-rational. Nevertheless, there is a way of considering the behavior performances based on them as *rational*. Based on this conclusion, while all animal behavior is certainly described as rational, at the same time it is still rudimentarily considered instinctual. The consequence is that every detailed hypothesis on animal intelligence and cognitive capacities is given up: it is just sufficient to acknowledge the general rationality of animal behavior. Let us illustrate in which sense we have to interpret this apparent paradox. I think the analysis of this puzzling problem can further improve knowledge about model-based and manipulative ways of thinking in humans, offering at the same time an integrated view regarding some central aspects of organisms' cognitive behavior.

Explanations in terms of psychological states obviously attribute to human beings propositional attitudes, which are a precondition for giving a *rational* picture of the explained behavior. These attitudes are a combination of beliefs and desires. Rational internal – doxastic – states characterize human behavior and are related to the fact that they explain why a certain behavior is appropriate on the basis of a specific relationship between beliefs, desires, and actions (cf. Magnani [1, chapter seven]). How can this idea of rationality be extended to nonlinguistic creatures such as human infants and several

---

[14] An illustration of the different spatial coordinate systems and their kinesthetic features in rat navigation skills (egocentric, allocentric, in terms of route in a maze space, etc.) is given in the classical Tolman, Ritchie, and Kalish [66], O'Keefe and Nadel [67], Gallistel [68].

types of animals, where the role of instinct is conspicuous? How can the inferential transformations of their possible internal thoughts be recognized when, even if conceivable as acting in their nervous systems, these thoughts do not possess linguistic/propositional features?

The whole idea of rationality in human beings is basically related to the fact we are able to apply *deductive* formal-syntactic rules to linguistic units in a truth preserving way, an image that directly comes from the tradition of classical logic: a kind of rationality robustly related to "logico-epistemological" ideals. The computational revolution of the last decades has stressed the fact that rationality can also be viewed as linked to ways of thinking such as *abduction* and *induction*, which can in turn be expressed through more or less simple *heuristics*. These heuristics are usually well-assessed and shared among a wide community from the point of view of the criteria of applicability, but almost always they prove to be strongly connected in their instantiation to the centrality of language. Indeed cognitive science and epistemology have recently acknowledged the importance of model-based and manipulative ways of rational thinking in human cognition, but their efficacy is basically considered to be strictly related to their hybridization with the linguistic/propositional level. Consequently, for the reasons I have just illustrated, it is still difficult to acknowledge the rationality of cognitive activities that are merely model-based and manipulative, like those of animals.

At the beginning of this section I said that, when dealing with rationality in nonlinguistic creatures, tradition initially leads us to a straightforward acknowledgment of the presumptive and intrinsic "rationality" of instincts. The background assumption is the seeming impossibility that something ineluctable like instinct cannot be at the same time intrinsically rational. Of course the concept of rationality is in this case paradoxical and the expression "rationality" has to be taken in a Pickwickian sense: indeed, in this case the organisms at stake "cannot" be irrational. A strange idea of rationality! Given the fact that many performances of nonlinguistic organisms are explainable in terms of sensory preconditioning (and so are most probably instinct-based – hard-wired – and without learnt and possibly conscious capacities which enable them to choose and decide), the rationality of costs and benefits in these behaviors is expressed in the "non-formal" terms of Darwinian "fitness". For example, in the optimal foraging theory, "rationality" is related to the animal's capacity – hard-wired thanks to evolution – to optimize the net amount of energy in a given interval of time. Contrarily to the use of some consciously exploited heuristics in humans, in animals many heuristics of the same kind are simply hard-wired and so related to the instinctual adaptation to their niches.

The following example provided by Bermúdez can further clarify the problem. "Redshanks are shorebirds that dig for worms in estuaries at low tide. It has been noticed that they sometimes feed exclusively on large worms and at other times feed on both large and small worms. [...] In essence, although a large worm is worth more to the red shank in terms of quantity of energy gained per unit of foraging time than a small worm, the costs of searching exclusively for large worms can have deleterious consequences, except when the large worms are relatively plentiful" [18, p. 117]. The conclusion is simple: even if the optimal behavior can be described in terms of a "rational" complicated version of expected utility theory, "[...] the behaviors in which it manifests itself do not result from the application of such a theory" (*ibid.*). We can account for this situation in our abductive terms: the alternatives which are "abductively" chosen by the redshanks are already wired, so that they follow hardwired algorithms developed through evolution, and simply instantiate the idea of abduction related to instincts present in Peircean insights.

The situation does not change in the case that we consider short-term and long-term rationality in evolutionary behaviors. In the case of the redshank we deal with "short-term" instinct–based rationality related to fitness, but in the case of animals that sacrifice their lives in a way that increases the lifetime fitness of other individuals we deal with "long-term" fitness. It has to be said that sometimes animals are also "hardwired" to use external landmarks and territory signs, and communicate with each other using these threat-display signals that consent them to avoid direct conflict over food. These artifacts are just a kind of instinct-based *mediators*, which are "instinctually" externalized and already evolutionarily stabilized[15].

4.2 Levels of Rationality in Animals

Beyond the above idea of "rationality" in animals and infants as being related to tropistic behaviors connected to reflexes and inborn skills such as imprinting or classical conditioning, the role of intermediary internal representations has to be clearly acknowledged. In this last case we can guess that a "rational" intelligence closer to the one expressed in human cognition, and so related to higher levels of abductive behavior, is operating. We fundamentally deal with behaviors that show the capacity to choose among different outcomes, and which can only be accounted for by hypothesizing learnt intermediate representations and processes. In some cases a kind of decision-making strategy can also be hypothesized: in front of a predator an animal can fight or flee and in some sense one choice can be more rational than the other. In front of the data, to be intended here as the "affordances" in a Gibsonian sense, provided through mere perception and which present various possibilities for action, a high-level process of decision-making is not needed, but choice is still possible. With respect to mere wired capacities the abductive behavior above seems based on reactions that are more flexible.

---

[15] These mediators are similar to the cognitive, epistemic, and moral mediators that humans externalize thanks to their plastic high-level cognitive capacities, but less complex and merely instinct-based. I have fully described the role of epistemic mediators in scientific reasoning in [7], and of moral mediators in ethics in [1]. See also the following section.

Bermúdez [18, p. 121] labels Level 1 this kind of rationality. It differs from "rationality" intended as merely instinct-based, expressed in immutable rigid behaviors (called Level 0). Level 1 rationality (which can still be split in short-term and long-term) is for example widespread in the case of animals that entertain interanimal interactions. This kind of rationality would hold when we clearly see ir-rational animals, which fail to signal to the predator and instead flee, thus creating a bad outcome for group fitness (and for their own lifetime fitness: other individuals will cooperate with them less in the future and it will be less probable for them to find a mate).

To have an even higher level rationality (Level 2) we need to involve the possibility of abductively selecting among different "hypotheses" which make the organisms able – so to say – to "explain" certain behaviors: a kind of capacity to select among different "hypotheses" about the data at hand, and to behave correspondingly. This different kind of "rational" behavior, is neither merely related to instincts nor simply and rudimentarily flexible, like in the two previous cases.

To make the hypothesis regarding the existence of this last form of rationality plausible, two epistemological pre-conditions have to be fulfilled. The first is related to the acknowledgment that model-based and manipulative cognitions are endowed with an "inferential" status, as I explained above when dealing with the concept of abduction, taking advantage of the semiotic perspective opened up by Peirce. The second relates to the rejection of the restricted logical perspective on inference and rationality I have described in the previous subsection, which identifies inferences at the syntactic level of natural and artificial/symbolic languages (in this last case, also endowed with the truth-preserving property, which produces the well-known isomorphism between syntactic and semantic/content level).

At this high-rationality level we can hypothesize in nonlinguistic organisms more than the simple selection of actions, seen as merely wired and operating at the level of perceptions like the theory of immediate affordances teaches, where a simple instrumental conditioning has attached to some actions a positive worth. Instead, in Level 2 rationality, complicated, relatively stable, internal representations that account for consequences are at work. In this case selecting is selecting – so to speak – for some "reasons": a bird that learns to press a lever in a suitable way to obtain food, which will then be delivered in a given site, acts by considering an association between that behavior and the consequences. A kind of instrumental pseudobelief about the future and about certain probable regularities is established, and contingencies at stake are represented and generalized in a merely model-based way. Then the organism internally holds representations with some stability and attaches utility scores to them: based on their choice a consequent action is triggered, which will likely satisfy the organism's desire. The action will be stopped, in a nonmonotonic way, only in the presence of out-coming obstacles, such as the presence of a predator.

Of course the description above suffers the typical anthropomorphism of the observer's "psychological" explanations. However, beliefs do not have to be considered explicit; nevertheless, some actions cannot be explained only on the basis of sensory input and from knowledge of the environmental parameters. Psychological explanations can be highly plausible when the goal of the action is immediately perceptible or when the distal environment contains immediately perceptible instrumental properties. This is obvious and evident in the case of human beings' abilities, but something similar occurs in some chimpanzees' behavior too. When chimpanzees clearly see some bananas they want to reach and eat, and some boxes available on the scene, they have to form an internal instrumental belief/representation on how to exploit the boxes. This "pseudobelief" is internal because it is not immediately graspable through mere perceptual content:

> Any psychological explanation will always have an instrumental content, but the component needs not take the form of an instrumental belief. [...] instrumental beliefs really only enter the picture when two conditions are met. The first is that the goal of the action should not be immediately perceptible and the second is that there should be no immediately perceptible instrumental properties (that is to say, the creature should be capable of seeing that a certain course of action will lead to a desired result). The fact, however, that one or both of these conditions is not met does not entail that we are dealing with an action that is explicable in non-psychological terms. [18, p. 129]

The outcomes are represented, but these "pseudorepresentations" lack in lower kinds of rationality. The following example is striking. A food source was taken away from chicken at twice the rate they walked toward it but advanced toward them at twice the rate they walked away from it: after 100 trials, this did not affect the creatures' behavior which failed to represent the two contingencies ([69] quoted in [18, p. 125]). Chicken, which do not retreat from a certain kind of action faced with the fact that a repeated contingency no longer holds, are not endowed with this high level "representational" kind of abductive rationality.

5 Artifactual Mediators and Languageless Reflexive Thinking

5.1 Animal Artifactual Mediators

Even if the animal construction of external *artifactual mediators* is sometimes related to instinct, as I have observed in the subsection 4.1, it can also be the fruit of plastic cognitive abilities strictly related to the need to improve actions and decisions[16]. In this case action occurs through the expert delegation of

---

[16] I have already stressed that plants also exhibit interesting plastic changes. In resource-rich productive habitats where the activities of the plants "generate"

cognitive roles to external tools, like in the case of chimpanzees in the wild, that construct wands for dipping into ant swarms or termite nests. These wands are not innate but highly specialized tools. They are not merely the fruit of conditioning or trial and error processes as is clearly demonstrated by the fact they depend on hole size and they are often built in advance and away from the site where they will be used.

The construction of handaxes by the hominids had similar features. It involved paleocognitive model-based and manipulative endowments such as fleeting consciousness, private speech, imposition of symmetry, understanding fracture dynamics, ability to plan ahead, and a high degree of sensory-motor control. I have already said in subsection 1.1 they represent one of the main aspects of the birth of *material culture* and technical intelligence and are at the root of what it has been called the process of a "disembodiment of mind" [16, 19].

From this perspective the construction of artifacts is an "actualization" in the external environment of various types of objects and structures endowed with a cognitive/semiotic value for the individual of for the group. Nonlinguistic beings already externalize signs like alarm calls for indicating predators and multiple cues to identify the location of the food caches, which obey the need to simplify the environment and which of course need suitable spatial memory and representations [71, 72]. However, animals also externalize complicated artifacts like in the case of Darwin's earthworms that I have illustrated in subsection 1.1.

These activities of cognitive delegation to external artifacts is the fruit of expert behaviors that conform to innate or learnt embodied templates of cognitive doing. In some sense they are analogous to the templates of epistemic doing I have illustrated in [7], which explain how scientists, through appropriate actions and by building artifacts, elaborate for example a simplification of the reasoning task and a redistribution of effort across time. For example, Piaget says, they "[...] need to manipulate concrete things in order to understand structures which are otherwise too abstract" [73] also to enhance the social communication of results. Some templates of action and manipulation, which are implicit and embodied, can be *selected* from the set of the ones available and pre-stored, others have to be *created* for the first time to perform the most interesting creative cognitive accomplishments of manipulative cognition.

Manipulative "thinking through doing" is creative in particularly skilled animals, exactly like in the case of human beings, when for example chimpanzees make a "new" kind of wand for the first time. Later on the new

---

various resources above and below ground that strongly modify the environment, plants themselves exhibit various kinds of, so-called, morphological plasticity – that is, the replacement of existing tissues [13, p. 300]. It is important to note that plant plasticity is particularly advantageous when responses are reversible rather than irreversible [70].

behavior can possibly be imitated by the group and so can become a shared "established" way of building artifacts. Indeed chimpanzees often learn about the dynamic of objects from observing them manipulated by other fellows: a process that enhances social formation and transmission of cognition.

5.2 Pseudological and Reflexive Thinking

Among the various ways of model-based thinking present in nonlinguistic organisms, some can be equated to well-known inferential functional schemes which logic has suitably framed inside abstract and ideal systems. There are forms of pseudological uses of negation (for example dealing with presence/absence, when mammals are able to discern that a thing cannot have simultaneously two contrary properties), of *modus ponens* and *modus tollens* (of course both related to the presence of a pseudonegation), and of conditionals (cf. Bermúdez [18, chapter seven]). Of course, these ways of reasoning are not truth preserving operations on "propositions" and so they are not based on logical forms, but it can be hypothesized that they are very efficient at the nonlinguistic level, even if they lack an explicit reference to logical concepts and schemes[17]. They are plausibly all connected with innate abilities to detect regularities in the external niche. In addition, forms of causal thinking are observed, of course endowed with an obvious survival value, related to the capacity to discriminate causal links from mere non-causal generalizations or accidental conjunctions[18].

It is interesting to note in prelinguistic organisms the use of both "logical" and fallacious types of reasoning. For example the widespread use of "hasty generalization" shows that poor generalizations must not only be considered – in the perspective of a Millian abstract universal standard – as a bad kind of induction. Even if hasty generalizations are considered bad and fallacious in the light of epistemological ideals, they are often strategic to the adaptation of the organism to a specific niche [77].

An open question is the problem of how nonlinguistic creatures could possess second-order thoughts on thoughts (and so the capacity to attribute thoughts to others) and first – and second-order – desires (that is desires when one should have a specific first-order desire). In human beings, self-awareness and language are the natural home for these cognitive endowments. Indeed,

---

[17] On the formation of idealized logical schemes in the interplay between internal and external representations cf. [74].

[18] Human prelinguistic infants show surprise in front of scenes when "action at a distance" is displayed (it seems they develop a pseudothought that objects can only interact causally through physical contact) [75]. Some fMRI experiments on "perceptual" causality are described in [76]: specific brain structures result involved in extracting casual frameworks from the world. In both children and adults these data show how they can grasp causality without inferences in terms of universality, probability, or casual powers.

it is simple to subsume propositions as objects of further propositions for ourselves and for others, and consequently to make "reflexive" thinking possible. This kind of thinking is also sensitive to the inferences between thoughts, which are suitably internally represented as icons of written texts or as representations of our own or others' external voices. In addition, the use of external propositional representations favors this achievement, because it is easy to work over there, in an external support, on propositions through other propositions and then internally recapitulate the results.

If it is difficult to hypothesize that animals and early infants can attribute beliefs and desires to other individuals without the mediation of language and of what psychologists call the "theory of mind", but it is still plausible to think that they can attribute goal-desires to other individuals. In this sense they still attribute a kind of intentionality, and are consequently able to distinguish in other individuals between merely instinctive and purposeful conducts[19].

In human beings, intentional attitudes are attributed by interpreters who abductively undertake what Dennett [80] calls the "intentional stance": they abduce hypotheses about "intentions". These attributions are "[...] ways of keeping track of what the organism is doing, has done, and might do" [81, p. 73]. However, animals too have the problem of "keeping track" of the behavior of other individuals. For example, it is very likely they can guess model-based abductive hypotheses about what other organisms are perceiving, even if those perceptions are not comprehended and made intelligible through the semantic effect produced by language, like in humans[20]. The importance of this capacity to monitor and predict the conduct of conspecifics and/or predators is evident, but other individuals are not seen as thinkers, instead they are certainly seen as doers.

Recent research has shown in animals various capacities to track and "intentionally" influence other individuals' behavior[21]. Tactical deception takes advantage of the use of various semiotic and motor signs in primates: for example, some females, by means of body displacements not seen by a dominant male, can cheat him when they are grooming another non-dominant male [82]. Ants, through externalized released pheromone, deceive members of other colonies: these signs/signals play the role of indirect exchanges of chemicals as units of cheating communication[22]. These activities of deception can be seen in the light of the ability to alter other individuals' sensory perceptions. The case of some jumping hunting spiders illustrated by Wilcox and Jackson

---

[19] Recent research on mirror neurons in primates and human beings support the neurological foundation of this ability [78, 79].

[20] On the encapsulation of perception in language in humans cf. subsection 3.2.

[21] Of course these capacities can be merely instinct-based and the fruit of a history of selection of certain genetic "programs", and consequently not learned in particular environmental contingencies, like in the cases I am illustrating here.

[22] Cf. Monekosso, Remagnino, and Ferri [83] that also illustrate a computational learning program which makes use of an artificial pheromone to find the optimal path between two points in a regular grid.

is striking. By stalking across the web of their prey, they cheat it, through highly specialized signals, also suitably exploiting aggressive mimicry. The interesting thing is that they plastically adapt their cheating and aggressive behavior to the particular prey species at stake, all this by using a kind of trial and error tactic of learning, also reverting to old strategies when they fail [84].

To conclude, it can be conjectured that, at the very least, emotions in animals can play a kind of reflexive role because they furnish an appraisal of the other states of the body, which arise in the framework of a particular perceptual scenario. This fact clearly refers to another kind of reflexivity, distant from the one that works in beings able to produce thoughts of thoughts, attribute thoughts to others (so possessing a "theory of mind"), monitor thoughts and belief/desire generation and engage in self-evaluation and self-criticism[23]. Also in adult humans emotions play this reflexive role, but in this case usually emotions are trained and/or intertwined with the effects produced by culture and thus language[24]. It seems researchers agree in saying that propositions/sentences are the only suitable mediators of second order thoughts. It is plausible to conclude that nonlinguistic creatures are excluded from many typically human ways of thinking, and it is plausible to guess that this reciprocally happens for humans, who do not possess various perceptual and cognitive skills of animals.

5.3 Affect Attunement and Model-Based Communication

An interesting extension of the model I have introduced in my recent [26], concerning "mimetic and creative representations" in the interplay between internal/external is furnished by the merely model-based case of some nonlinguistic and prelinguistic living beings. Human infants entertain a coordinated communication with their caregivers, and it is well known that many psychoanalysts have always stressed the importance of this interplay in the further development of the self and of its relationships with the unconscious states. Infants' emotional states, as "signs" in a Peircean sense, are displayed and put out into the external world through the semiotic externalization of facial expressions, gestures, and vocalizations. The important fact here is that this cognitive externalization is performed in front of a living external "mediator", the mother, "the caregiver", endowed with a perceptual system that can grasp the externalized signs and send a feedback: she cognitively and affectively mediates the initial facial expression and the interplay among the subsequent ones. The interplay above is also indicated as a case of human *affective attunement* [85].

In general an agent can expect a feedback also after having "displayed" suitable signs on a non living object, like a blackboard, but it is clear that

---

[23] Nevertheless, we have seen that nonlinguistic organisms "can" revise and change their representations.
[24] Cf. Magnani [1, chapter six].

in this last case a different performance is at play, which involves explicit manipulations of the external object, and not a mere exchange of – mainly facial-based – sensations, like in affective attunement. The external delegated representation to a non living object shows more or less complicated active responses, which are intertwined with the agent's manipulations. For example, a blackboard presents intrinsic properties that limit and direct the manipulation in a certain way, and so does a PC, which has – with respect to the blackboard – plenty of autonomous possibility to react: usually the interplay is hybrid, taking into account both propositional, iconic (in a Peircean sense), and of course motor aspects[25].

In affect attunement, the interplay is mainly model-based and mostly iconic (also taking advantage of the iconic force of gestures[26] and voice), meaningful words are also present, but the semiotic "propositional" flow is fully understood only on the part of the adult, not on the part of the infant, where words and their meanings are simply being learnt. The infant performs an "expressive" behavior based on appearances and gestures that are spontaneously externalized to get a feedback. Initially the expressions externalized are directly *mimetic* of the inner state but – through the interplay – where subsequent recapitulations of the mother's facial expressions are performed and are gradually, suitably picked up "outside" the mom's body, novel "social" expressions are formed. These expressions are shared with the mom and thus they are no longer arbitrary. Once stabilized, they constitute the expected affective "attunement" to the mom/environment, which is the fruit of a whole abductive model-based activity of subsequent "facial hypotheses". In this process, the external manifestation of the nonlinguistic organism is established as the quality of feelings that testify a shared affect. A new way of sharing affect is abductively *created*, which is at the basis of the further social expression of emotions.

In the case of externalization of signs in non-human animals, when the sharing of affect is not at play, we are, for example, faced with the mere communication of useful information. Many worker honeybees socially externalize dances that express the site where they have found food to inform the other individuals about the location:

> [...] the waggle dances communicate information about direction, distance, and desirability of the food source. Each of these three dimensions of variation is correlated with a dimension of variation in the dance. The angle of the dance relative to the position of the sun indicates the direction of food source. The duration of a complete

---

[25] It has to be noted that for Peirce iconic signs are generally arbitrary and flexible but there are some symbols, still iconic, which are conventional and fixed, like the ones used in mathematics and logic.

[26] Mitchell [86] contents infants need a connection between kinesthesis and vision. That is, without this connection the organism would not be able to connect the kinesthetic image it has of its own body with any visual image.

figure-of-eight circuit indicates the distance to the food source (or rather the flying time to the food source, because it increases when the bees would have to fly into a headwind). And the vigor of the dance indicates the desirability of the food to be found. [18, p. 152][27]

The externalized figures performed through movements are agglomerative[28] signs that grant a cognitive – communicative – *mediator* to the swarm. Through this interplay with other bees, the dancers can get a feedback from the other individuals, which will help them later on to refine and improve their exhibition. In the case of animals, which perform these kinds of externalizations on a not merely innate basis, the true "creation" of new ways of communicating can also be hypothesized, through the invention of new body movements, new sounds or external landmarks, which can be progressively provided, if successful, as a cognitive resource to the entire group.

Related to both the infant affect attunement and bee dances illustrated above an epistemological remark is fundamental. When we speak about internal and external representations in the abductive interplay we put ourselves in the perspective of the researcher, who "sees" two or more different agents in the sense of folk psychology. Nevertheless, in the two examples, the agents are not reified in the sense that "they" do not perceive "themselves" as agents, like we instead do. Rather, for instance in the case of affect attunement, it is the process itself that is responsible for the formation of the infant's agentive status. A clarification of this problem can be found in some cognitive results derived form neurological research, which I have described in a forthcoming paper [89].

6 Conclusion

The main thesis of this paper is that model-based reasoning represents a significant cognitive perspective able to unveil some basic features of abductive cognition in non-human animals. Its fertility in explaining how animals make up a series of signs and are engaged in making or manifesting or reacting to a series of signs in instinctual or plastic ways is evident. Indeed in this article I have illustrated that a considerable part of this semiotic activity is a continuous process of "hypothesis generation" that can be seen at the level of

---

[27] Bees would certainly find human communication very poor because we do not inform our fellows on the location of the closest restaurant by dancing!
[28] The theoretical distinction between agglomerative diagrammatic signs and discursive signs in sentential reasoning, together with many other fundamental clarifications of Peircean insights, also concerning mathematical reasoning, are given in Stenning [87]. On the cognitive advantages (and also disadvantages) – in humans – of diagrammatic dynamic reasoning over sentential reasoning cf. Jones and Scaife [88]: in a watcher/user/learner better cognitive offloading is allowed by external diagrammatic dynamic representations and their "hidden" dependencies.

both instinctual behavior and representation-oriented behavior, where non-linguistic pseudothoughts drive a "plastic" model-based cognitive role. I also maintain that the various aspects of these abductive performances can also be better understood by taking some considerations on the concept of affordance into account. From this perspective the referral to the central role of the externalization of artifacts that act as mediators in animal languageless cognition becomes critical to the problem of abduction. Moreover, I tried to illustrate how the interplay between internal and external "pseudorepresentations" exhibits a new cognitive perspective on the mechanisms underling the emergence of abductive processes in important areas of model-based inferences in the so-called mindless organisms.

The paper also furnished further insight on some central problems of cognitive science. I maintain that analysis of the central problems of abduction and hypothesis generation in non-human animals further clarifies other related topics in model-based reasoning, like pseudological and reflexive thinking, the role of pseudoexplanatory guesses in plastic cognition, the role of reification and beliefs, the problem of the relationship between abduction and perception, and between rationality and instincts, and the issue of affect attunement as a fundamental kind of model-based abductive communication.

In summary, in light of the considerations I outline in this paper it can be said that a considerable part of abductive cognition occurs through model-based activity that takes advantage of pseudoexplanations, reifications in the external environment, and hybrid representations. An activity that is intrinsically *multimodal*. This conclusion rejoins what I have already demonstrated in my recent article [90], from the perspective of distributed cognition: abductive hypothetical cognition involves a full range of various sensory modalities, which clearly stress its multimodal character.

References

1. Magnani, L.: *Morality in a Technological World. Knowledge as Duty*. Cambridge University Press, Cambridge (2007)
2. Gruen, L.: The morals of animal minds. In Bekoff, M., Allen, C., Burghardt, M., eds.: *The Cognitive Animal. Empirical and Theoretical Perspectives on Animal Cognition*. The MIT Press, Cambridge, MA (2002) 437–442
3. Darwin, C.: *The Descent of Man and Selection in Relation to Sex* [1871]. Princeton University Press, Princeton, NJ (1981)
4. Darwin, C.: *The Formation of Vegetable Mould, through the Action of Worms with Observations on their Habits* [1881]. University of Chicago Press, Chicago (1985)
5. Crist, E.: The inner life of eartworms: Darwin's argument and its implications. In Bekoff, M., Allen, C., Burghardt, M., eds.: *The Cognitive Animal. Empirical and Theoretical Perspectives on Animal Cognition*. The MIT Press, Cambridge, MA (2002) 3–8

6. Grau, J.W.: Learning and memory without a brain. In Bekoff, M., Allen, C., Burghardt, M., eds.: *The Cognitive Animal. Empirical and Theoretical Perspectives on Animal Cognition*. The MIT Press, Cambridge, MA (2002) 77–88
7. Magnani, L.: *Abduction, Reason, and Science. Processes of Discovery and Explanation*. Kluwer Academic/Plenum Publishers, New York (2001)
8. Tooby, J., DeVore, I.: The reconstruction of hominid behavioral evolution through strategic modeling. In Kinzey, W.G., ed.: *Primate Models of Hominid Behavior*. Suny Press, Albany (1987) 183–237
9. Pinker, S.: *How the Mind Works*. W.W. Norton, New York (1997)
10. Pinker, S.: Language as an adaptation to the cognitive niche. In Christiansen, M.H., Kirby, S., eds.: *Language Evolution*. Oxford University Press, Oxford (2003)
11. Novoplansky, A.: Developmental plasticity in plants: implications of non-cognitive behavior. *Evolutionary Ecology* **16** (2002) 177–188
12. Sachs, T.: Consequences of the inherent developmental plasticity of organ and tissue relations. *Evolutionary Ecology* **16** (2002) 243–265
13. Grime, J.P., Mackey, J.M.L.: The role of plasticity in resource capture by plants. *Evolutionary Ecology* **16** (2002) 299–307
14. Peirce, C.S.: *Collected Papers of Charles Sanders Peirce*. Harvard University Press, Cambridge, MA (1931-1958) vols. 1-6, Hartshorne, C. and Weiss, P., eds.; vols. 7-8, Burks, A.W., ed.
15. Peirce, C.S.: *The Charles S. Peirce Papers: Manuscript Collection in the Houghton Library*. The University of Massachusetts Press, Worcester, MA (1967) Annotated Catalogue of the Papers of Charles S. Peirce. Numbered according to Richard S. Robin. Available in the Peirce Microfilm edition. Pagination: CSP = Peirce / ISP = Institute for Studies in Pragmaticism.
16. Mithen, S.: *The Prehistory of the Mind. A Search for the Origins of Art, Religion and Science*. Thames and Hudson, London (1996)
17. Donald, M.: *A Mind So Rare. The Evolution of Human Consciousness*. W.W. Norton and Company, New York (2001)
18. Bermúdez, J.L.: *Thinking without Words*. Oxford University Press, Oxford (2003)
19. Magnani, L.: Mimetic minds. Meaning formation through epistemic mediators and external representations. In Loula, A., Gudwin, R., Queiroz, J., eds.: *Artificial Cognition Systems*. Idea Group Publishers, Hershey, PA (2006) 327–357
20. Saidel, E.: Animal minds, human minds. In Bekoff, M., Allen, C., Burghardt, M., eds.: *The Cognitive Animal. Empirical and Theoretical Perspectives on Animal Cognition*. The MIT Press, Cambridge, MA (2002) 53–58
21. Kuhn, T.S.: *The Structure of Scientific Revolutions*. University of Chicago Press, Chicago (1962)
22. Quine, W.V.O.: *Word and Object*. Cambridge University Press, Cambridge (1960)
23. Lakatos, I.: Falsification and the methodology of scientific research programs. In Lakatos, I., Musgrave, A., eds.: *Criticism and the Growth of Knowledge*, Cambridge, MA, MIT Press (1970) 365–395
24. Rivas, J., Burghardt, G.M.: Crotaloporphysm: a metaphor for understanding anthropomorphism by omission. In Bekoff, M., Allen, C., Burghardt, M., eds.: *The Cognitive Animal. Empirical and Theoretical Perspectives on Animal Cognition*. The MIT Press, Cambridge, MA (2002) 9–18

25. Fodor, J.: *Psychosemantics*. The MIT Press, Cambridge, MA (1987)
26. Magnani, L.: Semiotic brains and artificial minds. How brains make up material cognitive systems. In Gudwin, R., Queiroz, J., eds.: *Semiotics and Intelligent Systems Development*. Idea Group Inc., Hershey, PA (2007) 1–41
27. Dummett, M.: *The Origins of Analytical Philosophy*. Duckworth, London (1993)
28. Lévi-Bruhl, L.: *Primitive Mentality*. Beacon Press, Boston (1923)
29. Evans, C.S.: Cracking the code. Communication and cognition in birds. In Bekoff, M., Allen, C., Burghardt, M., eds.: *The Cognitive Animal. Empirical and Theoretical Perspectives on Animal Cognition*. The MIT Press, Cambridge, MA (2002) 315–322
30. Sherry, D.S.: Food storage, memory, and marsh tits. *Animal Behavior* **30** (1988) 631–633
31. Magnani, L., Bardone, E.: Sharing representations and creating chances through cognitive niche construction. The role of affordances and abduction. In Iwata, S., Oshawa, Y., Tsumoto, S., Zhong, N., Shi, Y., Magnani, L., eds.: *Communications and Discoveries from Multidisciplinary Data*, Berlin, Springer (2007) Forthcoming.
32. Heinrich, B.: Raven consciousness. In Bekoff, M., Allen, C., Burghardt, M., eds.: *The Cognitive Animal. Empirical and Theoretical Perspectives on Animal Cognition*. The MIT Press, Cambridge, MA (2002) 47–52
33. Gibson, J.J.: *The Ecological Approach to Visual Perception*. Houghton Mifflin, Boston, MA (1979)
34. Brunswik, E.: *The Conceptual Framework of Psychology*. University of Chicago Press, Chicago (1952)
35. Brooks, R.A.: Intelligence without representation. *Artificial Intelligence* **47** (1991) 139–159
36. Domenella, R.G., Plebe, A.: Can vision be computational? In Magnani, L., Dossena, R., eds.: *Computing, Philosophy and Cognition*, London, College Publications (2005) 227–242
37. Beers, R.: Expressions of mind in animal behavior. In Mitchell, W., Thomson, N.S., Miles, H.L., eds.: *Anthropomorphism, Anecdotes, and Animals*. State University of New York Press, Albany, NY (1997) 198–209
38. Tirassa, M., Carassa, A., Geminiani, G.: Describers and explorers: a method for investigating cognitive maps. In Nualláin, S.Ó., ed.: *Spatial Cognition. Foundations and Applications*, Amsterdam/Philadelphia, John Benjamins (1998) 19–31
39. Roitblat, H.L.: The cognitive dolphin. In Bekoff, M., Allen, C., Burghardt, M., eds.: *The Cognitive Animal. Empirical and Theoretical Perspectives on Animal Cognition*. The MIT Press, Cambridge, MA (2002) 183–188
40. Herman, L.M.: Exploring the cognitive world of the bottlenosed dolphin. In Bekoff, M., Allen, C., Burghardt, M., eds.: *The Cognitive Animal. Empirical and Theoretical Perspectives on Animal Cognition*. The MIT Press, Cambridge, MA (2002) 275–284
41. Ben Jacob, E., Shapira, Y., Tauber, A.I.: Seeking the foundation of cognition in bacteria. From Schrödinger's negative entropy to latent information. *Physica A* **359** (2006) 495–524
42. Rock, I.: Inference in perception. *PSA: Proceedings of the Biennial Meeting of the Philosophy of Science Association* **2** (1982) 525–540
43. Thagard, P.: *Computational Philosophy of Science*. The MIT Press, Cambridge, MA (1988)

44. Hoffman, D.D.: *Visual Intelligence: How We Create What We See*. Norton, New York (1998)
45. Magnani, L.: Disembodying minds, externalizing minds: how brains make up creative scientific reasoning. In Magnani, L., ed.: *Model-Based Reasoning in Science and Engineering, Cognitive Science, Epistemology, Logic*, London, College Publications (2006) 185–202
46. Turing, A.M.: Intelligente machinery [1948]. *Machine Intelligence* **5** (1969) 3–23. Edited by B. Meltzer and D. Michie.
47. Day, R.L., Laland, K., Odling-Smee, J.: Rethinking adaptation. The niche-construction perspective. *Perspectives in Biology and Medicine* **46(1)** (2003) 80–95
48. Donald, M.: Hominid enculturation and cognitive evolution. In Renfrew, C., Mellars, P., Scarre, C., eds.: *Cognition and Material Culture: The Archaeology of External Symbolic Storage*, Cambridge, The McDonald Institute for Archaeological Research (1998) 7–17
49. Whiten, A., Byrne, R.: Tactical deception in primates. *Behavioral and Brain Sciences* **12** (1988) 233–273
50. Whiten, A., Byrne, R.: *Machiavellian Intelligence II: Evaluations and Extensions*. Cambridge University Press, Cambridge (1997)
51. Byrne, R., Whiten, A.: *Machiavellian Intelligence*. Oxford University Press, Oxford (1988)
52. Gould, J.L.: Can honey bees create cognitive maps? In Bekoff, M., Allen, C., Burghardt, M., eds.: *The Cognitive Animal. Empirical and Theoretical Perspectives on Animal Cognition*. The MIT Press, Cambridge, MA (2002) 41–46
53. Cook, R.G.: Same-different concept formation in pigeons. In Bekoff, M., Allen, C., Burghardt, M., eds.: *The Cognitive Animal. Empirical and Theoretical Perspectives on Animal Cognition*. The MIT Press, Cambridge, MA (2002) 229–238
54. Schusterman, R.J., Reichmuth Kastak, C., Kastak, D.: The cognitive sea lion: meaning and memory in laboratory and nature. In Bekoff, M., Allen, C., Burghardt, M., eds.: *The Cognitive Animal. Empirical and Theoretical Perspectives on Animal Cognition*. The MIT Press, Cambridge, MA (2002) 217–228
55. Ramsey, F.P.: Facts and propositions. *Aristotelian Society Supplementary Volume* **7** (1927) 152–170
56. Dretske, F.: *Knowledge and the Flow of Information*. The MIT Press, Cambridge, MA (1988)
57. Roberts, W.A.: Spatial representation and the use of spatial code in animals. In Gattis, M., ed.: *Spatial Schemas and Abstract Thought*, Cambridge, The MIT Press (2001) 15–44
58. Freska, C.: Spatial cognition. In Mántaras, R.L.D., Saitta, L., eds.: *ECAI 2004. Proceedings of the 16th European Conference on Artificial Intelligence*, Amsterdam, IOS Press (2000) 1122–1128
59. O'Keefe, J.: Kant and sea-horse: an essay in the neurophilosophy of space. In Elian, N., McCarthy, R., Brewer, B., eds.: *Spatial Representation. Problems in Philosophy and Psychology*, Oxford, Oxford University Press (1999) 43–64
60. Raftopoulos, A.: Reentrant pathways and the theory-ladenness of perception. *Philosophy of Science* **68** (2001) S187–S189. Proceedings of PSA 2000 Biennal Meeting.
61. Raftopoulos, A.: Is perception informationally encapsulated? The issue of theory-ladenness of perception. *Cognitive Science* **25** (2001) 423–451

62. Fodor, J.: Observation reconsidered. *Philosophy of Science* **51** (1984) 23–43 Reprinted in [91, pp. 119–139]
63. Churchland, P.M.: Perceptual plasticity and theoretical neutrality: a replay to Jerry Fodor. *Philosophy of Science* **55** (1988) 167–187
64. Spelke, E.S.: The theory-ladenness of observation and the theory-ladenness of the rest of the scientific process. *Philosophy of Science* **68** (2001) S176–S186 Proceedings of the PSA 2000 Biennal Meeting.
65. Shanahan, M.: Perception as abduction: turning sensory data into meaningful representation. *Cognitive Science* **29** (2005) 103–134
66. Tolman, E.C., Ritchie, B.F., Kalish, D.: Studies in spatial learning II. Place learning versus response learning. *Journal of Experimental Psychology* **37** (1946) 385–392
67. O'Keefe, J., Nadel, S.: *The Hippocampus as a Cognitive Map*. Oxford University Press, Oxford (1978)
68. Gallistel, C.R.: *The Organization of Learning*. The MIT Press, Cambridge, MA (1990)
69. Hershberger, W.A.: An approach through the looking glass. *Animal Learning and Behavior* **14** (1986) 443–451
70. Alpert, P., Simms, E.L.: The relative advantages of plasticity and fixity in different environments: when is it good for a plant to adjust? *Evolutionary Ecology* **16** (2002) 285–297
71. Shettleworth, S.J.: Spatial behavior, food storing, and the modular mind. In Bekoff, M., Allen, C., Burghardt, M., eds.: *The Cognitive Animal. Empirical and Theoretical Perspectives on Animal Cognition*. The MIT Press, Cambridge, MA (2002) 123–128
72. Balda, R.P., Kamil, A.C.: Spatial and social cognition in corvids: an evolutionary approach. In Bekoff, M., Allen, C., Burghardt, M., eds.: *The Cognitive Animal. Empirical and Theoretical Perspectives on Animal Cognition*. The MIT Press, Cambridge, MA (2002) 129–134
73. Piaget, J.: *Adaption and Intelligence*. University of Chicago Press, Chicago (1974)
74. Magnani, L.: Abduction and cognition in human and logical agents. In Artemov, S., Barringer, H., Garcez, A., Lamb, L., Woods, J., eds.: *We Will Show Them: Essays in Honour of Dov Gabbay*, London, College Publications (2007) 225–258 vol. II.
75. Spelke, E.S.: Principles of object segregation. *Cognitive Science* **14** (1990) 29–56
76. Fugelsang, J.A., Roser, M.E., Corballis, P.M., Gazzaniga, M.S., Dunbar, K.N.: Brain mechanisms underlying perceptual causality. *Animal Learning and Behavior* **24(1)** (2005) 41–47
77. Magnani, L., Belli, E.: Agent-based abduction: being rational through fallacies. In Magnani, L., ed.: *Model-Based Reasoning in Science and Engineering. Cognitive Science, Epistemology, Logic*, London, College Publications (2006) 415–439
78. Rizzolatti, G., Carmada, R., Gentilucci, M., Luppino, G., Matelli, M.: Functional organization of area 6 in the macaque monkey. II area F5 and the control of distal movements. *Experimental Brain Research* **71** (1988) 491–507
79. Gallese, V.: Intentional attunement: a neurophysiological perspective on social cognition and its disruption in autism. *Brain Research* **1079** (2006) 15–24
80. Dennett, D.: *The Intentional Stance*. The MIT Press, Cambridge, MA (1987)

81. Jamesion, D.: Cognitive ethology and the end of neuroscience. In Bekoff, M., Allen, C., Burghardt, M., eds.: *The Cognitive Animal. Empirical and Theoretical Perspectives on Animal Cognition.* The MIT Press, Cambridge, MA (2002) 69–76
82. Tomasello, M., Call, J.: *Primate Cognition.* Oxford University Press, New York (1997)
83. Monekosso, N., Remagnino, P., Ferri, F.J.: Learning machines for chance discovery. In Abe, A., Oehlmann, R., eds.: *Workshop 4: The 1st European Workshop on Chance Discovery*, Valencia, Spain (2004) 84–93
84. Wilcox, S., Jackson, R.: Jumping spider tricksters: deceit, predation, and cognition. In Bekoff, M., Allen, C., Burghardt, M., eds.: *The Cognitive Animal. Empirical and Theoretical Perspectives on Animal Cognition.* The MIT Press, Cambridge, MA (2002) 27–34
85. Stern, D.N.: *The Interpretation World of Infants.* Academic Press, New York (1985)
86. Mitchell, R.W.: Kinesthetic-visual matching, imitation, and self-recognition. In Bekoff, M., Allen, C., Burghardt, M., eds.: *The Cognitive Animal. Empirical and Theoretical Perspectives on Animal Cognition.* The MIT Press, Cambridge, MA (2002) 345–352
87. Stenning, K.: Distinctions with differences: comparing criteria for distinguishing diagrammatic from sentential systems. In Anderson, M., Cheng, P., Haarslev, V., eds.: *Theory and Application of Diagrams*, Berlin, Springer (2000) 132–148
88. Jones, S., Scaife, M.: Animated diagrams. An investigation into the cognitive effects of using animation to illustrate dynamic processes. In Anderson, M., Cheng, P., Haarslev, V., eds.: *Theory and Application of Diagrams*, Berlin, Springer (2000) 231–244
89. Magnani, L.: Neuro-multimodal abduction. In: *Proceedings of the International Conference "Applying Peirce"*, Helsinki, Finland (2007) Forthcoming.
90. Magnani, L.: Multimodal abduction. External semiotic anchors and hybrid representations. *Logic Journal of the IGPL* **14(1)** (2006) 107–136
91. Goldman, A.I., ed.: *Readings in Philosophy and Cognitive Science.* Cambridge University Press, Cambridge, MA (1993)

Communicative Gestures Facilitate Problem Solving for Both Communicators and Recipients

Sandra C. Lozano and Barbara Tversky

Department of Psychology, Stanford University, Jordan Hall, Building 420, Stanford, CA 94305
scl@psych.stanford.edu, bt@psych.stanford.edu

Summary. Gestures are a common, integral part of communication. Here, we investigate the roles of gesture and speech in explanations, both for communicators and recipients. Communicators explained how to assemble a simple object, using either speech with gestures or gestures alone. Gestures used for explaining included pointing and exhibiting to indicate parts, action models to demonstrate assembly, and gestures used to convey narrative structure. Communicators using gestures alone learned assembly better, making fewer assembly errors than those communicating via speech with gestures. Recipients understood and learned better from gesture-only instructions than from speech-only instructions. Gestures demonstrating action were particularly crucial, suggesting that superiority of gestures to speech may reside, at least in part, in compatibility between gesture and action.

1 Introduction

The significance of gesture in communication has been noted for at least two millennia. The legendary Roman teacher of rhetoric, [36], analyzed gesture in detail in his 11-volume text on oratory. He catalogued many fine-grained aspects of gesture, including when specific hand shapes were appropriate and how to coordinate gesture with speech. The ancient Romans paid special attention to the hands' role in rhetoric. Orators were trained how to gesture to underscore points in their speeches. This practice continues. Contemporary politicians are coached on how to use their hands to make themselves appear honest, convincing, and compassionate.

Although it is generally agreed that gestures can play a number of functions in communication and in thought (see [20, 33]), much controversy surrounds the issue of whether gestures' primary function is to help communicators or recipients. According to one side of this debate, gestures' primary function is internal: gestures aid the communicator by facilitating thinking and speaking. One function gestures serve is to facilitate lexical access and increase fluency for communicators (e.g., [12, 26, 31]). When communicators sit on their hands, for example, their verbal fluency decreases, most notably for describing space [29]. When not allowed to gesture in a tip-of-the-tongue state, speakers suffer from disfluencies and impaired lexical access [15]. One hypothesis that these findings suggest is that gesturing facilitates lexical access by providing a cross-modal prime to find words [30, 33, 34].

Another possible internal function gestures serve for communicators is to help communicators organize their thinking for speaking [1, 28]. According to this viewpoint, gestures enable communicators to organize spatial representation into packages that are optimally suited for speaking. As such, gestures may reduce cognitive load for the communicator. For example, pointing improves children's performance on counting tasks [2]. Gestures can offer children and adults ways to express information that is difficult to express in speech. Notably, gestures often convey information that is different from or more sophisticated than that expressed in speech [20, 21].

According to the other side of the debate, gestures mainly serve an external function in communication: they facilitate comprehension and understanding for the recipient. This viewpoint holds that gestures have communicative functions in that they serve to make speech meaningful to recipients (e.g., [7, 14]). The strongest support for this view comes from studies on the effects of visual accessibility on gesturing. For example, when recipients are asked to guess the identity of a described but unnamed object (e.g., a fishing rod), they are faster when the communicator performs an illustrative gesture [38]. When recipients are asked to reproduce drawings described by a communicator, their reproductions are more accurate if they have interacted with a gesturing, rather than non-gesturing communicator [18]. A subsequent analysis of the discourses produced in the two conditions suggests that the advantage to recipients in the gesture condition likely comes from the illustrative quality of the hand gestures [19].

Communicators seem to believe that gestures aid conveying their messages. They gesture more if they can see their recipients than if they cannot [9, 10], they use more gestures that depict semantic content when they can see their recipients are visible than when they cannot [4]; and they change the spatial orientation of their gestures based on the location of their recipient [35].

One purpose of the present study is to shed light on this debate by reframing the question of whether gestures are communicative and considering an alternative solution in the debate, namely that gestures could help *both* communicators and recipients. In addition to seeking evidence that gestures benefit both communicators and recipients, the present study asks whether, in situations where gestures benefit both parties, it is the same or different gestures that do so. In other words when communicators gesture, is it the case that one type of gestures they produce helps them while another type helps their recipients, or do the same gestures help, and help in the same manner, for both parties?

Beyond the question of *whom* gestures help, a second purpose of the present study is to address the question of *why* gestures help. Traditionally it has been claimed that gesture is ancillary to language use [7, 33]. If a communicator's gestures change, it is because a communicator's language has changed [28, 32]. Thus, in situations where gestures benefit communication, whether for the communicator or recipient, it is because they are doing so *indirectly* – that is, they help because they augment or improve the quality of the speech they accompany [22, 27, 33].

Or do gestures help *directly* – for example, by conveying specific semantic content? Although it is possible that gestures help communication by improving speech, evidence from the literature on embodied cognition suggests an alternate possibility – that gestures help, not by virtue of how they change speech, but rather by in and of themselves. For example, theories of embodied cognition propose that knowledge is rooted in perceptual experience, which in turn, guides action in the world (e.g., [6, 16]). Many communicative gestures are like miniature actions and in this way can provide a communicator with motor experience that can guide knowledge acquisition and learning.

One piece of evidence supporting the idea that gestures benefit communication because they are miniature actions comes from research on the compatibility between motor actions and conceptual tasks: Observers were asked to judge whether a cup was right-side-up or upside down [42]. On half of the trials, the cup's handle was on the same side of the display as the hand a participant responded with, while for the other half of the trials the cup's handle was on the opposite side. Even though the handle position was irrelevant to the judgment, participants responded faster if the handle was on the same side they responded from. This suggests that potential motor actions, such as reaching for the cup, affect perceptual judgments, even when those actions are not directly relevant to the judgment.

Other tasks show effects of compatibility of actions to cognition. Participants made sensibility judgments on sentences that implied a particular direction of movement by pushing a lever toward or away from themselves [17]. Participants responded more quickly when sentence meaning was compatible with response direction (e.g., responding to "Open the drawer" by pulling a lever, or responding to "Close the drawer" by pushing a lever). This action-meaning compatibility effect occurred not only for physical actions but also for abstract ones. For example, participants were faster to make sensibility judgments for sentences like "Liz told you the story" by pulling a lever backward and for sentences like "You told Liz a story" by pushing a lever forward.

Yet another example comes from a study in which participants described action cartoons to listeners [37]. For half of the cartoon retellings, participants were not allowed to gesture. When gesturing was prevented participants were reliably slower to describe spatial content in the cartoons, but showed no impairment in describing nonspatial content. One interpretation is that gesturing provided a way of embodying the spatial content, so that preventing gesture impaired access to the spatial elements of the cartoon representation.

Collectively, these results highlight the ways in which gesture can directly, rather than indirectly, enhance knowledge and learning. Perceptually grounded knowledge guides action in the world, and gesture offers a more direct way than speech of conveying and acquiring this knowledge. Thus, the extreme prediction arising from theories of embodied cognition is that communicators will learn action better if forced to communicate through gestures only rather than communicating partially through gesture and partially through speech.

In order to test this prediction and to gain a better understanding of when and why gestures facilitate learning, the present studies begin by examining what types of gestures people produce in order to convey different semantic information about a task. They then turn to the questions of whether the gestures used in communication can facilitate task learning and performance and, if so, whether certain types of gestures are more beneficial than others to learning.

We chose a task designed to elicit explanatory gestures, namely, demonstrating how to put a piece of furniture together. People enjoy this task and can accomplish it in a reasonable amount of time, though they often make assembly errors [25]. Learning to assemble is representative of many tasks and complex concepts that people need to learn and explain in everyday life, such as how to operate things, how things work, how to put things together, and how to carry out a set of procedures. For all of these reasoning tasks, there is a set of parts arranged in meaningful way, and a set of actions or procedures that accomplish certain goals. Demonstrating assembly of a piece of furniture entails conveying spatial, structural information – what the parts are and how they fit together – as well as action information – how to attach the parts. Space and action descriptions are known to elicit copious iconic gestures (e.g., [33, 39, 40]).

Participants learned how to assemble the piece of furniture by using a photograph of the completed furniture as a guide. They then made an instructional video in which they reassembled the furniture and explained the assembly task to others. One group of participants received no constraints on how they gave their instructions; they were free to use both speech and gestures to communicate. Another group of participants, who thought they were making a demonstration for non-English speakers, was restricted to using only gestures, but no speech, to communicate. We expected that this manipulation would increase the number of gestures that participants produced, and more importantly, that it would provide insight into which gestures were most critical for communicating the assembly task. In other words, we expected that, relative to participants communicating through speech and gestures, participants restricted to only gestural communication would show a selective increase in the types of gestures that were most conducive to task performance and comprehension. This manipulation was also expected to help in answering the question of whether gestures help indirectly, by facilitating speech, or directly, by providing action information.

2 Experiment 1: Producing Demonstrations with Gesture or Speech

In Experiment 1, people learned how to perform an assembly task and then produced videos to teach the task to others. Specifically, participants first built a television cart using a photograph of a fully built one as a guide. Then they were divided into three groups, a control group that merely re-performed the assembly task or one of two experimental groups that produced videos instructing others how to assemble. One group of instructors was allowed to speak and gesture; the other group was restricted to gesturing on the pretext that the instructive video for was non-English speakers.

This paradigm addresses questions about the nature of speech and gestures used in explanations. What gestures are used to convey what meanings? How does communicating through gestures and speech differ from communicating through gestures only? What distinguishes the types of gestures produced by those restricted to only gestural communication? Which of these modes of communication, speech with gestures or gestures only, is more effective and why?

The paradigm also addresses questions about the effects of communication mode on the communicator, in particular, on reassembly performance. Are there effects of mode, speech with gesture vs. gesture alone vs. control reassembly, on the accuracy of second-time assembly? Two different but reasonable accounts give opposite predictions about who should perform best. An *indirect facilitation hypothesis* predicts that the *Speaker* group should perform reassembly best. According to this account, gestures help communication indirectly – they facilitate speech production and improve verbal fluency. However, the gestures themselves are not what promote information understanding and learning. Thus, this account predicts that participants who communicate through a combination of gestures and speech should outperform those restricted to communication through gestures only.

An account based on the *direct embodiment hypothesis* makes the opposite prediction: it predicts that participants restricted to gestures only should perform best. Explanatory gestures are, in effect, miniature actions; thus, the similarity and compatibility between these gestures and assembly actions ought to facilitate assembly performance for *Gesturer* participants. Thus, this account would predict that the best group should be the *Gesturer* demonstrators followed by the *Speaker* demonstrators, followed by the control group of reassemblers.

2.1 Method

Participants

Thirty-seven Stanford University undergraduates participated for course credit. One participant was excluded due to failure to follow instructions. The results reported below are based on the data of 19 males and 17 females.

Materials

Experimental materials consisted of the Vandenburg Mental Rotation Test [44] and two identical television carts made by Talon Systems Inc. ®. Each TV cart measured 17 × 25" × 21" in size, and consisted of two sideboards, a lower shelf, an upper shelf, a support board, pegs for attaching the support board, screws, screwdriver, and wheels. TV cart assembly performances were recorded with a digital video camera.

Design and Procedure

At the beginning of the experiment, all participants completed the Vandenburg Mental Rotation Test of spatial ability [44]. This test correlates with behavior on a broad range of spatial tasks, including assembly of the TV cart and production of diagrammatic instructions [23]. Next, all participants learned to build the television cart by using only a photograph of a fully built one as a guide (see Figure 1). Then, each participant was randomly assigned to one of three conditions for the second assembly of the cart: *Speakers*, *Gesturers*, or *Control*. Participants had not previously been informed that this reassembly phase would occur. The photograph of the TV cart was removed during reassembly and the experimenter was not present during either assembly or reassembly.

Participants in the two communicative conditions (Speakers and Gesturers) received instructions as follows: "Many people believe that they can learn a novel task best when they see someone else show them how to do it. Now that you are knowledgeable about how to assemble a TV cart, we would like you to please make a videotape in which you clearly demonstrate to someone else

Fig. 1. Illustration of the television cart that participants studied when initially learning the assembly task.

how to do this task". The instructions then diverged on the basis of whether a participant was in the Speakers or Gesturers condition. In the Gesturers condition, speaking was prohibited; these participants were instructed that "Since your demonstration might be viewed by a non-English speaker, the videotape will not have a soundtrack. You may use gestures and any strategies other than speech that you think will best communicate cross-linguistically how to assemble the TV cart". Speakers received the following instructions: "You may use speech, gestures, and any other strategies that you think will best communicate how to assemble the TV cart". In contrast, *Control* participants received the following instructions: "We are interested in how learning and experience can improve performance on an assembly task. Now that you have had practice building a TV cart and are knowledgeable about the task, we would like you to now assemble a new, identical TV cart".

Coding

Two independent coders counted and categorized all communicative hand gestures. Categorizations were based on the functions that gestures served and are elaborated in the following section. Agreement rates between coders ranged from 92%–98% across gesture types.

2.2 Results and Discussion

We first examine the types and frequencies of gestures produced by *Speaker* vs. *Gesturer* participants. Then we look at the different groups' relative success in reassembling the TV cart. Finally, we examine whether the types of gesture used in explanation predict assembly success.

Gestures

Gestures used to identify objects. Participants often either exhibited or pointed to cart pieces during assembly in order to draw attention to the pieces. Exhibiting was defined as "an action by which a person brings a thing into a conspicuous location and manifestly holds it there for inspection" [8]. In contrast to exhibiting, pointing occurred when a person used a finger to point to or to trace a piece sitting on the table or already in hand. Both gesture types were ways of indicating and highlighting the importance of individual pieces, as Figure 2 shows.

Exhibiting and pointing never occurred in the *Control* condition, but both kinds of gestures appeared often in the communicative conditions, as is evident from Figure 3. Communicators used exhibiting gestures to identify objects significantly more times ($M = 12.71$, $SEM = 1.17$) than pointing gestures ($M = 8.13$, $SEM = 1.84$), $F(1, 22) = 4.42$, $p < .05$. Overall, *Gesturers* exhibited cart pieces significantly more times ($M = 15.08$, $SEM = 1.07$) than did *Speakers* ($M = 10.33$, $SEM = .70$), $F(1, 22) = 4.73$, $p < .05$. Likewise, *Gesturers* pointed to objects to identify them more times

Fig. 2. Examples of exhibiting (top) and pointing (bottom) gestures used to identify TV cart pieces.

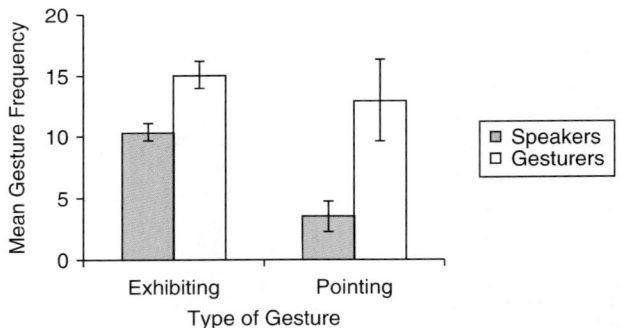

Fig. 3. Mean frequencies of pointing and exhibiting by Speakers and Gesturers. Error bars represent standard errors of means.

($M = 12.83$, $SEM = 3.34$) than did *Speakers* ($M = 3.42$, $SEM = 1.25$), $F(1, 22) = 6.04$, $p < .05$. Thus, *Speakers* used gestures to draw attention to objects that were used in assembly less often than *Gesturers* did. On the whole, communicators used exhibiting to draw attention to large moveable parts, such as the top, side, and support boards, whereas they used pointing to draw attention to small stationary ones, such as the holes for screws.

Gesturing to highlight action information. In order accommodate a viewer's perspective, participants had to perform actions in a way that was awkward for themselves. Would demonstrators value action information enough to accommodate a viewer's perspective during assembly action steps? We addressed this question by coding videotapes for the percentage of action steps that were made visible to the camera. Communicators showed assembly steps to the camera more often than Control participants, $F(2, 33) = 9.94$, $p < .01$. Also, Gesturers made reliably more steps visible to the camera ($M = 88\%$ of steps, $SEM = 3\%$), than Speakers did ($M = 59\%$ of steps, $SEM = 6\%$), $p < .01$. In providing descriptions of environments, large and small, people often make gestural models of the environment, using their hands to "sketch" a map of the space, locating landmarks within it or using their hands to "sketch" a route within a space [13]. In explaining how to assemble the TV cart, communicators also made gesture models of the actions to be performed as well as the structure of the TV cart. One technique was to provide previews of the action of each step, constructing gestural models using the hands and arms. Before starting a step, communicators often used gesture models to convey either information about actions that would occur in a step or about what the object structure would look like upon step completion. A string of gestures was coded as a model when three or more successive gestures were coordinated to portray either structure or action (see Figure 4 for examples).

Although *Speakers* ($M = 4.25$, $SEM = .68$) and *Gesturers* ($M = 3.00$, $SEM = .74$) did not differ in the total number of models that they built, $F(1, 22) = 1.57$, $p > .05$, the two groups did differ in the types of models they built. *Gesturers* modeled the structure of the cart, but even more often, they modeled the actions needed to accomplish each step. *Gesturers* used significantly more action models ($M = 1.82$, $SEM = .48$) than *Speakers*, who never gesturally modeled action ($M = 0$, $SEM = 0$), $F(1, 22) = 15.78$, $p < .01$.

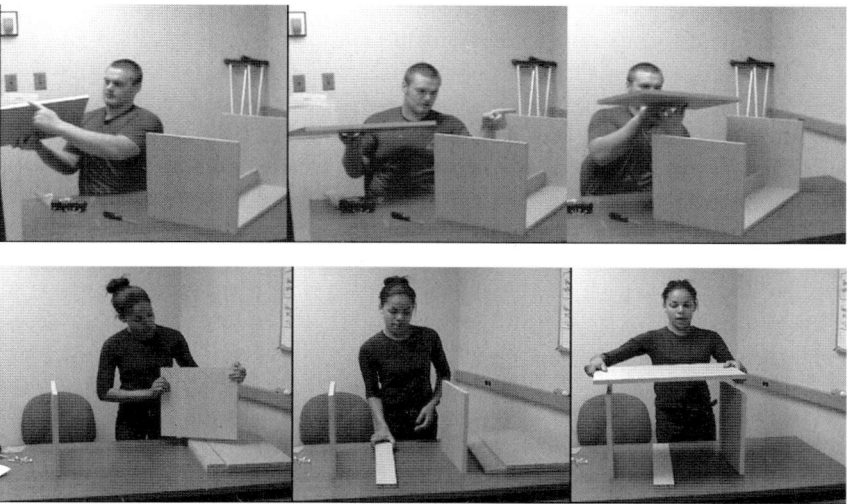

Fig. 4. Examples of a gestural action model (top) and a gestural structure model (bottom) used to communicate information prior to the beginning of actual assembly.

Thus, the major difference between the demonstrations of *Speakers* and those of *Gesturers* was in using gestures to convey action information. *Speakers* did use gestures to indicate parts, by exhibiting or by pointing. They also used gestures to indicate the structure of the TV cart. However, only *Gesturers* made action information explicit, by making sure their viewers could see the critical actions and by modeling the actions needed to perform the task. *Speakers* used words to convey this information. As we shall see, conveying action in gesture is more effective, both for communicator (this experiment) and for recipient (next experiment).

Imposing a narrative structure on demonstrations.

Unlike *Control* participants, communicators frequently imposed a narrative structure on their demonstrations. Exhibiting and pointing to parts initiated their demonstrations, much as ingredients introduce a recipe. Next, demonstrators often segmented the assembly process into discrete, meaningful action steps. For example, *Speakers* used verbal time markers to signal step initiation (e.g., "the next step is to attach the bottom shelf") or step completion (e.g., "now that these two boards are attached, this part of the cart is complete"). Although *Speakers* never used gestures to mark steps, *Gesturers* often did, in order to communicate the actions that grouped into each step (see Figure 5 for examples). Their gestures were coded for whether they signaled step initiation (e.g., holding up two fingers to signal starting the second step) or step completion (e.g., using the hand to make "okay" or "thumbs up" signs upon step completion). We then compared the frequency of usage of *Speakers'* verbal step markers to *Gesturers'* gestural step markers (see Figure 6). On average, *Speakers* marked 47% of steps ($SEM = 3\%$), while *Gesturers* marked only 31% of steps ($SEM = 4\%$), $F(1, 22) = 10.48$, $p < 0.01$. Apparently, it is easier to convey information about step time and order verbally than gesturally.

Fig. 5. Example of marking step initiation gesturally (left) and marking step completion gesturally (right).

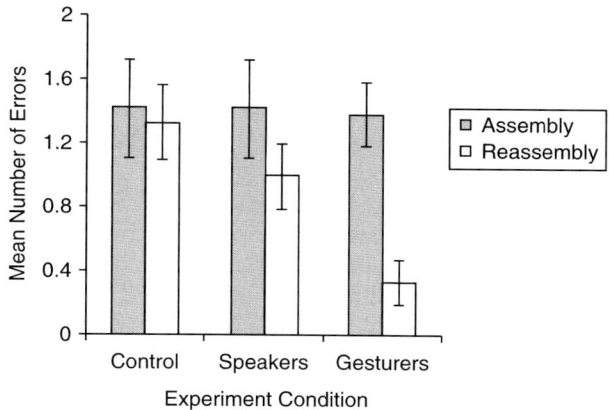

Fig. 6. Mean number of errors made during assembly and reassembly by Speakers, Gesturers, and Control participants. Error bars represent standard errors of means.

Finally, demonstrators signaled final completion of the entire assembly task. Speakers did this explicitly; 92% ($SEM = 2\%$) of Speakers signaled completion by saying things like "that's it", or "you're done". Of Gesturers, 100% ($SEM = 0\%$) signaled completion by showing or presenting the completed TV cart, usually by unfolding both hands and spreading them out toward the cart. Thus overall, Speakers and Gesturers did not differ in terms of the *types* of gestures that they produced; rather they differed with respect to the *number* of gestures produced and the *semantic content* their gestures conveyed; whereas Speakers primarily used gestures to convey structural information, Gesturers used gestures to highlight action information.

Assembly Performance

Videotapes of assembly performance were coded for total assembly time and total errors. Errors could take three forms: attaching pieces in the wrong order (e.g., building the entire cart before trying to insert the support board), attaching pieces that should not be connected to each other (e.g., attaching wheels to the top shelf), or attaching a piece in the wrong orientation (e.g., attaching the top shelf upside down).

All participants successfully assembled the TV cart using the box photograph as their only guide. The three groups made equally many mistakes when learning from the box photograph, $F(2,33) = .021$, $p > .05$, so there was no reason to think that any condition had more talented builders in it (see Figure 8). During first-time assembly, participants scoring higher in spatial ability on the Vandenburg MRT assembled the TV cart faster than those scoring low in spatial ability, $r(34) = -.41$, $p < 0.05$; this is consistent with findings from previous research [24]. Interestingly, spatial ability scores did not correlate with reassembly times or reassembly error rates, $p > .05$ for both. Assembly and reassembly times and errors also did not correlate with each other, $p > .05$.

For reassembly, *Control* participants' only job was to rebuild the cart, but communicators had the additional task of giving instructions to someone else. *Gesturers*, in particular, had to worry about giving instructions in a way that was very unfamiliar to them. This extra task could have interfered with cart building. Thus, a paired-samples t-test was conducted for each group to assess whether or not improvement in performance had occurred. Surprisingly, *Control* participants failed to improve during reassembly ($t(11) = .001, p > .05$); they made as many mistakes during reassembly ($M = 1.33, SEM = .23$) as during assembly ($M = 1.42, SEM = .31$). In contrast, *Speakers* made significantly fewer errors during reassembly ($M = 1.00, SEM = .21$) than during original assembly ($M = 1.42, SEM = .31$), $t(11) = 2.80, p < .05$. *Gesturers* also improved when rebuilding the cart ($M = .33, SEM = .14$ for reassembly vs. $M = 1.38, SEM = .20$ for assembly), $t(11) = 4.068, p < .01$; all except four of them performed reassembly perfectly, and those who still made mistakes all improved over their original assembly performance.

Although the three groups made equally many errors during original assembly, their error rates differed significantly during reassembly, $F(2, 33) = 6.70, p < .01$ (see Figure 6). *Gesturers* made reliably fewer errors ($M = .33, SEM = .14$) than *Control* participants ($M = 1.33, SEM = .23$), $p < .01$, and marginally fewer errors than *Speakers* ($M = 1.00, SEM = .21$), $p = .057$. The three groups did not differ in the total amount of time they took to build the television cart; but this likely attributable to differences in how the three groups distributed their time during the assembly task. Specifically, *Control* participants and *Speakers* spent much of their time staring at the cart pieces in confusion or correcting errors they had made. In contrast, *Gesturers* spent time in between assembly steps planning out the assembly process or trying to communicate with recipients. The superiority of the *Speakers* to the *Control* participants rules out verbal overshadowing [41] as an explanation because participants in the control condition neither gestured nor spoke. A verbal overshadowing account also cannot explain the direct relationship of specific gestures to performance, discussed next.

The Relation Between Gesture Use and Assembly Performance

If certain gestures, specifically action gestures, facilitate learning for the communicator, then frequency of these gestures should be associated with better assembly performance. In particular, greater use of action gestures should be associated with fewer assembly errors. Other kinds of gestures, such as those that convey structural information, would not necessarily correlate with assembly performance. One way communicators drew attention to action information was by making action steps visible. A second way was to create gestural models of action. Use of both correlated with assembly performance. The more action steps that communicators made visible to the

Table 1. Correlations Between Different Gesture Types and Assembly Performance.

	Action Step Visibility	Action Models	Structure Models	Exhibiting	Pointing
Action Models	$r = 0.39$ $p = 0.063$				
Structure Models	$r = -0.62$ $p = 0.001$	$r = -0.27$ $p = 0.21$			
Exhibiting	$r = 0.424$ $p = 0.039$	$r = 0.59$ $p = 0.03$	$r = -0.28$ $p = 0.33$		
Pointing	$r = 0.38$ $p = 0.067$	$r = 0.54$ $p = 0.006$	$r = -0.15$ $p = 0.50$	$r = -0.64$ $p = 0.001$	
Reassembly Errors	$r = -0.46$ $p = 0.024$	$r = -0.49$ $p = 0.015$	$r = 0.63$ $p = 0.001$	$r = -0.40$ $p = 0.052$	$r = -0.15$ $p = 0.49$

(N = 24 for all correlations reported in the above table)

camera, the fewer errors they made, $r(22) = -.46$, $p < .05$; and the more gestural action models that communicators used, the fewer assembly errors they made, $r(22) = -.49$, $p = .015$. In contrast, the more gestural structure models that communicators used the more assembly errors they made, $r(22) = .63$. Neither exhibiting nor pointing, which are ways of highlighting structural information, rather than action information, reliably predicted assembly performance ($p > .05$ for both). A full summary of the relations among frequency of different gesture types and assembly performance can be found in Table 1.

In sum, the present results support the *direct embodiment hypothesis*, and not the *indirect facilitation hypothesis*. Paying attention to action information was crucial for good assembly performance. Gestures can highlight, and indeed exemplify, action information: the more participants used gestures for this purpose, the better they performed the assembly task. Gestures that highlighted structural information did not benefit the communicator; only action-related gestures facilitated performance.

3 Experiment 2: Learning from Gestures or Speech

In the first study, participants learned and then explained how to assemble a TV cart using gestures and speech or gestures alone. Communicators who were restricted to gesture assembled the cart better than those who could communicate more naturally, using both gesture and speech. The second study is designed to shed light on two issues: Do gestures facilitate the performance of recipients, as well as of communicators; and if so, is it the same or different gestures that help recipients? Both *Speakers* and *Gesturers* used gestures in their explanations. What distinguished the *Gesturers* was that they used gestures to convey action; *Speakers* used gestures primarily to convey structure. Was the primary reason for the facilitation due to the gestural mode or to the action information, or to both? We examine those possibilities in the present experiment. Participants learned to assemble the TV cart from a video instruction, without the box photograph. The same demonstrator appeared in four videos. In half, she only spoke and in the other half, she only gestured. In half of each of those, she conveyed only structural information and in the other half, action information. Recipients watched one of the four videos and rated it for quality of explanation. Then they received a surprise recall task: they were asked to assemble the TV cart themselves.

3.1 Method

Participants

Forty-five Stanford University undergraduates participated for either course credit or monetary compensation. One participant's data was excluded due to computer failure during film viewing. The results reported below are based on the data of 21 males and 23 females.

Materials

Experimental materials consisted of the Vandenburg MRT, ratings questionnaires, demonstration films, and a TV cart. Ratings sheets contained questions about the effectiveness of instructions, ease of comprehension, and a manipulation check; ratings were made using 7-point Likert scales. All films were presented on a 21-inch, flat screen computer monitor, using a program written in PsyScope 1.2.5 [11]. The four versions of the test film presented steps and information in the same order, and all versions were equivalent in length. The only differences between films were the variables that we manipulated, namely instruction Modality and Information Type. Detailed film scripts can be found in Appendix A and Appendix B. The TV cart that appeared in the films and that Experiment 2 participants built after film viewing, was the same one as was used in Experiment 1. TV cart assembly performances were recorded with a digital video camera.

Design and Procedure

A 2 × 2 factorial between-subjects design was used. The manipulated factors were *Information Modality* (*Gesture-only* or *Speech-only*) and *Information Type* (*Action* or *Structure*). Each participant watched one of the four versions of the test film twice. Whereas participants in the *Speech-only* condition were told that they would be evaluating someone giving instructions to English speakers, participants in the *Gesture-only* condition were told that they would be evaluating someone giving instructions to non-English speakers. Aside from this, however, all participants received identical instructions. Subsequent to viewing the film twice, participants evaluated it using a rating questionnaire. When filling out their ratings questionnaires, participants were instructed to consider how easy it was to understand and learn from the demonstrator's instructions. After completing the questionnaire, all participants were presented with the surprise task of building the TV cart themselves. They were presented with all necessary assembly materials and were instructed to assemble the TV cart as quickly and as accurately as possible. None of the participants were provided with nor ever saw the box photograph. Their assembly performances were videotaped, and the experimenter was not present during assembly.

3.2 Results and Discussion

Demonstration Ratings

Would participants rate *Gesture-only* instructions as being better and easier to understand than *Speech-only* instructions? Would participants rate instructions providing *Action* information as being better and easier to understand than ones conveying *Structure* information? We addressed these questions by creating two composite scores from participants' questionnaire ratings, namely an *Effectiveness* score and a *Comprehension* score, both of which could range from 1 (low effectiveness/comprehensibility) to 7 (high effectiveness/comprehensibility). Effectiveness and Comprehension scores were positively related to each other, $r(42) = .40$, $p < 0.05$, suggesting that the two measures captured similar but non-overlapping aspects of video demonstrations. We then looked at whether either of these composite scores was influenced by *Information Modality* or *Information Type*. As can be seen in Figure 7, *Information Modality* did influence the *Effectiveness* score, $F(1, 40) = 32.75$, $p < .01$; participants rated the *Gesture-only* instructions as being more effective ($M = 6.36$, $SEM = .16$) than the *Speech-only* instructions ($M = 4.59$, $SEM = .26$). *Effectiveness* scores were not influenced by *Information Type*, $F(1, 40) = .78$, $p > .05$ and the interaction between *Information Modality* and *Information Type* was not significant, $F(1, 40) = .34$, $p > .05$. *Gesture-only* demonstrations consistently received higher *Effectiveness* ratings, regardless of what type of information was presented.

As Figure 7 also shows, *Comprehension* ratings were also influenced by *Information Modality*, $F(1, 40) = 30.82$, $p < .01$; participants rated the *Gesture-only* instructions as being easier to understand ($M = 6.05$, $SEM = .22$) than the *Speech-only* instructions ($M = 5.16$, $SEM = .36$). *Comprehension* scores were not influenced by *Information Type*, $F(1, 40) = .45$, $p > .05$ and the interaction between *Information Modality* and *Information Type* was not significant, $F(1, 40) = .18$, $p > .05$. Thus, assembly instructions were more effective and easier to understand if provided via gestures rather than speech. Participants did not report being better able to understand

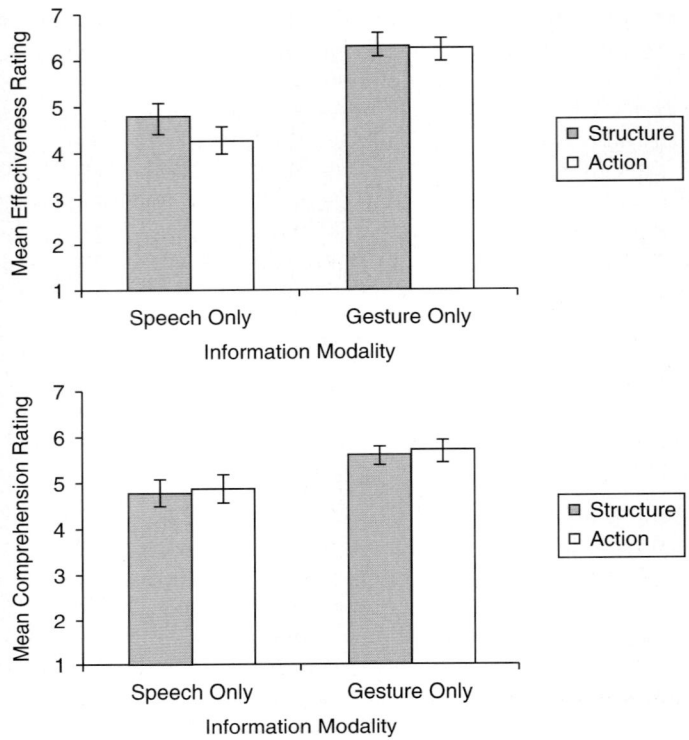

Fig. 7. Mean Effectiveness ratings (top) and mean Comprehension ratings (bottom), as a function of Information Modality and Information Type. Error bars represent standard errors of means.

Action than *Structure* instructions, but this did not necessarily preclude the possibility that they would learn differently from the two types of information.

Participants' questionnaires also contained free response sections in which they were supposed to comment on anything they especially liked or disliked in the demonstration that they saw. Their responses provide insight into not only whether they found a particular kind of demonstration useful, but also why the demonstration was (or was not) useful. Of the participants who saw a *Speech-only* demonstration, 82% reported that they did not like it because of the lack of gestures. For example, they listed complaints such as "She named the pieces but didn't point them out, which confused me"; "The beginning was confusing because she should have picked up the materials and shown them to me"; "Not everyone knows what words like parallel or perpendicular mean. Why didn't she just show me the pieces or at least point to them?" And 45% of the participants who viewed *Speech-only* films specifically commented that it was easier for them to learn the assembly task by just watching the actual assembly process than by listening to the verbal instructions. In contrast, only 9% of participants who viewed a *Gesture-only* film complained about comprehension difficulties due to the lack of speech.

Learning from Demonstrations

Do people learn the assembly task better from gestural instructions than from verbal instructions? Does the type of information (*Action* or *Structure*) influence their ability to learn the assembly task? Videotapes of assembly performance were coded for assembly time and number of errors. As in Experiment 1, there was no correlation between spatial ability and either errors or assembly time ($p > .05$ for both).

Information Modality did reliably affect assembly errors, $F(1, 40) = 61.60$, $p < .01$, and assembly time, $F(1, 40) = 6.91, p = .012$. As Figure 8 shows, participants who learned from a *Gesture-only* demonstration made fewer assembly errors ($M = .77$, $SEM = .15$) than participants who viewed a *Speech-only* demonstration ($M = 3.18$, $SEM = .29$; and they also assembled the TV cart

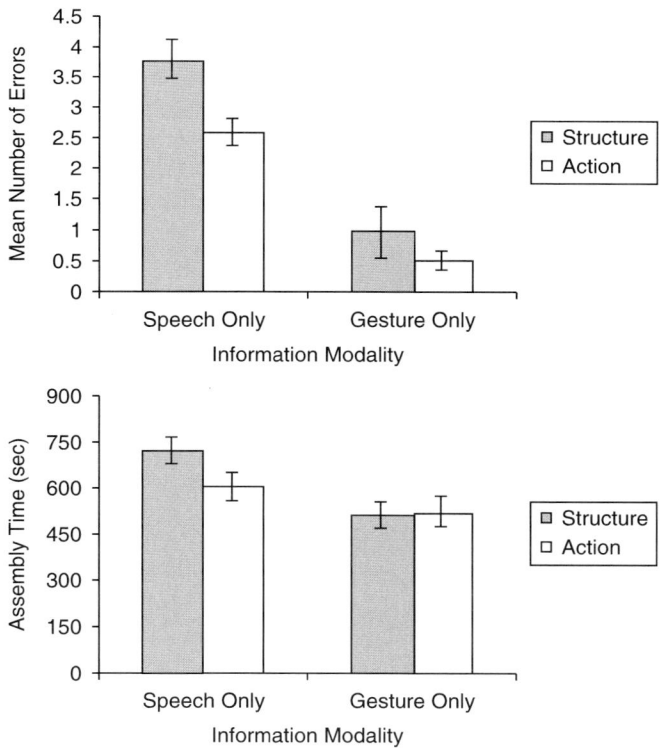

Fig. 8. Mean assembly time (top) and mean number of assembly errors (bottom), as a function of Information Modality and Information Type. Error bars represent standard errors of means.

faster ($M = 523.77$ sec, $SEM = 36.25$ sec) than did *Speech-only* participants ($M = 668.55$ sec, $SEM = 41.78$ sec). *Information Type* affected errors, $F(1, 40) = 6.34, p = .016$, but did not have a significant influence on assembly time, $F(1,40) = .90, p > .05$. Viewing a demonstration that contained *Action* information resulted in fewer assembly errors ($M = 1.59, SEM = .31$) than a demonstration that contained *Structure* information ($M = 2.36, SEM = .36$). There was not a significant interaction between *Information Modality* and *Information Type* for either errors, $F(1, 40) = 1.08, p > .05$ or for assembly time, $F(1, 40) = 1.45, p > .05$. In sum, *Gesture-only* films consistently produced better assembly performance than did *Speech-only* films; similarly, demonstrations with action information produced better assembly performance than demonstrations with only structural information.

4 General Discussion

Can communicative gestures facilitate problem solving and if so why? We investigated this in two experiments that involved explaining, learning, and evaluating explanations of a simple assembly task. Both experiments compared explanations using gesture alone or speech on performance; the first experiment examined explanations and performance of communicators and the second examined evaluations and performance of recipients. The task was assembly of a TV cart; this task is representative of a broad class of everyday problems that includes comprehending how to operate something, understanding how to put something together, and learning how something works. It is a task that students can accomplish in a reasonable amount of time, but frequently with errors. In the first experiment, people first assembled a TV cart using the photograph of the completed cart on the box as a guide. Then the photograph was removed and participants either had to simply re-perform the assembly task, or they had to re-perform it in a communicative way, to instruct someone else how to assemble the TV cart. Some communicators were allowed to use speech with gestures, while others were restricted to using gestures alone. Surprisingly, even though speaking is a more natural way to explain, those restricted to gesture, an unnatural task, reassembled the TV cart more accurately than those allowed to speak. These results supported the idea captured by the *direct embodiment hypothesis* – that gestures help communication because of the action information they directly convey – and ran counter to the idea captured by the *indirect facilitation hypothesis* – that gestures are themselves uninformative but they help communication because they facilitate speech.

In other words, gestures appeared to help communicators, not indirectly by facilitating speech, but by the more direct route of giving people valuable motor knowledge and experience. Verbally explaining while assembling appears to have induced a focus on objects used in assembly, rather than on assembly actions. The benefits of communication compatibility can be seen from the kinds of gestures both *Speakers* and *Gesturers* used, but especially in the gestures unique to *Gesturers*, namely models of action. Models of action are compatible with action, so communicating action with action models is especially beneficial, both to communicators and to recipients.

The gestures used by both *Gesturers* and *Speakers* explained the assembly task using several classes of gestures. They used deictic gestures, pointing to small parts and exhibiting large ones, to designate the parts to be assembled explicitly. They used sequences of related gestures to model and explain the desired structure of the object. Importantly, the *Gesturers* but not the *Speakers* used gestural action models to explain the actions needed for each assembly step. Finally, *Gesturers* and *Speakers* alike imposed a narrative structure on their assembly as they explained: they introduced the task by conveying the goal and showing the parts; they segmented the action into steps; with clear beginnings and ends; they ended their explanations by demonstrating the completed object. Although this narrative structure was intended for others, it seemed to benefit the communicators as well. Explaining for others is also explaining for self.

Clues to the second major effect of the first experiment, the superior reassembly of communicators restricted to gesture over those allowed the more natural task, unrestricted explaining using speech and gesture spontaneously, come from the kinds of gestures used by those who gestured without speaking. People who explained only with gestures gave better demonstrations than people who explained with both speech and gestures. *Gesturers'* demonstrations were better for three key reasons. First, *Gesturers* almost always went out of their way to accommodate a viewer's perspective but speakers did this much less often. Second, *Gesturers* made greater use of exhibiting and pointing to clearly identify and draw attention to TV cart parts. And third and most importantly, *Gesturers'* demonstrations focused on action information, whereas *Speakers'* demonstrations focused on structural and descriptive information. Of these differences, the third seems most influential. Accommodating the viewers' perspective presumably helps recipients more than it helps communicators. Clearly identifying parts may help both communicators and recipients, but it is something that is easily done in words. Moreover, *Speakers* and *Gesturers* alike used gesture to identify parts; the difference was one of degree. The third factor, using gestures to demonstrate assembly action, seems critical. *Gesturers* not only used more gestures than speakers, they also used gestures for a different purpose: *Gesturers* modeled the action of assembly whereas *Speakers* only modeled the structure of the object. Modeling the action accomplishes several ends. Modeling action requires *Gesturers* to plan actions and organize them hierarchically. Modeling action conveys the relation between actor and object for each step by showing, not by telling. Finally, modeling action is compatible with action; it is mini-actions. Speech explaining action bears no resemblance to action. Action gestures demonstrate and explain, and they also enact the actions that need to be performed.

Gesturing to explain doing, then, facilitates the doing for communicators. What about recipients? Do the gestures that communicators think will be beneficial to recipients actually help recipients? This question was addressed in the second experiment, first, by examining what information recipients believed was important and second, by examining what information actually facilitated their performance. In the second experiment, participants viewed a videotaped demonstration of how to assemble the TV cart, rated that demonstration, and then assembled their own TV cart. Four videotapes were made by the same actor, who carefully and deliberately provided step-by-step instructions as she assembled the TV cart. In two videos each, the actor used only gestures or only speech; this was crossed with structural or action information. Both gestures (as opposed to speech) and action information (as opposed to structural information) independently facilitated learning. Recipients preferred to learn the assembly task from instructions using gesture alone rather than from instructions using speech alone. They rated gesture-only instructions as being both more effective and easier to understand. Although they did not rate action information as being better or more comprehensible, recipients did learn better, making fewer assembly errors, after viewing action instructions than after viewing structural instructions, irrespective of communication modality. Learning was best when recipients viewed gesture-only instructions; speech-only instructions led recipients to perform the assembly task slower and with more errors. Thus, communicators' gestures do indeed benefit communicators and recipients alike; they seem to play the same helpful role for both parties. Let us now analyze in detail what kinds of gestures promoted performance, and why.

Why are gestures better than words, for both communicator and recipient? One proposal for the superiority of gestures to speech is that gestures are a form of embodied knowledge. Embodied approaches to cognition suggest that we acquire knowledge about the world around us by acting on it and by interacting with particular objects in particular situations [6, 16]. One of the earliest demonstrations of how perceptually grounded knowledge guides action came from Anderson and Ortony [5], who presented participants with sentences like "Pianos can be pleasant to listen to" or "Pianos are difficult to move". They found that in the former case, knowledge about pianos as musical instruments was highly accessible, whereas in the latter case knowledge about pianos' physical characteristics (e.g., their weight) was highly accessible. In other words, in the latter but not the former case, people seemed to engage in a simulation focused on action (i.e., moving a piano).

The present results are consistent with this finding. In the present studies, producing or listening to speech led participants to focus on what objects looked like and to ignore action information. Thus, they did not successfully learn how to perform the actions of TV cart assembly. In contrast using or seeing gestures made participants highly aware of what actions were important and how to perform them. Communicative gestures were like miniature task actions; they provided participants with the perceptual knowledge and motor experience needed to guide action learning.

A related but different proposal for why gestures alone are superior is that gestures serve to focus attention on particular aspects of a discourse, implicitly, the most important aspects of the discourse (e.g., [20]). Producing gestures in the absence of speech requires extra attention from communicators: they must figure out how to convey information in gestures that they would normally convey in speech; and they must make sure that no ambiguous information is carried by the gestures since they are the sole mode of communicating. Similarly, interpreting and learning from gestures without the support of speech requires added attention from recipients. Any complete account of the facilitation of gestures is bound to have more than one factor, and focusing attention may be one of them. However, it cannot be the only factor, as it is clear from the data that the kinds of gestures used are critical over and above the fact of gesturing. This suggests that gestures carry significant semantic content in and of themselves beyond emphasizing what is conveyed in the language.

What gestures carry that semantic content? Two types of gestures were prevalent in the explanations: deictic and iconic. Deictic gestures are traditionally defined as "finger points or other indications of concrete or imaginary objects or people", whereas iconic gestures "represent attributes, actions, or relationships of objects or characters" [33, p. 377]. In the present task, deictic gestures were primarily the points and exhibits and iconic gestures were primarily the models of structure and action. Which of these kinds of gestures were responsible for the gains for demonstrators and recipients? The deictic gestures seemed to play only a minor role in organizing and conveying the information essential for performing the task. The deictics refer, and words do an excellent job of referring. The parts had names, some created on the fly, which easily served that role. Rather than the deictics, the iconic gestures conveyed the essential information, but not all the iconics. It was a particular subset of the iconic gestures, namely, the action models, that facilitated assembly performance.

This specific facilitation to learning from a particular class of gestures suggests a further explanation of why gestures were superior to words, specifically that gestures were more effective because of their similarity to actual assembly actions. The gestures that people used to communicate, in particular the action models, were in effect, miniature task actions; by gesturing, people were, in effect, "practicing" the task they were about to perform. The action models had a clear advantage for communicators. Those who only gestured made models of action and assembled better than those who gestured and spoke; those who gestured and spoke made gesture models of structure rather than action. Furthermore, within those who only gestured, the number of action models predicted performance. For recipients, action information and gestures, rather than speech, augmented performance. Thus, it is a particular type of gestures, namely iconic gestures demonstrating action, that accounts for the superior performance of gesture over speech for both communicators and recipients. Gestures are uniquely suited to convey action as they are actions.

Communicating with gestures bears interesting comparisons to communicating with diagrams; they appear to have a common semantics and structure. Both modes have a modicum of iconicity, so that they can show rather then tell. A parallel project in which participants assembled the TV cart using the box photograph and then produced instructions, verbal, diagrammatic, or both for others to assemble revealed similar features in language and diagrams, features that also appear in gestures [25]. Like the gestural explanations, the descriptive and depictive explanations had a narrative structure, an introduction in which the parts were listed in words or pictures, a sequence of discrete steps, and an ending, the completed cart. The highly rated diagrams showed the perspective of action, just like the more effective gesturers. Highly-rated diagrams also indicated assembly actions, not just structure. Diagrams are uniquely appropriate for conveying structure; they use elements and spatial relations on paper to convey elements and spatial relations in the world (e.g., [43]). Explainers added arrows to diagrams to convey action. In other research, diagrams that added arrows to structure led to better understanding of mechanical systems [25] and to better assembly of the TV cart [23]. Gestures leave no trace, so although they can convey structure, they are less appropriate for conveying structure than diagrams as they rely on memory to combine parts. Gestures naturally enact so they are uniquely appropriate for conveying action.

Does the compatibility of gesture and action mean that the superiority of gestures over speech for both communicators and recipients is restricted to domains where action information is essential? We think not. Previous research has shown that specific gestures reflect the content of thought. For example, problem solvers using a discrete solution to an arithmetic problem use discrete gestures and problems solvers using a continuous solution use continuous gestures [3]. Other studies have shown that generic gestures can facilitate general word finding (e.g., [30]). Few studies have matched the content of gestures to the content of the thought that gestures facilitate, as in this task, where gesturing action facilitated performing action. But gestures do far more than demonstrate action. For one thing, gestures are effective at conveying structure; it is just that the present task we investigated depends on action. Since gestures can easily demonstrate elements and spatial relations, and language can only do this abstractly, we believe that gestures will be more effective than language alone at conveying structure. A considerable portion of information that people convey and learn depends on either structure or action or both. Think of the stories we tell each other at the end of the day, what happened where and when. Then there is metaphor. Another large portion of information that people convey is metaphorically based on structure or action, the relations among entities, and the ways that the entities and relations change. So there is every reason to believe that gesture goes a long way in human communication

Here, we have shown the powerful role that gesture plays in shaping mental representations of action and illuminating their contents, not just for communicators but also for recipients. The findings suggest that gesture might facilitate thought, performance, and learning in other domains as well.

A

The following is the instructional script used in the *Structure* information condition in Experiment 2. The speech used in *Speech-only* videos appears in regular font. The communicative gestures used in *Gesture-only* videos appear in bold italic font. The assembly actions seen by participants in all conditions appear in regular italic font.

1. You have four white pegs.
   Exhibits and counts pegs
   Places pegs in upper left corner of table in a line
2. You have eight silver screws.
   Exhibits and counts screws
   Places screws on table in a line below pegs
3. You have four black wheels.
   Exhibits and counts wheels
   Places wheels on table in a line below screws
4. And you have one screwdriver.
   Exhibits screwdriver
   Places screwdriver on table below wheels
5. You also have 2 square sideboards. They each have six holes on the inside and four holes on the outside.
   Exhibits each sideboard, points to each of the six holes on the inside, and then points to each of the four holes on the outside
   Places sideboards in lower right corner of table
6. You have a long thin support board. It is the smallest board and has two holes on each end.
   Exhibits all surfaces of support board and points to the two holes on each end
   Places support board on table
7. You have a large rectangular board, which is the bottom shelf.
   Exhibits all surfaces of bottom shelf
   Places bottom shelf on table above support board
8. And your largest rectangular board is the top shelf.
   Exhibits all surfaces of top shelf
   Places top shelf on top of bottom shelf
9. To begin assembly, the top shelf should lie flat in the center of the table. The first square sideboard stands upright and attaches perpendicular to the top shelf.

Exhibits top shelf and exhibits first sideboard; then poses sideboard perpendicular to top shelf; then exhibits a pair of screws and the screwdriver
Places top shelf upside down in the center of the table and attaches first sideboard to top shelf on her left side, starting with upper corner and then moving to lower corner

10. Both ends of the support board have holes for the white pegs. The first sideboard also has holes on the inside.
    Exhibits support board and poses it on top of top shelf; then exhibits the pegs
    Inserts 4 pegs to support board and then attaches it to first sideboard
11. Just like the first sideboard, the second sideboard has two holes on the inside.
    Exhibits second sideboard and poses it parallel to first sideboard
    Attaches second sideboard to support board
12. The second sideboard has two holes that match the remaining top shelf holes.
    Points to holes on second sideboard; then exhibits a pair of screws and the screwdriver
    Attaches second sideboard to top shelf on her right side, starting with upper corner and then moving to lower corner
13. The bottom shelf goes in between the two sideboards. It has two holes on one side that match with the remaining two holes of the second sideboard.
    Exhibits bottom shelf and poses it parallel to and above top shelf; then exhibits a pair of screws and the screwdriver
    Attaches bottom shelf to second sideboard, on her right side, starting with lower corner and then moving to upper corner
14. The bottom shelf has two holes on the other side that match with the remaining two holes of the first sideboard.
    Points to holes on bottom shelf; then exhibits a pair of screws and the screwdriver
    Attaches bottom shelf to first sideboard, on her left side, starting with lower corner and then moving to upper corner
15. There are four wheels and four holes on the bottom edges of the two sideboards.
    Exhibit the wheels
    Inserts wheels into holes, going from her upper left, to upper right, to lower right, to lower left
16. And here is the completed television cart.
    Waves open palms over top of cart to signal task completion
    Flips cart into upright position

B

The following is the instructional script used in the *Action* information condition in Experiment 2. The speech used in *Speech-only* videos appears in regular font. The communicative gestures used in *Gesture-only* videos appear in bold italic font. The assembly actions seen by participants in all conditions appear in regular italic font.

1. You have four pegs, which you will use to attach the small support board.
   Exhibits pegs and shows how/where they attach to support board
   Places pegs in upper left corner of table in a line
2. You have eight screws, which you will use to attach the top and bottom shelves to the sideboards.
   Exhibits screws and shows how/where they attach to the boards
   Places screws on table in a line below pegs
3. You have four wheels, which you will insert to the holes on the bottom edges of the sideboards.
   Exhibits wheels and shows how/where they attach to the boards
   Places wheels on table in a line below screws
4. And you have one screwdriver, which you will use for tightening all the screws.
   Exhibits screwdriver, points to the screws, and shows which direction to turn screwdriver
   Places screwdriver on table below wheels
5. You also have sideboards, to which you will attach the top shelf, support board, bottom shelf, and wheels.
   Exhibits all surfaces of sideboards and shows how/where they attach to top and bottom shelves
   Places sideboards in lower right corner of table
6. You have a support board, into which you will insert pegs and then attach to the inside holes of the sideboards.
   Exhibits all surfaces of support board and shows how/where it attaches to sideboards
   Places support board on table
7. You have a bottom shelf, which you will attach in between the sideboards.
   Exhibits all surfaces of bottom shelf and shows how/where it attaches to sideboards
   Places bottom shelf on table above support board
8. And you have a top shelf, which you will attach above the support board.
   Exhibits all surfaces of top shelf and shows how/where it attaches to sideboards
   Places top shelf on top of bottom shelf
9. To begin assembly, position the top shelf and first sideboard perpendicular to each other and so that their finished surfaces face the same direction. Line up the holes of the first sideboard and the top shelf. Insert screws to the aligned holes and then tighten each screw with your screwdriver.

Exhibits top shelf and first sideboard; traces finished edge of top shelf and points at camera; traces its unfinished edge and points behind her; traces its finished surface and points at table; traces finished surface of first sideboard and points at camera; points to its support board holes and then points toward inside of structure; exhibits screws and pantomimes inserting them to top shelf and sideboard
Places top shelf upside down in the center of the table and attaches first sideboard to top shelf on her left side, starting with upper corner and then moving to lower corner

10. Insert the pegs to the holes in the support board. Position the support board on top of the top shelf and so that its finished surface matches that of the structure. Slide the pegs in the support board into the holes on the inside of the first sideboard.
Pantomimes inserting pegs to support board; traces finished surface of support board and points at camera; traces unfinished surface and points behind her; pantomimes attaching support board to structure
Inserts 4 pegs to support board and then attaches it to first sideboard

11. Position the second sideboard parallel to the first sideboard and so that its finished surfaces match that of the structure. Attach it to the pegs in the support board.
Traces finished edge of second sideboard and points at camera; points back and forth between pegs and holes for them on second sideboard; pantomimes attachment of second sideboard and support board
Attaches second sideboard to support board

12. Line up the holes of the second sideboard and the top shelf. Insert screws to the aligned holes and then tighten each screw with your screwdriver.
Exhibits screws and pantomimes inserting them to top shelf and second sideboard
Attaches second sideboard to top shelf on her right side, starting with upper corner and then moving to lower corner

13. Position the bottom shelf in between the two sideboards and so that its finished surfaces match that of the structure. Line up the holes of the bottom shelf and the second sideboard. Insert screws to the aligned holes and then tighten each screw with your screwdriver.
Exhibits bottom shelf and pantomimes positioning it between sideboards; traces finished edge of bottom shelf and points at camera; traces its unfinished edge and points behind her; traces its finished surface and points at table; points back and forth between its holes and corresponding holes on second sideboard; exhibits screws and pantomimes inserting them to bottom shelf and sideboard

Attaches bottom shelf to second sideboard, on her right side, starting with lower corner and then moving to upper corner
14. Line up the holes on the other side of the bottom shelf with the first sideboard. Insert screws to the aligned holes and then tighten each screw with your screwdriver.
    Exhibits screws and pantomimes inserting them to bottom shelf and first sideboard
    Attaches bottom shelf to first sideboard, on her left side, starting with lower corner and then moving to upper corner
15. Insert the wheels into the four holes on the bottom edges of the two sideboards.
    Exhibits wheels and pantomimes attaching them to the holes in the sideboards
    Inserts wheels into holes, going from her upper left, to upper right, to lower right, to lower left
16. And here is the completed television cart.
    Waves open palms over top of cart to signal task completion
    Flips cart into upright position

Acknowledgement. We wish thank Bridgette Martin, Julie Heiser, Angela Kessell, Teenie Matlock, and Herb Clark for valuable advice and assistance. We also thank Jane Solovyeva and Karen Holmquist for expert help in executing the experiments and analyzing the data. Portions of this research were supported by Office of Naval Research, Grants Number NOOO14-PP-1-O649, N000140110717, and N000140210534 to Stanford University and a grant from the Edinburgh-Stanford Link at the Center for the Study of Language and Information at Stanford University. Reprinted from *Journal of Memory and Language* 55(1) 47-53 Copyright (2006), with permission from Elsevier.

References

1. Alibali, M.M., Kita, S., Young, A.: Gesture and the process of speech production: We think, therefore we gesture. *Language and Cognitive Processes* **15** (2000) 593–613
2. Alibali, M.W., DiRusso, A.A.: The function of gesture in learning to count: More than keeping track. *Cognitive Development* **14** (1999) 37–56
3. Alibali, M.W., Bassok, M., Olseth Solomon, K., Syc, S.E., Goldin-Meadow, S.: Illuminating mental representation through speech and gesture. *Psychological Science* **10** (1999) 327–333
4. Alibali, M.W., Heath, D., Myers, H.: Effects of visibility between speaker and listener on gesture production: Some gestures are meant to be seen. *Journal of Memory and Language* **44** (2001) 169–188
5. Anderson, R.C., Ortony, A.: (1975). On putting apples into bottles: A problem of polysemy. *Cognitive Psychology*, 7, 167–180
6. Barsalou, L.W.: Perceptual symbol systems. *Behavioral and Brain Sciences* **22** (1999) 577–609

7. Clark, H.H.: *Using language*. Cambridge: Cambridge University Press. (1996)
8. Clark, H.H., Krych, M.A.: Speaking while monitoring addressees for understanding. *Journal of Memory and Language* **50** (2004) 62–81
9. Cohen, A.A.: The communicative function of hand gestures. *Journal of Communication* **27** (1977) 54–63
10. Cohen, A.A., Harrison, R. P.: Intentionality in the use of hand illustrators in face-to-face communication situations. *Journal of Personality and Social Psychology* **28** (1973) 276–279
11. Cohen, J.D., MacWhinney, B., Flatt, M., Provost, J.: PsyScope: An interactive graphic system for designing and controlling experiments in the psychology laboratory using Macintosh computers. *Behavior Research, Methods, Instruments & Computers* **25** (1993) 257–271
12. Dittmann, A.T., Llewelyn, L.G.: Body movement and speech rhythm in social conversation. *Journal of Personality and Social Psychology* **23** (1969) 283–292
13. Emmorey, K., Tversky, B., Taylor, H.A.: Using space to describe space: Perspective in speech, sign, and gesture. *Journal of Spatial Cognition and Computation* **2** (2000) 157–180
14. Engle, R.A.: Not channels but composite signals: Speech, gesture, diagrams, and object demonstrations are integrated in multimodal explanations. In M.A. Gernsbacher & S.J. Derry, eds., *Spatial language: Cognition and computational perspectives*. Kluwer Academic, Dordrecht, The Netherlands (1998)
15. Frick-Horbury, D., Guttentag, R.E.: The effects of restricting hand gesture production on lexical retrieval and free recall. *American Journal of Psychology* **111** (1998) 43–62
16. Glenberg, A.M.: What memory is for. *Behavioral and Brain Sciences* **2** (1997) 1–55
17. Glenberg, A.M., Kaschak, M.P.: Grounding language in action. *Psychonomic Bulletin & Review* **9** (2002) 558–565
18. Graham, J.A., Argyle, M.: A cross-cultural study of the communication of extra-verbal meaning by gestures. *International Journal of Psychology* **10** (1975) 57–67
19. Graham, J.A., Heywood, S.: The effects of elimination of hand gestures and of verbal codability on speech performance. *European Journal of Social Psychology* **5** (1975) 189–195
20. Goldin-Meadow, S.: *Hearing gesture: How our hands help us think*. Belknap Press, Cambridge (2003)
21. Goldin-Meadow, S., Alibali, M.W., Church, R.B.: Transitions in concept acquisition: Using the hand to read the mind. *Psychological Review* **100** (1993) 279–297
22. Haviland, J.: Pointing, gesture spaces, and mental maps. In D. McNeill (Ed.), *Language and gesture*. Cambridge: Cambridge University Press (2000) 13–46
23. Heiser, J., Phan, D., Agrawala, M., Tversky, B., Hanrahan, P.: Identification and validation of cognitive design principles for automated generation of assembly instructions. *Proceedings of Advanced Visual Interfaces* (2004) 311–319
24. Heiser, J., Tversky, B.: Descriptions and depictions in acquiring complex systems. In *Proceedings of the 24th annual meeting of the Cognitive Science Society*. Fairfax, VA. (2002)
25. Heiser, J., Tversky, B.: Cognitive design principles for visualizations: Revealing and instantiating. In *Proceedings of the 25th Annual Meeting of the Cognitive Science Society*. Boston, MA. (2003)

26. Hewes, G.: Primate communication and the gestural origins of language. *Current Anthropology* **14** (1973) 5–24
27. Kendon, A.: *Gesture: Visible action in utterance.* New York: Cambridge University Press. (2004)
28. Kita, S., Özyürek, A.: What does cross-linguistic variation in semantic coordination of speech and gesture reveal?: Evidence for an interface representation of spatial thinking and speaking. *Journal of Memory and Language* **48** (2003) 16–32
29. Krauss, R.M.: Why do we gesture when we speak? *Current Directions in Psychological Science* **7** (1998) 54–59
30. Krauss, R.M., Hadar, U.: The role of speech-related arm/hand gestures in word retrieval. In L. Messing & R. Campbell (Eds.), *Gesture, speech, and sign.* New York, NY: Oxford University Press. (1999) 93–116
31. Krauss, R.M., Chen, Y., Chawla, P.: Nonverbal behavior and nonverbal communication: What do conversational hand gestures tell us? In M.P. Zanna (Ed.). *Advances in experimental social psychology* **28**. San Diego, CA: Academic Press. (1996)
32. Levinson, S.: *Pragmatics.* Cambridge: Cambridge University Press. (1983)
33. McNeill, D.: *Hand and mind: What gestures reveal about thought.* University of Chicago Press, Chicago (1992)
34. McNeill, D.: Analogic/analytic representations and cross-linguistic differences in thinking for speaking. *Cognitive Linguistics Special Issue: Language Acquisition* **11** (2000) 43–60
35. Özyürek, A.: So speakers design their cospeech gestures for their addressees. The effects of addressee location on representational gestures. *Journal of Memory and Language* **46** (2002) 688–704
36. Quintilian: In H.E. Butler (Ed. and Trans.), *Institutes of oratory.* London: William Heinemann Press. (1920)
37. Rauscher, F.H., Krauss, R.M., Chen, Y.: Gesture, speech, and lexical access: The role of lexical movements in speech production. *Psychological Science* **7** (1996) 226–231
38. Riseborough, M.G.: Physiographic gestures as decoding facilitators: Three experiments exploring a neglected facet of communication. *Journal of Nonverbal Behavior* **5** (1981) 172–183
39. Rogers, W.T.: The contribution of kinesic illustrators toward the comprehension of verbal behavior within utterances. *Human Communication Research* **5** (1978) 54–62
40. Rogers, W.T.: The relevance of body motion cues to both functional and dysfunctional communicative behavior. *Journal of Communication Disorders* **12** (1979) 273–282
41. Schooler, J.W., Engstler-Schooler, T.Y.: Verbal overshadowing of visual memories: Some things are better left unsaid. *Cognitive Psychology* **22** (1990) 36–71
42. Tucker, M., Ellis, R.: On the relations between seen objects and components of potential actions. *Journal of Experimental Psychology: Human Perception and Performance* **24** (1998) 830–846
43. Tversky, B.: Spatial schemas in depictions. In M. Gattis, ed., *Spatial schemas and abstract thought.* MIT Press, Cambridge (2001) 79–111
44. Vandenburg, S.G., Kuse, A.R.: Mental rotations, a group test of three-dimensional spatial visualization. *Perceptual and Motor Skills* **47** (1978) 599–604

The Concept of Fallacy is Empty
A Resource-Bound Approach to Error

John Woods

Department of Philosophy, University of British Columbia, Vancouver, Canada,
and Department of Computer Science, King's College, London, UK
woodsj@dcs.kcl.ac.uk

Summary. Do model-based approaches to reasoning have a stake in accounting for errors of reasoning? If mainstream logic is anything to go on, a theory of bad reasoning is wholly subsumed by a theory of good reasoning, with the former construed as the complement of the latter. In an older tradition (e.g., Mill's *System of Logic*), errors are best considered as a stand-alone component of any psychologically real approach to logic. Such is the assumption of this essay. Historically, logic's almost exclusive preoccupation with error is to be found in what it may chance to say about fallacies. In the tradition that has come down to us since Aristotle, fallacies are errors of reasoning that are attractive, widely-distributed enough to be called "universal", and difficult to correct, that is, possessed of significant levels of incorrigibility.

In what follows, I sketch a resource-bound model of cognitive agency, in which, among other things, errors of reasoning are relative to an agent's cognitive targets. From this a surprising result emerges. None of the standard *list* of the fallacies is a member of the extension of the traditional *conception* of them. This, "the negative thesis", is developed here with particular attention to *ad ignorantiam* and hasty generalization. We also note that the negative thesis, if true, throws considerable light on the fact that, for all its long history, fallacies have resisted theoretically deep explication. In other words, it helps answer the question, "Why is fallacy theory so difficult?"

1 The Origins of Fallacy Theory

In recent years model-based reasoning has achieved a certain prominence among logicians and cognitive scientists[1]. Its repute is deserved, notwithstanding that it has some vigorous rivals. Although both model-based and non-model-based systems aim at elucidations of good reasoning, there are certain issues that challenge them equally across the lines of their respective theoretical and methodological differences. One of these challenges is as old as the history of systematic logic itself. It is the challenge to identify, analyze and set conditions for the avoidance of fallacious reasoning. Aristotle defined a

---

[1] See, for example, [2, 6], and a special issue of the *Logic Journal of IGPL*, 2006.

fallacy as an argument that appears to be a syllogism but is not a syllogism in fact[2]. A syllogism is a (classically) valid argument, none of whose premisses is redundant and whose conclusion is neither identical to nor immediately inferrable from any single premiss, hence is derived without circularity. Aristotle further provides that syllogisms not have multiple conclusions; and it follows from the non-circularity requirement that syllogisms not have inconsistent premisses[3]. It is widely assumed by logicians of the present-day that syllogisms are of little relevance to the concerns of contemporary logic. This is a mistake. Aristotle's syllogistic is the first ever relevant, paraconsistent, intuitionist, nonmonotonic logic[4].

Syllogisms arise in the context of an attempt by Aristotle to discipline the distinction between genuine and merely apparent (or sophistical) refutations, although it is clear that syllogisms may also serve in demonstrations, instructional arguments and examination arguments. Fallacies likewise arise in the context of refutation-arguments. But here too Aristotle sees that they are easily committable in other contexts of reasoning and argument. In the case of refutations, the difference between a genuine and sophistical refutation is that the former embeds a syllogism and the latter embeds what only appears to be a syllogism. Since mistaking a non-syllogism for a syllogism is a fallacy, then mistaking a sophistical refutation for a refutation is also a fallacy[5].

Aristotle was plainly of the view that fallacies are a seductive and common fault. The large and dense *Topics* and the shorter and more accessible *On Sophistical Refutations* devote a number of passages to fallacies and to how they might be spotted and avoided. Notwithstanding the more formal treatment of syllogisms in the *Analytics*, in these earlier treatises Aristotle is much concerned with giving the actual reasoner on the ground practical instruction by which he might be guided in the transaction of his reasoning tasks. Indeed

---

[2] Aristotle defines syllogisms in *On Sophistical Refutations,* 165^a, 1–3, repeats the definition at several other places in the *Organon*: "For a syllogism rests on certain statements such that they involve necessarily the assertion of something other than what has been stated, through what has been stated".

[3] Suppose that, contrary to fact, $< A, B, A >$ is a syllogism. Then by an operation called "argumental conversion", $< \neg A, A, \neg B >$ is also a syllogism. But since conversion is syllogistity-preserving and $< A, B, A >$ is not a syllogism, neither is $< \neg A, A, \neg B >$.

[4] Its relevance is a strict version of Anderson-Belnap's "full-use" sense of the term [3]. Its paraconsistency flows from the consistency requirement. Its nonmotonicity is secured by the premiss-irredundacy condition. Its intuitionistic character arises from the requirement that a syllogism not have multiple conclusions. (See, regarding the link between non-multiple conclusions and intuitionism, [4]. See also [5].)

[5] Aristotle describes the tie between syllogisms and refutations at *On Sophistical Refutations* 171^a, 1–5: "... it is altogether absurd to discuss refutation without first discussing syllogisms; for a refutation is a syllogism, so that one ought to discuss syllogisms before describing false [i.e. sophistical] refutation; for a refutation of that kind is merely an apparent syllogism of the contradictory of a thesis."

we may say that the founder of logic was the first applied logician. Parts of the Aristotelian taxonomy have not been preserved in what is now regarded as the traditional conception of fallacy, but there can be no serious doubt that the tradition retains much of the flavour of Aristotle's original idea of them.

It cannot be said that Aristotle had much success with his fallacies programme. His original list of thirteen sophistical refutations is discussed scatteredly throughout *On Sophistical Refutations,* mainly in chapters 4, 5, 6 and 7. Altogether there are over thirty passages in which fallacies are considered. But no one thinks that any of this comes close to forming a comprehensive and credible account. It is possible that Aristotle abandoned the task of fallacy theory owing to the almost correct proof in the *Prior Analytics* of the perfectability thesis. This is the claim that every inapparent syllogism can be shown to be a syllogism using finite methods which themselves are entirely obvious. This is an amazing feat. It comes close to showing that the property of syllogisity is effectively recognizable. It is also possible that Aristotle worked up a comprehensive account of the fallacies in texts that have not survived. A third possibility is that Aristotle quit the fallacies project on account of its difficulty.

We now leap ahead to 1970, the year in which C.L. Hamblin published *Fallacies* [6]. In that work Hamblin excoriates his fellow logicians for having given up on the fallacies programme, and he traduces writers of the introductory textbooks of the period for restricting their accounts of fallacies to ludicrous caricatures and puerile definitions. Goaded by Hamblin's criticisms, there has been a kind of renaissance of the fallacies project, especially among informal logicians [7, 8], although contributions have also been forthcoming from a scattering of logicians who are more in the logical mainstream[6]. But here too it cannot be said that the efforts of the past nearly forty years have produced much by way of a settled theoretical consensus among logicians – as much a result of neglect as of doctrinal differences. It takes little reflection to see that this very fact constitutes one of the imperatives of fallacy theory itself. It can now be expected to answer the question, "Why is fallacy theory so difficult?"

2 The Traditional Conception of Fallacy

As it has evolved since Aristotle's day, the traditional conception of fallacies encompasses a rather loose grouping. In [13], the list has eighteen entries: the *ad baculum, ad hominem, ad misericordiam, ad populum, ad verecundiam,* affirming the consequent, denying the antecedent, begging the question, equivocation, amphiboly, hasty generalization, biased statistics, composition and division, faulty analogy, gambler's and *ignorato elenchi.* [14] discusses seventeen fallacies [15] eighteen, [16] twenty-eight, and [17] only eleven. While

---

[6] See, for example, [9–12].

all these lists are pairwise disjoint, there is nonetheless a considerable overlap among them. [18, ch. 1] light-heartedly baptized his list "the Gang of Eighteen".

It has come to be widely held that, on the traditional conception, a pattern of reasoning is fallacious when four conditions are met. (1) The reasoning is *erroneous*. (2) The reasoning is *attractive*; i.e., its erroneousness is inapparent. (3) The reasoning has *universal* appeal; i.e., it is widely resorted to. (4) The reasoning is *incorrigible*; i.e., levels of post-diagnostic recidivism are high. Let us call this the *EAUI* conception of fallacy (which has the attraction of being pronounceable "Yowee"). The *EAUI* conception has had a long history, originating, as we have said, with Aristotle. Fallacy theorists in this tradition have concentrated their attention on the error-condition, and have tended to regard the other three as more or less well-understood just as they stand. This is a regrettable turn of events. No account of fallacies can pretend to completeness as long as it leaves these three conditions in their present largely unexamined state. Nor can an account of an error of reasoning proceed in a principled way without taking into account what the human reasoner's *target* is and what *resources* are available for its attainment. More particularly, the relevant account of error must disarm the objection that satisfaction of the last three conditions is reason to believe that the first condition is *not* met. For a piece of reasoning that is attractive, universal and hard to do without suggests that it might not be an error after all.

In the case of the more or less traditional list of fallacies – the Gang of Eighteen, – the above pair of observations fall into an attractive kind of possible alignment. Arising from it is a certain model in which all four conditions manage to be satisfied. In it a piece of reasoning is erroneous in relation to a target that embeds a standard that it fails to meet. It is attractive, universal and hard to do without ("incorrigible") in relation to a more modest target, embedding a lesser standard which the reasoning does meet. The reason *is* erroneous (in relation to the higher standard) and *looks not* to be erroneous because, in relation to the lower standard, it *isn't* erroneous. Accordingly, the identity, and appropriateness, of a reasoner's target and its embedded standard precede any assessment of fallaciousness. And we must take seriously the possibility that the usual run of fallacies are errors only in relation to targets that reasoners don't usually set for themselves.

I want to see whether I can muster some support for two theses about the Gang of Eighteen.

> *Negative Thesis.* The fallacies listed in the Gang of Eighteen are either not mistakes or, if they are, are not mistakes which beings like us typically commit.
>
> *Positive Thesis.* Owing to the resource- and design-limitations under which individual reasoners must operate, a significant number of the Eighteen are rationally *acceptable scant resource-adjustment strategies*. As such, they are cognitive virtues.

Both theses bear on the question of why fallacy theory has made such little progress in spite of being on logic's agenda (albeit sometimes inertly) for more than two millennia. If the negative thesis is true, the difficulty of getting fallacies right is explained by our directing our efforts at things that aren't fallacies. It is rather like trying to analyze the genetic structure of radishes by directing one's probes to marshmallows. Similarly, if the positive thesis is true, the difficulty posed by fallacy theory can be explained by the fact that the Eighteen are attractive, universal and incorrigible because they aren't errors and are, rather, generally benign methods for the adjustment of our cognitive tasks to our actual interests and our actual cognitive capacities.

3 Resisting the EAUI-*Conception*

It is necessary to pause briefly to take note of a pair of challenges to the *EAUI*-conception. On the one hand, some logicians are of the view that it is a serious distortion of Aristotle's founding idea. On the other, there are those who think that, entirely aside from what Aristotle may have thought, it is better to understand the idea of fallacy in some or other non-*EAUI* fashion. In the first group one finds Jaakko Hintikka and (perhaps less insistently) Hamblin himself. Hintikka thinks that Aristotle's idea of fallacy was not the precursor of the *EAUI*-conception, that Aristotle did not think that fallacies were errors of reasoning or inference and that fallacies are actually mistakes committed in question-and-answer games [19]. Accordingly, if logic were understood in the general manner of the syllogistic (and, afterwards, first order classical logic), fallacies would not fall within its ambit. But, in fact, since logic is actually an inherently interrogative enterprise [19] and [20], fallacies do fall within the ambit of logic, provided that logic is taken in its proper sense. Hintikka's interpretation of Aristotle is examined and rejected in [21], with a reply from Hintikka [21]. In much the same spirit, [6, ch. 8] proposes that fallacy theory might better prosper within the precincts of modern revivals of the mediaeval logics of dialogue-games. I myself have a twofold opinion of these suggestions. One is that, apart from their intrinsic merits, they are not proposals that Aristotle would have accepted. The other is that their intrinsic merits are rather dubious. True, some of the Gang of Eighteen – notably begging the question – appear to be dialectical improprieties[7]. But the great bulk of the Eighteen resist such construal; consider, for example, hasty generalization, *post hoc ergo propter hoc*, false analogy, biased statistics, gambler's composition and division, affirming the consequent and denying the antecedent.

This is not the place to settle these differences. For present purposes it suffices that I show my hand. Although there are defectors here and there, the dominant view among logicians is indeed that the *EAUI*-conception is

---

[7] In fact, question-begging is not a fallacy either. See [22].

indeed the traditional idea of the fallacies[8]. Since my task is to investigate the Eighteen under this conception, nothing more need be said about these peripheral entanglements. But before quitting this point, it is also necessary to make mention of another – and somewhat related – rump in the fallacies research community. This second group is dominated the Amsterdam School of pragma-dialectics, according to which a fallacy is simply any violation of the discourse rules that govern critical discussions [26]. A similar view is taken in the post-1982 writings of Douglas Walton, who sees a fallacy as an illegitimate move in a conversational exchange which is designed to frustrate the rightful goals of the type of dialogue that the exchange instantiates [27]. My view is that the move to define fallacies in general, and the Eighteen in particular, as dialogical improprieties has the effect of substituting a stipulated definition of fallacy for the traditional conception. There is nothing wrong as such with stipulative re-definitions that bring about conceptual change. The concept of straight line had to be adjusted to fit relativity theory and the concept of particle had to be re-configured to accommodate quantum mechanics. But, as Quine has said of the non-classical logics, with stipulative definitions the returns had better be good. For a long time, I have thought that in the case of the dialogical re-definition of fallacy the returns aren't nearly good enough [18, ch. 9][9]. I shall return to this point.

4 Errors of Reasoning

It is noteworthy that logic's historical engagement with error bears dominantly on mistakes of reasoning or misinferences. The concept of error ranges far and wide, encompassing perceptual errors, mechanical errors, faulty memories, factual misinformation, and so on. But logic's interest does not extend so far. Examination of the fallacies literature discloses a striking complacency about the error-condition. It is taken as given that invalidity and inductive weakness are errors of reasoning just as they stand. This cannot be right, since it provides, among other things, that every inductively strong argument is an error of reasoning thanks to its invalidity. But beyond that, it is quite wrong to think of invalidity and inductive weakness as errors of reasoning in their own right. Not catching this is one of formal logic's more serious failings. One of the virtues of informal logic is its re-admittance of the concept of *agency*

---

[8] Support for the *EAUI*-conception is widespread. See [23, p. 172]: "By definition, fallacy is a mistake in reasoning, a mistake which occurs with some frequency in real arguments and which is characteristically deceptive". See also [24, p. 333]. "Fallacies are the attractive nuisances of argumentation, the ideal types of improper inference. They require labels because they are thought to be common enough or important enough to make the costs of labels worthwhile [...]". Such a view is also endorsed by [25, p. 1] as "the standard conception of a fallacy in the western logical tradition [...]".

[9] A reprint of [28].

as a load-bearing item of theory. Of course, informal logic is not alone in this. Much of computer science is agent-based, as are a good many non-classical systems of logic, such as epistemic and deontic logic, situation semantics, logics of defeasible reasoning, and practical logics of cognitive systems. For the most part, however, in all these areas the analysis of reason*ing* precedes the analysis of reason*ers*. In most cases there is little or no stand-alone investigation of what reasoners are actually like – of what they are interested in and what they are capable of. To a quite large extent reasoners are merely virtual in these studies. They are posited as beings or devices that implement the theory's rules, without regard to whatever else may be true of them. This gets things in the wrong order. In what the mediaevals called *ordo cognescendi*, a realistic theory of human reasoning requires a prior and independent account of the human reasoner. So I conjecture that

> *Difficulty.* One of the reasons that fallacy theory is so difficult is that theorists have not honored the conceptual priority of reasoners over reasoning.

We may say, then, that one of the clear advantages of an agent-based theory of reasoning is that what the theorist says about reasoning *can* be informed by what he makes it his business to learn about what reasoners are like. By my own lights, this is an opportunity that reasoning theorists ignore at their peril. For one thing, trying to sort out how reasoning works and the conditions under which it is good without due regard to what reasoners are actually like is a standing invitation to massive over-idealization[10].

Perhaps the first thing to notice about human individuals is the extent to which they are *cognitive* beings. They desire to know – they have a drive to know – what to believe and how to act. It is a drive structured in such a way that its satisfaction comes about only when the desiring subject is in an

---

[10] Let us also observe in passing that when agents are admitted to logical theory in a serious way, there are two consequences of particular note. One is that logic is *pragmaticized*; that is, it takes on the general colouration of the third member of C.W. Morris' trichotomy of syntax/semantics/pragmatics [29]. A second consequence is that the admittance of agents to one's logic has the effect of *psychologizing* the logic, since if you admit agents you admit them as they come, psychological makeups and all. Taken together, the pragmaticizing and psychologizing of logic give load-bearing significance to agents not just as language-users, not just as the performers of speech acts, but also as the subjects and manipulators of cognitive states. Disdained on both scores by mainstream logicians (after all, there are no people in model theory), this is not a luxury that agent-based theories of belief dynamics, default logics, defeasible logics and practical logics can afford. Whether the mainstream likes it or not, psychologism in logic is now a re-opened research question [30, 31, 31, 32]. Similarly, the pragmatic dimension of agent-based theories has growingly taken hold since the pioneering work of [10, 33–35], and onwards to the more recent [36] and [37].

appropriate cognitive state. At a minimum it is the state of taking some or other requisite proposition as known[11].

It has long been recognized that reasoning as an aid to cognition. This being so, a theory of reasoning is asking for trouble if it fails to take into account its cognitive orientation of reasoners. So there are two constraints on an account of reasoning that a would-be theorist ignores at her peril. She can ignore the fact that the nature and the goodness of reasoning are affected by what it is like to be a reasoner. She can also overlook that reasoning is intimately connected to the transaction of the reasoner's cognitive agendas.

All of this has a bearing on the notion of error. We may now say that something is an error only in relation to a cognitive target, and that, thus relativized, an error is a failure to meet an attainment standard embedded in that target.

> *The relationality of error.* An individual cognitive agent x commits an error M in relation to his cognitive target T iff x fails to meet an attainment standard S for T.

Here is an example. Harry wants to produce a sound demonstration of a proposition of topology. He works out a proof. The proof attains Harry's objective only if it is valid. Validity is the standard here. If the proof is invalid, Harry misses his target. His error is a violation of the validity standard. Another example: You and your team are running a drug approval trial in your lab at the Department of Health. Your target is experimental confirmation of the safety or otherwise of the drug. Experimental confirmation is a lofty target, necessarily so in the present case. It embeds a tough attainment standard. It is the standard of inductive strength, usually reckoned in terms of high conditional probability on suitably (and rigorously selected) random samples. If you fail the standard of inductive strength, you have missed your target. Your error in this case is a violation of that standard.

It takes only a moment's reflection to see that the topological example and the drug trial example are far from typical, to say nothing of canonical. If Harry's target were to decide whether to attend Sarah's rather dull annual Christmas party, he would be wasting his time looking for a truth-preserving proof of the proposition that he need not attend, or an experimentally impeccable projection to the same effect from some random sample. In the circumstances, he might be better-served by looking for considerations that give that proposition defeasible or plausibilistic support.

Given the state of play in present-day approaches to defeasible, presumptive and plausibilistic reasoning, it may safely be supposed that many empirically-minded theorists would grant that by a large margin the reasoning of real-life reasoners is rarely in response to targets set so high. Still, the view persists that truth-preservation is better than experimental confirmation, which in turn is better than the more practical targets set by individual

---

[11] Ignoring here sublinguistic, subconscious cognition. See, e.g., [38, 39] and [40].

reasoners on the ground. On this view, strictest is best and, concomitantly, reasoning that satisfies the attainment standards of the strictest targets is reasoning at its most perfect. On the contrary, when one takes into account the cognitive constitution of the real-life human reasoner, the strictest-is-best thesis loses all credibility.

5 Resource-Bound Agency

I have been saying that a decent theory of reasoning must be rooted in an account of the reasoning agent. Apart from having a cognitive orientation, what else about agency should the reasoning theorist take note of? Of paramount importance is an agent's *resource-boundedness*. As anyone who has actually been one will attest, individual agents must transact their cognitive agendas under press of scant cognitive resources – resources such as information, time, storage and retrieval capacity, and computational complexity. The classical literatures on theory change, belief dynamics and decision theory – as well as much of economics, theoretical computer science and cognitive psychology – are careless about giving this fact its due[12]. Even when they recognize these features of human performance on the ground, they marginalize it as a *de facto* failing, as something that debases an individual's rationality. Seen this way, the theorist's normative ambitions must now be prosecuted in the la-la-land of perfect information and logical omniscience, under an approximation relation that no one has ever bothered to define, much less demonstrate the existence of [36].

The limitations on an individual agent's cognitive wherewithal present themselves in two main ways, both important. They are the typical scantness of the individual's resources and the typical modesty of his cognitive targets. Scantness is a *comparative* quantity. Beings like us have less information, less time and less fire-power in the exercise of our cognitive agendas than is typical of institutional agents such as Nato or MI5. Scantness therefore does not strictly imply scarcity. There are instances in which beings like us have all the resources needed to attain our cognitive targets. But often – very often – we do not. It is here that scantness turns into scarcity. It is here that we must do the best we can with what we've got. It is a mistake to see this as an assault upon rational adequacy. The rational agent is not someone who is free of setbacks and disadvantages. He is someone who knows how to manage his setbacks and to adjust to his disadvantages appropriately, that is, *sensibly* and *productively*. So, then,

> *Individual agency.* It is a defining condition of individual agency that individual agents operate under press of comparatively scant resources.

---

[12] Important exceptions are [41–44]. See also [45]. For work done independently of these contributions see [46], [36, ch. 2]. See also [47, ch. 2] and [37, ch. 2].

If, as I say, an agent's rationality is intimately bound up with how he manages his limitations – if, in other words, the Dirty Harry Principle is a principle of rationality[13] – then the second aspect of an individual's cognitive limitations becomes apparent drops out. Just as the Olympic pole-vaulter does not go into training in order to follow the cow over the moon, the rational agent tends to set himself targets whose attainment in principle lies with his means to attain.

Proportionality of target selection. The individual cognitive agent sets targets that tend to lie within his cognitive reach.

Accordingly,

Target modesty. It is typical of individual agents to set cognitive targets of comparative modesty.

Jointly, then, someone is an individual, as opposed to an institutional, agent to the extent to which his cognitive targets are selected with a circumspection that reflects the comparative paucity and frequent scarcity of his cognitive resources[14].

We are now positioned to make a fundamental limitation claim about individual agents. In the transaction of their various cognitive agendas, individuals have little occasion to regard either truth-preservation or experimental confirmation as suitable targets. So, in the cognitive lives of individual agents on the ground, it is seldom the case that either validity or inductive strength is the requisite attainment-standard. As for validity, most of the things we desire to do know do not yield to sound demonstrations, and even where they did, finding them is typically beyond the reach of beings like us. We haven't the *time* and we haven't the *fire-power* and we haven't the *need*[15]. Inductive strength is similarly positioned. It is virtually never the case that individuals have the wherewithal to generate for a proposition the kind of support that a drug trial by Health Canada would provide (if done well) or a well-confirmed scientific theory would provide. This being so, inductive strength in the logician's technical sense of the term – is hardly ever the standard in play in an individual's reasoning[16]. So we must allow that

---

[13] "A man's got to know his limitations" – Clint Eastwood, playing Harry Callaghan in the 1971 movie *Dirty Harry*.

[14] This is not to overlook that institutional agents often are required to labor under press of scarce resources. But comparatively speaking, they typically have more of them, even when they are stretched, than individual agents. And typically, even when they are stretched, this enables them to select loftier targets than individual agents are typically capable of.

[15] Of course, there is deductivism to be considered. But not here.

[16] Consider here Mill's claim in *A System of Logic* that induction is not for individuals, but for *societies*. [48].

Lightening up on validity and inductive strength. As a default position, one must *not* attribute to an individual reasoner cognitive targets of which validity or inductive strength are the necessary attainment-standards.

Moreover,

Not judging them harshly. As a further default position, a finding that an agent's reasoning is either invalid or inductively decrepit is *not* as such a reason to judge it negatively.

At this point I imagine that readers can see where I'm headed. In order to get there quickly, let me simply declare myself on a further methodological point. We have already seen that something is an error only in relation to a cognitive target and its embedded attainment-standard. This works as a direct constraint on *assessment*.

Attribution precedes assessment. For any target T, before judging an individual's reasoning against a standard required for T-attainment, it must be determined first that T is indeed the target that the reasoner has set for himself.

6 Goodbye to the Eighteen

We have it now from the analysis of what reasoners are like together with the attribution-precedes-assessment principle, that except in the presence of particular considerations, to the contrary

Not typically errors. When performed by individual reasoners, invalid or inductively weak reasoning is *not* typically an error.

Consequently,

Not fallacies. Reasoning that is not typically an error is *not* a fallacy on the *EAUI* conception.

If we examine the large literature on the Gang of Eighteen, we find a striking consensus to the effect that the mistake embedded in most of these fallacies is either the error of invalidity or the error of inductive weakness. I leave it as an exercise to riffle through the list and tick those that qualify thus. This is a first step in support of the negative thesis that the Eighteen are either not mistakes or are not mistakes committed by us. It does not confirm it outright, in as much as there are items on the list for which, as we have seen, some theorists claim dialectical impropriety and nothing else (*petito* is perhaps the obvious example)[17]. But if the case that I have been trying to

---

[17] Even so, I say again that question-begging is not a fallacy. See [22].

build has merit, the Gang of Eighteen is in shambles and the negative thesis is broadly, if not wholly, confirmed.

I want now to switch our focus to the positive thesis, which says that a good many of the Eighteen are not only not errors, but they are rationally sound scant-resource adjustment strategies. I'll confine myself to two examples. One is hasty generalization. The other is the *ad ignorantiam* reasoning. I'll deal with these in reverse order.

7 Ad Ignorantiam *Reasoning*

In its modern[18] version the *ad ignorantiam* is a mistake in the form

1. It is not known that P
2. Hence, not-P.[19]

It is, of course, an invalid schema, and much railing against it has come from the informal logic community on grounds of (2)'s simply "not following" from (1). On the face of it, this is indeed a pretty hopeless kind of reasoning. But if attention is paid to the circumstances in which arguments of this form are actually presented, it is easy to see that they are typically enthymemes the missing premiss of which is an *autoepistemic conditional* in the form,

If P were the case, I (we) would know that it is.

Computer scientists have been long aware that premissses of precisely this sort are in play in negation-as-failure contexts. For example, if you consult the departure board at the Vancouver Airport and find no Tuesday flight to London after 9:00 p.m., you rightly conclude that there is no such flight. Your reasoning is autoepistemic.

i. If there were a flight after 9:00 we would know it (from the board).
ii. We don't know it (the board makes no such announcement).
iii. So, there is no such flight.

Of course, even this is an invalid argument (the board might have malfunctioned), and it certainly has nothing like the inductive support of drug trials. (For one thing, how big is your sample? What is your evidence that it is at all representative? Hearsay doesn't count). But it is a good argument all

---

[18] The term "ad ignorantiam" originated with Locke [49]. As Locke saw it, an *ad ignorantiam* move in an argument is one in which one of the parties, having presented his case, puts to the other the following challenge: "Accept my case or say something that betters it." Whatever one makes of such a demand, it is plainly not a *EAUI*-fallacy. Even if it were a mistake of some sort, it is not a mistake of reasoning.

[19] Alternatively,
    1. It is not known that $\sim P$
    2. Hence, P.

the same. Proposition (*iii*) is detachable as a default from premises (*i*) and (*ii*). Similarly, anticipating that he may be late for dinner, Harry's wife asks whether there will be a Department meeting today. Harry replies, "I take it not, for otherwise I would know".

It would be wrong to leave the impression that autoepistemic reasoning is error-free. Certainly it is easy enough to be mistaken about the missing autoepistemic premium. If Harry worked at the old IBM, there could well be a meeting without his knowing about it. Harry might have fallen victim to one of those infamous unannounced dismissals over the lunch hour, returning to find his office effects in the hall and the lock on his door changed. Even so, no one would say that it is typical of beings like us to be mistaken about our autoepistemic assumptions. Consequently, *ad ignorantiam* reasoning of the autoepistemic sort cannot be a fallacy on the *EAUI* model.

A little reflection shows how useful negation-as-failure reasoning is. It combines two essential cognitive virtues. One is that when we actually resort to it, it tends to be right, rather than wrong. The other is that it is efficient. It achieves striking economies of time and information. (After all, it is based on *lack* of information.) So I think that we may conclude the *ad ignorantiam* lends us support to both our theses, the positive as well as the negative.

8 Hasty Generalization

As traditionally conceived of, hasty generalization is a sampling error. It is the error of generalizing from an unrepresentative sample. In the classical literature, one of the standard marks of a sample's unrepresentativeness is its smallness. Traditional approaches to the fallacies seize on this factor, making hasty generalization the fallacy of mis-generalizing from an over-small sample. By these lights, any would-be analysis of this fallacy must take due notice of two factors. It must say what a generalization is. It must also say what is lacking in the relationship between an over-small sample and the generalization it fallaciously "supports". In traditional approaches, this is all rather straightforward. A generalization is a universally quantified conditional statement. And what the over-small sample fails to provide for it is inductive strength, or high (enough) conditional probability. Let us remind ourselves that fallacies are target-relative and resource-sensitive. So we must take care to observe that, even as traditionally conceived of, hasty generalization is not a fallacy *as such*. It is a fallacy only in relation to a cognitive target of which the production of an inductively well-supported universally quantified conditional is an attainment-standard. It is easy to see that for certain classes of institutional agents – think again of the Health Canada labs or, more expansively, of the whole sprawling project of experimental science – generalizing on over-small samples is indeed an error. It is so precisely when agents such as these set themselves targets that, short of meeting the standard of inductive strength, are unreachable.

It is also easy to see that when it comes to *individual* agents, the empirical record amply attests to two importantly linked facts. The first fact is:

The commonplaceness of haste. For beings like us, small-sample generalization is a widespread practice.

The second fact is:

The soundness of the practice. By and large, our track record as hasty generalizers is a good one. For the most part, the hasty generalizations we actually commit do not derange the Enough Already Principle.

The Enough Already Principle. Beings like us are right enough enough of the time about enough of the right things to survive, prosper and occasionally build great civilizations.

The empirical record discloses a third fact of importance to our enquiry. It is that:

The rarity of universal generalizations. When beings like us generalize, it is not typically the case that we generalize to universally quantified conditional propositions[20].

Why should this be the case? The principal reason is that universally quantified generalizations are *brittle*. They are felled by any single true negative instance. In contrast, a generic statement, e.g., "Ocelots are four-legged" are *elastic*. They can be true even in the face of some true negative instances. Ozzie, the ocelot, is three-legged. This topples "For all x, if x is an ocelot, then x is four-legged", but leaves standing "Ocelots are four-legged"[21]. This has a two-directional bearing on generalization. From the point of view of instantiation, a true negative instance of a universal conditional carries a twofold cost. One must give up the contrary of the negative instance and one must give up the generalization from which it was inferred. However, a true negative instance from a generic generalization carries only the first cost. We have to give up the contrary of the negative instance, but one needn't give up the generalization. So there is a striking economy in confining one's generalizations to the generic. From the other direction, i.e., in the move from sample to generalization, there is also an advantage that redounds to the generic. Take again a sample of ocelots, all of which except for Ozzie are four-legged. There is no non-trivial generalization of this that we know how to make. True, we might generalize to "For all x, if x is an ocelot and is not three-legged, then x is four-legged." But this is trivial. We might also generalize to "For all x, if x is an ocelot, then *ceteris paribus* x is four-legged." But this is not a

---

[20] For example: It never snows in Vancouver in April; Sarah always goes to bed before 10:00; Harry is never late in paying his bills; ocelots are four-legged; rhododendrons never do well in Toronto; and so on.

[21] Concerning generic statements, see [50].

generalization. It is a generalization-*schema,* for whose *ceteris paribus*-clause there is, as yet, no principled and suitably general an explication. In contrast, generalizing to generic statements, even from samples containing true negative instances, is decidedly more economical and truer to how beings like us perform on the ground.

Finally, in the cognitive economy in which resource-bound agents are required to perform, haste is a cognitive virtue. It is not always a virtue if it leads one to error[22], needless to say, but in the case of hasty generalization it bears repeating that by and large when we *resort to it,* we get it right, not wrong. So in the form in which we typically do perform it, it cannot be a fallacy. Such is the providence of the *EAUI* conception.

9 Fallibilism

I said at the beginning that anyone seeking to produce a logic of fallacious reasoning on the *EAUI*-conception would have to focus on the four defining conditions of it: error, attractiveness, universality and incorrigibility. I suggested that one explanation of why the Eighteen appear to satisfy the last three of these conditions can be set out as follows. For individual reasoners, the Eighteen are either not errors, or if they are, are not typically committed by beings like us. It is possible, of course, that for institutional agents, who have the resources to meet targets of considerable loftiness, at least some of the Eighteen are indeed errors. This helps explain why the Eight *appear* to be fallacies. They appear to be errors because they are errors for institutional agents. They are attractive because they are not errors for us or not errors that we typically commit. They are attractive to us for the same reason. They are universal because evolution has kitted us out to reason similarly in similar situations. And they are incorrigible because they present nothing that requires correction or across the board suppression.

Attractive though the suggestion may have seemed initially, it is doubtful that we can now persist with it. Consider the following two examples of institutional agency. Let M be the community of number theorists since, say, 1900. Let E be the community of experimental and statistical scientists since that same year. It is clear that truth-preservation is a dominant target in M and that validity is an attainment-standard. Similarly, the target of experimental confirmation is dominantly present in E and with it, the standard of inductive strength. Consider now the traditional fallacies of affirming the consequent and denying the antecedent, each plainly an invalid bit of reasoning. If, as in M, truth-preservation is the target, then these would be errors in M. But are they fallacies? One could hardly think so. These are not mistakes that are committed in M with anything like the frequency required to make

---

[22] As [51] points out, sometimes a more efficient way to learn is to make "natural" mistakes which admit of speedy and reliable correction.

them fallacies. Nor are they attractive or incorrigible in M. So again they can't be fallacies. Mathematicians make mistakes. But they don't make *those* mistakes.

Much the same must be said of the so-called inductive fallacies – e.g., hasty generalization, *post hoc ergo propter hoc* and the gambler's fallacy. These are errors only in relation to the target of experimental confirmation or probabilistic projection. E is an institutional agent for which this is a dominant standard. Although these are errors in E, they are not fallacies. Scientists don't commit them with any frequency to speak of. They are not attractive and not incorrigible to scientists in their white coat moments. Scientists make mistakes. But they don't make *those* mistakes.

This leaves the Gang of Eighteen in pretty forlorn shape. We are having a difficult time in finding agents who commit them in fulfillment of the conditions that make them fallacies. This lends support to our earlier suggestion of a serious "radish problem" for the Eighteen on the $EAUI$-conception. We seek for an analysis of radishes but we channel our investigations to marshmallows. Not only do we not get radishes right, we also end up with a ludicrous theory of marshmallows. As regards the fallacies, this puts massive pressure on the Gang of Eighteen or the $EAUI$-conception, or both. If we assume that the $EAUI$-definition is sound, we must be ready for the possibility that the Eighteen aren't in its extension (which is precisely the purport of the negative thesis). On the other hand, perhaps the $EAUI$-conception itself is where the problem lies. Perhaps it is the case not only that the Eighteen don't satisfy the four $EAUI$-conditions, but also that *nothing* does. I am not ready at present to assert as a fact that the concept of fallacy is empty. But I admit to being much drawn to the idea. It is an interesting possibility that does not merit dismissal out of hand.

There are, of course, some obvious objections to consider, beginning with fallibilism. Fallibilism is an epistemological doctrine which in all its variations honors the empirical fact that

Error abundance. Beings like us make errors, lots of them.

Of course, fallibilism is not scepticism. It does not purport that we are *always* mistaken. Indeed it honors the further fact that

Knowledge abundance. Beings like us have knowledge, lots of it.

On the face of it, the two abundance theses stand to one another in an uneasy tension. But that tension appears to dissipate, or anyhow to diminish, once we throw the Enough Already Principle into the mix. Notwithstanding that we commit lots of errors, these are not in the aggregate errors of sufficient moment to deny us our collective survival and prosperity. We are right enough enough of the time about enough of the right things. The point at hand is this. If we commit lots of errors, then we do lots of things that appear not to be errors. Moreover, fallibilists are not of the position that with human individuals errors are just one-off miscues. Their view rather is that error is persistent and recurrent in the human species. This is getting to be rather close to satisfying the *EAUI*-conditions on fallacious reasoning. So might it not be the case that although the Eighteen aren't in the extension of the *EAUI*-conception, lots of our other errors are? If so, would it not fall to the fallacy theorist to seek out the identities of those errors, assigning them suitable names and organizing them in appropriate taxonomies?[23]

Granted that we commit errors on a grand scale, how might it come to pass that there are no fallacies? I lack the space to consider this question with the care and detail it deserves[24]. But it is possible to sketch out a possible answer, which strikes me as meriting a certain consideration. To that end, let us repeat that the errors associated with the *EAUI*-conception of fallacies are errors of *reasoning* – in particular, errors affecting inferences or the reaching of conclusions.

(1) Then the first thing to say is that most of our errors are not errors of reasoning. Rather they are mechanical errors, perceptual errors, errors of forgetfulness, errors arising from misinformation, and the like. Since these are not errors of reasoning, the question of fallaciousness does not arise for them.

(2) A second point has to do with the structure of defeasible reasoning. Virtually everyone agrees that accepting a proposition α is an error should α turn out to be false. Picking a false α is "getting the wrong answer". Yet virtually everyone also believes that the inference pattern

1. α
2. $\prime\prime^{def} \beta$

in which α is the premiss, β the conclusion and $\prime\prime^{def}$ is the defeasible therefore-sign, is one which can be sound even though α is true and β turns out to be false. For concreteness let α = "Ocelots are four-legged" and β = "Ozzie the ocelot is four-legged". Let it be the case that, as before, Ozzie is in fact three-legged. The qualification "def" on? indicates that β is a default drawn from the generic α. It is a good inference, even though it is not truth-preserving. It is a good inference even though its conclusion chances to be false, and even though it is an inference that would have to be given up once this fact came to light. This reminds us that, as long as β's falsity is not known to him, an agent might reasonably conclude it from a true α by defeasible inference. When he does so defeasibly, drawing that default is *not* an error of reasoning, although persisting with it once β's falsity became known would be a mistake. We draw default inferences with great frequency, almost as naturally as we breathe. In lots of these cases, we make a mistake, the mistake, namely, of "picking a β that is the wrong answer". But in so doing, we have not committed an error

---

[23] For example, what of the conjunction fallacy of Kahneman and Tversky? For arguments that this, too, is not fallacy, see [43] and [52].
[24] This is undertaken in [53].

in *reasoning*. So here too is a large class of cases in which the error we commit is not even a candidate for fallaciousness.

Needless to say, defeasible inference is not alone in allowing for the possibility of reasoning correctly to a false conclusion. Inductive inference has the same feature. Consider a simplified schema.

1. $\Pr(P) = n$
2. Evidence E exists.
3. $\Pr(P/E) = n + m$ (for suitably high n and m)
4. Therefore, P.

Or, in an assertoric variation,

4'. Therefore, probably P.[25]

Here it is understood that assertorically modified or not, the inference to P requires only a suitable degree of inductive strength, and that this strength is present when the values of n and m are high enough. But nowhere is it required for the inference to be inductively sound that P be true. One can be right in the inference one draws and yet also, in picking P, get the wrong answer.

Much the same is true of abductive reasoning. Consider another simplified schema. Let T be an explanation-target that an agent X cannot attain on the basis of what he now knows (K). Assume further that X lacks the wherewithal to repair his ignorance in a timely way, and yet he wishes his ignorance not to paralyze actions of the kind that T-attainment would provide a basis for. Suppose also that a proposition H, whose truth-value is not known to X, is such that, *were* it true, then K updated by H *would* explain T. Then, X abduces H in two steps. First, he conjectures that H. Secondly, on that basis, he activates H; that is, he releases it for premissory work in inferences and decisions relevant to his interest in T in the first place. Summarizing,

1. K doesn't attain T.
2. If H were true, $K(H)$ would attain T.
3. Therefore, H is a reasonable conjecture.
4. Therefore, H.[26]

As with the other cases, the therefore-operator is not intended to be truth-preserving, nor as with inductive inference, need it here be probability-enhancing. Abductive inference is typically weaker than that. But the main point remains untouched. A reasonable abduction of H is compatible with H's falsity.

Defeasible and abductive inference dominates the reasoning of the ordinary individual. They both allow for the compatibility of good reasoning and false

---

[25] For a reminder of the use of "probably" as an assertion-modifier see [54].
[26] For more detailed examinations of abduction, see [37, 55, 56] and [57].

conclusions. Although individuals avail themselves of it less, inductive reasoning has the same character. This tells us something of the first importance. Let f be the frequency with which we reason reasonably to false conclusions, that is, with which we reason in an error-free way to an error. The higher the value of f, the greater the likelihood that the Error Abundance Thesis is confirmed by errors that aren't errors of *reasoning*. Everything that we so far know about these things suggests that as a matter of fact the value of f is rather high. Accordingly,

> *The unlikelihood of fallacy.* The likelihood is high that the errors that lend greatest confirmation to the Error Abundance Hypothesis are not errors of a type that qualify as fallacies.

(3) We come now to a third point. Let Σ be a set of priors, i.e., a database, a set of given facts, a knowledge-base, a set of premisses, or some such thing. Things like Σ we happen upon endlessly as we make our way through the minutes and hours of our engagement with the world. In a rough and ready way, we can note that these Σs are informative for us in two ways. They carry information directly (by way of "what the facts say") and they convey information by inference (by way of "what the facts mean"). When his Σs bears on him relevantly, a cognitively competent individual will have a generally good track-record discerning the information that Σ carries, as well as the information that can be inferred from it. Clearly there are variations within these competencies; and, *in extremis*, they may break down. Consider the case in which an agent is reasonably good at reading what a Σ says, but no good at all at discerning what it means, that is, what should be inferred from it. What we see here is the absence of reasoning rather than bad reasoning. The person who doesn't know what to make of his Σs is in a bad way. There is something wrong with his reasoning skills. There is a deficit of reasoning here, a failure of reasoning. But it would be going too far to call it an error. Of course, it is *vacuously* true to say that this is error-free reasoning, but there is no comfort in saying so. What we learn from this is that there is more to learn about deficiencies of reasoning than can be found in any theory of reasoning-errors, or in any theory that requires fallacies to be errors of reasoning.

We now have the means to say that in their failure to engage the factor of error in a robust way, theories of agent-based reasoning are asking for trouble, and theories of fallacious reasoning are guaranteed to get it[27]. For the present, it may be that we have now said enough to lend some support to the conjecture of several paragraphs ago.

> *No fallacies.* When our errors are *bona fide* errors of reasoning, they occur with neither the frequency, the attractiveness, nor incorrigibility required to qualify them as fallacies of any kind on the *EAUI*-conception. More briefly, there are no such fallacies.

---

[27] A first attempt at subduing the concept error may be found in [58]. Also relevant, in addition to [22], are [59], and [60].

I do not say that the thesis of the emptiness of the *EAUI*-concept is now *fait accompli*. Even so, it now enjoys some backing of non-trivial weight. If the thesis is right, it leaves fallacy theory in considerable disarray. If so, there is a biting irony to it. As I mentioned in section 5, for years I have complained that jettisoning the traditional concept of fallacy by pragma-dialecticians, in favour of a stipulated successor that dances to the provisions of their preferred theory of argument, was solving an honourable and difficult problem by changing the subject. If, as pragma-dialecticians say, a fallacy is any deviation from a rule of civilized discourse, then fallacy theory is no more difficult than the problem of specifying those rules[28]. My position all along has been that the pragma-dialectical solution of the fallacies problem has, in Russell's words about another thing, all the virtues of theft over honest toil. But if, as I now conjecture, the traditional concept of fallacy is indeed empty, it is much harder to persist with one's dissatisfactions with the Amsterdam School. Of course, it by no means follows from the emptiness of the *EAUI*-concept, that fallacies must be conceptualized in the Amsterdam way. There are two rival possibilities to keep in mind. Either fallacies are properly conceptualizable, but not in the Amsterdam way. Or fallacies are like phlogiston. The trouble with phlogiston was not that it was misconceived. It was that there wasn't any. The concept "phlogiston" was empty. No one thinks on that account that we must now find an extension for it by getting it to mean something different.

All this, of course, is rather tentative – one might even say "defeasible". One thing is clear, however, the fallacies project is still a wide-open question for the logical theory of the 21st century.

References

1. Magnani, L., Nersessian, N., eds.: *Model-Based Reasoning in Scientific Discovery, Technological Innovation and Values.* Kluwer, New York (2002)
2. Magnani, L., ed.: *Model-Based Reasoning in Science and Engineering.* College Publications, London (2006)
3. Anderson, A., Belnap Jr., N.: *Entailment: The Logic of Relevance and Necessity.* Princeton University Press, Princeton, NJ (1975)
4. Shoesmith, D., Smiley, T.: *Multiple-Conclusion Logic.* Cambridge University Press, Cambridge (1978)
5. Woods, J.: *Aristotle's Earlier Logic.* Hermes Science Publications, Oxford (2001)
6. Hamblin, C.: *Fallacies.* Methuen, London (1970)
7. Hansen, H., Pinto, R.: *Fallacies: Classical and Contemporary Readings.* The Pennsylvania State University Press, University Park, PA (1995)
8. Johnson, R.: *The Rise of Informal Logic.* Vale Press, Newport News Va (1996)
9. Barth, E., Krabbe, E.: *From Axiom to Dialogue: A Philosophical Study of Logic and Argumentation.* De Gruyter, Berlin and New York (1982)
10. Hintikka, J.: *Knowledge and Belief.* Cornell University Press, Ithaca, NY (1962)

---

[28] Even this is harder than it may first appear.

11. Woods, J., Walton, D.: *Fallacies: Selected Papers 1972-1982.* Foris de Gruyter, Berlin and New York (1989)
12. Mackenzie, J.: Four dialogue systems. *Studia Logica* **XLIX** (1990) 567–583
13. Woods, J., Irvine, A., Walton, D.: *Argument: Critical Thinking, Logic and the Fallacies.* Prentice Hall, Toronto (2004) 2nd Revised Edition.
14. Copi, I.: *Introduction to Logic.* MacMillan, New York (1986)
15. Carney, J., Scheer, K.: *Fundamentals of Logic.* Macmillan, New York (1980)
16. Schipper, E., Schuh, E.: *A First Course in Modern Logic.* Henry Holt, New York (1959)
17. Black, M.: *Critical Thinking.* Prentice-Hall, New York (1946)
18. Woods, J.: *The Death of Argument. Fallacies in Agent-Based Reasoning.* Kluwer, Dordrecht and Boston (2004)
19. Hintikka, J.: The fallacy of fallacies. *Argumentation* **1** (1987) 211–238
20. Hintikka, J., Sandu, G.: Game-theoretical semantics. In van Benthem, J., ter Meulen, A., eds.: *Handbook of Logic and Language.* Elsevier (1997) 361–410
21. Hintikka, J.: What was aristotle doing in his early logic, anyway? a reply to woods and hansen. *Synthese* **113** (1997) 241–249
22. Woods, J.: Begging the question is not a fallacy (2007) to appear.
23. Govier, T.: Reply to massey. In Hansen, H.V., Pinto, R.C., eds.: *Fallacies: Classical and Contemporary Readings.* The Pennsylvania State University Press, University Park PA (1995) 172–180
24. Scriven, M.: Fallacies of statistical substitution. *Argumentation* **1** (1987) 333–349
25. Hitchcock, D.: Why there is no argumentum ad hominem fallacy (2006) www.humanities.mcmaster.ca/~hitchckd/. Accessed on 16/07/06.
26. Eemeren, F.v., Grootendorst, R.: *Argumentation, Communication and Fallacies. A Pragma-Dialectical Perspective.* Lawrence Erlbaum Associates, Hillsdale, NJ and London (1992)
27. Walton, D.: *A Pragmatic Theory of Fallacy.* University of Alabama Press, Tuscaloosa (1995)
28. Woods, J.: Pragma dialectics: A radical departure in fallacy theory. *Bulletin of the International Society for the Study of Argumentation* (1989) 5–15
29. Morris, C.: *Writings on the General Theory of Signs.* Mouton, The Hague (1971)
30. Jacquette, D.: *Psychologism the philosophical shibboleth.* In Jacquette, D., ed.: Philosophy, Psychology and Psychologism: Critical and Historical Readings on the Psychological Turn in Philosophy. Kluwer (2003) 245–262
31. Pelletier, F., Elio, R., Hanson, P.: The many ways of psychologism in logic: History and prospects (2007) Studia Logica, to appear.
32. Gabbay, D., Woods, J.: *The Ring of Truth: Towards a Logic of Plausibility.* Volume 3 of A Practical Logic of Cognitive Systems (2007)
33. Searle, J.: *Speech Acts. An Essay in the Philosophy of Language.* Cambridge University Press, Cambridge (1969)
34. Stalnaker, R.: Presupposition. *Journal of Philosophical Logic* **2** (1973) 447–457
35. Stalnaker, R.: Assertion. *Syntax and Semantics* **9** (1978) 315–332
36. Gabbay, D., Woods, J.: *Agenda Relevance: A Study in Formal Pragmatics,* vol. 1 of A Practical Logic of Cognitive Systems. North-Holland, Amsterdam (2003)
37. Gabbay, D., John Woods, J.: The practical turn in logic. In Gabbay, D.M., Guenthner, F., eds.: *Handbook of Philosophical Logic, 2nd revised edition.* Kluwer (2005) 15–122

38. Churchland, P.: *A Neurocomputational Perspective; The Nature of Mind and the Structure of Science.* MIT Press, Cambridge, MA (1989)
39. Bruza, P., Widdows, D., Woods, J.: A quantum logic of down below. In Engesser, K., Gabbay, D.M., eds.: *Handbook of Quantum Logic.* Elsevier, Amsterdam (2007)
40. d'Avila Garcez, A., Gabbay, D., Ray, O., Woods, J.: Abductive neural networks: Abductive reasoning in neural-symbolic systems (2007) *Topoi*, to appear.
41. Simon, H.: *Models of Man.* John Wiley, New York (1957)
42. Simon, H.: *Models of Bounded Rationality: Behavioral Economics and Business Organization.* The MIT Press, Cambridge and London (1982)
43. Gigerenzer, G.: *Adaptive Thinking: Rationality in the Real World.* Oxford University Press, New York (2000)
44. Gigerenzer, G., Selten, R.: *Bounded Rationality: The Adaptive Toolbox.* MIT Press, Cambridge MA (2000)
45. Stanovich, K.: *Who is Rational? Studies of Individual Differences in Reasoning.* Erlbaum, Mahwah, NJ (1999)
46. Woods, J., Johnson, R., Gabbay, D., Ohlbach, H.: Logic and the practical turn. In D.M. Gabbay, R.M. Johnson, H.O., Woods, J., eds.: *Studies in Logic and Practical Reasoning volume 1 of the Handbook of the Logic of Argument and Inference.* Elsevier/North Holland (2002) 1–39
47. Gabbay, D., Woods, J.: T*he Reach of Abduction: Insight and Trial, volume 2 of A Practical logic of Cognitive Systems.* North-Holland, Amsterdam (2005)
48. Mill, J.: *A System of Logic.* Longman's Green, London (1961) Originally published in 1843.
49. Locke, J.: *An Essay Concerning Human Understanding.* Clarendon Press, Oxford (1975) edited by Peter H. Nidditch, Originally published in 1690.
50. Carlson, G., Pellerier, F., eds.: *The Generic Book.* Chicago University Press, Chicago (1995)
51. Gigerenzer, G.: I think, therefore i err. *Social Research* **1** (2005) 195–217
52. Woods, J.: Probability and the law: A plea for not making do (2007) to appear.
53. Gabbay, D., Woods, J.: *Seductions and Shortcuts: Fallacies in the Cognitive Economy, volume 4 of A Practical Logic of Cognitive Systems* (2007)
54. Toulmin, S.: *Stephen Toulmin, The Philosophy of Science: An Introduction.* The Hutchinson University Library, London (1982)
55. Magnani, L.: *Abduction, Reason and Science: Processes of Discovery and Explanation, Dordrecht.* Kluwer Plenum, New York (2001)
56. Gabbay, D., Woods, J.: Advice on abductive logic. *Logic Journal of the IGPL* **14** (2006) 189–219
57. Aliseda, A.: *Abductive Reasoning: Logical Investigation into the Processes of Discovery and Evaluation.* Springer, Amsterdam (2006)
58. Woods, J.: *Error* (2007) to appear.
59. Woods, J.: Why is fallacy theory so difficult? (2007) to appear.
60. Woods, J.: Lightening up on the ad hominem (2007) to appear.

Abductive Reasoning, Information, and Mechanical Systems

Maria Eunice Quilici Gonzalez[1], Mariana Claudia Broens[2], and Fabricio Loffredo D'Ottaviano[3]

[1] Philosophy Department, UNESP, São Paulo, Brazil
   gonzalez@marilia.unesp.br
[2] Philosophy Department, UNESP, São Paulo, Brazil
   mbroens@marilia.unesp.br
[3] Pediatrics Department, UNIFESP, São Paulo, Brazil
   loffdotta@gmail.com

Summary. We investigate, from a philosophical perspective, the relation between abductive reasoning and information in the context of biological systems. Emphasis is given to the organizational role played by abductive reasoning in practical activities of embodied embedded agency that involve meaningful information. From this perspective, meaningful information is provisionally characterized as a self-organizing process of pattern generation that constrains coherent action. We argue that this process can be considered as a part of evolutionarily developed learning abilities of organisms in order to help with their survival. We investigate the case of inorganic mechanical systems (like robots), which deal only with stable forms of habits, rather than with evolving learning abilities. Some difficulties are considered concerning the hypothesis that mechanical systems may operate with meaningful information, present in abductive reasoning. Finally, an example of hypotheses creation in the domain of medical sciences is presented in order to illustrate the complexity of abduction in practical reasoning concerning human activities.

1 Introduction

The relationship between abductive reasoning and meaningful information (exemplified in coherent action) constitutes our main subject of investigation. Based upon Charles Sanders Peirce's hypotheses on the logical nature of abduction, and Gilbert Ryle's characterization of *abilities* (as distinct from *habits*), we investigate the nature of meaningful information in the domain of abductive practical reasoning. In this domain, we distinguish between simbolic meaningless information, conceived from a mechanistic view, and meaningful information implicit in systemic coherent agency, arguing in defence of the following hypotheses:

1. Living organisms, differently from inorganic systems, deal with meaningful information. This, in turn, can be understood as a self-organizing process of patterns generation which provides sufficient conditions for the development of coherent action.
2. Coherent action, in contrast to purely habitual movements, involves learning abilities developed in accordance with a process of self-organization.
3. Abductive inference constitutes one of the main organizational sources of coherent actions emergent from self-organizing learning processes.

Hypotheses 1–3 are investigated in two steps: Section 2, *Abductive Reasoning and the Debate between Mechanistic and Systemic Orders*, introduces the main properties of abductive reasoning, as originally proposed by Charles S. Peirce, and addresses the debate regarding its (non) mechanical nature. Section 3, *Can Mechanical Systems Develop Abductive Reasoning?*, focuses on the distinction between habits and abilities in the domain of systemic coherent action. Inspired by Ryle, we argue that abilities, differently from mere habits, involve abductive reasoning and learning processes which unify different properties of events in specific practical contexts. A brief inquiry into the possibility of including robots in the realm of agency that requires learning and an ability to deal with meaningful information is, then, developed. Finally, an example of abductive reasoning, in the area of medical sciences, is presented to illustrate the non-mechanical aspect of abduction in practical human activities.

2 Abductive Reasoning and the Debate between Mechanistic and Systemic Orders

Abductive reasoning can be characterized as a form of practical reasoning that constitutes one of the basic pillars of learning abilities. These abilities present an unifying property which allows patterns to be identified in the domain of action.

As originally formulated by Peirce, abduction is a way of reasoning in which one infers to the best explanation of a problem, an anomaly or a surprising event. This type of reasoning can be described in four steps that initiate with, (1), the perception of an anomaly or a surprising event and develops with, (2), the search for hypotheses (H_1, H_2, H_n) for the purpose of explaining the anomaly. Once reasonable hypotheses are elaborated, (3), inductive testing of these hypotheses is undertaken. Finally, (4), deductive confirmation of the plausibility of the new hypotheses indicates that they could predict and explain the original perceived anomaly which initiated the abductive reasoning in question [12].

Peirce argues that this type of reasoning applies both in science and in everyday situations whenever a surprising or anomalous fact, F, is observed. He stresses that the search for an explanation of a surprising fact, F, which interrupts a chain of well established habits, is guided by the supposition that:

i. If a hypothesis H (concerning the possible nature of F) were true, F would be a matter of course;
ii. Hence, there is reason to suspect that H is true [12].

The validity of abductive reasoning is the object of great discussion between logicians and philosophers, given that it only provides useful guidelines to the "admissibility" of explanatory hypotheses. Differently from inductive or deductive inferences, abductive reasoning is fallible in that no guarantee is provided about its correctness; it can, nevertheles, be useful as a creative way of selecting hypotheses and structuring an organism's actions, as we are going to explain in Part 3.

To be properly characterized, the above preliminary description requires an understanding of Peirce's ontology, according to which the universe is in a process of continuous expansion, acquiring and modifying habits – or tendencies to repetition. On the basis of this principle of habit formation, the cosmos is conceived as being structured by a continuous flow of information available to all existent beings. He emphasizes the organizational role played by abductive reasoning in the expansion of knowledge related to practical activities of embodied agents rooted in the cosmos. Furthermore, according to his consideration of the law of habit formation – which describes the tendency for repetition of already existent events –, minds are everywhere: Wherever the possibility of habit formation exists, minds are going to start their continuous evolutionary path that is also governed by chance and failures. From Peirce's perspective, beliefs are chains of stable habits that constraint action. Given the dynamic character of these beliefs, sometimes they have to be abandoned or radically transformed in order to deal with the intricacies of the world.

It is in the context of a complex universe of habit formation and transformations that abductive reasoning applies whenever chance and novelties occur, disrupting previously existent stable beliefs. Guided by the principle of fallibilism (which emphasizes the epistemic relevance of chance, failure and spontaneity in the process of knowledge acquisition) Peirce argues that, even though abductive reasoning can be expressed logically, in terms of the mentioned rules of inference it is not a mechanical procedure:

> Thus, the universe is not a mere mechanical result of the operation of blind law. The most obvious of all its characters cannot be so explained. It is the multitudinous facts of all experience that show us this; but that which has opened our eyes to these facts is the principle of fallibilism. Those who fail to appreciate the importance of fallibilism reason: We see these laws of mechanics; we see how extremely closely they have been verified in some cases. We suppose that what we haven't examined is like what we have examined, and that these laws are absolute, and the whole universe is a boundless machine working by the blind laws of mechanics. [12]

One important reason underlying Peirce's rejection of the mechanistic view is that it underestimates the role of chance, spontaneity and error in the dynamic of novelty production in the universe. Absolute chance – which is not governed by laws – constitutes the very possibility of initiating, in general, an abductive reasoning. Criticizing Spencer's mechanistic view of evolution, Peirce stresses that:

> [...] mechanical law, which the scientific infallibilist tells us is the only agency of nature, mechanical law can never produce diversification. That is a mathematical truth – a proposition of analytical mechanics; and anybody can see without any algebraical apparatus that mechanical law out of like antecedents can only produce like consequents. It is the very idea of law. So if observed facts point to real growth, they point to another agency, to spontaneity for which infallibilism provides no pigeon-hole. [12]

Thus, it is in the background of a non-mechanistic ontology that Peirce's conception of abductive reasoning applies: In the always-changing universe, spontaneity and failure launch the seeds that will break the chain of well-established habits initiating new forms of organization.

This very brief summary of Peirce's non-mechanistic view help us to understand the organizational role played by abductive reasoning in practical activities of embodied embedded agents. Rooted in a spontaneous and ever growing universe, agents may come across surprising events or anomalies that force them to modify their beliefs (or stable habits), generating innumerable tracks of information. As mentioned, in these circumstances they look for possible hypotheses or diverse ways of expanding their habits in order to maintain coherence in the set of usual actions. This process can be considered as a part of evolutionarily developed learning abilities of organisms in order to help with their survival. According to our Hypothesis 3, abductive inference constitutes one of the main organizational sources of coherent actions from emergent self-organizing learning processes.

As pointed out in [6] and [9], the label "self-organization" refers to a process through which new forms of organization may emerge *spontaneously* from the dynamic interaction established between elements that are initially independent. The spontaneous characteristic of a self-organizing process requires that no *a priori* plan or central controller should direct the development of the process in question: Its organization, when it happens, should result mainly from the exclusive dynamics of the interaction between its elements.

Triggered by basic needs of survival, organisms developed informational meaningful paths, which constrain their actions creating habits that will be further developed in order to fulfill less basic needs. From this perspective, meaningful information can be characterized as a self-organizing process of pattern generation that constrains coherent action [8].

In this context, information differs from symbolic mechanical information (processed by machines) in at least one fundamental aspect: It cannot be reduced to pure physical stimuli that need to be represented and interpreted according to specific rules or programs in order to become meaningful; its meaning is generated and developed by means of the relevant action of organisms situated in their natural niche [7]. Information could be named here, following the notion of Gibson, *systemic ecological information* in that it is a systemic property of the pair "organism-environment", which does not need to be interpreted in order to acquire meaning.

Taking into consideration the above hypotheses, the main question that interests us here at the moment is: Could inorganic mechanical systems like robots, guided by mechanical laws, evolve in such a way that it would make sense for them to reason in an abductive way? In that which follows, we are going to inquire into this possibility.

3 Can Mechanical Systems Develop Abductive Reasoning?

According to our preliminary characterization of abductive reasoning as introduced in Section 2, this form of reasoning involves self-organization and meaningful information in the domain of action. We claimed that abductive reasoning constitutes one important way of organizing learning abilities that go beyond mere habits or mechanical repetition. Here, in Section 3, we are going to provide evidence for this hypothesis focusing on the distinction between mechanical habits and learning abilities, such as developed by Ryle in his formative work *The Concept of Mind*.

Even though the approaches of Ryle and Peirce belong to different philosophical contexts, both of them address the distinction between pure habits and abilities or *creative intelligent capacities*: Habits involve some kind of automatism that is incompatible with the performance of "able actions". Ryle suggests that:

> [...] the distinction between habits and intelligent capacities could be illustrated by the parallel distinction between the methods used for inculcating the two sorts of second nature [or acquired dispositions]. We build up habits by drill, but we build up intelligent capacities [or abilities] by training. [13]

While habits result from repetitions, abilities involve attention and training towards constant adjustments to the best situation. However, specific abilities can be transformed into habits, and the creation of habitual actions could require attention in unusual contexts. Thus, for example, walking is a physiological potentiality for certain animals which has to be actualized by exercises. Humans learn how to walk initially at the expense of great amounts of energy, but when this ability is acquired it becomes a habit. After acquiring walking abilities, healthy adults walk without expending too much energy, but in specific circumstances, such as walking near a precipice, attention and adjustments may be necessary. Thus, the characterization of habits and abilities requires considerations about the context to which they belong.

The cognitive relevance of practical abilities is further developed throughout Ryle's characterization of what he calls *knowing how* in contrast to *knowing that*. To *know how* to walk, for example, is a practical, non-propositional, kind of knowledge. This kind of knowledge can be as relevant as *knowing that*, which corresponds to the classical theoretical conception of propositional knowledge developed in accordance with general rules of inference.

Although Ryle is not an explicit pragmatist, he suggests that intelligent action involves the ability to deal with *novelties* in order to fulfill certain convenient needs. Practical reasoning expressed by means of different forms of *knowing how* (to walk, to sing, to play, to cook, etc) is embodied in a certain sense; we do not need to formalize an inference in order to know how to act, for instance, when jumping out of the way of a runaway truck that heads in our direction. This does not mean, however, that such an action would be irrational or the product of exclusive chance; dispositions, which create some form of structuring, would probably be incorporated in this action.

In short, *knowing how* to do something does not require two steps: Mentally represented plans and the action itself. As stressed by Ryle, to *know how* consists of a special kind of array of complex dispositions, acquired by training, which are continuously actualized:

> Knowing how, then, is a disposition, but not a single-track disposition like a reflex or a habit. Its exercises are observances of rules or canons or the application of criteria, but they are not tandem operation of theoretically avowing maxims and then putting them into practice. [13]

Dispositions are relational properties of events, objects or chemical phenomena, like the solubility of salt in water (without water, how could salts be soluble?), the breakability of glass and affordances of different types of food to specific species. In the case of organisms, dispositions are biological properties (and embodied tendencies or potentialities) which might be actualized in order to generate patterns of action in evolutionarily developed contexts. These dispositions can be improved thanks to attention and self-organizing processes of adjustment that give rise to an embodied memory of relationships established between organisms and their niches. This relational character of the (complex) notion of dispositions may help us to elucidate the role of abductive reasoning in the structuring of humans' embodied abilities in so far as it provides elements for the emergence of systemic actions.

From Peirce's perspective, abductive reasoning is not a specific human ability – it is spread all around nature wherever transformations occur. It is not our aim to discuss this claim here, but rather to explain our Hypothesis

(3) that abductive inference constitutes one important organizational source of coherent actions emergent from self-organizing learning processes.

Given that the pragmatic view of coherent actions focuses on the systemic evolutionary character of contextually situated agency, it should not be analyzed from the formal perspective of classical logic. If it was, we might become trapped in the debate about fallacies of the following kind:

> Human beings are mortal.
> Plants are mortal.
> Ergo, human beings are plants. [4]

As is well known, this type of fallacy has for many centuries been considered a source of error and deception. Thus Aristotle, in "On Sophistical Refutations", stresses that:

> The refutation which depends upon the consequent arises because people suppose that the relation of consequence is convertible. For whenever, suppose A is, B necessarily is, they then suppose also that if B is, A necessarily is. This is also the source of the deceptions that attend opinions based on sense-perception. [1]

However, from a systemic pragmatic perspective, this form of reasoning can be legitimized in specific contexts involving patterns of action available to living beings. In this sense, Bateson [3–5] argues that humans and plants share common properties (such as that of being mortal) that allow both of them to be classified as organisms in terms of common patterns that connect their evolutionary path.

In a similar way, Magnani and Belli claim that fallacies may have a cognitive relevance for practical activities related to the survival of organisms [10]. They argue that hasty induction, for instance, could be used to protect living beings in dangerous situations. Thus, if we have a painful experience with fire, a "strategic rationality" can be created in the general form "fire is always dangerous". Even though this reasoning is not valid from the classical logic perspective, it could be biologically relevant to organize actions related to an ability to deal with fire.

Assuming that abductive reasoning is a type of pragmatic agent-based strategy, then its formal properties, initially regarded as a kind of fallacy from the classical perspective, can be re-evaluated. As mentioned in Section 2, one way of expressing abductive reasoning according to Peirce's view is:

> A surprising fact, C, is observed,
> But if H were true, C would be a matter of course;
> Hence, there is reason to suspect that H is true. [12]

As a tool to help with the formulation and selection of possible hypotheses to explain surprising events, abduction can be interpreted as a powerful means to understand creative thinking that initiates with surprise and ends with a well-structured elucidation of a set of events. Given that innumerable hypotheses could be formulated in different contexts, the power of creative abductive reasoning seems to depend on the embodied embedded circumstances in which agents are immersed.

Pragmatically considered, this way of reasoning "to the best explanation" may help embodied embedded agents to organize their beliefs (or strong habits) creating dispositions and possible able actions, contributing to the production of "survival strategies" [10]. This type of strategy can, and often does, lead to errors but, as stressed by Peirce: "It resembles instinct too in its small liability to error: For though it goes wrong oftener than right, yet the relative frequency with which it is right is on the whole the most wonderful thing in our constitution" [12].

Peirce's reference to instinct in his analogical explanation of the power of abductive reasoning brings us back to our initial question concerning the possibility of mechanical systems (like robots) being able to incorporate abductive reasoning; i.e.: Can abductive reasoning be properly characterized in terms of mechanical processes of information? A first answer to this question, based on the mentioned peircean anti-mechanicist conception of evolution of the cosmos, would be no. However, considering the nuances and complexities of contemporary mechanical systems this straightforward answer could be mitigated. After all, what might be the main difficulties concerning the *real* possibility that sophisticated robots could deal with abductive reasoning?

One difficulty (concerning the possibility of mechanical systems operating with abductive reasoning) is that mechanistic perspectives, such as those of weak and strong Artificial Intelligence, do not give value to the fundamental relationship between abductive reasoning and the agents' ability to be surprised. As pointed out in [9], even though Turing argues, in his well-known paper "Computing Machinery and Intelligence", that machines very often surprised him with their performances [14], he did not investigate the possibility that machines themselves could be surprised.

Peirce investigated (passive or active) surprise from a logical perspective as an initiator of abductive reasoning in terms of conflict of expectations. We believe that a similar role could be attributed to surprise as characterized by Aristotle who considered surprise the origin of philosophical activity [2].

Another difficulty involves the role of chance and failures in the development of abductive reasoning. The pragmatic fallibilist approach to knowledge seems to be underestimated in mechanical modeling of abductive reasoning. As indicated, in this sort of reasoning chance and failures may make sense in specific contexts, particularly those related to shared collective history. However, when investigated from a purely formal logical point of view errors can be seen as generators of fallacies, while in an agent-based perspective they may be of great value to the understanding of the origin and systemic history of survival strategies. Given that robots are not alive, it is not clear how they could deal with failures and surprises in this context.

A third and more central difficulty is related to the importance of criteria of relevance that are usually applied by embodied agents to distinguish relevant from non-relevant information amongst the immense amount of data available in the environment. When trying to search for evidence that could support an initial hypothesis, agents frequently explain an anomaly or a persisting problem, searching for the best explanation (in accordance with abductive reasoning) thanks to their embodied dispositions.

To conclude this paper, we are going to present an example of hypotheses creation, via abductive reasoning, in the domain of medical sciences (undertaken by one of the authors) in order to illustrate the complexity of abduction in practical reasoning. This example illustrates the importance of criteria of relevance in the selection of data in the domain of human activities: Without them we seem to be lost in a sea of meaningless information, in which differences would make no difference.

4 A Bone Metabolism Hypothesis Creation

During work in the Pediatric Polyclinic of the Federal University of São Paulo (UNIFESP-EPM) we could observe a considerable number of asthmatic patients with bone loss. As is well known, the gold standard treatment given is based on inhaled corticosteroids (IC), which are anti-inflammatory drugs that act directly in the lungs[1].

Although these medicines are not found to lead to bone mass loss, we had a strong belief that they do this. So, we have a problem that consists of providing a good explanation for our belief. In order to do that, we initiate an abductive reasoning, elaborating the following hypothesis (H_1): IC are absorbed by lungs and gastro-intestinal tract, and may cause as many collateral effects as systemic corticosteroids do [11]. These effects include a decrease of calcium absorption and an increase of calcium excretion; inside the bone matrix the osteoblasts, which rebuild the reabsorbed tissue, are inhibited, whereas the activity of osteoclasts leads to bone reabsorption.

In order to inductively test our hypothesis, we had 34 patients (half from each gender) who had been using IC for asthma for at least 12 months. All patients were treated in our clinic from February 2005 to February 2006, and were 4 to 14 years of age with a median of 11 years and 6 months. At the end of the period we found that according to their bone densiometry 5 (29%) of the males and 6 (35%) of the females had low bone mass for chronological age [15].

---

[1] National Institutes of Heath, National Asthma Education Program. Expert panel report 2: guidelines for the diagnosis and management of asthma. Bethesda, MD: National Institutes of Health, National Heart, Lung, and Blood Institute, 1993; 1–153. Available at www.nhlbi.nih.gov/guidelines/asthma/asthgdln.htm. Accessed November 19, 2005.

Although we found a high prevalence of bone loss, it was not yet possible to correlate IC use and bone loss in a conclusive way. Furthermore, it was not even possible to correlate a cumulative dose of IC that would presuppose bone loss. Chi square tests and Fisher's exact tests were used for statistical analysis.

As the above example illustrates, in order to successfully complete our abductive reasoning, longer studies with a higher number of patients would be necessary for more definitive conclusions. The difficulty present in this abductive reasoning is mainly due to the fact that we are dealing with organisms understood as complex systems and, as such, their analysis involves a variety of perspectives. These perspectives include living conditions, individual and collective history, genetic factors, and environment, among others.

Given the complexity underlying the development of the above practical exercise, our question is: How could a mechanical system select meaningful information that could be relevant to complete our abductive reasoning? Are we facing a similar impasse as in the case of the well-known *Frame problem* which highlights the difficulty of finding criteria of relevance to select information available for practical actions in contrast to pure random movements? Our belief is that embodied embedded learning abilities should be developed in more specific contexts in order to complete this abductive inference.

5 Final Comments

Our main purpose here was to investigate the relation between abductive reasoning and meaningful information supposedly unfolded in coherent action. Our study, which included a practical clinical case investigation on bone metabolism, was based on Peirce's conception of abductive reasoning (by which one infers to the best explanation of a problem, an anomaly or a surprising event), and Ryle's distinction between *habits* (that involve some sort of *automatism*) and *abilities* (performed by organisms possessing pragmatic competence to satisfy basic needs in dynamical contexts). We distinguished between information, conceived from a mechanistic view, and meaningful information implicit in systemic agency.

From this pragmatic point of view, we claim that: organisms, embedded in a context, are able to perceive patterns that connect them to their niches and organize their actions in coherent ways; by frequently incorporating abductive reasoning, they learn how to interact competently in an always challenging environment.

As for our hypothesis concerning the dynamics of bone metabolism, our collective evolutionary history may indicate new meaningful informational patterns that may help us to see the relevance of the data available in our niche.

Finally, a provisional answer to our initial question of whether mechanical systems, like robots, can develop these abilities involving abductive reasoning is: Mechanical systems would develop these abilities if they could perform actions that presuppose embodied competence to deal with meaningful information (involving self-organization as previously characterized). At the moment, they seem to be at the first step of the habit formation process, given that they can only learn how to execute certain specific tasks from the mechanical perspective, which allows little scope for self-organization and historically rooted competence.

Acknowledgements. We would like to thank Vivian and Artur Gonzalez for their helpful suggestions, and to Andrew Allen for the English revision. We also would like to thank our colleagues of the research groups: Cognitive Science, Unesp, and Self-Organization, Unicamp, for their comments and suggestions.

References

1. Aristotle: On sophistical refutations. In: Aristotle. *The Complete Works of Aristotle:*, the revised Oxford's translation. Edited by J. Barnes. 1 v. Princeton University Press, Princeton (1971)
2. Aristotle: Metaphysics. In: Aristotle. *The Complete Works of Aristotle:* the revised Oxford's translation. Edited by J. Barnes. 1 v. Princeton University Press, Princeton (1971)
3. Bateson, G.: *Mind and Nature: A Necessary Unity.* Bantam, New York (1980)
4. Bateson, G.: Men are like plants. In: Thompson, W.I. *Gaia - A Way of Knowing: Political Implications of the New Biology.* Lindisfarne Press, New York (1987)
5. Bateson, G.: *Steps to an Ecology of Mind. Collected Essays in Anthropology, Psychiatry, Evolution, and Epistemology.* University Of Chicago Press, Chicago (1972)
6. Debrun, M.A.: A idéia de auto-organização. In M. Debrun, M.E.Q. Gonzalez amd O. Pessoa Jr. eds.: *Auto-organização: Estudos Interdisciplinares*, vol. 18, Coleção CLE, 1–23, Campinas, Brasil (1996)
7. Gibson, J.J.: *The Ecological Approach to Visual Perception.* Houghton-Miffin, Boston (1986)
8. Gonzalez, M.E.Q.: Information and mechanical models of intelligence. *Pragmatics & Cognition* **13** (2005) 565–582
9. Gonzalez, M.E.Q. and Haselager, W.F.G.: Creativity and self-organization: contributions from cognitive science and semiotics. *S.E.E.D. Journal – Semiotics, Evolution, Energy, and Development* **3** (2003) 61–70
10. Magnani, L. and Belli, E.: Agent-based abduction: being rational through fallacies. In: Magnani L., ed.: *Model-Based Reasoning in Science and Engineering. Cognitive Science, Epistemology, Logic.* College Publications, London, (2006) 415–439
11. Mortimer, K.J., Harrison, T.W. & Tattersfield, A.E.: Effects of inhaled corticosteroids on bone. In: *Annals of Allergy, Asthma and Immunology* **94(1)** (2005) 15–22
12. Peirce, C.S.: *Collected Papers of Charles Sanders Peirce.* Harvard University Press, Cambridge, MA, vols. 1-6, Hartshorne, C. and Weiss, P., eds.; vols. 7–8, Burks, A. W., ed. (1931–1958)

13. Ryle, G.: *The Concept of Mind*. Penguin, London (2000)
14. Turing, A.: Computing machinery and intelligence. *Mind* **50** (1950) 433-450
15. Van der Sluis, I.M. and Muinck Keizer-Schrama S.M.: Osteoporosis in childhood: bone density in children in health and disease. *Journal of Pediatric Endocrinology and Metabolism* **14/(7)** (2001) 817–32

Automated Abduction in Scientific Discovery

Oliver Ray

Department of Computing, Imperial College, London, UK and
Department of Computer Science, University of Cyprus, Nicosia, Cyprus
or@doc.ic.ac.uk,oliver@cs.ucy.ac.cy

Summary. The role of abduction in the philosophy of science has been well studied in recent years and has led to a deeper understanding of many formal and pragmatic issues [1–5]. This paper is written from the point of view that real applications are now needed to help consolidate what has been learned so far and to inspire new developments. With an emphasis on computational mechanisms, it examines the abductive machinery used for generating hypotheses in a recent *Robot Scientist* project [6] and shows how techniques from *Abductive Logic Programming* [7] offer superior reasoning capabilities needed in more advanced practical applications. Two classes of abductive proof procedures are identified and compared in a case study. Backward-chaining logic programming methods are shown to outperform theorem proving approaches based on the use of contrapositive reasoning.

1 The Role of Abduction in Scientific Discovery

Abduction was first recognized as an fundamental form of logical inference alongside deduction and induction by C.S. Peirce just over a century ago [8]. His key insight was that abductive inference could be formally characterized by a simple rearrangement of the Aristotelian syllogism. He illustrated the idea by means of an example concerning the color of some beans taken from a bag [8, 2.623]. If one considers a deductive syllogism in which the conclusion (or *result*) "These beans are white" is derived from the major premise (or *rule*) "All the beans from this bag are white" and the minor premise (or *case*) "These beans are from this bag", then the corresponding abductive inference is obtained by exchanging the minor premise and conclusion so that the case is now abduced from the rule and result. In this way abduction provides a plausible explanation or hypothesis for why some individuals (these beans) share a certain characteristic (being white) indicative of some particular class (the beans in this bag). Of course, the abduced explanation is a suggestion rather than a certainty; and the hypothesis "These beans are from this bag" would have to be withdrawn if it were later to emerge that these beans were in fact taken from another bag of mixed color beans.

In his later writings Peirce shifted the emphasis from the syllogistic form of abduction to its pragmatic function within the philosophy of science. He began to identify abduction with the process of scientific hypothesis formation and saw it cooperating with deduction and induction within an incremental cycle of knowledge development. As illustrated in Figure 1, the cycle begins with the discovery of an *anomaly* or a "surprising fact" [8, 5.188] not explicable by one's existing knowledge. A plausible *hypothesis* or "flash of insight" [8, 5.181] must therefore be sought to account for this fact. In other words, one must adopt a hypothesis which is "likely in itself, and renders the facts likely" [8, 7.202]. This process of hypothesis is what Peirce calls abduction. He argues that testable *predictions* must then be found that would follow if the hypothesis were true and these must be compared against the results of experimental *observations*. Adequate confirmation may justify the tentative acceptance of the hypothesis as part of one's growing knowledge base, but insufficient support may eliminate one hypothesis in favor of another. The process of drawing up predictions is what Peirce calls *deduction*, while the evaluation of hypotheses is what he refers to as *induction*. New anomalies may be discovered along the way and thereby trigger further instances of the cycle. The next section considers a recent attempt to automate scientific methodology using specially developed hardware and software platforms.

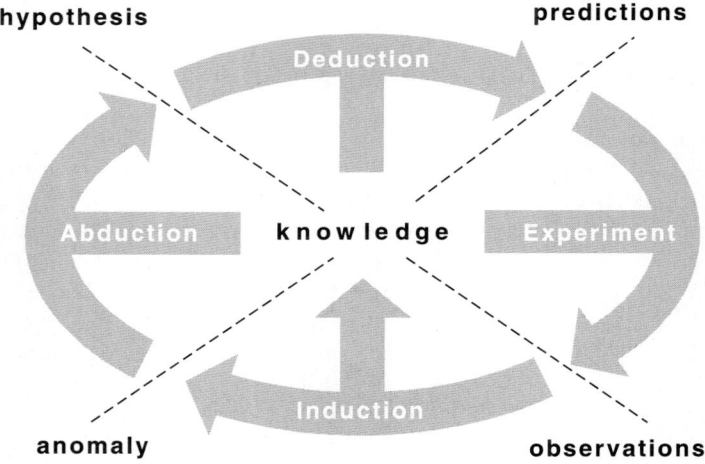

Fig. 1. Peirce's cycle of scientific knowledge development (copyright © 2005, Oliver Ray, reprinted by permission).

2 The Automation of Scientific Method

Robot Scientist [6] is an ongoing project which aims to automate scientific process by integrating advanced laboratory robotics and *Artificial Intelligence* (AI). According to its creators, 'The system automatically originates hypotheses to explain observations, devises experiments to test these hypotheses, physically runs the experiments using a laboratory robot, interprets the results to falsify hypotheses inconsistent with the data, and then repeats the cycle' [6]. A prototype Robot Scientist has been demonstrated in a highly regarded biological case study [6], where it was shown to cost-effectively reconstruct parts of a known biosynthesis pathway. An improved robot platform, illustrated in Figure 2 is now being built in a collaboration between the University of Wales and Caliper Life Sciences Inc [9].

The new Robot Scientist is and is based around the Caliper Sciclone i1000 liquid-handling workstation supported by peripheral freezing, incubating and plate reading equipment. When primed with a set of frozen cellular strains and a supply of nutrients, the system is able to revive selected strains from the freezer, incubate them in nutrient controlled plates and measure their respective growth rates over time. With no human assistance, the system can perform more than 1,000 experiments per day, logging in excess of 100,000 data measurements. Special heuristics have also been developed to allow the Robot Scientist to select a series of experiments that minimize the expected cost of converging to a correct theory [10].

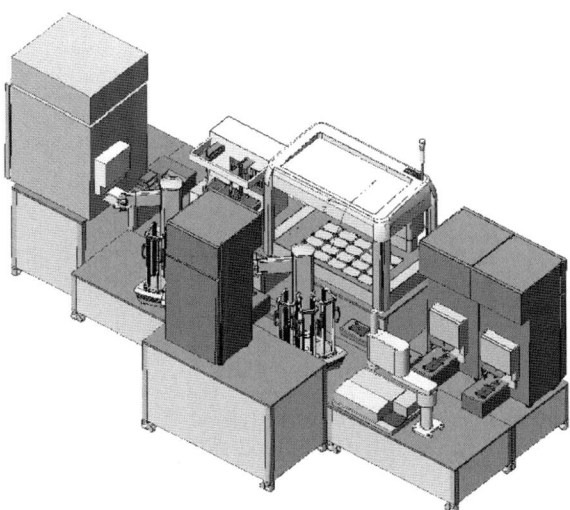

Fig. 2. Schematic overview of the Robot Scientist (copyright © 2005-2006, University of Wales Aberystwyth, reprinted by permission).

The "brain" of the Robot Scientist is a software system called *Progol5* [11] which uses a form of *contrapositive reasoning* [12] to generate logical hypotheses that explain a set of observed facts with respect to a prior theory. In contrast to many other machine learning systems, this allows Progol5 to learn hypotheses for theoretical predicates which cannot be observed directly and presented as examples. A recent analysis of Progol5, in [13], has shown that its use of contrapositives can be viewed as a restricted form of abduction. In the next section this mechanism is described in more formally and contrasted with an alternative backward-chaining approach based on an extension of the Prolog [14] logic programming paradigm.

3 Computational Models of Abductive Inference

Abduction is the task of finding a hypothesis H that *explains* a goal G with respect to a theory T. To automate this process, it is necessary to formalize the notion of explanation. At the risk of neglecting important philosophical concerns relating to causality and creativity, it is usual to equate explanation with deductive logical entailment and employ a naive hypothetico-deductive view of abduction where it is required to find a consistent H that, together with T, entails G (i.e. $T \wedge H \models G$). Computationally it is convenient to regard the theory T as a logic program (i.e. a universally quantified conjunction of Horn clauses of the form $a \leftarrow b_1, \ldots, b_n$) and the goal G as a query (i.e. an existentially quantified conjunction of atoms written? $g_1, \ldots, g_n$).

Epistemic considerations mean that the hypothesis is usually taken as a conjunction of ground atoms declared in advance as being abducible and often subject to domain-dependent integrity constraints. For convenience, the abducibles can be compactly specified by a set, AB, of ground terms, called *schemes*; which, as explained in [15], can contain special terms of the form *#pred*, where # is a special symbol called a *place-marker* and *pred* is an (optional) predicate called the *type* of that place-marker. Intuitively, each placemarker is seen as a holder for a constant of the appropriate type.

Proof procedures for automating abduction date back to the 1970's and can be partitioned into two categories according to whether they reason backwards from the goal [16–18] or forwards from the negation of the goal [19–21]. The former *backward-chaining* approaches augment Prolog-style reasoning techniques with a facility for assuming ground abducible atoms necessary to prove the goal, while the latter *contrapositive reasoning* approaches use theorem proving methods to infer negated hypotheses from the theory and the negation of the goal. Formally, the equivalence of the these approaches is ensured by the principle of *inverse entailment* which states that

$$T \wedge H \models G \text{ iff } T \wedge \neg G \models \neg H$$

The backward-chaining approach is exemplified by the *Abductive Logic Programming* (ALP) [7] procedure of Kakas and Mancarella [22], which extends Prolog inference with abduction. This is done by replacing non-abducible subgoals in the query by their resolvents [23] with theory clauses and assuming abducible subgoals where necessary to derive the empty clause. Contrapositive reasoning is used by Progol5 [11] to compute the negative ground literals entailed by the negated goal. Intuitively, the contrapositives of each theory clause are formed by swapping the head atom with one of the body atoms and then negating the swapped atoms in order to provide a logically equivalent clause with a negated body atom in the head.

The distinction can be illustrated by the task of inferring the explanation $H = \{b(1)\}$ for the goal $G =?a(1)$ given the theory $T = \{a(X) \leftarrow b(X)\}$ and the abducible $AB = \{b(\#)\}$. Contrapositive reasoning computes H by looking for those atoms of whose negation there is a proof from the negation of G and the contrapositives of T – i.e. it finds a proof of $\neg b(1)$ from $\neg b(X) \leftarrow \neg a(X)$ and $\neg a(1)$. Backward-chaining computes H by looking for those atoms which must be assumed so as to ensure the existence a proof of G using the clauses in T – i.e. it finds a proof of $a(1)$ from $a(X) \leftarrow b(X)$ by assuming $b(1)$. In the rest of this paper, these approaches are described in more detail and illustrated by means of a case study.

3.1 Contrapositive Reasoning (Progol5)

In classical logic, the contrapositive of a conditional $A \leftarrow B$ refers to the logically equivalent formula $\neg B \leftarrow \neg A$ obtained by negating and interchanging the consequent and antecedent. Similarly, the contrapositives of a Horn clause $A \leftarrow B_1, \ldots, B_n$ are defined as the n rules of the form $\neg B_i \leftarrow \neg A, B_1, \ldots, B_{i-1}, B_{i+1}, \ldots, B_n$ obtained by negating and interchanging the head atom with each of the body atoms in turn. By treating such negated predicates $\neg p$ as new predicates in their own right, contrapositives can be viewed as standard Horn clauses and used in Prolog computations to check whether a *negative* literal is implied by a Horn theory [11, 12].

To ensure computed atoms satisfy the required language bias, Progol5 adds a type atom back into the body of each contrapositive where B_i has an abducible predicate. This atom is obtained by replacing the predicate p of B_i with a new predicate $*p$ defining the well-typed instances of p. The result is denoted $\neg B_i \leftarrow \neg A, B_1, \ldots, B_{i-1}, *B_i, B_{i+1}, \ldots, B_n$. For each abducible scheme s, one type clause is added to the theory of the form $*p(t_1, \ldots, t_n) \leftarrow q_1(X_1), \ldots, q_m(X_m)$ where $p(t_1, \ldots, t_n)$ is the atom obtained from s by replacing each place-marker $\#q_i$ by a fresh variable X_i. In this way, abducibles are always grounded with terms of the correct type [11].

Given a Horn theory T, a ground atomic goal G, integrity constraints IC and abducibles AB, Progol5 will try to compute an atomic hypothesis H by showing $T \wedge \neg G \models \neg H$. To do this, it will augment T with all of its contrapositives along with the negation of G and the type rules generated by AB. It will then query the negation of each scheme s in AB, in turn, after replacing all place-markers by fresh variables. Any successful ground instances of these queries are un-negated and returned as possible abductive explanations. Each such abducible is added to the theory and checked for consistency with IC by ensuring the atom $\bot$ fails as a query [11].

To avoid an unnecessary proliferation of contrapositives, Progol5 uses an optimization based on the so-called *predicate dependency graph*: whose nodes are the predicates of T and where there is a edge from p to q iff there is a clause in T having the predicate p in its head and the predicate q in its body. Specifically, Progol5 uses this dependency graph to suppress the formation of any contrapositives $\neg B_i \leftarrow \neg A, B_1, \ldots, B_{i-1}, *B_i, B_{i+1}, \ldots, B_n$ where the predicate of B_i does not lie on a path from the predicate of G to an abducible. Since it can be shown that none of these contrapositives can participate in a successful computation [13], they can be safely omitted in order to reduce the size of the Progol5 search space [11]. In the next section, an alternative approach is considered that uses backward-chaining instead of contrapositives to implement abduction.

3.2 Backward-Chaining (ALP)

Backward-chaining refers to the Prolog-like strategy of replacing subgoals in the query by their resolvents with clauses in the theory until the empty clause is derived. Abductive reasoning can be realized by extending this approach with a facility for assuming atoms needed to complete a proof and for checking the consistency of those assumptions against any integrity constraints. One of the best known abductive techniques is the ALP procedure of Kakas and Mancarella [7] which interleaves two types of derivation: namely, abductive derivations, in which abducibles are assumed, and consistency derivations, in which integrity is checked.

Given a theory T, goal G, abducibles AB, the Kakas-Mancarella procedure starts an abductive derivation by progressively resolving the G against the clauses in T until an abducible atom a is selected. A consistency derivation is then invoked to determine if this atom a can be added to the (initially empty) hypothesis H without violating any integrity constraints that may be optionally specified to restrict which combinations of abducibles may be assumed at the same time.

To ensure ALP explanations satisfy the language bias, it is convenient to employ the same type clauses as Progol5 by adding one clause $*p(t_1, \ldots, t_n) \leftarrow q_1(X_1), \ldots, q_m(X_m)$ to the theory T for each head-declaration scheme in AB (as described in the previous section). At the same time is also convenient to replace each abducible predicate p with a new predicate p' by adding to T one bridging clause of the form $p(X_1, \ldots, X_n) \leftarrow *p(X_1, \ldots, X_n), p'(X_1, \ldots, X_n)$ for each abducible predicate p of arity n. These transformations simplify the design of ALP procedures by ensuring that abducibles do not appear in the heads of any clauses and are grounded by well-typed constants [24].

4 A Case Study in Computational Biology

Backward-chaining and contrapositives are now compared using a biological case study involving a metabolic regulatory mechanism of E. coli [25], a single-celled bacterium which lives in the human gut. Its preferred food is the sugar *glucose*, but it can metabolize *lactose* if it produces two enzymes: a *permease*, which imports lactose into the cell; and a *galactosidase*, which breaks it down. These enzymes are coded by two genes, *lac(y)* and *lac(z)*, which lie on a cluster of genes called the *lac-operon*. These genes are expressed when two regulators are present: *allolactose*, which is a derivative of lactose that is present when lactose is present; and *cAMP*, whose formation is inhibited by glucose and which is present when glucose is absent.

As shown in Fig. 3, cAMP enables an activator molecule (a.1) to bind to the operon (a.2) which induces polymerase to bind also (b.3). Allolactose then disables a repressor molecule (c.4-5) which allows polymerase to express the genes (d.6-8). In this way, E. coli avoids producing these enzymes if the preferred food source (glucose) is available or if the alternative food source (lactose) is not. A simplified model of this process is given by the theory T below. It states E. coli can metabolize lactose if permease and galactosidase are present. Like all enzymes, they are present if a gene which codes them is expressed. These enzymes are coded by the genes lac(y) and lac(z), which are expressed when allolactose and cAMP are present. In turn, these are present when the sugars lactose and glucose are present and absent, respectively.

$$T = \begin{cases} metabolise(\_, lact) \leftarrow present(perm), present(gala) \\ present(Enz) \leftarrow codes(Gene, Enz), express(Gene) \\ express(lac(\_)) \leftarrow present(allo), present(cAMP) \\ present(allo) \leftarrow present(lact) \\ present(cAMP) \leftarrow absent(gluc) \\ codes(lac(y), perm) \\ codes(lac(z), gala) \\ sugar(lact) \\ sugar(gluc) \end{cases}$$

$$G = ?metabolise(ecoli, lact)$$

$$AB = \begin{cases} present(\#sugar) \\ absent(\#sugar) \end{cases}$$

The goal G to be explained is the observation that E. coli is metabolizing lactose. The abducibles AB allow assumptions of the form $present(s)$ and $absent(s)$ where s is a sugar. Note that, for brevity, some constant names have been truncated in the obvious way (e.g. "lactose" to "lact"). Note also this example employs standard Prolog conventions whereby constants start with lowercase letters and variables start with uppercase letters or an underscore. The following sections detail the result of applying backward-chaining (as used in ALP) and contrapositives (as used by Progol5) to this example.

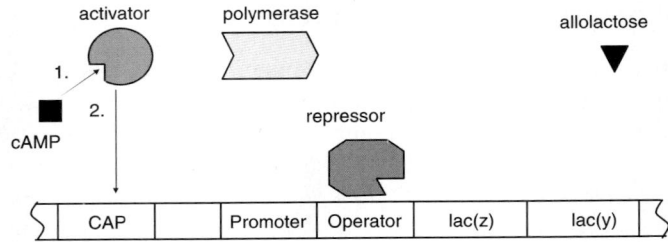
(a) Activator being enabled by cAMP and binding to CAP site

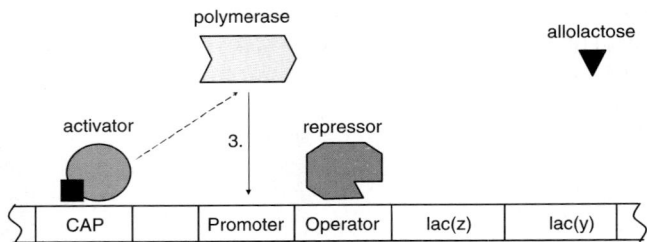
(b) Activator inducing polymerase to bind to promoter site

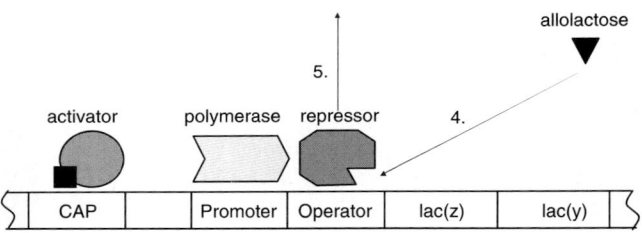
(c) Repressor being disabled by allolactose and detaching from operator

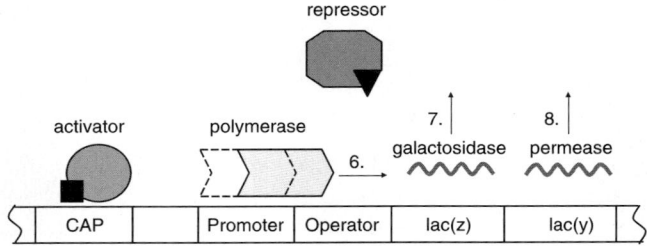
(d) Polymerase transcribing genes

Fig. 3. Biological Regulation of the LAC-Operon.

4.1 Application of Backward-Chaining

First consider the ALP methodology of backward-chaining from the goal. The full set of typing and bridging clauses added by ALP to the original theory T are shown below. The goal G results in one query for which the corresponding computation is shown in Figure 4. The query is shown at the top and each line corresponds to one resolution step. Note that, for brevity, both predicate and constant names are truncated (e.g. "present" to "pres"). Backtracking points are shown as side branches (with the number of resolution steps before failure given in brackets). Abduced atoms are underlined at the point they are assumed. The computed answer, obtained by removing the primes from the abduced atoms, is $H = \{present(lactose), absent(glucose)\}$. The search space is finite and comprises 1 successful branch, 12 failed branches and 45 resolution steps. The computed hypothesis is easily seen to be correct.

$$*present(X) \leftarrow sugar(X)$$
$$*absent(X) \leftarrow sugar(X)$$
$$present(X) \leftarrow *present(X), present'(X)$$
$$absent(X) \leftarrow *absent(X), absent'(X)$$

4.2 Application of Contrapositives

Next consider the Progol5 methodology of querying the negated abducibles. The full set of typing and contrapositive clauses added by Progol5 to the theory T are shown below. Note that, in the actual syntax and terminology of Progol5, the abducibles AB are represented by the so-called *head-declarations* $modeh(*, present(\#sugar))$ and $modeh(*, absent(\#sugar))$ and the goal is identified by the Progol5 *directive observable(metabolise/2)*. The first two clauses below are type clauses, the next seven are contrapositives, and the last is the negation of the goal. Notice also that the dependency graph optimisation used by Progol5 suppresses the contrapositive $\neg codes(Gene, Enz) \leftarrow \neg present(Enz), express(Gene)$.

$*present(X) \leftarrow sugar(X)$
$*absent(X) \leftarrow sugar(X)$
$\neg present(perm) \leftarrow \neg metabolise(\_, lact), *present(perm), present(gala)$
$\neg present(gala) \leftarrow \neg metabolise(\_, lact), present(perm), *present(gala)$
$\neg present(allo) \leftarrow \neg express(lac(\_)), *present(allo), present(cAMP)$
$\neg present(cAMP) \leftarrow \neg express(lac(\_)), present(allo), *present(cAMP)$
$\neg present(lact) \leftarrow \neg present(allo), *present(lact)$
$\neg express(Gene) \leftarrow \neg present(Enz), codes(Gene, Enz)$
$\neg absent(gluc) \leftarrow \neg present(cAMP), *absent(gluc)$
$\neg metabolise(ecoli, lact)$

The abducibles result in two queries shown in Figure 5, where finitely failed branches are denoted ■ and infinitely failed branches are denoted ∞. Both

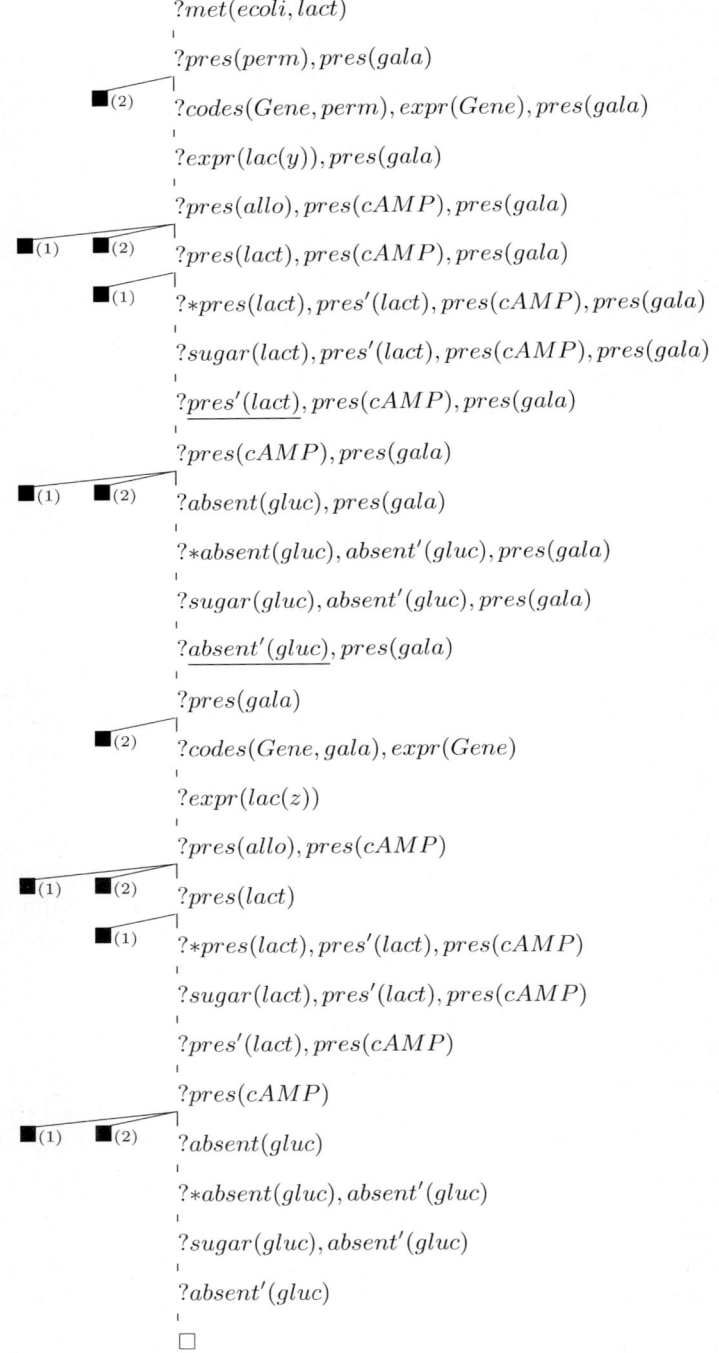

Fig. 4. Successful Backward-Chaining Computation.

?¬absent(S)
?¬pres(cAMP), ∗absent(gluc)
?¬exp(lac(G)), pres(allo), ∗pres(cAMP), ∗absent(gluc)
?¬pres(E), codes(lac(G), E), pres(allo), ∗pres(cAMP), ∗absent(gluc)

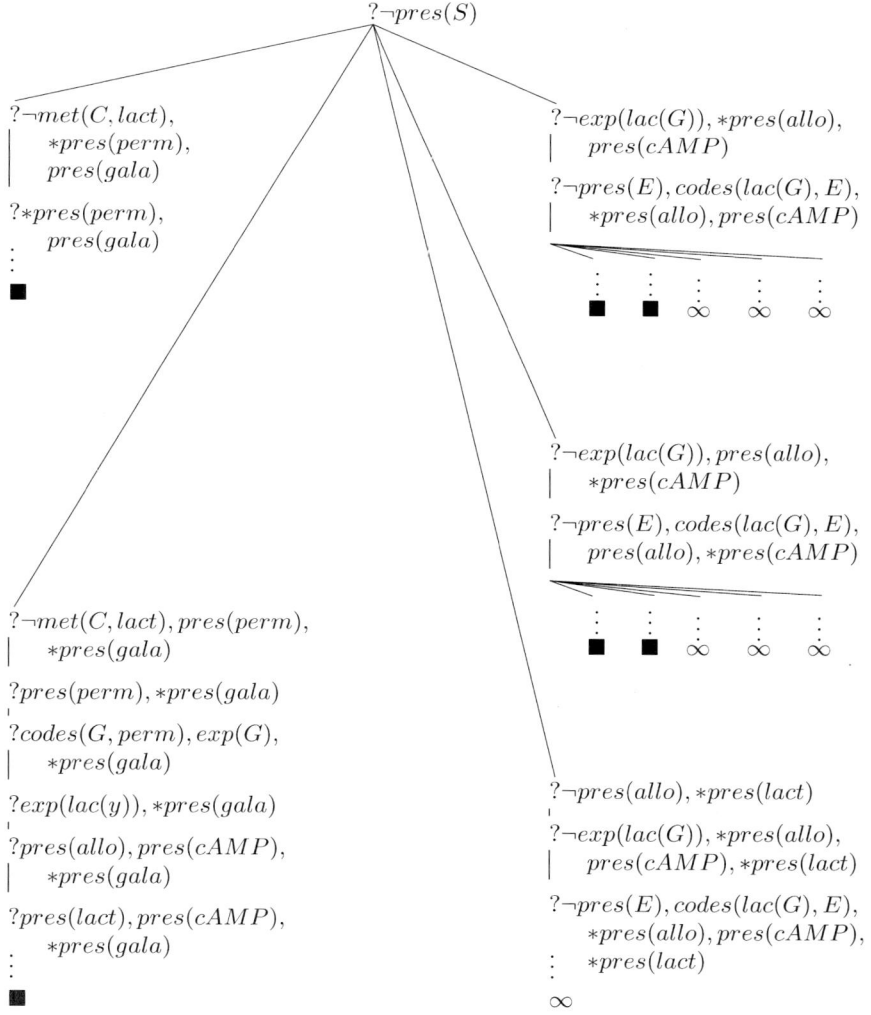

Fig. 5. Failed Contrapositive Computation.

computations are infinitely failed. Since the latter occurs as a subgoal in the former after 3 resolution steps, it suffices to analyze the latter computation. There are no successful branches and, at the default Progol5 depth-bound of 30 resolution steps, this search space comprises 710,645 finitely failed branches 8,199,009 non-terminated branches and contains 21,465,271 resolution steps.

5 Conclusion

It is apparent from the above analysis that contrapositives and backward-chaining are dual methods in the sense that the latter extends a given theory to perform abduction using standard Prolog inference, whereas the former extends Prolog inference to perform abduction on the given theory. Although they are closely related by the principle of inverse entailment, the case study shows there are important practical differences. Most obvious is the fact that, unlike ALP, contrapositives cannot practically be used to compute hypotheses containing more than one atom (since the number of potential queries would grow exponentially with the number of atoms and would require the introduction of non-Horn theorem proving techniques). In the above case study, the first failure of Progol5 is its inability to compute more than one atom by contrapositives, when two atoms are needed; which highlights the importance of previous work which shows how multiple-atom abductive explanations can be efficiently exploited in non-OPL systems [13].

The inability of contrapositives to compute hypotheses with more than one atom is not the only reason for the failure of Progol5 in this case study. This can be shown by adding the atom *absent(glucose)* to the theory T so that the unique minimal hypothesis comprises the single atom *present(lactose)*. While this hypothesis is trivially computed by ALP, once again, Progol5 fails to find a solution. The shortcoming is a previously noted limitation of Progol5 [26] which means it cannot compute atoms like *present(lactose)* that are needed more than once in an Prolog proof of the goal [27]. In the above example each branch of the Progol5 computation fails on the very atom *present(lactose)* required as a hypothesis – indicating for the first time the practical significance of this incompleteness in non-trivial applications. Even though this incompleteness can be overcome by the use of non-Horn theorem proving methods [12], this would exacerbate the already poor performance of contrapositives compared with backward-chaining.

Although the automation of abductive reasoning and scientific method has proven to be a beneficial exercise, several philosophical and practical issues have been necessarily avoided. The logical approaches discussed in this paper assume the initial theory is correct and they do not provide methods to identify and withdraw existing rules that are subsequently found to be incorrect. Even though recent progress has been made integrating ALP with inductive learning [13] to allow the inference of more expressive logic programs with *negation* [28], truly creative scientific reasoning has yet to be reliably automated.

In addition, the Robot Scientist is limited to performing a rather narrow class of experiments. Attempts to overcome these restrictions are important directions for future work that are likely to shed new light on old philosophical arguments and break new ground in the application of AI.

Acknowledgements

The author is grateful to Ross King for helpful discussions on the Robot Scientist and for his kind permission to reproduce Figure 2. Thanks also go to Stephen Muggleton and Antonis Kakas for discussions regarding the Progol5 and Kakas-Mancarella proof procedures. Valuable suggestions were also made by Alessandra Russo and Krysia Broda.

References

1. Magnani, L.: *Abduction, Reason, and Science. Processes of Discovery and Explanation.* Kluwer (2001)
2. Aliseda, A.: Abductive Reasoning: Logical Investigations into Discovery and Explanation. Volume 230 of *Synthese Library*. Springer (2006)
3. Gabbay, D., Woods, J.: *The reach of abduction: Insight and trial.* Elsevier (2005)
4. Flach, P., Kakas, A., eds.: Abduction and Induction: essays on their relation and integration. Volume 18 of *Applied Logic Series*. Kluwer (2000)
5. Josephson, J., (Eds.), S.J.: *Abductive inference: computation, philosophy, technology.* Cambridge University Press (1994)
6. King, R., Whelan, K., Jones, F., Reiser, P., Bryant, C., Muggleton, S., Kell, D., Oliver, S.: Functional genomic hypothesis generation and experimentation by a robot scientist. *Nature* **427** (2004) 247–252
7. Kakas, A., Kowalski, R., Toni, F.: Abductive Logic Programming. *Journal of Logic and Computation* **2**(6) (1992) 719–770
8. Hartshorne, C., Weiss, P., Burks, A., eds.: *Collected papers of Charles Sanders Peirce.* Harvard University Press (1931-1958)
9. King, R., Benway, M.: Robot Scientist: an Autonomous Platform for Systems Biology Discovery. A poster at the 12th Annual Conference and Exhibition of the Society for Biomolecular Sciences (2006)
10. Bryant, C., Muggleton, S., Oliver, S., Kell, D., Reiser, P., King, R.: Combining Inductive Logic Programming, Active Learning and Robotics to Discover the Function of Genes. *Electronic Transactions on Artificial Intelligence* **5 (Section B)** (2001) 1–36
11. Muggleton, S., Bryant, C.: Theory Completion Using Inverse Entailment. In: *Proceedings of the 10th International Conference on Inductive Logic Programming.* Volume 1866 of Lecture Notes in Computer Science. Springer Verlag (2000) 130–146
12. Stickel, M.: A Prolog technology theorem prover: A New Exposition and Implementation in Prolog. *Theoretical Computer Science* **104** (1992) 109–128
13. Ray, O.: *Hybrid Abductive-Inductive Learning.* PhD thesis, Department of Computing, Imperial College London, UK (2005)

14. Lloyd, J.: *Foundations of Logic Programming*. Springer Verlag (1987)
15. Muggleton, S.: Inverse Entailment and Progol. *New Generation Computing, Special issue on Inductive Logic Programming* **13**(3-4) (1995) 245–286
16. Jr., H.P.: On The Mechanization of Abductive Logic. In: *Proceedings of the 3rd International Joint Conference on Artificial Intelligence*, William Kaufmann (1973) 147–152
17. Finger, J., Genesereth, M.: *Residue: a deductive approach to design synthesis*. Technical Report STAN-CS-85-1035, Stanford University, USA (1985)
18. Stickel, M.: A Prolog-like inference system for computing minimum-cost abductive explanations in natural-language interpretation. *Annals of Mathematics and Artificial Intelligence* **4** (1991) 89–106
19. Morgan, C.: Hypothesis generation by machine. *Artificial Intelligence* **2**(2) (1971) 179–187
20. Cox, P., Pietrzykowski, T.: Causes for Events: Their computation and Application. In Siekmann, J., ed.: *Proceedings of the 8th International Conference on Automated Deduction*. Volume 230 of Lecture Notes in Computer Science., Springer (1986) 608–621
21. Inoue, K.: Linear resolution for consequence finding. *Artificial Intelligence* **56**(2-3) (1992) 301–353
22. Kakas, A., Mancarella, P.: Database Updates through Abduction. In: *Proceedings of the 16th International Conference on Very Large Databases*, Morgan Kaufmann (1990) 650–661
23. Robinson, J.: A Machine-Oriented Logic based on the Resolution Principle. *Journal of the ACM* **12**(1) (1965) 23–41
24. Ray, O., Kakas, A.: Prologica: a practical system for abductive logic programming. In: Dix, J., Hunter, A., eds.: *11th International Workshop on Nonmonotonic Reasoning*. IFL Technical Report Series, Clausthal University of Technology (2006) 304–312
25. Jacob, F., Monod, J.: Genetic regulatory mechanisms in the synthesis of proteins. *Journal of Molecular Biology* **3** (1961) 318–356
26. Ray, O., Broda, K., Russo, A.: Hybrid Abductive Inductive Learning: a Generalisation of Progol. In: Horváth, T., Yamamoto, A., eds.: *Proceedings of the 13th International Conference on Inductive Logic Programming*. Volume 2835 of Lecture Notes in Artificial Intelligence., Springer Verlag (2003) 311–328
27. Ray, O.: The need for Ancestor Resolution when answering queries in Horn clause logic. In: Gabbrielli, M., Gupta, G., eds.: *Proceedings of the 21st International Conference on Logic Programming*. Volume 3668 of Lecture Notes in Computer Science., Springer Verlag (2005) 410–411
28. Ray, O.: Using abduction for induction of normal logic programs. In: *Proceedings of the ECAI'06 Workshop on Abduction and Induction in Artificial Intelligence and Scientific Modelling*. (2006) 28–31

Abduction, Medical Semeiotics and Semioethics
Individual and Social Symptomatology from a Semiotic Perspective

Susan Petrilli

Dipartimento di Pratiche Linguistiche e Analisi di Testi, University of Bari, Bari, Italy
s.petrilli@alice.it

Summary. My paper starts from Thomas A. Sebeok (1920-2001) and his conception of semiotics integrated with Charles Sanders Peirce's pragmatic concept of "abduction", disregarded by Sebeok in his own original reformulation of Peircean theory. Sebeok was not interested in logic just as he was not interested in critiquing today's social system. Sebeok denominated his particular approach to semiotics as "global semiotics". Taking global semiotics as our starting point, we propose to develop it in the direction of semioethics, which presupposes Sebeok's interpretation of ancient medical semeiotics or symptomatology as an initial phase in the history of semiotics.

1 Abduction, Icon, and Agapasm from a Synechist Perspective

Abduction is the name of a given type of argumentation, of progression from one interpretant to another. Abduction is foreseen by logic, but (especially in its more risky expressions) supersedes the logic of identity and develops through argumentative procedures that are risky, that is, eccentric, creative, or inventive. By contrast with induction and deduction, in abduction the relation between interpreted sign and interpretant sign is regulated by similarity, attraction and reciprocal autonomy. Abduction is grounded in the logic of otherness, substantial dialogism, and creativity. It proceeds through relations of fortuitous attraction among signs and is dominated by iconicity.

We have made the claim that abductive argumentative procedure is risky; in other words, it advances through arguments that are tentative and hypothetical, leaving a minimal margin to convention and to mechanical necessity. To the extent that it transcends the logic of identity and equal exchange among parts, abduction belongs to the side of excess, exile, *dépense*, giving without a profit, the gift beyond exchange, desire. It proceeds more or less according to the "interesting", and is articulated in the dialogic and disinterested relation among signs – a relation regulated by the law of creative love, so that abduction is also an argumentative procedure of the agapastic type.

In Charles S. Peirce's philosophical system (see, for example, his writings collected in the volume *Chance, Love and Logic* [1]), "chance", "love" and "necessity" indicate three modes of development. The connection between love, or what Peirce calls *agapasm*, is particularly interesting from the perspective of the present paper. In his renown paper "On a New List of Categories" [2, 1.545-559], Peirce elaborates his doctrine of categories in terms of the triad firstness, secondness, and thirdness, which are always co-present, interdependent and irreducible. The doctrine of categories constitutes the foundation of his ontology and cosmology, therefore it is connected to his ontological-cosmological triad with the distinction between agapasm, anancasm, tychasm, to his typology of inferential logic with the triad abduction, induction, and deduction, and to his sign typology, in particular the triad icon, index, and symbol.

The connection between agapasm and abduction offers a platform to discuss Victoria Welby's (1837-1912) work on signifying processes in relation to Peirce. In a reconstruction of the history of semiotics, Welby is a name to remember along with others like Mikhail M. Bakhtin and Emmanuel Levinas, for a better understanding of the sign in theoretical terms.

Understood as development through the forces of affinity and sympathy and with reference to the Peircean triad, icon, index and symbol, we may claim that agapasm is strongly iconic by virtue of the force of attraction, that is, the relation of similarity or affinity among interpretants. In agapastic evolution, chance (tychasm) and necessity (anancasm) are also operative. However the forces of attraction, affinity, freedom, and fortuitousness dominate in the relation among interpretants forming the continuous (synechetic) flow of infinite semiosis, just as iconicity dominates over indexicality and symbolicity.

The concept of continuity or synechism involves regularity. As emerges from her own philosophy of the sign processes permeating the entire signifying universe, Welby too believed that development is articulated in a continuous sign structure, and that continuity presupposes relational logic grounded in otherness. The logic of otherness is a "dia-logic" – that is, logic that recovers the dimension of *dialogicality*, as understood by Bakhtin. In other words, following both Peirce and Bakhtin, dialogicality is considered as a modality of semiosis, which may or may not involve verbal signs and may or may not take the form of dialogue. Dialogicality thus described is determined by the degree of opening towards otherness. And agapastic evolution is achieved through the law of love; creative and altruistic love, love founded on the logic of otherness, as would say authors like Welby, Bakhtin and Levinas.

In tychastic development – connected to symbolicity in semiotic terms and to induction in argumentative terms, new interpretive routes with unpredictable outcomes, some of which are fixed in "habits", are determined by chance. In anancastic development – connected to indexicality and deduction, new interpretive routes are determined by necessity, by causes that are both internal (the logical development of ideas, of interpretants that have already been accepted) and external (circumstances) with respect to consciousness, with the possibility of making predictions concerning eventual results. Instead, in agapastic development, deferral from one interpretant to the next is iconic and abductive. Development is neither regulated by chance, nor by blind necessity, but, as Peirce says, "by an immediate attraction for the idea itself, whose nature is divined before the mind possesses it, by the power of sympathy, that is, by virtue of the continuity of mind" [2, 6.307]. As an example of agapasm, the evolution of thought according to the law of creative love, Peirce cites the *divination* of genius, the mind affected by the idea before it is comprehended or possessed, by virtue of the force of attraction that the idea exercises upon him in the context of relational continuity among signs in the great semiosic network of the universe or semiosphere.

Paradoxically, in tychastic development chance generates order. In other words, the fortuitous result engenders the law, and the law in turn finds an apparently contradictory explanation in terms of the action of chance. This is the principle that informs the work of Charles Darwin on the origin of the species and natural selection. In Peirce's opinion, one of the reasons why Darwin won so much favor was that the values informing his research – represented by the principle of the survival of the fittest – measured up to dominant values of the time, which, in the final analysis, were founded on the assertion of the logic of identity and could be summed up in the word "greed".

Logic understood in the strict sense as necessary cause is connected with anancastic development. The limitation to this kind of development is the conviction that only one kind of logical development is possible. By maintaining that the conclusion derived from the premises could not be anything different, all other argumentative modes, and consequently the possibility of free choice, are excluded [2, 6.313]. However, while the mechanisms of construction, contingency, and mechanical necessity effectively dominate over the relation between interpreted and interpretant signs in anancastic inferential procedure, this does not exclude the effect of other interpretive modalities, active even when the anancastic procedure prevails. In semiotic terms, in the case of anancasm the relation between the intepreted and interpretant is predominantly indexical; in argumentative terms, it is deductive. The relation between the conclusion and its premises is regulated by reciprocal constriction and operates at low degrees of otherness and dialogism.

The goal of agapastic development is development itself (of the cosmos, language, thought, the subject), continuity of semiosis, of the development of an idea. According to Peirce, in a universe regulated by the principle of continuity and relational logic, where no single fact, datum, idea or individual exists in isolation, creative evolution (beaten out at the rhythm of hypotheses, discoveries, and qualitative leaps) is achieved through the combined effect of agapasm and synechism. Therefore, shifting our attention to the problem of subjectivity, the self as a communicating entity, far from being solitary in essence, has its roots in *agape*. By virtue of the continuity of thought, and therefore of relational logic, the main force ruling over the deferral among signs from an evolutionary perspective is that of *agapic or sympathetic comprehension and recognition*. And the simultaneous and independent occurrence of a genial idea to a number of individuals not endowed with any particular powers – a consequence of belonging to the same great semiosphere – might be considered as a demonstration of this [2, 6.315-316]. As a sign, in Peirce's view, the self too develops according to the laws of inference [2, 5.313].

2 Medical Semeiotics and Semioethics: from Sebeok's Biosemiotics and Beyond

Let us now consider Sebeok's "global semiotics". According to Sebeok, semiotics is not only *anthroposemiotics* but also *zoosemiotics, phytosemiotics, mycosemiotics, microsemiotics, endosemiotics, machinesemiotics, environmental semiotics*.

Sebeok's interests cover a broad range of territories ranging from the natural sciences to the human sciences. Consequently, he deals with theoretical issues and their applications from as many angles as the disciplines called in question: linguistics, cultural anthropology, psychology, artificial intelligence, zoology, ethology, biology, medicine, robotics, mathematics, philosophy, literature, narratology, and so forth. Even though the initial impression might be of a rather erratic mode of proceeding as he experiments various perspectives and embarks upon different research ventures, in reality Sebeok's expansive and seemingly distant interests find a focus in his "doctrine of signs" and in the fundamental conviction subtending his general method of enquiry that the universe is perfused with signs, indeed, as Peirce hazards, may be composed exclusively of signs.

As a fact of signification the entire universe enters Sebeok's "Global Semiotics". Semiotics is the place where the "life sciences" and the "sign sciences" converge, therefore the place where consciousness is reached of the fact that the human being is a sign in a universe of signs.

Sebeok has propounded a wide-ranging vision of the study of signs that coincides with the study of the evolution of life. After Sebeok's work both the conception of the field of semiotics (the sign science) and its history have changed noticeably. Thanks to him semiotics at the beginning of the new millennium presents a far more enlarged view than that of the first half of the 1960s. Sebeok has extended the boundaries of traditional semiotics or more correctly semiology – which is restrictively based upon the verbal paradigm and is vitiated by the *pars pro toto* error. He tagged this conception of semiotics the "minor tradition" and promoted, instead, what he called the "major tradition", as represented by Locke and Peirce and early studies on signs and symptoms by Hippocrates and Galen. Semiotics, therefore, is at once recent if considered from the viewpoint of the determination of its status and awareness of its wide-ranging possible applications, and ancient if its roots are traced back at least, following Sebeok [3] to the theory and practice of Hippocrates and Galen.

By virtue of this "global" or "holistic" approach, we can immediately associate Sebeok's inquiry into the "life of signs" with his concern for the "signs of life". In Sebeok's view, *semiosis and life converge.*

The boundaries of the history of semiosis have also been extended as a result of the links forged between semiosis and the biological sciences, a consequence of including not only communication but also signification and symptomatization in our conception of semiosis or sign process or sign situation.

Moreover, the boundaries of the history of semiotics are further expanded by awareness of the interdisciplinary character of semiotics, including its relation to the natural and biological sciences.

Semiotics as a field of study is far older than is usually presented. Its origins are in fact ancient, notwithstanding superficial descriptions in a phonocentric and glottocentric key. Semiotics grew out of symptomatology, or medical semeiotics, which Galen considered one of the principal branches of medicine. In this sense, Hippocrates and Galen, with their early studies on signs and symptoms, were among the very first semioticians.

If semiotics is concerned with life over the whole planet (given that life and semiosis converge), and if the original motivation for the study of signs is the "health" of semiosis, it may well follow that an important task for semiotics today is to care for life in all its diversity – a task that is especially urgent in the era of globalization. Semiotics is capable of understanding the entire semiosic universe and of discussing the various forms of separatism, technicalism, and overspecialization that afflict it. All of this should result in awareness of ethical responsibility.

Semioethics is the term we have proposed for this particular unbounded trend in semiotics, as announced in my book with Augusto Ponzio entitled *Semiotics Unbounded*. This title is intended to underline the totalizing and yet open character of semiotics, with what we may call its "detotalizing method".

Semiotics is an extremely wide-ranging field that crosses into many disciplines. Precisely because of this, and given that we must keep account of progress in the sciences – "human", "natural" and "logico-formal" sciences –, semiotics must be ready to renew itself and to interrogate the very methods and categories it employs. Semiotics is unbounded, and so is the object of its studies – the sign network. This leads to the subtitle of our book: *Interpretive routes through the open network of signs.*

In a paper of 1989, "Semiosis and Semiotics: What Lies in Their Future?", included in *A Sign Is Just a Sign* [4, pp. 97–99], Sebeok introduced a second meaning of the term "semiotics", beyond indicating the general science of signs. This new meaning refers to *the specificity of human semiosis*, the human capacity for metasemiosis, and is of vital importance for a *transcendental founding of semiotics* as a doctrine of signs. Says Sebeok:

Semiotics is an exclusively human style of inquiry, consisting of the contemplation – whether informally or in formalized fashion – of semiosis. This search will, it is safe to predict, continue at least as long as our genus survives, much as it has existed, for about three million years, in the successive expressions of Homo, variously labeled – reflecting, among other attributes, a growth in brain capacity with concomitant cognitive abilities – *habilis, erectus, sapiens, neanderthalensis,* and now *s. sapiens.* Semiotics, in other words, simply points to the universal propensity of the human mind for reverie focused specularly inward upon its own long-term cognitive strategy and daily manoeuvrings. Locke designated this quest as a search for "humane understanding"; Peirce, as "the play of musement". [4, p. 97]

According to this second meaning, "semiotics" is the human species-specific capacity for *metasemiosis*. In other words, human semiosis is capable of reflecting on signs, therefore not only interpretation undistinguished from direct response to signs, but also interpretation understood as reflection on signs, as suspension of response and deliberation.

Beyond Aristotle who at the beginning of his *Metaphysics* claimed that man tends by nature to knowledge, we can now claim that man tends by nature to semiotics. Human semiosis, anthroposemiosis, also presents itself as *semiotics* in the double sense just described and as such is free to venture across the entire semiosic universe. All the same, however, we must guard against the *pars pro toto error* (made in the history of ideas), that is, the oversimplification of reducing semiosis in general to anthroposemiosis.

In another paper included in *A Sign Is Just a Sign* [4, pp. 83–96], Sebeok explains the correspondences that exist between the various branches of semiotics and different types of semioses, from the world of micro-organisms to the Superkingdoms and the human world. Human semiosis is characterized as semiotics thanks to a modeling device specific to humans, called "language" (it is virtually certain that *Homo habilis* was endowed with language, but not speech).

In another important paper included in [4], entitled "In What Sense is Language a 'Primary Modeling system'?", Sebeok describes language as a "modeling device". Every species is endowed with a model that produces its own world, and "language" is the name of the model belonging to human beings. The human modeling device, or language, is completely different from the modeling devices of other life forms. Its distinctive feature is what the linguists call "syntax". Thanks to syntax hominids do not just experience one "reality", one world, but can also frame an indefinite number of possible worlds. This capacity is unique to human beings. Thanks to syntax human language is like Lego building blocks, it can reassemble a limited number of construction pieces in an infinite number of different ways. As a modeling device language can produce an indefinite number of models; in other words, the same pieces can be taken apart and put together to construct an infinite number of different models. Thanks to language not only do human animals produce worlds as do other species, but, as says Leibniz, human beings can also produce an infinite number of possible worlds. This leads to the "play of musement", a human capacity which Sebeok considers particularly important for scientific research and all forms of investigation as well as for fiction and all forms of artistic creation.

The exquisitely human propensity for musement implies the ability to carry out such operations as predicting the future or "traveling" through the past, the ability, that is, to construct, deconstruct and reconstruct reality, inventing new worlds and interpretive models. (The happy expression "the play of musement" is used by Sebeok, interpreter of Peirce, as the title of his book of 1981 [5].

Speech appeared with *Homo sapiens* as an adaptation, like language, but *for the sake of communication*, and much later than language. Consequently, language too ended up becoming a communication device; and speech developed out of language as a derivative *exaptation* [6]. Exapted for communication, first in the form of speech and later of script, language enabled human beings to enhance the nonverbal capacity with which they were already endowed. On the other hand, speech was *exapted* for modeling and eventually functioned as a *secondary modeling system*. In addition to increasing the communication capacity, speech also increased the capacity for innovation and "play of musement". Such aspects as the plurality of languages and "linguistic creativity" (Chomsky) testify to the capacity of language understood as a primary modeling device, for producing numerous possible worlds.

Sebeok uses the concept of modeling as proposed by the so-called Moscow-Tartu school (A.A. Zaliznjak, V.V. Ivanov, and V.N. Toporov. Ju. M. Lotman) where it is used to denote natural language ("primary modeling system") and other human cultural systems ("secondary modeling systems"). However, differently to the Moscow-Tartu school, Sebeok goes further to extend the concept of modeling beyond the domain of anthroposemiosis. With reference to the biologist J. von Uexküll and his concept of *Umwelt*, Sebeok's interpretation of model may be translated as an "outside world model". On the basis of research in biosemiotics, the modeling capacity is observable in all life forms [7, pp. 49–58] [8, pp. 117–127]. *The Forms of Meaning. Modeling Systems Theory and Semiotic Analysis*, co-authored by Sebeok with Marcel Danesi, published in 2000, further develops the fundamental notion of "model" [9].

The study of modeling behavior in and across all life forms requires a methodological framework developed in the field of biosemiotics, what Sebeok in his research on the interface between semiotics and biology called "modeling systems theory". Modeling systems theory studies semiotic phenomena as modeling processes [9, pp. 1–43].

In the light of semiotics conceived in terms of modeling systems theory, semiosis – which is a vital characteristic of all life forms – may be defined as "the capacity of a species to produce and comprehend the specific types of models it requires for processing and codifying perceptual input in its own way" [9, p. 5].

The applied study of modeling systems theory is called *systems analysis*, which distinguishes between primary, secondary and tertiary modeling systems.

The primary modeling system is the innate capacity for simulative modeling, in other words, it is a system that allows organisms to simulate something in species-specific ways [9, pp. 44–45]. The primary modeling system specific to *Homo* is also called "language". The secondary modeling system is the system that subtends both indicational and extensional modeling processes. The nonverbal form of indicational modeling has been documented in various species, whereas extensional modeling is a uniquely human capacity given that it presupposes *language* (human primary modeling system) as distinguished from *speech* (human secondary modeling system) [9, pp. 82–85]. The tertiary modeling system subtends highly abstract, symbol-based modeling processes. Tertiary modeling systems are human cultural systems [9, pp. 120–129].

Syntactics, deconstruction and reconstruction, production of many possible worlds, semiotics, with the ensuing capacity for evaluation, responsibility, inventiveness, planning, criticism, are all prerogatives of language. With language the being of communication finds its *otherwise*. Insofar as man is endowed with language, insofar as he is a semiotic animal, human behaviour cannot be circumscribed to communication, being, ontology. From this perspective man reveals his capacity for *otherness*. He can present himself as other and propose other possibilities beyond the *alternatives* forseen by "being" as the latter emerges in communication society.

Proceeding beyond Sebeok we may claim that the human being as a unique semiotic animal, that is, the only animal capable of reflection upon signs and communication, has a singular responsibility towards life (which is made of signs and communication) – which also means the quality of life. More than *limited responsibility*, the type of responsibility involved is *unlimited responsibility*, as understood by Levinas, *responsibility without alibis*, *absolute responsibility*.

3 Logic, Ethics, and Social Symptomatology

The human capacity for abduction, that it, for hypothesis, is the condition for critical and dialogic totalization. It implies that the ability to grasp the *reason* of things cannot be separated from the capacity for *reasonableness*. The issue at stake may be stated in the following terms: given the risks inherent in social reproduction today for semiosis and for life, *human beings must at their very earliest transform from rational animals into reasonable animals.*

Referring to the problem of subjectivity in the framework of his pragmaticism, therefore to self considered as a set of actions, practices, habits, Peirce identified "power" as opposed to "force" as one of its fundamental characteristics. Power is not "brute force" but the "creative power of reasonableness", which, by virtue of its agapistic orientation rules over all other forms of power [2, 5.520]. We could say that power, that is, the ideal of reasonableness, is the capacity to respond to the attraction exerted upon self by the other; power and reasonableness are related to the capacity for response to the other and the modality of such response is dialogue.

It is significant that Peirce turned his attention specifically to the normative sciences in the final phase of his research. He linked logic to ethics and to esthetics: while logic is the normative science concerned with self-controlled thought, ethics is the normative science that focuses on self-controlled conduct, and esthetics the normative science devoted to ascertaining the end most worthy of our espousal. In this context, Peirce took up the question of the ultimate good, *summum bonum*, or ultimate value which he refused to identify with either individual pleasure (hedonism) or with a societal good such as the greatest happiness for the greatest number of human beings (English utilitarianism). Instead, he insisted that the *summum bonum* could only be defined in relation to the "evolutionary process", that is, to a semiosic process of growth. Specifically, he identified the highest good in the continuous "growth of concrete reasonableness".

Our responsibilities towards life in the global communication-production phase of development in today's society are enormous, unbounded. Indeed, when we speak of life the implication is not only human life, but all life throughout the whole planetary ecosystem, from which human life cannot be separated. As the study of signs, semiotics cannot evade this issue.

Originally, semiotics was understood as "semeiotics" (a branch of the medical sciences) and was focused on symptoms. Nowadays the ancient vocation of semiotics as it was originally practiced for the "care of life" must be recovered and reorganized in what we propose to call "semioethic" terms. This issue is particularly urgent in the present age in the face of growing interference in communication between the historical-social sphere and the biological sphere, between the cultural sphere and the natural sphere, between the semiosphere and the biosphere.

Proceeding beyond Sebeok's perspective, we are not simply alluding to the capacity for being otherwise with respect to being, but to the capacity that is specific to humans for being *otherwise than being*, that is, *otherwise than today's being as shaped in communication*, or so-called globalization. The capacity for otherwise than being in fact subtends all possibilities of being otherwise. This capacity is characteristic of the semiotic animal and consists of the capacity to transcend being and today's global communication world. This capacity renders the semiotic animal completely responsible not only for social reproduction, but also for life over the whole planet, the two things of course being inseparable. The capacity for *otherwise than being* denies the semiotic animal all possible alibis offered by interpretation, response, action, choices and standpoints whose sense and scope are limited to social communication in the world as it is, to alternatives made available by the world as it is. Instead, the semiotic animal is endowed with, indeed characterized by a capacity for *otherness*, for *otherwise than being*.

Semioethics is the result of two thrusts: one is *biosemiotics*, the other *bioethics*. Insofar as it is connected with medical semeiotics, semioethics is *listening*, it is oriented to listening, not only in the sense of the general theory of signs subtending semiotics, but in a medical sense; semiotics must be listening in the sense of medical semeiotics or symptomatology. In other words, semiotics must listen to the symptoms of today's globalized world and identify the different expressions of unease and disease – in social relations, in international relations, in the life of single individuals, in the environment, in life generally over the entire planet.

On the basis of abductive inferential processes the aim is to make a diagnosis, a prognosis and to indicate possible therapies for the future of globalization and the health of semiosis, by contrast with a globalized world tending towards its own destruction.

The semiotician today must be ready to interpret the symptoms of semiosis and its malfunctioning as produced by globalization in today's global communication-production society.

What we propose to call "semioethics" must take the current phase in historico-social development as its starting point and proceed to analyzing today's society rigorously and critically, today's communication-production social structures, the communication-production relationships forming the presentday world.

If semiotics is to meet its commitment to the "health of semiosis" and to cultivate its capacity for understanding the entire semiosic universe, it must continuously refine its auditory and critical functions, that is, its capacity for listening and critique. To accomplish such tasks we believe that the trichotomy that distinguishes between (1) cognitive semiotics, (2) global semiotics, and (3) semioethics, is no less than decisive not only in theoretical terms but also for reasons of a therapeutic order.

4 Symptomatology of Global Communication Today from a Semioethic Perspective

If we consider the contribution made by global semiotics to semioethics in relation to present day global communication, sign practitioners are faced with an enormous responsibility, that of evidencing the limits of today's communication-production society. Semiotics must now accept the responsibility of denouncing incongruencies in the global system with the same energy, instruments and social possibilities produced by the global communication-production system itself – it is time to denounce the dangers involved in this system for life over the entire planet.

The current phase in the development of today's production system is "global communication". This expression may be understood in at least two senses: that communication in the present day and age is characterized by its *planetary extension*, and that communication is *realistic in the sense that it accommodates the world as it is*. Globalization implies the *omnipresence* of communication in production and characterizes the entire productive cycle: not only is globalization present at the level of the market, of exchange, as in earlier phases in socio-economic development, but also at the level of production and consumption. Globalization is tantamount to heavy interference by communication-production not only in human life, but life in general over the planet.

For an understanding of world-wide global communication-production we need a view that is just as global. The special sciences taken separately are unable to provide this. On the contrary, a global view is offered by the general science of signs or *semiotics* as it is taking shape today on the international scene thanks to the approach fostered by Sebeok and his ongoing work for further development.

A full understanding of the current phase in global communication implies a full understanding of the risks involved by global communication, including the risk that *communication itself may come to an end*. This risk, however, is not simply the risk of "incommunicability", theorized and represented in film and literature, a subjective-individualistic malady ensuing from the transition to communication in its current forms (and which cannot be separated from production). Instead, when we speak of the "risk that communication may come to an end", we are referring above all to identification between communication and life, and therefore to the risk that life may come to an end on the planet, considering the enormous potential for destruction in today's society by contrast with all other earlier phases in the development of today's dominant social system.

Therefore, the expression *global* communication-production does not only refer to the expansion of the means of communication and of the market at a world-wide level, but also to the fact that all human life is englobed into the communication-production system: whether in the form of development, well-being and consumerism or of underdevelopment, poverty and the impossibility to survive; health or sickness; normality or deviation; integration or emargination; employment or unemployment; transfer of people functional to the work-force characterizing emigration, or transfer of people whose request of hospitality is denied, characteristic of migration; whether in the form of the traffick and use of legal or illegal merchandize, from drugs to human organs, to "non-conventional" weapons. Indeed, this process of englobement is not limited to human life alone. All life over the entire planet is now irremediably involved (even compromised and put at risk) in the communication-production system.

Reflection on problems relevant to semioethics today in the context they in fact belong to, the context of globalization, requires an approach that is just as global: an approach that is not limited to considering partial and sectorial aspects of the communication-production system according to internal perspectives functional to the system itself. Therefore on an empirical level this approach is not limited to psychological subjects, subjects reduced to parameters imposed by the social sciences, subjects measurable in terms of statistics. Global communication-production calls for a methodological and theoretical perspective that is as global as the phenomenon under observation. In other words, it calls for a perspective that understands the logic of global communication-production, and therefore can proceed to a *critique* of the system it subtends.

Social reproduction in the global communication-production system is destructive. Reproduction of the *productive cycle* itself is destructive. It destroys: a) machines, which are continuously substituted with new machines – not because of wear, but for reasons connected with competetivity; b) jobs, making way for automation which leads to an increase in unemployment; c) products on the market where new forms of consumerism are elicited, completely ruled by the logic of reproducing the productive cycle; d) already existing products which once purchased would otherwise exhaust the demand, and which in any case are designed to become outdated and obsolete almost immediately, as new and similar products are continuously introduced on the market; e) commodities and markets which are no longer able to resist competition in the global communication-production system.

Communication-production destroys natural environments and life forms. It also destroys different economies and cultural differences eliminated by the processes of homologation operated by market logic: nowadays, not only are habits of behavior and needs but even desires and the imaginary homologated and rendered identical (though the possibility of satisfying needs and desires is never identical). Communication-production also destroys traditions and cultural patrimonies that contrast with or obstacle or are simply useless, non functional to the logic of development, productivity and competition. Communication-production destroys productive forces that escape the limits of production systems that penalize intelligence, inventiveness and creativity by subjecting them to "the reason of the market" (which of course cannot be avoided in the current phase of necessary investment in "human resources"). The destructive character of today's production system is also manifest in the fact that it produces growing areas of underdevelopment as *the very condition for development*, areas of human exploitation and misery to the point of non-survival. This logic is behind the expanding phenomenon of *migration* which so-called "developed" countries are no longer able to contain owing to objective internal space limits – far more serious than in earlier phases in the development of social systems.

The principle of exploiting other people's work is destructive, work costs less the more it produces profit: with the help of global communication developed countries turn to low cost work in underdeveloped countries ("stay where you are and we will bring work to you"). The disgrace of the communication-production world is manifest in the spreading exploitation of child labor that is heavy and even dangerous (much needs to be said and done about children as today's victims of underdevelopment – in misery, sickness, war, on the streets, in the work-force, on the market).

The destructive character of global communication-production is evidenced by war which is always a scandal. Global communication-production is the communication-production of war. War requires ever new markets for the communication-production of conventional and unconventional weapons. War also demands to be acknowledged as just and necessary, a necessary means of defence against the menacing "other", of achieving respect for the rights of one's "own identity", "one's own difference". The truth is that identities and differences are not at all threatened or destroyed by the "other", but rather by today's social system which while encouraging and promoting identity and difference renders them *fictitious* and *phantasmal*. And this is what pushes us to cling to such values so passionately, being a logic which fits the communication-production of war to perfection.

With the spread of "bio-power" (Foucault) and the controlled insertion of bodies into the production apparatus, world communication goes hand in hand with the spread of the concept of the individual as a separate and self-sufficient entity. The body is perceived as an isolated biological entity, as belonging to the individual, as a part of the individual's sphere of belonging. This has led to the quasi total extinction of cultural practices and worldviews based on intercorporeity, interdependency, exposition and opening of the body (what remains is the expression of a generalized tendency to museumification; mummified remains studied by folklore analysts, archeological remains preserved in ethnological museums and in the histories of national literatures).

Technologies of separation are applied to human bodies, interests, to individual and collective subjects and are functional to production and to making production and consumption converge, which is a characteristic of presentday production systems. Instead, thanks to its ontological perspective global semiotics (or semiotics of life) can if nothing else oppose a whole series of signs showing how each instant of individual life is interrelated, even compromised with all other life forms over the planet.

To acknowledge such interrelatedness, such compromise involves a form of responsibility which transcends positive rights and limited responsibilities, responsibilities with alibis. And to acknowledge interrelatedness is ever more urgent the more the reasons of production and global communication functional to it impose ecological conditions that obstacle and distort communication between our bodies and the environment.

References

1. Peirce, C.S.: *Chance, Love and Logic*. Harcourt, New York (1923) edited by Morris R. Cohen.
2. Peirce, C.S.: *Collected Papers of Charles Sanders Peirce*. The Belknap Press Harvard University Press, Cambridge (Mass.) (1931-1966) edited by eds.

C. Hartshorne, P. Weiss, and A. W. Burks, 8 Vols., (References are to CP, followed by volume and paragraph numbers).
3. Sebeok, T.A., ed.: *How Animals Communicate.* Indiana University Press, Bloomington, IN (1979)
4. Sebeok, T.A.: *A Sign Is Just a Sign.* Indiana University Press, Bloomington (1991)
5. Sebeok, T.A.: *The Play of Musement.* Indiana University Press, Bloomington, IN (1981)
6. Gould, S.J., Vrba, E.: Exaptation. *Palebiology* **8(1)** (1982) 14–15
7. Sebeok, T.A.: *American Signatures. Semiotic Inquiry and Method.* Oklahoma Press, Norman-London (1991) edited by I. Smith.
8. Sebeok, T.: *Signs. An Introduction to Semiotics.* Toronto University Press, Toronto (2001)
9. Sebeok, T.A., Danesi, M.: *The Forms of Meaning. Modeling Systems Theory and Semiotic Analysis.* Mouton de Gruyer, Berlin (2000)

Abduction and Modeling in Biosemiotics and Sociosemiotics

Augusto Ponzio

Dipartimento di Pratiche Linguistiche e Analisi di Testi, University of Bari, University of Bari, Bari, Italy
augustoponzio@libero.it

Summary. Semiosis may be interpreted as the capacity with which all life-forms are endowed to produce and comprehend the species-specific models of their worlds. Primary modeling is the innate capacity for simulative modeling in species-specific ways. The primary modeling system of the species *Homo* is language. Secondary and tertiary modeling systems presuppose language and consequently they are uniquely human capacities. The secondary modeling system is verbal language or *speech*. Tertiary modeling systems are all human cultural systems. There is a connection between language and abduction In abduction the relation between premises and conclusion is *iconic* and is dialogic in a substantial sense, in other words, it is characterized by high degrees of dialogism and inventiveness as well as by a high risk margin for error.

1 Modeling

Our definition of semiosis or sign process is centered around the notion of *interpretant*. The interpretant mediates between *objectum* (from *obicio*), that is, *solicitation* (interpretandum) and *response* (sign behavior). In Peirce's view, such mediation distinguishes a semiosis from a mere dynamical action – "or action of brute force" – which takes place between the terms forming a pair. Instead, semiosis results from a triadic relation which involves a cooperation of three subjects, a sign, its object, and its interpretant, and is not "in any way resolvable into action between pairs" [1, 5.484] (unless otherwise stated) the numbers in brackets in this chapter refer to *Collected Papers*, by Charles S. Peirce.

The interpretant does not occur in physical phenomena or in nonbiological interactions, but only in the organic world.

The terms "model" and "modeling" are used in the present text as understood by Thomas A. Sebeok and his global semiotics. "Modeling" is a pivotal notion used in Sebeok's global semiotics to explain life and behavior among living entities conceived in terms of semiosis. Therefore, global semiotics also involves modeling systems theory.

Modeling is the foundation of communication in all living beings. Communication necessarily occurs within the limits of a world and its characteristics as modeled by a given species, a world that is species-specific. Jakob von Uexküll speaks of invisible worlds to indicate the domain which englobes all animals according to the species they belong to. What an animal perceives, craves, fears and predates is relative to its own world. Human communication is the most complex and varied form of communication in the sphere of biosemiosis – the human animal is capable of modeling a potentially infinite number of possible worlds. Sebeok develops the concept of modeling from the so-called Moscow-Tartu school, though he enriches it by relating it to the concept of *Umwelt* as formulated by Jakob von Uexküll.

Semiosis may be interpreted as the capacity with which all life-forms are endowed to produce and comprehend the species-specific models of their worlds. Primary modeling is the innate capacity for simulative modeling in species-specific ways.

The so-called Moscow-Tartu school limits the concept of modeling to the human sphere (Lotman's "semiosphere") and distinguishes between the "primary modeling system", an expression used to denote natural language, and the "secondary modeling system", used for all other human cultural systems. Instead, Sebeok extends the concept of model beyond the domain of anthroposemiosis, and connects it to the concept of *Umwelt* as elaborated by the biologist Jakob von Uexküll, which in Sebeok's interpretation may be translated as "outside world model".

On the basis of research in biosemiotics, we now know that the modeling capacity is operative in all life forms. All life forms are endowed with a capacity for semiosis, therefore the capacity to produce and comprehend the species-specific models of their worlds. Primary modeling is the innate capacity of organisms for simulative modeling in species-specific ways.

The primary modeling system of the species *Homo* is what we may call, with Sebeok, *language,* which should not be confused with verbal language, as in the Moscow-Tartu school and in Noam Chomsky.

The Moscow-Tartu school, specifically Lotman, considered verbal language, and consequently any particular natural language, as a primary modeling system.

But a distinction must be made between *language* understood as "verbal language", that is, *as a communication system and a secondary modeling system,* and *language* understood *as a species-specific primary modeling device.* All animal species have a specific modeling device. But the species-specific trait of the human being is a modeling device capable of inventing *many* worlds – as Leibniz say infinite possible worlds – differently from other animals. Instead any other animal species has only one world. Human beings were endowed with this particular modeling device when they first appeared in evolutionary development; and this specific device conditions evolution determining the capacity for articulate speech in *Homo sapiens* and *Sapiens sapiens.*

With Peirce, Sebeok calls this species-specific human modeling capacity (a capacity for the "play of musement") "language" and distinguishes this from "speech": originally language was mute. As a modeling device capable of modeling different worlds, it allowed for communication. Therefore, language thus understood is at the basis of the different historical natural languages. This is the *"lingua mutola"* (mute language) discussed by Vico: originally all nations were mute and expressed themselves through acts or body language [2, p. 434].

The capacity for abduction is strictly connected with this species-specific human modeling.

Secondary and tertiary modeling systems presuppose language understood as a modeling device, therefore, these too indicate uniquely human capacities. In Sebeok's terminology, the secondary modeling system is verbal language or, *speech*; while tertiary modeling systems indicate all human cultural systems, symbol-based modeling processes grounded in language and speech.

Sebeok's tripartite distinction is fundamental in order to distinguish between modeling and communication, as well as to demonstrate the foundational character of modeling with respect to communication.

Language as a modeling device is related iconically to the universe it models. This statement is on the same line of thinking as Peirce and Jakobson who both stressed the importance of iconic signs. An equally important connection can be made with Ludwig Wittgenstein's *Tractatus*, particularly with the notion of "picturing".

The mind as a sign system or model represents what is commonly called the surrounding world or *Umwelt*. The model is an icon, a kind of diagram, where the most pertinent relations are of a spatial and temporal order. These relations are not fixed once and for all but can be fixed, modified and fixed again in correspondence (a resemblance relation) with the *Innenwelt* (inner world) of the human organism. On the basis of this model which may be compared to a diagram or a map, the human mind shifts from one node to another in the sign network, choosing each time the interpretive route considered most suitable.

2 Dialogism

In light of Sebeok's biosemiotics, the concept of dialogism may be extended beyond the sphere of anthroposemiosis and applied to all communication processes. But all communication processes are based not only on *modeling* (Sebeok), but also on *dialogism* (Bakhtin). We believe that modeling and dialogism are at the basis of all communication processes.

Already in *semiosis of information or signification* (Thure von Uexküll) interpretation is dialogical, here an inanimate environment acts as a "quasi-emitter" – or, in our terminology, the *interpreted* becomes a *sign* only because it receives an interpretation (an *interpretant*) by an interpreter which is necessarily a living being. Dialogue also subsists in *semiosis of symptomatization* (Thure von Uexküll), where the interpreted too is an interpretant response (*symptom*). Differently from *semiosis of information or signification*, in *semiosis of symptomatization* the interpreted does not arise for the sake of being interpreted as a sign. Obviously dialogue subsists in *semiosis of communication* (Thure von Uexküll) where the interpreted itself, before being interpreted as a sign by the interpretant is already an interpretant response calling for interpretation as a sign.

Jakob von Uexküll's "functional cycle" is a model for semiosic processes. As such, it has a dialogic structure and involves inferences of the "if... then" type which may even occur on a primitive level, as in Pavlovian semiosis or as prefigurations of the type of semiosis (where we have a "quasi-mind" interpreter) taking place during cognitive inference.

In the "functional cycle", the interpretandum produced by the "objective connecting structure" becomes an interpretatum and (represented in the organism by a signaling disposition) is translated by the interpretant into a behavioural disposition which triggers a behaviour into the "connecting structure". Uexküll does not use a dialogic model. However, the point is that in the "functional cycle" thus described, a dialogic relation is established between interpreted (interpretandum) and interpretant (interpreted by another interpretant, and so forth). The interpretant does not limit itself to identifying the interpreted, but rather establishes an interactive relationship with it.

Vice versa, not only does the "functional cycle" have a dialogic structure, but dialogue in communication understood in a strict sense, may also be analyzed in the light of the "functional cycle". In other words, the dialogic communicative relationship between a sender who intends to communicate something about an object and a receiver may be considered, in turn, on the basis of the "functional cycle" model. The type of dialogue discussed here corresponds to the processes described by the "functional cycle" as presented, in Thure von Uexküll's terminology, neither in *semiosis of information or signification*, nor in *semiosis of symptomatization*, but rather in *semiosis of communication*. In this case, the interpreted itself is already an interpretant response before it is interpreted as a sign by the interpretant, and this interpreted is addressed to somebody to be identified and to receive the required *interpretant of answering comprehension*.

The *Handbook of Semiotics* by Winfried Nöth [3] lacks the entry "Dialogue". However, this term is listed in the "Index of subjects and terms", which informs us that this subject is treated in a chapter titled "Communication and semiosis" (Part 3). Here the "functional cycle" is also mentioned [3, 176-180]. This shows the implications of Uexküll's biosemiosic "functional cycle" concerning the relation between dialogue and communication. Different communication models are discussed in order to show how biological models (such as the models proposed by Maturana, Varela, and Thure von Uexküll) (according to which communication is self-referential, autopoietic and a semiotically closed system) have a dialogic structure and are radically opposed both to the linear (Shannon and Weaver) and the circular (Saussure) paradigms. As reported by Nöth [3, p. 180], Thure von Uexküll [4, p. 14] demonstrates that Jakob von Uexküll's biosemiosic functional cycle [4, p. 8] is characterized by autonomous closure and therefore reacts to its environment only according to its internal needs.

The theory of autopoietic systems is incompatible with dialogism only if dialogism is understood reductively in terms of a communication model describing communication as a linear causal process. This is a process moving from source to destination. Similarly, there is incompatibility between autopoietic systems and dialogism, when dialogue is conceived as based on the conversation model governed by the turning around together rule. Moreover, the autopoietic system calls for a new notion of creativity. Another question remains, the question of how the principle of autonomous closure is compatible with dialogue conceived as the inner structure of the individual, therefore with creativity and learning.

As Maturana [5, pp. 54–55] suggests, another form of dialogic exchange may be conceived. This is different from communication conceived as a linear process from source to destination, or as a circular process in which participants take turns in playing the part of sender and receiver. From this point of view, the dialogic model conceived by Maturana may be described in terms of "pre- or anticommunicative interaction".

In light of the Bakhtinian notion of dialogism, dialogue is neither the communication of messages, nor an initiative taken by self. On the contrary, self is always in dialogue with the other, that is to say, with the world and with others, whether it knows it or not; self is always in dialogue with the word of the other. Identity is dialogic. Dialogism is at the very heart of the self. The self, "the semiotic self" [6], is dialogic in the sense of a species-specifically modeled involvement with the world and with others. The self is implied dialogically in otherness, just as the "grotesque body" [7] is implied in the body of other living beings. In fact, in a Bakhtinian perspective *dialogue* and *intercorporeity* are closely interconnected: there cannot be dialogue among disembodied minds, nor can dialogism be understood separately from the biosemiotic conception of sign.

As we have already observed [8], we believe that Bakhtin's main interpreters such as Holquist, Todorov, Krysinsky, and Wellek have all fundamentally misunderstood Bakhtin and his concept of dialogue. This is confirmed by their interpretation of Bakhtinian dialogue as being similar to dialogue in the terms theorized by such authors as Plato, Buber, Mukarovsky.

According to Bakhtin dialogue is the embodied, intercorporeal, expression of the involvement of one's body (which is only illusorily an individual, separate, and autonomous body) with the body of the other. The image that most adequately expresses this idea is that of the "grotesque body" [7] in popular culture, in vulgar language of the public place, and above all in the masks of carnival. This is the body *in its vital and indissoluble interconnectedness with the world and the body of others*. With the shift in focus from identity (whether individual, as in the case of consciousness or self, or collective, as in the case of, a community, historical language, or a cultural system at large) to alterity, a sort of *Copernican revolution* has been accomplished. Bakhtinian critique conducted in terms of dialogic reason not only interrogates the general orientation of Western philosophy, but also the dominant cultural tendencies that engender it.

The "Copernican revolution" operated by Bakhtin in relation to the conception of self, identity, and consciousness involves all living beings and not only mankind. Consciousness implies a dialogic relation that includes a witness and a judge. This dialogic relation is not only present in the human world but also in the biological. Says Bakhtin:

> When consciousness appeared in the world (in existence) and, perhaps, when biological life appeared (perhaps not only animals, but trees and grass also witness and judge), the world (existence) changed radically. A stone is still stony and the sun still sunny, but the event of existence as a whole (unfinalized) becomes completely different because a new and major character in this event appears for the first time on the scene of earthly existence – the witness and the judge. And the sun, while remaining physically the same, has changed because it has begun to be cognized by the witness and the judge. It has stopped simply being and has started being in itself and for itself... as Well as for the other, because it has been reflected in the consciousness of the other."
> ("From notes made in 1970–71, [9, p. 137])

At this point, a possible connection may be pointed out between Sebeok's biosemiotic conception and Bakhtin's dialogic conception. These two authors seem very distant from each other. In reality, this is not true. Bakhtin himself was seriously interested in biology. And, in fact, he developed his own conception of dialogue in close relation to the biological studies of his time, and particularly to the totalizing perspective delineated by Vernadsky and his conception of biosphere. For both Sebeok and Bakhtin, all living beings on the planet Earth are closely interrelated and interdependent, whether directly or indirectly, in spite of their apparent autonomy and separation.

Bakhtinian dialogue is not the result of an attitude that the subject decides to take towards the other. On the contrary, dialogue is the expression of the living being's condition of the biosemiosic impossibility of closure and indifference towards its environment, with which it constitutes a whole system named *architectonics* by Bakhtin. In human beings, architectonics becomes an "architectonics of answerability", semiotic consciousness of "being-in-the-world-without-alibis". Architectonics thus described may be limited to a small sphere – that is to say, the restricted life environment of a single individual, one's family, professional work, ethnic, religious group, culture, contemporaneity. Or, on the contrary, as consciousness of a "global semiotic" order (Sebeok), which may be extended to the whole world in a planetary or solar or even cosmic dimension (as auspicated by Victoria Welby). Bakhtin distinguishes between "small experience" and "great experience". The former is narrow-minded experience. Instead

> [...] in the great experience, the world does not coincide with itself (it is not what it is), it is not closed and finalized. In it there is memory which flows and fades away into the human depths of matter and of boundless life, experience of worlds and atoms. And for such memory the history of the single individual begins long before its cognitive acts (its cognizable "Self"). (Notes of 1950 [10, p. 99])

It must not be forgotten that in 1926 Bakhtin authored an article entitled "Contemporary Vitalism", in which he discusses problems of the biological and philosophical orders. This article was signed by the biologist Ivan Ivanovich Kanaev, and is an important tessera for the reconstruction of Bakhtin's thought since his early studies. Similarly to the biologist Jakob von Uexküll, Bakhtin too begins with an interest in biology, specifically in relation to the study of signs.

This article by Bakhtin on vitalism was written during a period of frenzied activity, the years 1924-29, in Petersburg, then Leningrad. In this productive period of his life, Bakhtin actually published four books on different subjects (Freud, Russian Formalism, philosophy of language, Dostoevsky's novel). He only signed the last with his own name while the others (together with several articles) were signed by Voloshinov or Medvedev.

In Petersburg Bakhtin lived in Kanaev's apartment for several years. Kanaev contributed to Bakhtin's interest in biology as well as to the influence exerted by the physiologist Ukhtomsky on his conception of the "chronotope" in the novel. Jakob von Uexküll is also quoted in Bakhtin's text on vitalism.

In "Contemporary Vitalism", Bakhtin criticizes vitalism, that is to say, the conception that theorizes a special extramaterial force in living beings as the basis of life processes. In particular, his critique is directed against the biologist Hans Driesch who interpreted homeostasis in the organism in terms of total autonomy from its surrounding environment. On the contrary, in his own description of the interaction between organism and environment, Bakhtin opposes the dualism of life force and physical-chemical processes and maintains that the organism forms a monistic unit with the surrounding world. The relation between body and world is a dialogic relation where the body responds to its environment modeling is own world.

The category of the "carnivalesque" – as formulated by Bakhtin and the role he assigns to it in his study on Rabelais – can be adequately understood only in the light of his global (his "great experience") and biosemiotic view of the complex and intricate life of signs.

The title of Bakhtin's book on Rabelais, literally *The Work of François Rabelais and Popular Culture of the Middle Ages and Renaissance*, stresses the intricate connection between Rabelais's work, on the one hand, and the view of the world as elaborated by popular culture (its ideology, its *Weltanschauung*) in its evolution from Ancient Greek and Roman civilization into the Middle

Ages and Renaissance, on the other, which in Western Europe is followed by the significant transition into bourgeois society and its ideology.

Bourgeois ideology conceives bodies as separate and reciprocally indifferent entities. Thus understood, bodies only have two things in common: firstly, they are all evaluated according to the same criterion, that is to say, their capacity for work; secondly, they are all interested in the circulation of goods, including work, to the end of satisfying the needs of the individual. Such ideology continued into Stalinist Russia, which coincides with the time of Bakhtin's writing, and into the whole period of real socialism where work and the productive capacity were the sole factors taken into serious considered as community factors. In other words, work and productivity were the only elements considered as what links individuals to each other. Therefore, beyond this minimal common denominator, individual bodies were considered as being reciprocally indifferent to each other and separate.

The carnivalesque participates in the "great experience" which offers a global view of the complex and intricate life of bodies and signs. The Bakhtinian conception emphasizes the inevitability of vital bodily contact, showing how the life of each one of us is implicated in the life of every other. Therefore, in what may be described as a "religious" (from Latin *religo*) perspective of the existent, this conception underlines the bond interconnecting all living beings with each other.

Furthermore, the condition of excess is emphasized, of bodily excess with respect to a specific function, and of sign excess with respect to a specific meaning: signs and bodies – bodies as signs of life – are ends in themselves. On the contrary, the minor and more recent ideological tradition is vitiated by reductive binarism, which sets the individual against the social, the biological against the cultural, the spirit against the body, physical-chemical forces against life forces, the comic against the serious, death against life, high against low, the official against the non-official, public against private, work against art, work against non official festivity. Through Rabelais, Bakhtin recovered the major tradition and criticized the minor and more recent conception of the individual body and life inherent in capitalism as well as in real socialism and its metamorphoses. Dostoevsky's polyphonic novel was in line with the major tradition in *Weltanschauung*, as demonstrated by Bakhtin in the second edition (1963) of his book of 1929 [11].

The self cannot exist without memory; and structural to both the individual memory and social memory is otherness. In fact, the kind of memory we are alluding to is the memory of the immediate biosemiotic "great experience" (in space and time) of indissoluble relations with others lived by the human body. These relations are represented in ancient forms of culture as well as in carnivalized arts: however, the sense of the "great experience" is anaesthetized in the "small", narrow-minded, reductive experience of our time.

Let us resume. Modeling and dialogism are pivotal concepts in the study of semiosis. Communication is only one kind of semiosis that (together with the semiosis of information or signification and the semiosis of symptomatization) presupposes the semiosis of modeling and dialogism. This emerges clearly if in accordance with Peirce and his reformulation of the classic notion of *substitution* in terms of *interpretation*, we consider the sign firstly as an interpretant, that is to say, as a dialogic response foreseen by a specific type of modeling. Moreover, Bakthin's concept of dialogue also contributes considerably to extending this concept beyond the human world connecting dialogism with semiosis from Sebeok's biosemiotic perspective, namely according to the point of view of *global* semiotics.

3 Abduction

Generally one maintains that dialogue commences with signaling behavior from a sender intending to communicate something about an object. But what is not taken into account is that the "if... then" inference, hypothesis formation, and a "chain of thought" are dialogic forms in themselves.

In inference, in the hypothetical argument, and in the chain of interpreted and interpretant thought signs generally, dialogue is implied in the relation itself between the interpreted and the interpretant. In inference, in hypothetical argument, and in the chain of thought generally which consists of the relation between interpreted and interpretant signs, dialogue is implied in the relation itself between interpreteds and interpretants. We shall now analyze dialogism in the "if... then" inference, in hypothesis formation, and in any "chain of thought".

The connection between semiosis and interpretation implies the connection between sign and argument, and therefore the connection between semiotics and logic.

Taking Peirce's viewpoint into consideration, we may say that the problem of the connection between identity and alterity in the sign is not just a problem of semiotics, but also concerns logic as theory of argumentation. In Peirce this problem directly concerns logic which as theory of argumentation also involves the problem of dialogue. Considered from the point of view of its relation to the object, the sign is a symbol insofar as it involves mediation of an interpretant; from the perspective of its relation to the interpretant, the sign-symbol is an *Argument*. This is true if the sign-symbol distinctly represents the interpretant which it determines as its Conclusion through a proposition that forms its Premise or, more generally, its premises [1, 2.95]. Depending on the type of sign relation that comes to be established in the argument between premise and conclusion, three kinds of arguments are possible: Deduction, Induction and Abduction.

In *deduction* where the relation between the premises and the conclusion is *indexical*, the degree of dialogism is minimal: here, once the premises are accepted the conclusion is obliged.

In *induction* which is also characterized by an unilinear inferential process, the conclusion is determined by habit and is of the *symbolic* order: identity and repetition dominate though the relation between premises and conclusion is no longer obliged.

In *abduction* the relation between premises and conclusion is *iconic* and dialogic in a substantial sense. In other words, it is characterized by high degrees of dialogism and inventiveness as well as by a high margin of risk for error. Claiming that abductive argumentative procedures are risky means that they are mostly tentative and hypothetical with just a minimal margin for convention (symbolicity) and mechanical necessity (indexicality). Therefore, abductive inferential processes engender sign processes at the highest levels of otherness and dialogism. These are the theses we intend to demonstrate in this paper.

The special relation that exists between sign (interpreted) and interpretant as understood by Peirce is a *dialogic* relation. Peirce evidenced *the dialogic nature of the sign and semiosis*.

In *semiosis of information or signification* (cf. Th. von Uexküll, "Biosemiosis", 1997, and "Varieties of Semiosis" 1991) where an inanimate environment acts as "quasi-emitter" – or, in our terminology, where the *interpreted* becomes a *sign* only because it receives an interpretation by the interpretant which is a response – receiver interpretation is dialogic. Not only does dialogue subsist in *semiosis of communication* (Th. von Uexküll) where the interpreted is already itself an interpretant response oriented to being interpreted as a sign before it is effectively interpreted as a sign by the interpretant, but dialogue also subsists in *semiosis of symptomatization* (Th. von Uexküll) where the interpreted is an interpretant response (*symptom*) which is not oriented to being interpreted as a sign, as well as in *semiosis of information or signification*.

Dialogue does not commence with signaling behavior from a sender intending to communicate something about an object. The whole semiosic process is dialogic. "Dialogic" may be understood as *dia-logic*. The *logic* of semiosis as a whole and consequently of Krampen's semiosic matrix (see his article entitled "Model of semiosis" in Posner, Robering, and [12, p. 248]) is *dia-logic*. The interpretant as such is "a disposition to respond", an expression used by Krampen [12, p. 259] to describe the dialogic interaction between a sender and receiver.

Krampen's semiosic matrix confirms the connection established between dialogue and semiosis insofar as it shows that the two terms coincide, not only in the sense that dialogue is semiosis but also in the sense that semiosis is dialogue – the latter being an aspect which would seem to escape Krampen. The dialogue process presented in the semiosic matrix is similar to the "if. . . then" semiosic process, to hypothesis formation, chain of thought, and *functional cycle* after Jakob von Uexküll. In Krampen's article the semiosic matrix illustrates dialogue with two squares which represent two partners, the sender and receiver, where each has its own rhombus representing the interpretant. Despite this division, the graphic representation of dialogue is not different from the author's diagrams representing other types of semiosis. It could be the model, for example, of an "if. . . then" semiosis in which the two distinct interpretants are the premises and the conclusion of an argument in a single chain of thought.

In deduction the relation to the interpretant is of the *indexical type*; in induction it is *symbolic*; in abduction, *iconic*.

In inference the relation between premises and conclusion may be considered in terms of *dialogue* between *interpreted sign* (premises) and *interpretant sign* (conclusion).

The two speakers among whom an argument is hypothetically divided are connected, in deduction, by a relation of reciprocal dependence and constraint. In the deductive argument the premiss determines the conclusion, that is, the precedent determines the consequent with the same force of compulsion with which the past imposes itself upon the present. The conclusion must passively acknowledge the premiss which has already been formulated like a *fait accompli* [1, 2.96].

In induction, on the other hand, the conclusion is not imposed by the premiss and is susceptible to modification. Despite this, however, the inductive argument is merely repetitive and quantitative, given that its sphere of validity remains that of the fact, that is, of the totality of facts on whose basis alone can it infer the future. As in deduction, the inductive process is unilinear and moves in a precise order of succession from the point of departure to the point of arrival without interruption, reversal or retroaction as opposed to abduction which, as we will see, moves backwards from the consequent to the antecedent.

In abduction, the thought-sign (the minor premise) and the thought-interpretant are connected by a dialogic relation which is not pre-determined by the pre-dialogic selection of a law. Retroaction of the interpretant on the premiss to the point that interpretation determines the major premise is precisely what causes us to define this type of reasoning as *retroduction* or *abduction*.

The dialogic division between the parts enables us to take into account the level of dialogic complexity, that is, of alterity, differentiation, distance and novelty established in the argument between the sign and the interpretant. The dialogic character of logic is discussed in a medieval tractatus on logic entitled *Summule logicales* by Peter of Spain (an author known to Peirce). It is not incidental that Peirce should have used the term *Speculative Rhetoric* to designate *transuasional logic* [1, 2.93], the doctrine of the general conditions whereby symbols and other signs refer to and determine the interpretants. In fact, the term Rhetoric implies reference to the addressee, the interlocutor, and recalls such expressions as to converse, to argument, to convince and to account for. Furthermore, it represents a break in the conception of reason and reasoning originated from Descartes, and therefore it alludes to the uncertain, probabalistic, and approximative nature of human knowledge. Peircean logic is presented as dia-logic.

The iconicity of abduction consists in establishing a relation between that which originally and naturally is not related: imaginative representation attempts an approach to that which is given as *other* in order to lead it back to a relation of similarity. Similarity is rightly listed by Peirce together with all that we associate with the category of obsistence; in fact, originality or firstness is surpassed by secondness or obsistence when whatever exists autonomously is related to something else. To have an understanding of alterity in a certain sense means to exceed it. The innovating, creative, displacing capacity of abduction so much in its exhibiting an image which draws that which seems to evade all constraints nearer, as in the capacity to direct does not consist itself towards the autonomously other.

In the abductive process we run the risk of surpassing the datum, thus developing an interpretant that has *its own alterity and autonomy* in so far as it is not motivated, justified or compensated by the object-datum it specifically refers to. Such self-sufficiency of the abductive interpretant, that is, its iconicity and originality presents a challenge, a provocation with regards to the concept of identity and totality. It thus questions even that which seemed settled and definitive, and exhibits an image which can neither be incorporated nor accounted for whether through immediate reference to the fact or datum, or on the basis of a system of pre established laws. With a logic that goes beyond the logic of exchange and equilibrium, it is possible for an argument to actualize firstness, originality, or alterity in the very core of the symbolic, of the law, of the transuasional.

In the succession deduction – induction – abduction the *degree of alterity* increases. Says Peirce:

> Abduction is the process of forming an explanatory hypothesis. It is the only logical operation which introduces any new idea; for induction does nothing but determine a value, and deduction merely evolves the necessary consequences of a pure hypothesis. Deduction proves that something *must* be; induction shows that something *actually is* operative; abduction merely suggests that something *may be*. [1, 5.172]

Abduction is the inferential process by which the rule that explains the fact is hypothesized through a relation of similarity (iconic relation) to that fact. The rule acting as general premise may be taken from a field of discourse that is close to or distant from that to which the fact belongs, or it may be invented *ex novo*. If the conclusion is confirmed, it retroacts on the rule and convalidates it (ab- or retro-duction). Such retroactive procedure makes abductive inference risky, exposing it to the possibility of error; at the same time, if the hypothesis is correct, abduction is innovative, inventive and sometimes even surprising.

4 Critique of Dialogic Reason

If we do not take into account Bakhtins's global (see his notion of "great experience") and biosemiotic view towards the complex and intricate life of signs, we will not understand the role in his work of the relation between "dialogism" and "carnivalesque". The latter is formulated in Bakhtin's study on Rabelais and then also used in the revised edition (1963) of his book on Dostoevsky.

In *Rabelais* Bakhtin tells us what carnival means for him. He refers to that complex phenomenon, existing in all cultures, formed by the system of attitudes, conceptions and verbal and nonverbal signs according to a carnivalesque and joyful idea of living. Carnival, therefore, does not only concern Western culture, nor the Russian spirit, but any culture of the world insofar as it is human.

Rabelais occupies a place of central importance in Bakhtin's overall conception. By contrast with oversimplifying and suffocating interpretations of Marxism, Bakhtin instead develops Marx's idea that the human being is fully realized when "the reign of necessity ends". Consequently, a social system that is effectively alternative to capitalism is one that considers *free time* and not labour time as the measure of *real* social wealth (see Marx, *Grundrisse*, 1857–61, Eng trans.: 708), in Bakhtin's terminology the "time of non official festivity", which is closely connected to what he calls the "great time" of literature.

Today, we are witness to the worldwide spread through global communication of the ideology of production and efficiency. This is in complete contrast with a carnivalesque vision of life. The difference also concerns individualism, which is exasperated by the ideology of production connected with the logic of competitivity. But even when the logic of production, individualism and efficiency is dominant, it cannot eliminate the constitutive inclination of the grotesque body, insofar as it is grounded in dialogism and intercorporeity, for involvement with the world and the body of others. Mankind's inclination for the "carnivalesque" endures, and this is testified, for example, by literary writing. In Orwell's 1984 [13], the greatest resistance against a social system based on the ideology of production and efficiency is in fact represented by literature. In this sense we may say that literature (and art, in general) is, and always will be, carnivalized.

Bakhtin's fundamental contribution to semiotics and "philosophy of language" is his *critique of dialogic reason*, a critique connected to Kant and Marx. Bakhtin inaugurates a "critique of dialogic reason" by contrast with Kant's "critique of pure reason" and Sartre's "critique of dialectic reason".

Bakhtin privileges the term "metalinguistics" for his own approach to the study of sign, utterance, text, discourse genre, and relations between literary writing and nonverbal expressions in popular culture, such as the signs of carnival. Bakhtin's critique of dialogic reason focuses on the concept of *responsibility without alibis*, a nonconventional idea of responsibility, which concerns existential "architectonics" in its relation with the I, the world and others.

Bakhtinian critique of dialogic reason is a critique of the concept of autonomy among individual bodies: in fact, autonomy is an illusion. Consequently, Bakhtin's critique is a critique of individual identity (such as consciousness or self) and of collective identity (such as community, historical language, or cultural system) where identity is conceived in terms of separation from the other following dominant ideological tendencies.

The problem of the critique of dialogic reason leads to the problem of the centrality of dialogue and dialogism in logic, in argumentation and in the biosemiosic universe.

Bakhtin (in his notes of 1970–71) evidences how unilaterality, ossification, rectilinear and unilateral dialectics derives from sclerotized dialogue. Monologic, unilinear and totalizing dialectics is necessarily orientated towards a synthesis and a conclusion. As such it calls for a "critique of dialogic reason". From this point of view Bakhtin is a milestone because all his research, including his latest paper of 1974 on the methodology of human sciences, focuses on the same problem faced by Sartre in *Critique de la Raison Dialectique* [14]: that is, whether human knowledge and understanding not only imply a specific method but also a *New Reason*. However, this problem cannot be adequately understood appealing to Sartre's belief in terms of a new relationship between *thought* and its *object*. In fact, Sartre's dialectics remains wholly inside the limits of monologic dialectics for he reduces the relation of otherness to a relation of identity and of reciprocal objectification: dialectics between *for self* and *for others* is dialectics in a totalizing consciousnesses, where the tendency is to assert one's own objectifying view.

Critique of dialogic reason is critique of the category of Identity which is dominant in Western thought and praxis. From the perspective of identity, sense coincides with partial and limited interests and engenders mystification: and this happens whether we are speaking of the identity of individual, group, nation, language, cultural system, or of a macro-community such as the European Community, the Western world, and so forth.

The category of Identity dominates today's world because of the *concrete abstractions* constructed upon it forming the *Reality* we experience: these concrete abstractions are "internal" to today's overall system of social reproduction. They include Individual, Society, State, Nation, Truth, Knowledge, Work, Trade Equality, Justice, Freedom, limited Responsibility, Need, Equal exchange, etc. However, it is not only a question of concrete abstractions ensuing from the system. Even more radically, the system itself is grounded in the category of Identity which is asserted structurally and constitutively as a Universal in the worldwide and global processes of Production, Exchange on the Market and Consumption. The logic orienting concrete abstractions in today's processes of social reproduction is the logic of Identity. And the categories of Individual and its rights, obligations, responsibilities, of Society and its interests, of State and its Politics (which reflect Reality as closely as possible), of Equal exchange and its demands, all obey the logic of Identity.

The places of argumentation internal to the order of discourse are the places of the logic of identity. Reason includes "the reason of war" even if in the form of *extrema ratio*, which presents war as legitimate, just and legal. Reason includes elimination of the other – from emargination and segregation to extermination. Reason is the Reason of Identity. Its logic is asserted by barricading, isolating, expelling or exterminating the other thereby laying the conditions for the construction of the concrete abstractions mentioned above. As anticipated, these concrete abstractions include the category of Individual which must firstly sacrifice its otherness to self in order to assert self as identity.

The Critique of Reason and Argumentation thus understood requires a *point of view that is other*. This approach calls for preliminary *recognition of the other*, or, better, recognition of the fact that recognition of the other is an *inevitable imposition*. Recognition of the other here is not conceived as a concession, a free choice made by the Individual, the Subject, the Same, but as a necessity imposed by alienation, the loss of sense, by the situation of *homo homini lupus*. The situation of *homo homini lupus is consequent* and not mythically antecedent (the allusion is to Hobbes's fallacy!) to such concrete abstractions as State, Politics, Law. A global semiotic perspective that keeps account of today's socio-economic context in terms of global communication evidences that the human individual, as a living body, is interconnected with all other forms of life over the whole planet thanks to the condition of diachronic and synchronic intercorporeity.

A global and detotalizing approach in semiotics demands availability towards the other, to an extreme degree, a disposition to respond, to listen to others in their otherness, a capacity of opening to the other, where such opening is measured in quantitative terms (the omnicomprehensive character of global semiotics), as well as in qualitative terms. All semiotic interpretations by the semiotician (especially at a metasemiotic level) cannot leave the dialogic relationship with the other out of consideration. Dialogism is, in fact, a fundamental condition for a semiotic approach in semiotics which though oriented globally, privileges the tendency to open to the particular and the local rather than to englobe and enclose. Accordingly this approach privileges the tendency towards detotalization rather than totalization.

As shown by Emmanuel Levinas, otherness obliges the totality to reorganize itself always anew in a process related to what he calls "infinity", and which (to use a phrase associated with Peirce) we could also relate to the concept of "infinite semiosis". This relationship to infinity is far more than cognitive: beyond the established order, beyond the symbolic order, beyond our conventions and habits, it tells of a relationship of involvement and responsibility with the other. This relationship with infinity is a relationship with what is most refractory to the totality, therefore it implies a relationship to the otherness of others, of the other person, not in the sense of another self like ourselves, another *alter ego*, an I belonging to the *same community*, but of an other in its extraneousness, strangeness, diversity, difference towards which we cannot be indifferent despite all the efforts and guarantees offered by the identity of the I.

Such considerations orient semiotics according to a plan that does not belong to any particular ideology. This kind of semiotics concerns human behavior as it ensues from the awareness of human being's radical responsibility as a "semiotic animal". Properly understood, the "semiotic animal" is a responsible actor capable of *signs of signs*, of mediation, reflection, and awareness in relation to semiosis over the whole planet. In this sense global semiotics must be adequately founded in cognitive semiotics, but it must also be open to an other dimension that is the ethical. This is the pivotal object of what Susan Petrilli and myself propose the to call *"semioethics"*. Its commitment is to the "health of semiosis", that is, *of life* on the Planet; and to cultivate an understanding of the entire semiosic universe, semiotics, understood as *semioethics*, must continuously refine its auditory and critical capacity, that is, its capacity for listening and criticism.

Translation from Italian by Susan Petrilli

References

1. Peirce, C.S.: *Collected Papers of Charles Sanders Peirce*. The Belknap Press Harvard University Press, Cambridge (Mass.) (1931-1966) edited by eds. C. Hartshorne, P. Weiss, and A. W. Burks, 8 Vols., (References are to CP, followed by volume and paragraph numbers).
2. Vico, G.: *Principi di Scienza Nuova*. Einaudi, Torino (1977) edited by F. Nicolini, 3 Vols.
3. Nöth, W.: *Handbook of Semiotics*. Indiana University Press, Bloomington (1990)
4. Uexküll, T.v.: The sign theory of Jacob von Uexküll. In Krampen, M., Oehler, K., Posner, R., Sebeok, T., Uexküll, T.V., eds.: *Classic of Semiotics*. Plenum (1981) 147–170
5. Maturana, H.N.: Biology of language: The epistemological reality. In Miller, G.A., Lenneberg, E., eds.: *Psychology and Biology of Language and Thought*. Academic Press (1978) 27–63
6. Sebeok, T., A.Petrilli, S., Ponzio, A.: *Semiotica dell'io*. Meltemi, Rome (2001)
7. Bakhtin, M.M.: *Rabelais and His World*. The MIT Press, Cambridge (1968) translated by H. Iswolsky and edited by K. Pomorska.
8. Ponzio, A.: *I segni e la vita. La semiotica globale di Thomas A. Sebeok*. Spirali, Milan (2003) in collab. with S. Petrilli.
9. Bakhtin, M.M.: *Speech Genres & Other Late Essay*. University of Texas Press, Austin (1986) edited by C. Emerson and M. Holquist.
10. Bakhtin, M.M.: *Sobranie sochinenij [Collected works]*, Vol. 5 (1940-1960). Russkie Slovari, Moskow (1996) edited by S. G. Bocharov and L. A. Gogotishvili.
11. Bakhtin, M.M.: *Problemy Tvorchestva Dostoevskogo (Problems of Dostoevsky's work)*. Priboj, Leningrad (1929)
12. Posner, R., Robering, K., Sebeok, T.: *Semiotics. A Handbook on the Sign-Theoretic Foundations of Nature and Culture,* 4 Vols. Walter de Gruyter, Berlin (1998)
13. Orwell, G.: *Nineteen Eigthy-Four. A Novel*. Penguin, New York (1949)
14. Sartre, J.P.: *Critique de la Raison Dialectique*. Gallimard, Paris (1960)

Reason out Emergence from Cellular Automata Modeling

Leilei Qi[1] and Huaxia Zhang[2]

[1] School of Public Administration, South China Normal University, Guangzhou, P. R. China
   qillei@126.com
[2] Department of Philosophy, Sun Yat-sen University, Guangzhou, P. R. China
   hsszhx@sysu.edu.cn

Summary. In this paper, we first construct a descriptive definition for emergence based on multilevel ontology, and then use Cellular Automata Modeling to simulate some classical emergent processes, such as Conway's game of life and virtual ants building highway, which shows how the emergent phenomena arise, how the emergence of system at higher levels is derived from the simply basic interaction rules of system elements and their initial conditions at lower-levels. Although those inferences are deducible, they are not analytic. They are "bottom-up" synthetic methods based on computer simulation. There are three conditions that must be met when an emergent phenomenon can be reasoned out from its low-level elements and their interaction rules: (1) They must be simulatable, namely, that the elements and their operation rules can be constructed. (2) They must be computable, at least computable in principle. (3) They are necessary configuration function at the high level and auxiliary hypotheses. These indicate the limitation of the method of "derived from simulation" for understanding emergence. Moreover, most of system emergent properties cannot be definitely predicted because of complexity, hierarchy, uncertainty and adaptability in the development of systems. That is to say, in fact, we are using a new reducible method to prove that it is insufficient to understand emergent phenomena only with reducible method.

1 Emergence

In both natural and social world, surprising new matter emerges in an endless way, such as the appearance of life with its own metabolism and self-reproduction, nervous systems with their own intelligent nature, mind with marvelous thoughts and the function of imagination and social culture with its diversity and cohesive force. It is unimaginable and incredible that these phenomena would appear, thinking of its lower levels objects and processes. We regard them as common things only after we have encountered them repeatedly. Emergence does not refer to the appearance of any kind of new matter, but it is an interlevel new phenomenon, referring to new phenomena properties, functions and laws of the whole complex system comprised of its parts. Therefore, we conclude there are four emergent features as follows:

(1) Wholeness. Wholeness is the idea that the whole possesses some kinds of properties which can not been provided with by and even no meaningful of its pre-existing components. Namely, the properties and system behaviors as a whole cannot be explained by behaviors of the parts. For example, atoms are lifeless, molecules are inelasticity and elemental particles are non-color. Wholeness presupposes different levels and it is "a gold medal" trademark of emergence.

(2) Novelty. The simple rules can cause surprising new phenomena, and thus generates numerous and complicate complex systems via continuous iteration and diachronic evolution. This characteristic is remarkable.

(3) Unpredictability. Even if initial conditions and evolved rules are known, future behaviors and structures of the system still cannot be predicted, as complex systems are complicated and non-linear. This "unpredictability" refers to "definite unpredictability", namely, actual states of emergence cannot be predicted definitely.

(4) Downward Causation. The whole at the higher level is comprised of elements at the lower level. While the lower level elements in turn are constrained by the higher level whole, and they act according to the rules which the higher level whole obeys. In other words, the system emergent properties have downward causation to its component elements. We can see the characteristic as a powerful weapon to argue against traditional reductionism. If the whole is reduced to its elements interaction at the lower level, the downward causation is neglected, and that is incomplete to understand the whole.

It is thus clear from the above characteristics that we should make observation from different aspects and levels when we try to understand the emergent properties. The world is of different levels with each level having its own specific entities, phenomena, properties and laws. That is the multilevel ontology. The existence of the different hierarchy as well as points of observation is the original concept, from which we can formally define emergence:

Let $\{S_i\}$ ($i = 1, 2, 3, \ldots, n$) being a set of the elements or "agents", I_{ij}^1 being the interaction rules between elements. Where ij means S_j acts on S_i or the input of S_j to S_i, while 1 denotes at the first level.

Then let's assume F^1 being the representation of the configuration function. $F^1(S_i^1)$ is the characteristics obtained from observation or measurement of the element S_i at the first level.

If the interaction between elements in $\{S_i\}$ can form a new mutually stable structure or system S^2, S^2 is called emergent structure or entity, which is,

$$S^2 = R(S_i^1, F^1, I_{ij}^1)$$

And if there is $P \in F^2(S^2)$, but $P \notin F^1(S_i^1)$, this P is regarded as P_e^2, which means the emergent properties at the second level can be defined as:

$$P_e^2 = \{P \mid P \in F^2(S^2) \wedge P \notin F^1(S_i^1)\}$$

Please note that the emergent structure R and emergent properties P defined in this way is a unified definition from an ontological and epistemological point of view. The configuration function $F(\ )$ is epistemological, but its independent variable, namely object of observation itself and S in $F(S)$ is ontological commitment. It indicates the existence of an observed complex system S.

The emergent concepts can date back to ancient Greek philosopher Aristotle who says: "The whole is more than the sum of its parts", while the concept is recounted by British Emergentists in the 20th century, such as S. Alexander [1], C. L. Morgan [5] and C. D. Broad [3] who think that the emergent properties at higher-level B "cannot be deduced from (lower-level) A-properties and the structure of the B-complex by law of composition which has manifested itself at lower levels", and it characterized "non-deducible properties" [3, pp. 77–78].

Morgan [5] believes that the emergent properties are unpredictable before their appearance. He says: "What is it that you claim to be emergent? The brief reply is: Some new kind of relation... It may still be asked in what distinctive sense the relations are new. The reply is that their specific nature could not be predicted before they appear in the evidence, or prior to their occurrence... In like manner we think that, on the level of physicochemical events, there could be no knowledge on the basis of which vital relatedness could be foreseen before it came. And so, too, at a later stage with mind as an emergent quality which expresses new relatedness of the conscious order" [5, pp. 64–65]. Alexander [1] sums up the ideas of British Emergentism. He says: "The higher quality emerges from the lower level of existence and has its roots therein, but it emerges therefrom, and it does not belong to that level, but constitutes its possessor a new order of existent with its special laws of behavior. The existence of emergent qualities thus described is something to be noted, as some would say, under the compulsion of brute empirical fact, or, as I should prefer to say in less harsh terms, to be accepted with the 'natural piety' of the investigator. It admits no explanation" [1, pp. 46–47].

Our understandings of emergence are similar to the British Emergentists', namely we all think emergence is the phenomenon produced at different levels. We cannot understand it completely at certain level. But the British Emergentists were too absolute and think it cannot be deduced, predicted, or even explained. They think that the appearance of emergence is a fact, but how it appears is a black box. We can only see the input at the lower-level and the output at higher-level, while the internal mechanism is undetectable. However, in the last 10 to 20 years of the 20th century, due to the development of complex sciences and computer science, basing on the high-speed computer, new mathematical methods, especially discrete dynamics and chaos dynamics, it is possible for us to open the black-box and explore the emergent process, mechanisms and structures based on interactions between agents of the low-level, and to derive emergence from simulation.

2 Simulation and Emergence Derived from Simulation

Generally speaking, the model in mathematics is a kind of homomorphic mapping. For example, the map (map or model) system is a homomorphic mapping of real physiognomy (prototype systems), and differential equation is the homomorphic mapping of electromagnetic movement of physical circuit. And that simulation in this paper is an iterated homomorphic mapping. It emphasizes two points: (1) the simulated (prototype) and the simulation system (model) are both dynamical systems. The simulation system describes dynamic process of the simulated system through updating their state. (2) Unlike traditional mathematical models, a simulation system does not have analytical solutions. The equation cannot be worked out with a solution by means of algorithm, such as addition, subtraction multiplication, division and extraction.

Thus, it requires the following three essential factors to simulate a natural or social real complex system:

(1) Elements
$$S_i = S_i(t), \; i = 1, 2, 3, \ldots, n \tag{1}$$

S_i is a model of system component elements. Where S_i can be viewed as ith element or the internal state of object and it changes with the change of time. Therefore, J. H. Holland described it as a dynamic mechanism in his famous book Emergence.

(2) Interaction Rules
$$I_{ij} = I(S_i, S_j) \tag{2}$$

Where I_{ij} is the representation of interaction rules or laws between elements. To be concrete, it is the jth object's action rules on the ith objects and it is the simulations to interactions between elements.

(3) Update Function U. An object update functional U is the state transition of systems and their elements state which transit in term of time sequence t.
$$S_i(t+1) = U(S_i(t), I_{ij}(t)), \; i = 1, 2, 3, \ldots, n \tag{3}$$

According to these three elements, we can build the system simulation $\sum s$, It refers to iterated update collection of all the objects state during calculation to evolution on time and aggregation on space. Namely,

$$\sum s = \sum_{t=1}^{m} \sum_{i=1}^{n} (S_i(t+1)) = \sum_{t} \sum_{i} (U(S_i(t), I_{ij}(t))),$$
$$\text{here } t = 1, 2, 3, \ldots, m; \; i = 1, 2, 3, \ldots, n \tag{4}$$

On basis of system simulation, if there are appropriate initial conditions C in place of "i" and "t", we can infer emergent conditions:

When S^2, namely $R(S_i^1, F^1, I_{ij}^1) \in \sum_s^1$, we say there is emergent structure derived from system modeling $\sum_s^1(S_i^1, F^1, I_{ij}^1)$ at level L_1, i.e. $\sum_s^1 \& C \vdash S^2$, iff $S^2 = R(S_i^1, F^1, I_{ij}^1) \in \sum_s^1$; when $P \in \sum_s^1(S_i^1, F^1, I_{ij}^1)$, we think there is emergent property P^2 derived from $\sum_s^1$, but here is an additional condition F^2, namely $P^2 = F^2(S^2)$, which is to say there must be a configuration function F^2 at the higher-level, and then P_e^2 can be derived.

$$\sum_s^1 \& F^2 \& C \vdash P_e^2 \qquad (5)$$

Note that the computer-based simulation we use must have physically implemental mechanisms, i.e. the machine of iteration procedure which implements $\sum s$, $\sum s$ is normally some kind of physically digital computers.

Next, we take the computer as the platform, using cellular automata as simulation methods of discrete dynamic systems to analyze several typical cases of life emergence.

3 Reason out "Living Organism" Emergence from Game of Life

"Game of Life" was invented by the mathematician John Conway in 1970. He first presents his grid construction (cell spaces) which is divided into many smaller lattices, as chessboard. Each lattice is called a cell and any adjoining cell is considered as its "neighbor", including diagonals. Because there are many ways to design neighbor, the way he used is called Moore neighborhood. In game of life, "life organisms" are "mobile" on the grid. Every cell of "life organisms" is element $S_i = S_i(t)$, which is described in the previous section. Each of these cells could have two states: 0 or 1, (it can be live or dead, active or quiescent) and they change states with time. Each cell updates simultaneously and independently in discrete time. The new state of a cell only depends on its own actual state itself and on its eight closest neighbors' state. Its update rules I_{ij} mentioned above is described as the following three transition rules or life rules:

1. A dead cell (0) with exactly three live neighbors becomes a live cell ("it's born").
2. A live cell with two or three live neighbors remains alive (survival).
3. In any other case, a cell dies or remains dead (overcrowding or loneliness).

If each grid is filled with 0 or 1 at random and allows this configuration to iterate according to the life rules, in other words, acting an update function U on all cells, after a period of time, the cell on grid may be dead, disappeared, quiescent as a stable pattern (stability), or becoming an oscillating pattern (blinker). But what interests Conway most is that it generates another particular pattern, R-pentomino (Figure 1(a)) [9].

The R-pentomino, its figure looks just like the letter "r", consisting of five non-zero states. It is thought to be a minimal initial configuration that

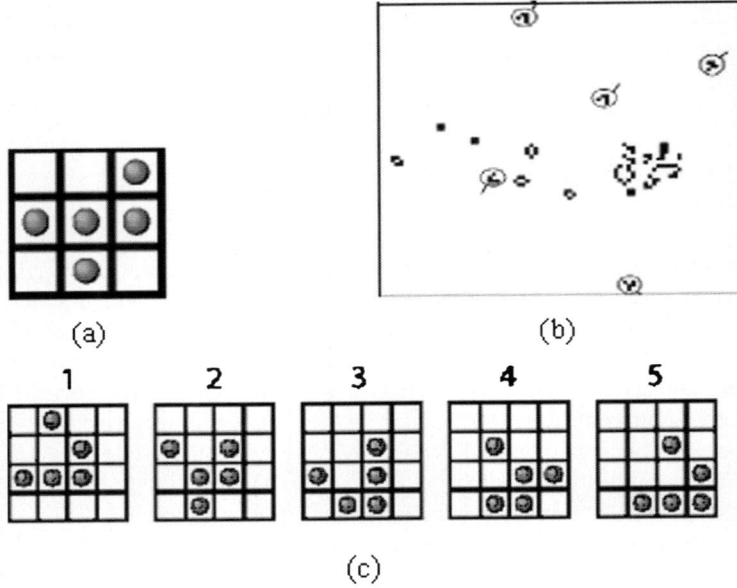

Fig. 1. The glider and its historical precursor. (a) R-pentomino. (b) Configuration developed from the R-pentomino, at the 224th step of evolution; the gliders are encircled, and their velocity vectors are sketched. (c) The glider.

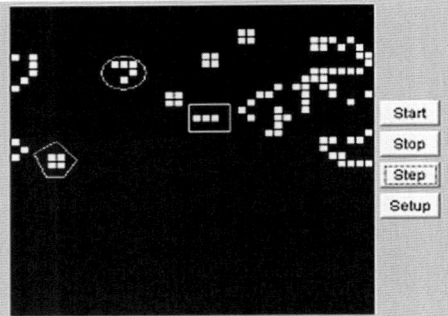

Fig. 2. The life forms.

generates unpredictably developing patterns. In the R-pentomino's evolution, it generates a cluster of small stable or oscillating patterns, and the most important one is called glider.

The behavior of each cell on cellular Spaces looks very simple, but it produces drastic activity and all kinds of unpredictably developing patterns (Figure 2) [8]. Figure 2 shows the configuration developed from R-pentomino's initial state after running 92 time step. Clicking "Start", the R-pentomino on grid is changing with time and gradually generates stability (encircled with a Pentagon), blinker (encircled with a rectangular), glider (encircled with an oval) and other life forms. These forms of life are at the edge in certain way, at the edge of growth and degeneration and at the edge of order and chaos. Sometimes, some stable and oscillating patterns are activated when they are influenced by other cells, and then they are likely to evolve into another form of life. The whole cellular spaces stabilize eventually only after 1103 generations. At this time, the other cells stay stable state, except for a blinker.

It is seen that the interesting game of life is the same as life. Simple component elements and rules create rather complex life forms in an unpredictable way. Therefore Conway has always firmly believed that "life organisms" in Game of Life look like the ones in the real world, moving, growing, reproducing, evolving, and even thinking. According to this, he designs life glider (in Figure 2, the glider is encircled with an oval) and other "life forms" in order to reveal the truth of life.

At present because the computer screen is not long or wide enough, the movement of certain kind of configurations is displayed only with the limited grid like checkerboard. When the glider and other "life forms" move to one side of the lattice, they are required to crawl out from the opposite side. The designing idea is just the same as Einstein's finite but boundless cosmological model. After clicking "Start", they are moving uniformly along the diagonal just like a reptile, and its velocity equal to $1/4$ of c, here c denotes the velocity from one cell to the next one in the grid . That is their macroscopic behavior rules and configuration behaviors that we can explore. In other words, it is a kind of phenomenon that we have seen at the higher-level, namely emergent hierarchy. That is $F^2(S^2)$) that we have seen after using the configuration function F^2 mentioned above. Note that if we click "step" to decompose the whole evolution, just like doing "everything in one action" in military training, we will clearly see that the transition rules of "Game of Life" only limit each cell to be "dead" or "alive", but do not limit the moving direction of every "life-organisms" and do not show how different forms of gliders shift across the screen. In other words, the position of each cell on the lattice is actually unchanged, and they do not "glide" themselves.

The process of glider "gliding" we see is merely the process of cells appearing and disappearing, being dead or alive, which only shows the cells updating and is a phenomenon observed to use configuration function F^1 at low-level. It is these phenomena that make us think of our lives. Macroscopically our life is persistent, but in fact, it is only the process of delivery or transition of atoms and molecules as well as metabolism of cells in our body. A glider is not just a set of cells. The cells of each generation are replaced completely. The process that the component atoms of your body is updating all the time after you were born and the component parts of a glider work in the same way. In view of dynamics, the application of game of life to simple rules causes "life glider" which is the dynamic, coherent and independent structure of mysterious phenomenon. This shows how the simple rules results in the complex structure similar to life and behavior. The phenomenon of life emergence is deduced from component elements at the low-level, simple rules and additional function $F^2(\ )$.

Game of life which belongs to the artificial system can perfectly explain how the complexity of life emerges from simple configurations and rules via cellular automaton simulation. The following is an example of the simulation of natural phenomena via cellular automata, which can further explain the emergent process of complexity.

4 Virtual Ants Building Highway

Ants are complex in physiology, whereas we will consider a virtual ant created by Chris Langdon as a simple agent S_i. It is able to produce a different type of complex behaviors according to simple rules I_{ij}. First, this ant is located on one of the grids (cells) that are painted either black or white. These grids are theoretically finite but boundless, which we have talked about in the last section. At each time step, the ant is always facing in one of four directions: north, south, east or west and acts according to the following rules ("reptilian rules"):

1. If the ant is now standing on a white cell, it paints the cell black and turns 90 degrees clockwise. Then the ant moves onto the cell.
2. If the ant gets onto a black cell, it paints the cell white and turns 90 degrees counterclockwise. Then the ant moves onto the cell.

According to "reptilian rules", Figure 3 [4, p. 265] illustrates eight steps that an ant would take starting from an initially blank grid. At each time step, the pictures are arranged from left to right and from top to down.

Based on Figure 3 for another 10,000 time steps or so, the ant will indeed form a chaotic-looking mess that has little or no structure. But after another 250 or more steps, the ant will start to build its "highway systems" (This phenomenon was discovered by James Propp and he calls a highway) (Figure 4) [4, p. 266].

Figure 4 shows an ant highway created from an initially blank configuration. The ant takes a step forward strictly obeying the "reptilian rules". After

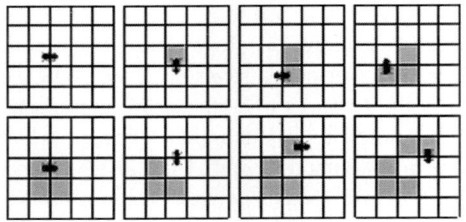

Fig. 3. Langton's virtual ant.

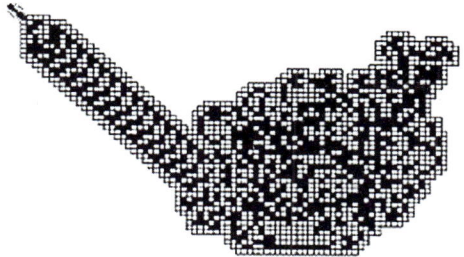

Fig. 4. A virtual ants building a highway.

extremely "chaotic" state, it finally builds a highway stretching towards the north-west. As an ant can build a highway that flows in any of the four diagonal directions, this one is only a chance in necessity. That is to say, an ant builds a highway ultimately according to "reptilian rules", but what is different is the direction. Clearly, if the ant's universe consists of an infinite cellular space that is initially blank, then the ant will happily build the highway forever. If it is in wraparound universe, the highway must eventually intersect a place in the ant's grid space where the ant has been before. As shown by Gary William Flake [4] such a "bump in the road" will usually force the ant back into a chaotic-like behavior, but it will often spontaneously start to build another highway, and so on. It is a process from chaos to order and then chaos cut of order, which is explained by the simulation to the virtual ant.

Virtual ants building highway proves not only the iteration of simple rules which can cause suddenly new phenomenon, but also the fact that the new phenomenon cannot be explained by the sum of its component parts alone, i.e. what Aristotle (350 B.C.) said: "The whole is more than the sum of its parts". Undoubtedly, it is an emergent phenomenon for an ant to build a highway, and that emphasis "the gold medal" feature of emergence-system wholeness, i.e. an emergent structure is a whole one. Note that it is unwitting for the ant to build the highway. Maybe there is no concept of "highway" in ants' world and they only obey inherent laws to take a step forward. Contrastively, the reason why we define such an outspread structure of the grid as a highway is our possession of knowledge of highway. In other words, we cannot discover such an emergent phenomenon until we have configuration F^2 at the higher-level.

Ian Stewart [4, p. 270] has made an interesting observation regarding our lack of knowledge about the long-term behavior of some of these virtual ants. To paraphrase, for any of these ants we know their Theory of Everything, in that all of the "physical" laws that govern the ant's universe are simple and known to us. We also know the initial configuration of the ant's universe. Yet we are helpless to answer a simple question in an analytical way: how does the ant build the highway? Does it ever build a highway? But with using cellular automata iterative modeling, it is possible to answer that. In the simulation of the virtual ant, we have seen it extremely simple that the virtual ant takes each step forward, but it produces unpredictable complex configuration step by step through iteration or recursion which is exactly the arising process of the emergence of complexity. And we have known, "building highway" or something like that is not virtual ant's exclusive feature. Similar patterns have been found in other fields. This shows it is feasible to use virtual ant and other multiagents-based simulation to illustrate the emergence of complexity.

5 Features and Conditions Derived from Simulation

According to the mathematical analysis of emergence above and the simulated case using cellular automata, we can see it is inappropriate for the British Proto-Emergentists, especially Alexander to claim emergence is completely unexplainable. So long as some simple elements and interaction rules could be found, various strange emergent phenomena will arise unpredictably from the higher-level. That is like playing chess – with simple rules, you can have quite different results. Moreover, the cellular automaton modeling shows how emergence springs up step by step and explains the emergent process and mechanism.

The so-called explanation of a kind of phenomenon is to clarify the mechanism of its generation and to illustrate how it is logically deduced. We have arrived at a conclusion through the above analysis: emergent phenomena are explainable and can be deduced from simulation. Let us first account for the methodological characteristics that the simulation of emergence can be deduced.

(1) It is not analytically deduced but synthetically simulated. So called analytical solution is a classical mathematical analysis, in which methods is used in order to solve a problem or to explain a phenomenon. First, it is necessary to set a mathematical equation and add some initial condition and certain parameters, then work it out via limited computation of the limited numbers by means of such algorithms as addition, subtraction multiplication, division and extraction. In this way, the known phenomenon can be explained exactly and the future behavior of system can be predicted. But it cannot solve the problems as complexity, emergence and nonlinearity. The so-called "derived from simulation" does not mean to get an analytical solution from decomposing the whole system to establish the equation. Its main purpose is to demonstrate and compute how local casual effects make use of aggregation in space and iteration in time to show the whole phenomenon of emergence, after modeling the system. And thereby, it is not a reasoning process from the whole to the parts, but from the parts to the whole. In fact, it is a synthetic method. Methodologically, R. K. Sawyer [7] regards it as a conversion from equation-based modeling to agent-based modeling [7, p. 263].

(2) It is partially rather than wholly reducible. Emergence arises from the high-level system. Moreover, its characters are observed and measured

by configuration function F^2 and expressed by proposition P^2 and theory T^2 of the high-level. There are no F^2, P^2 or T^2 in configuration function F^1, proposition P^1 and theory T^1 of lower level. For example, in chemistry, such phenomena and concepts as freezing point, boiling point, heat of solution, heat of sublimation of water are at the higher level, but they cannot be reasoned out completely by the quantum physics, because there are no such notions in quantum physics. Certainly, when the heat of sublimation of ice is 12.2 kilocalories, three-quarters of energy can be considered to destabilize the hydrogen bond between molecules. However, there is no term of hydrogen bond in quantum mechanics. Thus it can not be fully derived from quantum mechanics of the low-level. Only by adding the hypothesis about hydrogen bond which does not exist in quantum mechanics, can the heat of sublimation of ice be deduced. This is reasoning of partial reduction.

Similarly, life emergence is derived from cellular automata modeling, which is also partially deduced. The configuration, moving direction and speed of life gliders as well as glider guns and so on cannot be deduced from the initial conditions of cells state and their interaction rules. As long as we add function of configuration of the higher level to the reasoning, the emergent phenomenon of that level can be derived by simulation of interaction of elements at the lower level, namely, that the formula $\sum_s^1 (S_i^1, F^1, I_{ij}^1) \& C \vdash P^2$ is not tenable, while, $\sum_s^1 \& F^2 \& C \vdash P^2$ is tenable, where $F^2()$ is a function of configuration of the high-level. If the method of "reasoned out emergence by simulation" is considered as reductionism, it is only partial reduction. This method itself indicates some emergent properties themselves are irreducible, or at least partially irreducible. Therefore, the reductive method of "reasoned out emergence by simulation" just proves that it is not enough to understand emergence only by means of reductionism. At least two (high and low) kinds of configuration functions would be used at the same time. Emergence can be fully understood by the analysis and the synthesis of both the higher level and the lower level. In addition, the macroscopic emergent phenomenon may be composed of multiple emergent microscopic mechanisms. Certain microscopic mechanism is only an example among many explanations or illustrations to explain the macro-emergence, and this mechanism is always incomplete, especially in the case that multiple emergences are not meaningfully related.

In this way, if an emergent property can be derived from simulation, it requires to meet the following three conditions at least:

(1) It must be simulatable. Emergent phenomena and properties are sometimes very complicated. It is quite difficult or even impossible to simulate a system when their subsystems are indistinguishable or subsystems' interrelations entangled. Therefore, a system is simulatable if and only if two conditions are met. Condition I: There exists $S_i \in M$, where M is a model, namely some crucial features of the simulated system were abstracted as elements of the simulation system. Condition II: Some update functions U exist and they distribute all over S_i. These two conditions are not generally available,

due to the existence of inseparable systems and system properties. So the existence of these conditions themselves also indicated the limitation of reasoning out emergence by simulation. When you abstract elements S_i from a system, whether this S_i reflects the real properties of the system, or whether the update function U you define can show the development of systems is still a problem. In the best case, S_i is always simplified and that indicates inaccuracy and uncertainty.

(2) It must be computable. For example, in fractal theory, recursion or mapping of Mandelbrot Sets $Z_{n+1} = Z_n^2 + C$ is simulatable, but both Z and C are complex numbers. Because most C is approximate to Julia Sets, this recursion is instable equilibrium, and then this mapping is incomputable because computation is defined in real number field. Similarly, it is inevitable to extend to complex number when we use iteration of various functions to work out imaginary number. We should say it is of non-calculability in principle. As for computable emergent process in principle, if the amount of S_i is too large or the update function U too complicated, it will go beyond the really computable confine, it is actually incomputable and does not satisfy the conditions of simulation.

(3) It must possess necessary configuration function F^2 of the high-level. As is mentioned above, if the necessary function F^2 is not given, emergence can not be derived from simulation. The reasoned out emergent conditions from simulation show us that because of the complexity, uncertainty, sensitivity to initial conditions emerging in complex systems as well as the limitation of simulation, computation and observational states, the emergence of systems cannot be predicted accurately. Because of the lack of F^2, some real instances such as financial crisis, political situation abruptness and earthquake are considered unpredictable, especially when we saw them for the first time. On the other hand, we also notice that the emergence of system is not completely unpredictable. When we possess F^2, it is likely to infer the result by means of simulation. It is a good example that the warning and serving system of disasters has been putting into practice in our country. It can be seen that our emergence theory of complex systems is a farewell to that of British Emergentists, but in some extent a kind of return to the idea of the British Emergentists.

References

1. Alexander, S.: *Space, Time, and Deity*. Macmillan, London (1920)
2. Bedau, M. A.: Downward Causation and the Autonomy of Weak Emergence. *The International Directory of On-Line Philosophy Papers*, http//philosophy.hk/paper/index (1997)
3. Broad, C. D.: *The Mind and Its Place in Nature*. Routledge & Kegan Paul, London (1925)
4. Flake, G. W.: *The Computational Beauty of Nature: Computer Explorations of Fractals, Chaos, Complex System, and Adaptation*. The MIT Press (2001)

5. Morgan, C. L.: *Emergent Evolution*. Henny Bolt and Company (1925)
6. Rasmussen, S., Barret, C.L.: Elements of a Theory of Simulation. In *ECAL 95 Lecture Notes in Computer Science*. Springer-Verlag (1995)
7. Sawyer, R. K.: The Mechanisms of Emergence. *Philosophy of The Social Science* **34** (2004) 260-282
8. http://llk.media.mit.edu.contents/projects/emergence
9. http://www.rennard.org/alife/CollisionBasedRennard.pdf

Belief Ascription and *De Re* Communication

Yuan Ren

Department of Philosophy & Institute of Logic and Cognition, Sun Yat-Sen University, Guangzhou, P.R. China
hssreny@mail.sysu.edu.cn

Summary. Direct reference theorists and Fregeans have different opinions on how to explain belief reports of sentences containing proper names. In this paper I suggest an alternative way to understand how successful de re communication is possible, based on which I give an explanation of belief ascription that seems to avoid the shortcomings of both camps.

1 Background

The so-called "New Theory of Reference", which was advocated by Donnellan, Kripke, Putnam, Kaplan and Perry in the seventies of 20th century, has now become the new orthodoxy of analytic philosophy. The New Theory in fact mainly encompasses three parts: direct reference theory, historical-causal theory of reference, and semantic or mental content externalism. The main thesis of direct reference theory is a semantic one, namely, there exist some kinds of expressions (e.g. demonstratives, indexicals and proper names) which are directly referential, and the semantic contribution of such expressions to the sentence in which they occur are just their referents. On the other hand, the historical-causal theory of reference works mainly on epistemological level. The purpose of such a theory is to give an account of the epistemic mechanism of designation of some singular terms (in particular, proper names). Finally, mental content externalism wants to show that propositional attitude content does not supervene on local properties of individual persons but indexically dependent on the physical or social environment.

Do the above three parts coherently manifest a new scene of our understanding of reference? Kripke [13] thinks that the historical-chain picture of reference favors the Millian view of names, while Devitt [6] makes use of causal theory of reference to argue against direct reference theory. Evans [9] provides a hybrid account in a nice way which combines causal theory of reference with a semantics that associates names with a certain mode of presentation. On the other hand, it is often claimed that causal theory of reference is some kind of externalism (in the sense that links between names and objects are social and hence external to the individual's cognitive ability), and semantic externalism can be derived from direct reference theory (since sentences containing direct referential expression would express a singular or object-dependent proposition, propositional attitude content does not supervene on cognitive individuals but is dependent on elements of environment which contains the existence of object as a part of it).

Various aspects are involved in our talking about the notion of reference and theories of reference. Before we examine the debate between two camps, direct reference theorists and Fregeans, we may characterize theories of reference on four different levels:

(1) *Epistemological level.* In terms of which institution, mechanism or rules, could a singular term designate an object or fix its referent? What it is for a competent speaker of a language to understand a sentence containing referential expressions?

(2) *Semantic level.* What is said when a speaker utters a sentence containing a referential expression? Or, what is the semantic content of an utterance of a sentence containing a referential expression?

(3) *Psychological level.* On this level we mainly care about mental reference, that is, problems about mental representation and problems about propositoinal attitude and *de re* belief. What is the structure of a thought or a belief?

(4) *Pragmatic level.* On this level we consider referential communication as speech act. So the main question is how is it possible to obtain a successful referential communication?

2 The Fregean Sense and Direct Reference

In order to solve the puzzle of identity statement and that of substitution in the propositional context, Frege introduced the concept of Sinn. But as Dummett said: "Frege contented himself with laying down certain principles about sense and never attempted a specific account of the sense of any particular expression" [8, p. 136]. The only explicit formulation Frege made about the notion of sense of a referring expression is that it is the mode of presentation of the referent. That is to say, sense is the epistemological channel through which we contact the object and fix the referent. On the other hand, according to Frege, the sense of a referring expression is its contribution to the truth condition of the sentence in which it occurs. In this way, sense constitutes the semantic content of the expression. Finally, in order to solve the puzzle of substitution in propositional attitude context, Frege took the referent of a sentence in an indirect context as its normal sense. As a result, the sense of a sentence becomes the object of propositional attitude. This is the ground on which we explain the cognitive attitude and behavior of the subject. Thus sense functions as mental content. In summary, Frege's notion of sense in fact includes the first three levels of sense that we have characterized previously. In this way, Frege answered the question of reference concerning different levels in a unified way.

However, the Fregean sense faces with many difficulties, one of which is how to individuate the sense properly. It is difficult for us to find a suitable entity that can serve as the candidate for the notion of sense that satisfies all the above three levels. If we take the sense of a referring expression as some definite description or the identifying property of the referent, it is strongly challenged by Donnellan and Kripke via modal argument and argument from ignorance and error. Or we might take the sense of the referring expression as the linguistic meaning of the expression or linguistic convention that govern the expression, but again it faces serious problems: in the case of indexicals the difficulty is that the semantic value of an indexical is context-dependent whereas its linguistic meaning is stable. Kaplan [12] has successfully argued that the linguistic meaning of indexicals and demonstratives cannot enter the truth condition of the sentences in which they occur. Perry on the other hand expounds from the perspective of entertaining belief and explaining behavior, that essential indexicals cannot be semantically reduced to a cluster of equivalent definite descriptions. These challenges show that Fregean sense is actually incoherent, for the epistemological aspect and semantic aspect of Fregean sense have to be characterized by different criteria, which cannot obtain in a unified formulation.

In response to these challenges, neo-Fregeans would like to shift their interpretation of sense from the semantic aspect to the psychological aspect, substituting descriptive sense with non-descriptive sense, and interpreting the mode of presentation in terms of the way of thinking. But in so doing, they weakened the principle of objectivity of the Fregean sense. At the same time the following problem arises: if the sense of a referring expression varies with the cognitive subject, how is successful referential communication possible? For example, how is the communication of first person thought possible? We have noticed that Frege's solution to the information problem of identity statement aims at answering this question: How is it that the identity statement "Hesperous is Phosphorous" is empirical while "Hesperous is Hesperous" is a priori? Here being empirical is certainly different from the way an individual cognitive subject thinks of the object. On the other hand, Frege characterized sense as the content of the subject's propositional attitude. Later Evans gave the criterion of individuation of the notion of sense, which he called "Intuitive Criterion of Difference" for thoughts. We may call these two aspects of sense the epistemic content and cognitive content respectively.

Before we discuss the relation between these three levels of sense, let's first of all turn to what direct reference theory has to say. Direct reference theory has different versions. It's core is a semantic thesis that the semantic contribution of a direct referring expression to the sentence in which it occurs is its referent. Thus, an utterance of a sentence containing such an expression expresses a singular proposition. This thesis implies that direct referential expressions are *de jure* rigid designators, in contrast to *de facto* rigid designators which are characterized in terms of possible world truth conditions. According to this core thesis, proper names, indexicals and demonstratives are all direct referential expressions, whose semantic contents are merely their referents. However, this core thesis of direct reference theory can be combined with different epistemological theses. Some moderate direct reference theorists, such as the early Kaplan and Perry, think that indexicals and demonstratives fix their referents in terms of the linguistic meaning associated with them, i.e., their character. In other words, "direct means unmediated by any propositional component, not unmediated simpliciter" [12, p. 561]. So there still exists some form of epistemological mediator between a referential expression and its referents, but such a mediator does not enter into the truth condition of the utterance of the sentence containing that expression. Moreover, Kaplan and Perry use character to explain the behavior of the rational subject. Thus it plays the role of cognitive content of the expression. On the other hand, the radical direct reference theorists, or Millianists, such as Salmon and Wettstein, insist that a proper name is only a tag of the referent and there is no need of any epistemological mediator between a name and its referent, whether it is descriptive or non-descriptive. However, Salmon does not deny that there may be some information about the referent which is associated with the proper name. But he not only rejects that such information can be epistemological mediator between a name and its referent or can be used as referent determinant, but also that such information can enter the propositional content of the sentence that contains the name.

Both moderate and radical direct reference theorists hold the same semantic core: an utterance of a sentence containing direct referential expression expresses a singular proposition, and the semantic content of a referring expression is its referent. Direct reference theorists therefore must solve the puzzle which Frege did by introducing the notion of Sinn or mode of presentation. The typical strategy of direct reference theorists such as Salmon and Soames resorts to the distinction between semantically expressed information and pragmatically conveyed information of a sentence, in terms of the Gricean distinction between what is said and what is implicated. They thus claim that what is said, or semantic contents, of the utterances of sentences that contain different co-referential direct referential expressions, are the same, while what is different between the two utterances is transferred by a pragmatic process. However, direct reference theorists generally fail to offer a complete account explaining how this pragmatic process functions so as to communicate the epistemic or cognitive content among different speakers. In explaining the substitution of co-referential terms in propositional attitudes context, though direct reference theorists have got rid of the Fregean sense from belief content, they still have to explain belief itself as a *ternary* relation between the believer, the proposition and the way of taking the proposition. Moreover, they seem to have no way to give an adequate account of the third element.

This is why Evans and Forbes would consider direct reference theory to be a "notional variant" of the Fregean theory. Some philosophers thus reach the conclusion that sense is indispensable. What is at issue is simply whether it is descriptive or non-descriptive, or whether it is internal or external to the proposition the sentence expresses.

3 Explaining Belief Ascriptions

We now focus on the puzzle of substitution in propositional attitude context, which is said to be an objection to direct reference theory. It is widely held that the following two sentences have different truth values, (1) being true and (2) being false, supposing that Joey has no idea of the knowledge of identity between Hesperus and Phosphorus:

(1) Joey believes that Hesperus is Hesperus.
(2) Joey believes that Hesperus is Phosphorus.

There are different explanations for the above phenomenon. Fregeans would like to read the two sentences respectively as

(1f) Joey believes that the star visible in the morning sky is the star visible in the morning sky.
(2f) Joey believes that the star visible in the morning sky is the star visible in the evening sky.

Thus Fregeans draw the conclusion that the sentences (1) and (2) in fact express two different propositions (1f) and (2f), and the two propositions have different truth values.

On the other hand, Russellians or direct reference theorists would think both (1) and (2) in fact express one and the same proposition, that is,

(1r) Joey believes that Venus is Venus.

Besides, Russellians hold that (1) and (2) have the same truth value, for they express the same proposition. Russellians would like to claim that "although Joey does not seem to believe that the star visible in the morning sky is the star visible in the evening sky, he in fact believes that Venus is Venus". But now Russellians must afford the burden of giving an explanation of the apparent difference between (1) and (2).

Generally, direct reference theorists have different ways to say that. One comes from Salmon (and Braun), who takes belief ascription relation not as a traditional binary relation between subject and propositions but as a ternary relation among subject, propositions and the way of believing. Thus the difference between (1) and (2) is something like (1s) and (2s):

(1s) Joey believes in way_1 that Venus is Venus.
(2s) Joey believes in way_2 that Venus is Venus.

According to this explanation, there exist different ways of believing or ways of grasping propositions, such that a believer might believe the same singular proposition in different ways. That means sentences (1) and (2) might have different cognitive significance though they semantically express the same proposition (1r).

Another explanation comes from Soames (and Thau), who distinguishes carefully between the propositions "semantically expressed" and "pragmatically conveyed" by a sentence. According to Soames, "the proposition semantically expressed by $tv's$ that s is true with respect to a context c, world w, and assignment f of objects to variables, iff in w, the referent of t (with respect to c, w, and f) bears the relation expressed by attitude verb v (with respect to c) to the proposition semantically expressed by s (with respect to c and f)" [20, p. 140]. That is, we intuitively think that (1) is true and (2) is false because when we consider (1r), what we actually think might take the form of (1) or (2), which in fact are pragmatically enlarged propositions compared with (1r), and can be rewritten as (1p1) and (2p1):

(1p1) Joey believes that Venus (named with Hesperus) is Venus (named with Hesperus).

(2p1) Joey believes that Venus (named with Hesperus) is Venus (named with Phosphorus).

Or (1r) might be pragmatically enlarged in some other ways, that is (1p2) and (2p2):

(1p2) Joey believes that Venus (the star visible in the morning sky) is Venus (the star visible in the morning sky).

(2p2) Joey believes that Venus (the star visible in the morning sky) is Venus (the star visible in the evening sky).

The main difference between (1f), (2f) and (1p2), (2p2) is that Fregeans would think that (1f) and (2f) are semantically expressed by sentences (1) and (2) while Russellians would think that (1p2) and (2p2) are pragmatically conveyed by sentences (1) and (2). In other words, in some Russellians' opinion, sentences containing names (indexicals, demonstratives, etc.) express single propositions but communicate descriptive or general propositions whereas Fregeans think they both express and communicate Fregean thoughts, whether or not these sentences are simple sentences or embedded in propositional attitude context. Note that both sides insist on some kind of "semantic innocence", i.e., "the utterance of the embedded sentences in belief reports express just the proposition they would if not embedded, and these propositions are the contents of the ascribed beliefs" [5, p. 686].

The third explanation comes from Perry and Crimmins, maybe including Schiffer, who hold the so called "hidden-indexical theories"(HIT) view of belief ascription. But different from other direct reference theorists, they don't think that sentences (1) and (2) have the same truth condition and the same truth value. Because it rejects such "semantic innocence", we may take this view as a hybrid view. According to HIT view, (1) is true iff Joey believes the singular proposition "Venus is Venus" under certain mode of presentation which is determined contextually. The modes of presentation are "hidden" because they are referred to implicitly and they are "indexical" because they vary with the context of utterance. So the difference between sentences (1) and (2) lies in their expressing the different propositions involving different contextually specified modes of presentation.

Now we may summarize the four kinds of different explanations of belief ascriptions as below. For sentence:

(3) S believes that x is F.

there are different ways of reading according to different positions.

(3a) S believes $<$ the sense of $x, F >$ (Fregeans' reading)
(3b) S believes $< x(w), F >$ (Soames' reading)
(3c) S believes-in-way-$w < x, F >$ (Salmon's reading)
(3d) S believes $< x$ in $m(c), F >$ (HIT's reading)

As we have mentioned above, the mode of presentation of the referent is preserved in some sense in all these readings. But the important difference is that in (3a) the mode of presentation is internal to the semantic content of the sentence while in (3b) and (3c) the modes of presentation is external to the semantic content of the sentence. As to (3d), the mode of presentation is partially internal to the semantic content of the sentence because of its indexical feature.

4 What Does Referential Communication Mean?

As stated above, neither direct reference theories nor the Fregeanism is satisfactory for our purpose. The Fregean sense in fact incoherently includes several aspects, whereas direct reference theories emphasize that the semantic content of a sentence is a singular proposition, without satisfactorily explaining how the semantic content of a sentence correlates with cognitive significance and with communication. We instead attempt to construe such correlation by expounding on the structure of referential communication. Referential communicative activities are a type of speech-acts that involve the speaker and the audience, who are both rational agents. It's main characteristic lies in the fact that the speaker and the audience communicate an object-dependent thought, or *de re* thought, in the speech-act. The speaker utters a sentence containing a referring expression on a particular occasion, with the intention that the audience will think of a certain object, and identify it, so that she will obtain a certain belief which leads to an intending behavioral response. A successful referential communication means two points: (1) that the audience properly understands the speaker's intention, followed by the response intended by the speaker; (2) that the audience get the knowledge of referent from the speaker, as Evans once pointed out, "communication is essentially mode of transmission of knowledge ... If the speaker has the knowledge of x to the effect that it is F, and in consequence utters a sentence in which he refers to F, and says that it is F, and if his audience A understand the utterance and accept it as true (and there are no defeating conditions), then A himself thereby comes to know of x that it is F" [9, p. 310–311].

There are now two questions. One is, in order to reach a successful referential communication, what conditions must be satisfied? In other words, what does it mean by saying that the audience has properly understood the speaker's intention in a referential communication? The other is, if we define the semantic content of an utterance as what remains stable in a successful communication, and is common to the cognitive contents the speaker and the audience attach to the utterance, then what do we expect this semantic content to be?

In a referential communication, when a speaker utters a sentence containing a referring expression, several elements are related to the referring expression: (1) the referent of the referring expression on that particular occasion; (2) the linguistic meaning of the referring expression, or conventions related to it; (3) the epistemic content about the referent; (4) the cognitive contents of the speaker and the audience about the referent; (5) the speaker and the audience's common belief about the context; (6) conversational implicature. The meanings of (1) and (2) are explicit, and that of (5) and (6) rather clear, even though they must rely on the context for further clarification. What we have to explain are (3) and (4). As explained previously, epistemic content comes from the mode of presentation of the object, by which Frege had hoped to solve the identity puzzle. It's characteristic is that it contains information about the object and involves epistemic knowledge which is experienced and obtained by a certain linguistic community, shared by different subjects in the community. Moreover, by way of this knowledge, members of the linguistic community can successfully identify the relevant object in the world. Cognitive content was used by Frege to explain the puzzle of substitution of co-referential names in prepositional attitudes, that is, for different subjects, the same referring expression may have different cognitive value, which, according to Evans' "Intuitive Criterion of Difference", is the content of the belief formed when the speaker accepts as true a certain sentence containing a referring expression. Accordingly, the cognitive content of a referring expression, i.e. the way a subject thinks of the object, turns out to be different psychological modes of presentation of the object by different cognitive subjects. Forbes and some other philosophers use "dossier of information" or "file" to characterize cognitive content. In terms of this kind of content, we are able to explain the causal effects of the intentional mental states of different individuals. The key point here is what is the relation between epistemic and cognitive content, and how they bear on the semantic content of a sentence? In the following, we will answer this question by characterizing the structure of referential communication.

There are several different communication models which explain the sufficient and necessary conditions of successful referential communication. Type 1 model (Fregean style) holds that successful communication means, when a speaker utters a sentence containing a referring expression, the thought entertained by the audience is exactly the one held by the speaker when she utters the sentence. Type 2 model (Rusellian style) holds that communication is successful when the speaker and the audience share the same singular proposition. Type 3 model holds that successful communication requires certain correspondence or relevance between the thought held by the speaker and that held by the audience.

The difficulty of type 1 model is apparent: here the thought held by the speaker and that held by the audience contain different cognitive contents about the referent of the expression that are different for the speaker and the audience. We cannot expect to identify them. For example, the first person thought of the speaker can in no way be accessed by the audience, since it contains the self-conscious way of thinking of the speaker. Type 2 model seems reasonable. The speaker and the audience hold different thoughts. A singular proposition is attained by screening the subjective part of the speaker's thought (called "interpreting up" by Perry), then adding to it the audience's subjective understanding is formed the audience's thought (called "interpreting down" by Perry). Though this model seems more plausible than type 1 model, it in fact fails to characterize the actual picture of referential communication. Accepting type 2 model means that in forming the understanding of the speaker's thought, the audience first identifies the object the speaker intended to talk about, before she is able to think of the object from an audience's perspective. This is clearly inappropriate, for it is the result of a successful communication, not its starting point, that the audience identifies the object and thus obtains the same singular proposition held by the speaker. The idea of type 3 model can be expounded in McDowell's words: "Frege's troubles about "I" ... result from the assumption, ... that communication must involve a sharing of thoughts between communicator and audience. The assumption is quite natural, and Frege seems to take it for granted. But there is no obvious reason why he could not have held, instead, that in linguistic interchange of the appropriate kind, mutual understanding – which is what successful communication achieves – requires not shared thoughts but different thoughts which, however, stand and are mutually known to stand in a suitable relation of correspondence" [16, p. 290]. But it is not easy to explain what is suitable relation of correspondence. Bezuidenhout [3] proposes to explain successful communication between the speaker and the audience in terms of the similarity of different psychological modes of presentation. Apparently it is difficult to give the criterion of similarity. For example, in the communication of first person thought, it is hard to say what is the similarity between the first person thought held by the speaker and the third person thought held by the audience.

5 Conditions of *de Re* Communication

Here let's summarize the relevant conditions of successful referential communication: first, it is necessary to maintain the referent, otherwise the speaker's true belief cannot be transferred, and in turn the behavioral response that conforms to the speaker's intention cannot be produced. This also means that the mere understanding of the literal meaning of an utterance of a sentence cannot count as successful communication. Next, only maintaining the referent is not sufficient for successful communication, as shown by the examples in Loar [15][1]. More specifically, even if the audience has correctly attained the referent intended by the speaker, the audience might not have properly understood the speaker, for the belief might be obtained by accident. This indicates that we cannot judge that the audience has understood the speaker because she has grasped the right object simply by any arbitrary means. There must be some restriction on the relation between the ways of thinking or cognitive contents that are held by the speaker and the audience respectively about the object that is talked about. That is to say, successful communication does not only mean that the result of communication is to share the same referent, but also means to restrain the route by which the audience identify the referent. In Richard Heck's words, what is transferred is not only a true belief, but also knowledge, namely, justified true belief; and what understanding maintains is not only the referent, but also knowledge about the referent.

Furthermore, as indicated by McDowell quoted above, this restriction need not only be as strong as to request both parties to hold the same cognitive content about the object. Some (like Evans) proposed that the linguistic mode of presentation shared by the speaker and the audience can be considered the route to identify the referent. In this way, the linguistic convention connectes to the referring expression plus the referent form a dyadic group that can serve as the conditions for successful referential communication. But we note that this plan still fails to pass the test of Loar's example mentioned above, because even if the speaker and the audience both understand the linguistic sense of a referring expression and grasp the right referent, the audience will still not be able to fully understand the information the speaker intends to transfer, if the audience cannot understand in a proper way how the speaker identifies the referent. For example, in the case of demonstratives, the audience needs

---

[1] The example is: "Suppose that Smith and Jones are unaware that the man being interviewed on television is someone they see on the train every morning and about whom, in the latter role, they have just been talking. Smith says 'He is a stockbroker', intending to refer to the man on television; Jones takes Smith to be referring to the man on the train. Now, as it happens, Johns has correctly identified Smith's referent, since the man on television is the man on the train; but he has failed to understand Smith's utterance. It would seem that, as Frege held, some 'manner of presentation' of the referent is essential, even on referential use, to what is being communicated." (Loar [15, p. 357], also cf. Recanati [18])

to know by what means the speaker refers to the object, or, the audience needs to follow the speaker's demonstration in order to reach the referent. In addition, the speaker's demonstrative way of thinking is included in the knowledge transferred to the audience by the speaker's intention. In the case of indexicals, the audience needs to know who is the user of the indexical. This knowledge is not entirely included in the linguistic sense of the indexical. In the case of proper names, if there is no overlap between the "information dossiers" that the speaker and the audience have about the referent, the audience can in no way communicate according to the speaker's intention.

So far our conclusion is, successful referential communication requires that: (1) the audience needs to identify the referent in the way intended by the speaker, not in any arbitrary way; (2) there is something in common between the speaker's thought and the audience's, for example, the referent at least. Then how does (1) relate to (2)? Identifying the referent can mean two things, the one is to isolate a certain object form among many others, such like picking up the object that the speaker intends to talk about from the background. The other is to know a certain object to be the same one under different spatial-temporal situations or under different modes of presentation. (This is what Strawson [23] called identification and reidentification.) A possible case is that, for example, supposing that in certain case of demonstrative reference, neither the speaker nor the audience knows the object presented referentially from different aspects is one and the same object. In such a case, even if the audience can recognize a certain object from the background and identify it to be the one that the speaker is talking about, this identification of the object is still insufficient for successful communication, because the audience might just "happen" to get the right referent. That is, if there are two different modes of presentation m_1 and m_2 about the same object O, and neither the speaker nor the audience realizes that their modes of presentation are about the same object, then we cannot say that the audience has understood the speaker when the speaker talks about O in the way of thinking m_1 whereas the audience thinks of O in the way of m_2. Therefore, successful referential communication in fact requires:

(S1) the audience gets the right referent from the speaker, and
(S2) the audience knows the object the speaker is thinking of and that the audience is thinking of are identical (even if they are thinking of the referent in different ways).

These two conditions are sufficient for successful referential communication, for as long as the speaker and the audience are talking about the same object, and they know that they are thinking about the same object, successful referential communication is reached. Condition (S2) is the restriction of the so called "cognitive dynamics", and it indicates that the audience can justify the fact that she gets the right referent. Condition (S2) means that the cognitive contents of the speaker and the audience are about the same object and the audience knows this. Note that it does not require that the speaker and the audience have the same or even similar way of thinking of the referent.

Successful referential communication requires that the ways the speaker and the audience think of the object have certain relevance, characterized by the condition (S2). But how can (S2) be satisfied? Following Evans [9] and Bach [2], we divide the *de re* thought into three kinds, i.e, perception-based, memory-based and communication-based. In the case of demonstrative reference (typically using demonstratives), the speaker indicates the audience to fix the referent by demonstration. But because of the indeterminacy of demonstration in reference-fixing, the speaker and the audience often think of the referent in different ways. If there is no enough common knowledge or background to ensure the satisfaction of (S2), even if (S1) is satisfied, the communication may still fail. This is shown by Loar's example. In the case of using indexicals, for example in the communication of first person thought, normal contexts would generally provide enough background knowledge to make the audience know that "the speaker is normally using the indexical 'I' to refer to the speaker herself". (S2) is thus satisfied. In the case of proper names, how does the audience know the referents that the speaker and the audience think of with the proper name N are identical? This seems a bit complicated. For example, suppose the cognitive contents about London that the speaker and the audience have are totally different, without any overlap, when the speaker says "London is pretty", in what sense can we say the audience understands this utterance? In this case it seems impossible for the audience to understand the speaker, since the audience is in no position to know what she and the speaker think of is the same referent. The audience cannot understand the speaker until she learns in some way the referents she and the speaker are talking about are identical. For example, the audience finds that the London that she is thinking of turns out to be the same the speaker is thinking of, instead of two different cities with the same name.

Therefore, successful referential communication requires the satisfaction of (S2), that is, there must be certain identifying content that enables the audience to know what the speaker and the audience talk about with the referential expression is the same referent. Considering "the symmetry of the paths" for both parties in successful communication, this means there must be certain identifying content that enables both the speaker and the audience to know what they talk about with the referential expression is the same referent. This is what is required by successful communication that is more than merely getting the right referent. This means, in successful communication, singular proposition is not the only thing in common in the cognitive contents of the speaker and the audience. The content that enables the speaker and the audience to know they are thinking of the same object is the epistemic content about the referent, namely the epistemic content manifested in the identity statement $m_1 = m_2$.

6 Propositonal Attitudes and Semantic Content Revisited

From the previous analysis we can conclude that, successful referential communication not only maintains the referent, but involves knowledge about the relevance of the ways of thinking that the speaker and the audience have about the referent, that is, they can know as well as believe that they are thinking of the same referent. Apart from this, successful communication does not involve any more specific knowledge about the modes of presentation about the referent. One might still ask, how to explain how the speakers know that they are thinking of the same object, or how can one speaker know that the object she thinks of at different times is one and the same object? Is there any reductive explanation about condition (S2), so that the relevance of different ways of thinking about the same referent is finally involved in it? This is certainly one of the profoundest metaphysical problems, but it is not the burden of this paper to resolve it. What we have to examine here is, whether this relevance constitute a part of the semantic content of the referential utterance.

There is certain formal semantic characterization about what is said or semantic content of an utterance in terms of possible world semantics, taking semantic content as truth conditions in possible worlds, like what Kaplan has done. Another concept of what is said comes from Grice's approach, that is, what is said mainly involves the truth-value of the utterance and makes sense of semantic reasoning, whereas what is implicated mainly involves pragmatic reasoning. Grice gets what is said from deleting the implicature from what is communicated. But he does not give a positive formulation about what is said, simply indicating what is not what is said. Recanati [18] takes what is said to be sentence meaning (literal meaning) plus contextual ingredients of what is said. But as we have shown, in successful referential communication, although literal meaning of a term makes contribution to determining what is said, it is not a part of what is said itself.

In any case, we must determine the semantic content of an utterance, which is taken as the mediator of successful communication. However, it is still not clear as to how we can systematically explain the correlation between the speaker's and the audience's thoughts, for such correlation obviously relies on particular context. Just as Heck said recently: "We can define a notion of 'what is common' to the various Thoughts speakers associate with a given utterance and, if we like, call that the utterance's "meaning". Maybe the singular proposition that is determined by all those Thoughts will even turn out to be what is common... If one really wants to find something to call the meaning of an utterance, then perhaps what is common to the cognitive values the utterance has for different speakers is as good a choice as any" [11, p. 31].

Let's consider a simple fact. In referential communication, the speaker and the audience can ensure that they are thinking of the same referent in terms of explicit verbal communication. In the case of demonstrative reference, the speaker can tell the audience the unique characteristic of the demonstrated referent so as to eliminate the indeterminacy in reference fixing via demonstration. In the case of proper names, the speaker can provide more specific identifying content for the audience to verify that they are talking about the same referent. Of course, if the common knowledge between the speaker and the audience is abundant enough, the above clarification is unnecessary. For example, if both the speaker and the audience are graduates of philosophy department, they don't have to worry about the failure of referential communication when they utter the name "Immanuel Kant" in a normal context. Likewise, if both the speaker and the audience know that Cicero is a man, sentence

(4) Cicero is unmarried; and sentence
(5) Cicero is a bachelor

communicate the same thing to them in any normal context. Similarly if both the speaker and the audience know that Cicero is Tully, then sentence (2) Cicero is a bachelor and sentence

(6) Tully is a bachelor

mean the same to them in any normal context. Stanley [22] recently suggested the following principle of communication-expression that two sentences express the same proposition semantically iff they communicate the same proposition in normative context class C.[2]

Combined with our discussions above, we have this: if in a context both the speaker and the audience know that Cicero is Tully, then (2) Cicero is a bachelor and (3) Tully is a bachelor communicate the same proposition to them. Thus we propose the principle of communication-expression based on presupposition:

PECP: in normal context C, if the speaker and the audience share the presupposition P about relevant singular term t, then sentences S(t) and S'(t)

---

[2] **Stanley's Expression-Communication Principle (ECP):**
For all S, S', c, c', such that c and c' agree on all contextual feature relevant for determining what is said by S and S', S, relative to c, and S', relative to c', express the same proposition if and only if an utterance of S would communicate the same thing as an utterance of S' in every context c'' meeting the following four conditions:

(a) c'' agree with c and c' on assignments to all contextual sensitive items in S and S.
(b) It is common knowledge that all participants understand the items in S and S' and know the values of the context-dependent elements in S and S' relative to c''.
(c) It is common knowledge that each lexical items in S and S' would be intended to be used in accorded with its usual literal meaning.
(d) It is common knowledge that that the speaker would be perspicuous (i.e. not flout the maxim of Manner) [22, p. 329].

express the same proposition relative to context C, iff they always communicate the same proposition under presupposition P.

Here we use the notion of presupposition in the sense of Stalnaker's, which means the common belief of the speaker and the audience, and can be seen as a set of propositions that are 'taken for granted' by speakers when an utterance is made.

How the speaker knows $m_1 = m_2$ and the fact that the speaker has already known $m_1 = m_2$ are two different things. Their difference is that between "know how" and "know that". Explaining the former requires the construction of a mechanism relevant to "cognitive dynamics", while the latter is used to explain the success of referential communication. When $m_1 = m_2$ is already considered as common knowledge, sentences "m_1 is F" and "m_2 is F" express the same proposition to the speaker and the audience.

Therefore, our conclusion is, whether the mode of presentation or way of thinking about the referent enters the proposition expressed by the sentence that contains the referring expression relies on the common knowledge or presupposition of the speaker and the audience. In successful referential communication, in order to know that they are thinking of the same object, the speaker and the audience must have some common knowledge about the referent. This common knowledge is not any specific way of thinking and mode of presentation of the referent, but that the knowledge of both parties is about the same object. It is this common knowledge that constitutes the "sense" of the referring expression. Thus we may take the sense of a referring expression as the presupposition associated with it. Obviously this notion of sense has a reflexive characteristic.

Considering the common knowledge of the whole linguistic community, the reflexive sense of the proper name N is "'N' is the bearer of N". The reflexive sense of a token demonstrative will include certain perceptual elements about the context. Accordingly, the semantic content of an utterance of a sentence containing such a referring expression is not only the singular proposition, but includes the reflexive sense of the referring expression as a part. But it does not include any particular mode of presentation that is associated with the referent.

Finally, we can get a way out of Kripke's puzzle [13] along this line. Recall Kripke's story about Paderewski. Peter learned in some day that the musician Paderewski was good at violin. The other day he came to know the politician Paderewski was an excellent orator. But Peter had no way to know Polish politician Paderewski is also a good musician. So the following two sentences seem both true:

(7a) Peter believes that Paderewski had musical talent.
(7b) Peter believes that Paderewski had no musical talent.

But Peter is a rational human being, how could he believe a contradiction?

We have seen various solutions based on different semantic positions to the puzzle in section 4. According to the Fregean position, Peter actually believes two different Fregean thoughts, which do not constitute a contradiction. According to Russellian position, Peter believes the contradictory beliefs in different ways, or he believes in fact two pragmatically enlarged propositions, which seems similar to Fregean thought. I don't want to discuss the criticisms of these solutions now. I just sketch my way to the puzzle.

Supposing that Peter talks to himself about Paderewski, that is, he is a speaker as well as an audience. To obtain a successful communication to himself, according to the above analysis, not only must the referent be preserved (condition S1), but some identity knowledge of the referent should be kept in mind as well (condition S2). The latter would be some kind of presupposition for the communication. Since Peter has no way to know that politician Paderewski is the same as musician Paderewski, the presupposition now is musician Paderewski is not politician Paderewski. And when we consider the belief ascription sentences (7a) and (7b), in fact they should be read as:

(7c) Peter believes that Paderewski had musical talent and musician Paderewski is not politician Paderewski.

(7d) Peter believes that Paderewski had no musical talent and musician Paderewski is not politician Paderewski.

Notice that (7c) is not in conflict with (7d). There is nothing contradictory here.

Generally, when we consider the explanations of belief ascriptions of the form

(3) S believes that x is F.

we have an alternative reading according to our approach, that is, (3) could be read as

(3e) S believes that x is F and certain presupposition P.

Now we suggest an alternative way to explain the intensionality of belief context: the content of a speaker's belief that m_1 is F might be different from the content of her belief that m_2 is F, even if m_1 is m_2, where m_1 and m_2 are direct referential expressions.

(3f) S believes that m_1 is F.
(3g) S believes that m_2 is not F.

Since in some context (3f) and (3g) can be read as:

(3f1) S believes that x is F and $m_1 \neq m_2$.
(3g1) S believes that x is not F and $m_1 \neq m_2$.

Or

(3h1) S believes that x is F and presupposition P.
(3h2) S believes that x is not F and presupposition P.

where x is the referent of m_1 and m_2, since they are co-referential. Now the content of belief in (3h1) is $<x$ is F and presupposition $P>$, which is contradictory to $<$ not $<x$ is F and presupposition $P>>$, but is not always contradictory to $<x$ is not F and presupposition $P>$, which is content of belief in (3h2). Moreover, the content $<x$ is F and presupposition $P>$ is contradictory to $<x$ is not F and presupposition $P>$ if and only if the presupposition P expresses a truth, which means that the identity statement $m_1 = m_2$ is true and known by the speakers.

7 Conclusion

Neither direct reference theories nor Fregeanism is satisfactory in explaining belief ascription. We instead have attempted to construe how the semantic content of a sentence correlates with cognitive significance and with communication by expounding on the structure of *de re* communication. We claim that a successful referential communication requires two conditions (S1) and (S2). Then we may take the sense of a referring expression as the presupposition associated with it, which is included in conditions (S2), thus providing an explanation of belief ascription.

References

1. Almog, J., Perry, J., Wettstein, H., eds.: *Themes from Kaplan*. Oxford University Press (1989)
2. Bach, K.: *Thought and Reference*. Oxford University Press (1987)
3. Bezuidenhout, A.: The communication of de re thought. *No.* **31**–2 (1997) 197–225
4. Braun, D.: Understanding belief reports. *Philosophical Review* **107** (1998) 555–595
5. Crimmins, M., Perry, J.: *The Prince and Phone Booth: Reporting Puzzling Beliefs*. In the *Journal of Philosophy* **86** (1989) 685–711
6. Devitt, M.: Against direct reference. *Midwest Studies in Philosophy* **14** (1989) 206–240
7. Donnellan, K.: Proper Names and Identifying Descriptions. In D. Davidson and G. Harman, eds., *The Semantics of Natural Language*. Reidel, Dordrecht (1972)
8. Dummett, M.: *The Logical Basis of Metaphysics*. Harvard University Press (1991)
9. Evans, G.: *The Varieties of Reference*. Oxford University Press, Oxford (1982)
10. Heck, R.: The Sense of Communication. *Mind* **104** (1995) 79–106
11. Heck, R.: Do Demonstratives Have Senses? *Philosophical Print* **2** (2002) 1–33
12. Kaplan, David.: Demonstratives, In *Themes from Kaplan*. (1989) 481–563.
13. Kripke, S.: A Puzzle about Belief. In A. Margalit, ed., *Meaning and Use*. Reidel, Dordrecht (1979) 239–275
14. Kripke, S.: *Naming and Necessity*. Harvard University Press, Cambridge, MA (1980)

15. Loar, B.: The Semantics of singular terms. *Philosophical Studies* **30** (1976) 353–377
16. McDowell, J.: De Re Sense. *Philosophical Quarterly* **136** (1984) 283–294
17. Perry, J.: *The Problem of the Essential Indexical and Other Essays.* CSLI, Stanford, CA (2000)
18. Recanati, F.: *Direct Reference: From Language to Thought.* Blackwell, Oxford (1993)
19. Salmon, N.: *Frege's Puzzle.* MIT Press, Cambridge MA (1986)
20. Soames, S.: *Beyond Rigidity.* Oxford University Press, New York (2002)
21. Stalanker, R.: Reference and Necessity. In B. Hale and C. Wright, eds., *Blackwell Companion to Philosophy of Language.* (1997)
22. Stanley, J.: Modality and What is Said. *Philosophical Perspectives* **16** (2002) 321–344
23. Strawson, P.: *Individuals.* Routledge, London and New York (1959)
24. Thau, M.: *Consciousness and Cognition.* Oxford University Press, Oxford (2002)

Multiagent-Based Simulation in Biology
A Critical Analysis

Francesco Amigoni and Viola Schiaffonati

Dipartimento di Elettronica e Informazione, Politecnico di Milano, Milan, Italy
{amigoni,schiaffo}@elet.polimi.it

Summary. In this paper we critically analyze the use of multiagent systems for performing simulations of biological processes. From the one hand, the possibility of associating different elements of a biological process to independent computing entities, called agents, makes multiagent systems a powerful and flexible tool for simulation. From the other hand, the weak validation of the results obtained makes multiagent-based simulations hard to trust. We discuss these issues by referring to a specific example, the simulation of a signal transduction pathway.

1 Introduction

Computer simulations [1] are an important tool to investigate the properties of biological systems that are difficult to study in more traditional ways, for example with *in vivo* or *in vitro* experiments. Multiagent systems [2], which are composed of different interacting computing entities called agents, provide an interesting way to design and implement simulations of biological systems.

In this paper, we start from reviewing the state of the art relative to multiagent-based simulations of biological processes and then we critically analyze the resulting scenario. In particular, we concentrate on a specific example, the multiagent simulation of the MAPK pathway, one of the best documented signal transduction pathways. By referring to this example, we outline both the potentialities and the limitations of using multiagent systems to simulate biological processes. Among the potential advantages, the most prominent one is the possibility to define a new category of experiments, called *in virtuo* experiments, that are simulations that can be perturbed at run-time, modifying the entities involved. The most limiting aspect of the currently available multiagent biological simulators is the lack of any convincing validation of the results they produce.

The original contributions of this paper are relative to the critical analysis of a number of multiagent systems for biological simulation that have been taken from literature. The discussion of these systems will abstract away from many technological issues that are of little relevance for the purpose of this paper.

Before starting with the main content of the paper, we deem appropriate to discuss the relation between computer simulation and biology with an eye to the philosophy of science scenario. Computer simulations have been playing a central role in the field of biology from the very last years. However, when philosophers of science refer to simulation, physics is usually the only source of examples on which the analysis is based. Simulation is adopted in physics when no analytic method can be used for exploring the properties of the mathematical models representing phenomena or their portions. This amounts to say that, in order to be solved, some differential equations need to be transformed into difference equations, whose solution can be calculated by means of computers. Moreover, in physics, well-understood models or theories of the simulated phenomena are usually present. For this reason, the application of simulation procedures to derive accurate solutions for the equations is also a way to test the underlying model or theory. The following quotation shows how simulations, as they are intended in physics, can turn analytically intractable problems into computationally tractable ones:

> A computer simulation is any computer-implemented method for exploring the properties of mathematical models where analytical methods are unavailable. [3, p. 501]

Besides physics, however, other disciplines should be taken into account to picture the current status of science, also when investigating what a computer simulation is and how it works. In biology, for example, compact and elegant theories of the sort familiar in physics are rare. Explanations of phenomena are typically expressed by natural language narratives and are not always based on well-defined and complete paradigms [4]. Because of the complexity of biological systems and the lacking of satisfactory theories for their explanation, simulation in biology strongly participates in setting theoretical frameworks and contributes in a central way in constructing the theoretical knowledge. For this reason, the use of simulation in biology reveals something more than the use of simulation in physics. Computer simulation in biology is not just used for calculation or for re-modeling the original problem using a computer, as stated in [5]. Computer simulation is a dynamic process resulting from the execution of a computational model representing the behavior of a system. In this sense, a computer simulation provides access to the computational model by reproducing the system's behavior. According to this view, the behavior of complex systems, like those of biology, does not need a conclusive theory to be analyzed, but it can be examined as a phenomenon that *emerges* during computer simulations [6]. In the following of this paper, we discuss in more detail simulations in biology, when these simulations are performed exploiting multiagent systems. We explicitly note that in this paper we are mainly interested in multiagent-based simulations of biological processes rather than in the more general area of computer simulation of biological processes. This means, for example, that we do not consider simulations performed with high-performance computing technologies and simulations based on cellular automata.

This paper is structured as follows. The next section surveys multiagent systems and their applications to simulation in biology, in particular to simulation of the MAPK signal transduction pathway. Sections 3 and 4 critically analyze the pros and cons of using multiagent systems to simulate biological systems with specific reference to the simulation of the MAPK pathway. Finally, Section 5 concludes the paper.

2 Multiagent-Based Simulation in Biology

In this section, we briefly introduce multiagent systems and their roles in biology. Next, we describe in more detail how multiagent systems can be used to simulate a specific biological process.

2.1 Multiagent Systems and Biology

Multiagent systems are a powerful paradigm and technology developed at the intersection between artificial intelligence and distributed computing [2]. A multiagent system is composed of a number of interacting computing entities, called *agents*, that exhibit some degree of autonomy in their behavior. As a *paradigm*[1], multiagent systems enable the appropriate modeling of complex distributed systems [7]. As a *technology*, multiagent systems allow the effective implementation of distributed computing systems [8]. Usually, a system designed according to the multiagent paradigm is implemented using a multiagent technology, even if this is not mandatory.

Multiagent systems are becoming widely employed in biology, basically with two different roles. On the one hand, multiagent systems are used to support information gathering, processing, and integration [9]. For example, some multiagent systems have been proposed to perform automated annotation of genomes, collecting heterogeneous data from distinct locations [10, 11]. On the other hand, multiagent systems are used to simulate the behavior of biological systems [12]. For example, some multiagent systems have been proposed to simulate the behavior of *Escherichia coli* [13], the apoptosis of B-CD5 cells [14], and protein folding [15]. In this paper, we focus on this second role relative to simulation. In general, multiagent systems are adopted for simulation when the individual variability of a system's parts cannot be neglected and the whole behavior of the system results from the

---

[1] Here we are referring to a *programming paradigm*, that is an abstract view the programmer has of the execution of a computer program. Besides the multiagent paradigm, other well-known programming paradigms of computer science include the object-oriented and the procedural paradigms.

interaction of its components which can have different natures and structures. Hence, multiagent systems are suitable to simulate biological systems that can be decomposed in several independent but interacting entities, each one represented by an agent.

Since a comprehensive review of the state of the art on the simulation of biological processes by multiagent systems is outside the scope of this paper, in the following section we focus on a specific example, relative to the simulation of a signal transduction pathway.

2.2 Multiagent Systems Simulating the MAPK Pathway

Signal transduction pathways are cellular processes by which cells can detect, convert, and internally transmit information about the external environment. Signal transduction pathways are organized in sequences (cascades) of concurrent biochemical reactions that are activated by receptors. Basically, a signal transduction pathway operates in the following steps [16]:

(a) a signalling molecule arrives from outside the cell,
(b) a receptor on the surface of the cell interacts with the signalling molecule,
(c) the receptor interacts with intracellular pathway components, triggering a cascade of protein interactions within the cell,
(d) the signal arrives at destination and elicits a functional response (e.g., a gene transcription).

The MAPK (Mitogen-Activated Protein Kinase) pathway, which intervenes in different cellular functions, is one of the most understood signal transduction pathways [17, 18], even if some of its details are not yet firmly established. A number of multiagent systems have been developed to simulate the complexity and the concurrency of the MAPK pathway. We present some of these systems in order to provide a specific example of the efforts made in the last years in simulating biological processes with multiagent systems.

In the multiagent system presented in [19], every biochemical reaction is represented by an agent. An agent simulates the corresponding biochemical reaction in three steps: perception, decision, and action. During the first step the agent perceives from the environment the concentration of the elements involved in the reaction. Then, it calculates the reaction speed and the quantity of each reagent (decision step). Eventually, during the action step, the concentration of the reagents and of the products in the environment is updated. The parallel processing of the agents and their interactions "produce" a model of the MAPK pathway. The agents do not explicitly communicate with each other by exchanging messages. Their interactions are carried out only through (are mediated by) the environment.

In the multiagent system proposed in [20], the main components of the MAPK pathway are considered as agents. For example, agents are associated to receptors, non-catalytic proteins, and enzymes. The agents communicate through a shared data structure, called blackboard, that stores the intracellular signals the components exchange. An agent executes an action of its repertoire (e.g., a phosphorylation) when certain conditions are satisfied (e.g., when certain signals are stored in the blackboard). Also the spatial organization of the components (agents) is important in this system and it is modeled by organizing the blackboard in different abstraction levels and by separating agents in different groups.

Finally, in the multiagent system proposed in [21], the molecular species (e.g., Raf and Ras proteins, that are a serine/threonine-specific kinase and a regulatory guanosine triphosphate hydrolase, respectively) are considered as agents. The reactions are conducted by explicitly exchanging messages between agents. An agent a can start a reaction by sending a message to another agent b representing a potential reactant. In turn, agent b can participate in the reaction by responding to the message from agent a.

The three systems described above differ in what their agents represent. In the first case, agents represent biochemical reactions; in the second case, agents represent intracellular components; and in the third case, agents represent molecular species. This shows the flexibility of using the multiagent paradigm in modeling biological processes and in designing their simulators. Another kind of flexibility, provided by using multiagent technology to implement biological simulators, is that, in order to modify the model, it is sufficient only to add or remove agents. For example, adding a new biochemical reaction to the model proposed in [19] is as easy as adding a new agent (representing the new biochemical reaction) to the system. This kind of operations is usually well supported by the programming languages used to implement the multiagent systems. For example, the dynamic programming language (ORIS) used in [19] allows an agent to be created, destroyed, or modified during the simulation process. Hence, it is possible to disturb the cascade of the MAPK pathway during simulation and to observe the effects on the global behavior of the model. As another example of the flexibility of the multiagent technology, the system proposed in [21] is implemented using DECAF, a JAVA-based framework that allows high modularization and easy management of the agents composing the system.

3 Using Multiagent Systems for Simulation in Biology: The Pros

Multiagent systems can be effectively employed for simulating biological processes due to their flexibility relative both to their paradigm and to their technology. In this section we argue that a computer simulation can be regarded as an *experiment*. After having discussed the relationships between simulations and experiments, we claim that adopting multiagent systems as tools for simulation enables a new powerful type of experiment.

3.1 Simulation and Experimentation

Computer simulations present several interesting properties both at the applicative level and at the epistemological one. Probably the most significant one is that simulation may help scientists to investigate situations that cannot be explored by traditional experimental means. In this sense, simulation can be seen as a particular type of scientific experiment. The examples in the previous section offer an insight on how simulation plays the role of experiment. The MAPK pathway is investigated by simulating the behavior of its components and their interactions, thus supporting scientists in exploring situations that cannot be easily investigated with traditional experimental means. Moreover, the underlying theoretical model, representing the whole phenomenon, is shaped according to the results of simulation, as it happens when a theory in modeled in dependence of experiments. To claim that a computer simulation is a type of experiment means to consider simulation used as an *experimental tool* [22], a technique for conducting experiments on a computer. Basically, simulations can be exploited to test theoretical hypotheses or to provide additional knowledge about a phenomenon under investigation in a way very similar to traditional scientific experiments. But why using computer simulations as experiments instead of traditional experimental means? This question triggers a multifaced answer. The most common reason to use simulation is for predicting the behavior of complex systems. For example, simulation is employed for solving the three body problem, namely the problem of computing the mutual gravitational interaction of three masses, widely recognized as particularly difficult. Moreover, simulation is used to investigate systems otherwise inaccessible. For this reason, simulation can be seen as a *substitute* for an experiment impossible to make in reality, where impossibility can be either *theoretical* or *pragmatical* [23]. A theoretically impossible experiment is the analysis of counterfactual situations like, for example, to investigate the possibility of having the values of some fundamental constants (e.g., the charge of the electron) different from reality. A pragmatically impossible experiment is the study and manipulation of objects which we cannot access like, for example, the inner structure of a star.

When shifting from physics to biology the importance to consider theoretical scenarios which are not accessible by real experiments is enhanced. A computer simulation does not simply provide more speed in carrying out a large number of operations keeping track of many variables, but also it contributes with its results to set the theoretical framework, that in biology is usually weaker than in physics. In this sense, it is possible to say that computer simulation is a *new tool* for science, intermediate between theory and empirical methods [24]. It is a new tool since it allows *theoretical model experiments*, a sort of *thought experiments*, of a scope and richness far exceeding anything present before the introduction of computer simulation in science.

The experiments taking place as computer simulations are usually named *in silico* experiments. This name refers to the fact that these experiments are performed by running a computer program [25]. The concept of *in silico* experiment is an alternative that goes in parallel with the traditional experimental modes of empirical sciences like biology, namely *in vivo* and *in vitro* experimentations. The multiagent systems for simulating the MAPK pathway of Section 2.2 are examples of *in silico* experiments.

In the following, we summarize the different functions of *in silico* experiments with particular attention to the case of biology. The first one is to perform several accelerated experiments, that is to say that simulation is exploited to force the normal course of the events. This is possible when the model, or the group of models, on which the simulation is based has been already validated by real experiments (*in vivo* or *in vitro* experiments). As we have seen, this is a typical situation in physics where well-grounded theories can be usefully exploited in designing simulation tools and *in silico* experiments. The scenario of biology is quite different: complete and elegant theories similar to those of physics are rare. In this case, since the theoretical framework is incomplete, simulations are used to test hypotheses and to set the theoretical framework. This is the second function *in silico* experiments can cover. The multiagent systems described in Section 2.2 are examples of this second function: the MAPK pathway is modeled by embedding some of its elements in the agents and by letting agents interact with each other to reproduce the behavior of the MAPK pathway. Even if the MAPK pathway is one of the best documented signal transduction pathways, at the moment it does not exist any complete theory able to explain precisely all the elements of this process and their interactions. The results of the simulation process can provide new insights for the theoretical description of the MAPK pathway: the experimental results (obtained in simulation) can play a central role in shaping theoretical knowledge and in setting up a more complete description of the MAPK pathway.

3.2 From *In Silico* Experiments to *In Virtuo* Experiments

As already emphasized, the advantage of adopting multiagent systems for simulating biological processes lies in the flexibility offered by these tools, with respect both to the paradigm and to the technology. If a generic computer-based simulation can be regarded as a special type of experiment, a multiagent-based simulation can be regarded as an extremely flexible type of experiment. In [19], for example, in order to modify the model representing the MAPK pathway, it is sufficient only to add or remove an agent representing a biochemical reaction, without the need of changing the whole model[2]. Similarly, it is easy to change the models of [20] and [21], exploiting

---

[2] We stress that we consider computer simulation as the process resulting from the execution of a computational model that represents the behavior of a system.

the fact that multiagent systems are composed of independent agents that can be added and removed without affecting the other agents.

By specifically considering the example reported in [19], it is possible to observe that adding and removing agents at run-time allow to modify the structure of the experiment while it is being run. In other words, it is possible to disturb the cascade of the MAPK pathway during simulation and this is reflected on the behavior of the simulated biological process. In this way, the effects of changes on the global behavior of the model of the pathway can be immediately detected and observed. This high flexibility is supported by the multiagent technology, namely by the dynamic programming language used in the implementation, that allows an agent to be created, destroyed or modified during the simulation process.

These considerations lead to the concept of *in virtuo* experiment [26], which shows higher experimental potentialities than an *in silico* experiment. In *in silico* experimentations, simulations replace experiments and observations as source of data about the world. In *in virtuo* experimentations, which are special *in silico* experimentations, it is also possible to virtually manipulate the simulated system producing the data, even when this is not possible in reality. Note that, while it is a typical property of all computer simulations (and of *in silico* experiments) to easily change the values of the parameters characterizing simulations, it is specific of multiagent-based simulations to easily modify the structure of the experiment itself giving rise to *in virtuo* experiments. Hence, an *in virtuo* experiment is defined as an experiment taking place as a computer-based simulation, but with the possibility to disturb the model that is being run. This concretely means that it is possible to dynamically change the limit conditions and the constraints of the simulation (for example, by changing the values of some parameters) and also to add and remove elements during simulation. *In virtuo* experiments, similarly to *in silico* experiments, are necessary when it is difficult or impossible to make direct experiments for whatever reasons. *In virtuo* experiments, differently from *in silico* experiments, allow for testing at run-time the reactivity and adaptability of the model guiding the simulation [27]. We note that, although *in virtuo* experiments are in principle possible also without using agents, multiagent systems provide a particularly adequate technological support for their implementation.

4 Using Multiagent Systems for Simulation in Biology: The Cons

Besides the potential advantages of adopting multiagent systems as simulation tools, some problems can arise, in particular when considering *in virtuo* experiments. In this section we discuss one of these problems, namely why the process of validating simulation results is problematic and how a possible solution can be envisaged.

4.1 The Validation Problem

In virtuo experiments represent a new and powerful way of experimentation. Multiagent systems provide the technological support to realize the flexibility required by these experiments, namely they offer the possibility to intervene directly during the simulation process and to dynamically manipulate the structure of the simulation, even when this is impossible in reality. This allows to fully explore the model lying behind the simulation process, to investigate its properties, and to find out its limits. Besides these clear advantages, the main problem is the extent to what the results of an *in virtuo* experimentation can be trusted. This is the so called *validation problem* and is particularly urgent in the case of multiagent simulations in biology. The main reason for this urgency is that the credibility of such simulations does not derive from a governing theory that, in many cases, is lacking. Moreover, a direct comparison between the results obtained in simulation and data measured in reality is sometimes difficult, since experimental data are sparse.

To be more precise, the validation problem presents two aspects. The first aspect is the general problem of having a methodology for validating the results of simulations, which is typical of all scientific disciplines. For example, in biology, the definition of a more rigorous methodology to ground the trust in *in virtuo* experiments is central in order to adopt them with full awareness of their potential risks. The second aspect is specific to the scenario described in this paper and regards the problem of demonstrating that multiagent-based simulations in biology are more flexible and useful that other kinds of simulation techniques (as widely declared by their partisans). In this case it is necessary to prove the supposed superiority of the results obtained by using multiagent systems with those obtained with other techniques, like systems of differential equations or cellular automata. A more detailed comparison between different simulation methods is required; this means to develop quantitative comparisons to measure the performances of different techniques, to define what should be confronted, and which metrics to use.

In the literature on multiagent systems for biological simulation (including those relative to the MAPK pathway in Section 2.2) the validation problem is only peripherally tackled. As regarding the second aspect above, results from multiagent-based simulations are compared against other results obtained with more traditional simulation approaches, like for example systems of differential equations [19, 21], but not against real data. The only information shown is that, from a qualitative point of view, multiagent-based simulations are at least comparable to other approaches. Even if some researchers (see for example [19, 21]) recognize the importance to verify these results in a stronger way, the evaluations are currently still qualitative and not yet quantitative [14].

If this problem is basically a practical one that could be solved by introducing better testing techniques in the current scenario, the first aspect above (the general methodology for validation) requires a deeper theoretical insight.

The process of confronting results from simulation with experimental data can result in a very difficult task. For this reason a fully new methodology is required. The definition of this methodology is all the more urgent to fully show the potentialities of multiagent systems for simulation in biology. To move toward a possible solution, in the next section, we propose to consider a new conceptual framework in which to discuss the problem.

4.2 *In Virtuo* Modeling, *In Vitro* Validation: A Possible Solution?

Biologists often run simulations of systems about which data are very difficult to get from real experiments. In some cases simulations can completely substitute real experiments and observations. Some of these simulations, however, are successfully used in applications and their results are trusted (for example, those related to protein folding). Given that the models behind simulations are not always determined by a theory, what is the source of credibility of these results? Ideally, it could be possible to directly compare the results of predictions with real results and to observe their discrepancies. Unfortunately, a direct comparison is not always possible in practice, due to time or space constraints and to the fact that simulations are actually used when scientists want to learn about systems about which data are difficult to collect. Moreover, even in the case of a direct comparison between data from simulations and real data, the problem of the source of credibility still remains. Data need to be interpreted and this is not always a plain task.

A set of strategies is involved with validating simulations. The most used strategy is to consider how much a prediction is successful, namely the degree of its adherence to reality. Success, however, is an ambiguous and slippery concept which usually implies a commitment to truth. The best explanation for success lies in the fact that models are considered to *truthfully* describe parts of reality. We claim, rather, that the concept of *reliability* [28] in this technical context is more appropriate than that of truth. This concretely means that the source of credibility for simulation lies in the prior successes of the model building techniques adopted, without any commitment to how the simulation results reflect reality. Reliability can be characterized in terms of being able to produce outcomes fitting well with previously accepted data, observations, intuitions, without forgetting the capability of making successful predictions and of producing practical accomplishments.

The choice to think in terms of reliability instead of truth deals with the desire to stress the lack of guarantees implied in simulation procedures. Simulations are *fallible* in the same way that [29] has pointed out for experiments. Reliability provides, thus, a set of strategies that give good reasons to believe in the results of simulations, even if there is no guarantee that these results are conclusively and permanently correct. There can be situations in which these strategies are applied, but whose results are shown later to be incorrect. Computer simulation, when used with experimental purposes in mind, is fallible. No single strategy, neither a mix of them, guarantees the validity of an experimental result. This is not to be read as a poor success for science or as an evidence of its weakness; rather, it is a proof of its intrinsic complexity that should be always kept in mind to avoid the fallacies and ingenuities of thinking of simulations as conclusive experiments.

A possible way to overcome the validation problem, also in the field of multiagent-based simulations in biology, is to move toward a new methodology [30] in confirming results, which takes into account the warnings just stated when using computer simulation as an experimental procedure. This methodology should be double-sided. From the one side, *in virtuo modeling* should be the way to discover new scientific results by exploiting all the opportunities of a very flexible and dynamic methodology of experimentation. From the other side, *in vitro validation* should confirm the new discoveries in the most rigorous and precise way by a direct comparison with real data. Hence, if on the virtual side experimenters have the option of fully exploring all the potentialities of a model, on the validation side the purpose is to ground and test all possible results in accordance with real data, when possible, or with careful validation strategies, otherwise. We believe that a precise quantitative comparison between real data and results produced by multiagent-based and other simulation techniques in biology is the first step to tackle the validation problem.

5 Conclusions

In this paper we have provided a critical outlook on multiagent-based simulations in biology. The discussion moved from some multiagent systems proposed in literature to simulate the MAPK pathway in order to evaluate the current trends in the field. From the one hand, multiagent systems offer great flexibility in simulation of biological systems, both because the multiagent paradigm allows to associate agents to different entities of the domain and because the multiagent technology allows to alter the simulation conditions at run-time, promoting the new concept of *in virtuo* experiments. From the other hand, simulations conducted with multiagent systems are currently not validated against real data and this raises concerns about their credibility.

Future research should address, at the domain level, the development of new multiagent systems to simulate other biological processes. In this perspective, we think that interesting contributions could come from associating agents not only to the elements of a biological process, but also to alternative partial models of a biological process [31]. The composition of different alternative partial models can lead to a more accurate model of the biological process. Moreover, these multiagent simulators should be evaluated by biologists to assess their strengths, weaknesses, and roles. Finally, at the epistemological level, the new methodology of validation proposed here needs to be fully developed. This will represent the first step in the direction of a complete epistemology of computer simulation in which the ability of experimenting is just one of the properties a computer simulation exhibits.

References

1. Regev, A., Shapiro, E.: Cells as computation. *Nature* **419** (2002) 343
2. Wooldridge, M.: *An Introduction to Multiagent Systems*. John Wiley and Sons, Chichester, UK (2002)
3. Humphreys, P.: Computer simulations. In Fine, A., Forbes, M., Wessels, L., eds.: *Proceedings of the 1990 Biennal Meeting of the Philosophy of Science Association*, East Lansing, MI, USA (1991) 497–506
4. Andy, D., Brent, R.: Modelling cellular behaviour. *Nature* **409** (2001) 391–395
5. Hughes, R.I.G.: The Ising model, computer simulation, and universal physics. In Morgan, M., Morrison, M., eds.: *Models as Mediators*. Cambridge University Press, Cambridge, UK (1999) 97–145
6. Bedau, M.: Can unrealistic computer models illuminate theoretical biology? In: *Proceedings of the Genetic and Evolutionary Computation Conference Workshop Program*. (1999) 20–23
7. Jennings, N.: An agent-based approach for building complex software systems. *Communications of the ACM* **44** (2001) 35–41
8. Luck, M., Ashri, R., D'Inverno, M.: *Agent-Based Software Development*. Artech House, Boston MA, USA (2004)
9. Karasavvas, K.A., Baldock, R., Burger, A.: Bioinformatics integration and agent technology. *Journal of Biomedical Informatics* **37** (2004) 205–219
10. Bazzan, A.L.C., Duarte, R., Pitinga, A.N., Schroeder, L.F., Souto, F.D.A.: ATUCG - An agent-based environment for automatic annotation of genomes. *International Journal of Cooperative Information Systems* **12** (2003) 241–273
11. Decker, K., Khan, S., Schmidt, C., Situ, G., Makkena, R., Michaud, D.: Biomas: A multi-agent system for genomic annotation. *International Journal of Cooperative Information Systems* **11** (2002) 265–292
12. Cannata, N., Corradini, F., Merelli, E., Omicini, A., Ricci, A.: An agent-oriented conceptual framework for systems biology. In Merelli, E., González Perez, P.P., Omicini, A., eds.: *Transactions on Computational Systems Biology*. Volume III of LNBI. Springer (2005) 105–122
13. Jonker, C.M., Snoep, J.L., Treur, J., Westerhoff, H.V., Wijngaards, E.C.A.: BDI-modelling of intracellular dynamics. In: *Proceedings of the First International Workshop on Bioinformatics and Multi-Agent Systems*. (2002) 15–23
14. Ballet, P., Pers, J.O., Rodin, V., Tisseau, J.: A multiagent system to simulate an apoptosis model of B-CD5 cells. In: *Proceedings of the IEEE International Conference on Systems, Man, and Cybernetics*. (1998) 3799–3803
15. Bortolussi, L., Dovier, A., Fogolari, F.: Multi-agent simulation of protein folding. In: *Proceedings of the First International Workshop on Multi-Agent Systems for Medicine, Computational Biology, and Bioinformatics*. (2005)
16. Lodish, H., Berk, A., Zipursky, L., Matsudaira, P., Baltimore, D., Darnell, J.: *Molecular Cell Biology*. W.H. Freeman and Company, New York, NY, USA (2000)

17. Kolch, W., Calderc, M., Gilbert, D.: When kinases meet mathematics: The systems biology of MAPK signalling. *FEBS Letters* **579** (2005) 1891–1895
18. Orton, R.J., Sturm, O.E., Vyshermirsky, V., Calder, M., Gilbert, D.R., Kolch, W.: Computational modelling of the receptor-tyrosine-kinase-activated MAPK pathway. *Biochemical Journal* **392** (2005) 249–261
19. Querrec, G., Rodin, V., Abgrall, J.F., Kerdelo, S., Tisseau, J.: Uses of multi-agents systems for simulation of mapk pathway. In: *Proceedings of the Third IEEE Symposium on Bioinformatics and Bioengineering (BIBE03)*. (2003) 421–425
20. Gonzáles, P.P., Cárdenas, M., Camacho, D., Franyuti, A., Rosas, O., Lagúnez-Otero, J.: Cellulat: An agent-based intracellular signalling model. *BioSystems* **68** (2003) 171–185
21. Khan, S., Makkena, R., McGeary, F., Decker, K., Gillis, W., Schmidt, C.: A multiagent system for the quantitative simulation of biological networks. In: *Proceedings of the Second International Joint Conference on Autonomous Agents and Multiagent Systems (AAMAS2003)*. (2003) 385–392
22. Naylor, T.H.: *Computer Simulation Techniques*. John Wiley, New York, NY, USA (1966)
23. Hartmann, S.: The world as a process: Simulations in the natural and social sciences. In Hegselmann, R., Mueller, U., Troitzsch, K.G., eds.: *Simulation and Modeling in the Social Sciences from the Philosophy of Science Point of View. Theory and Decision Library*. Kluwer, Dordrecht, The Netherlands (1996) 77–100
24. Rohrlich, F.: Computer simulation in the physical sciences. In Fine, A., Forbes, M., Wessels, L., eds.: *Proceedings of the 1990 Biennal Meeting of the Philosophy of Science Association*, East Lansing, MI, USA (1991) 145–163
25. Zhao, J., Stevens, R.D., Wroe, C.J., Greenwood, M., Goble, C.A.: The origin and history of in silico experiments. In: *Proceedings of the UK e-Science All Hands Meeting*. (2006)
26. Desmeulles, G., Querrec, J., Redou, P., Kerdlo, S., Misery, L., Rodin, V., Tisseau, J.: The virtual reality applied to biology understanding: The in virtuo experimentation. *Expert Systems with Applications* **30** (2006) 82–92
27. Tisseau, J.: *Virtual reality, in virtuo autonomy*. Accreditation to Direct Research, University of Rennes 1 (2001)
28. Winsberg, E.: Models of success vs. success of models: Reliability without truth. *Synthese* **152** (2006) 1–19
29. Hacking, I.: *Representing and Intervening*. Cambridge University Press, Cambridge, MA, USA (1983)
30. Smallwood, R.H., Holcombe, W.M.L., Wlaker, D.: Development and validation of computational models of cellular interaction. *Journal of Molecular Histology* **35** (2004) 659–665
31. Amigoni, F., Schiaffonati, V.: A multiagent approach to modelling complex phenomena. *Foundations of Science* (forthcoming)

Mathematics through Diagrams: Microscopes in Non-Standard and Smooth Analysis

Riccardo Dossena[1] and Lorenzo Magnani[2]

[1] Department of Philosophy, University of Pavia, Pavia, Italy
   riccardo.dossena@istruzione.it
[2] Department of Philosophy, University of Pavia, Pavia, Italy and Sun Yat-sen University, Guangzhou, P.R. China
   lmagnani@unipv.it

Summary. Diagrams play an important role in the construction of mathematical concepts, mainly in (some) "limit" situations, like in the case of the mental representation of geometric tangent lines. They have many properties and can be viewed as particular *epistemic mediators*. Further, they are able to provide a better understanding of some mathematical concepts because they can be manipulated. In this paper we investigate how a particular kind of diagram (microscope) can serve to obtain two different and interesting visual representations of how a real function appears in small neighborhoods of its points.

1 The Role of Mathematical Diagrams

Sometimes, mathematical constructions (in particular, geometrical constructions) present curious or "limit" situations. For example, the concept of tangent line in terms of the limit of the differential quotient $\Delta x/\Delta y$ needs further "dynamic" explanation through diagrams: a graph of a function in which a secant line "becomes" a tangent line. This is the classical procedure adopted by a teacher to explain the concept of derivative to his students.

When manipulative aspects of external models prevail, like in the case of manipulating diagrams on the blackboard, we face what we call *manipulative abduction* (or action-based abduction) [1]. Manipulative abduction happens when we are thinking *through* doing and not only, in a pragmatic sense, about doing: diagrams offer various contingent ways of epistemic acting, like looking from different perspectives, comparing subsequent appearances, discarding, choosing, re-ordering and evaluating. Moreover, they present some features, typical of the so-called abductive *epistemic mediators*: simplification of the task and the capacity to obtain visual information otherwise unavailable.

Diagrams play an important role because they can be manipulated. In mathematics diagrams play various roles in a typical abductive way. Two of them are central:

- they provide an intuitive and mathematical *explanation* able to enhance the understanding of concepts that are difficult to grasp or that appear obscure and/or epistemologically unjustified. We will present some new diagrams in the following sections (microscopes within microscopes), which provide new mental representations of the concept of the tangent line in infinitesimally small regions.
- they help *create* new previously unknown concepts, as illustrated in the case of the discovery of non-Euclidean geometry in [2].

In the construction of mathematical concepts many external representations are exploited, both in terms of diagrams and symbols. We are interested in our research in diagrams which play an *optical* role – microscopes (that look at the infinitesimally small details), a *mirror* role (to externalize rough mental models), and an *unveiling* role (to help create new and interesting mathematical concepts, theories, and structures)[3].

Optical diagrams play a fundamental explanatory (and didactic) role in removing obstacles and obscurities and in enhancing mathematical knowledge of critical situations. They facilitate new internal representations and new symbolic-propositional achievements. In the example studied in the following section focusing on the calculus, the extraordinary role of optical diagrams in the interplay between standard/non-standard analysis is emphasized. Some of them could also play an unveiling role, providing new light on mathematical structures: it can be hypothesized that these diagrams can lead to further interesting creative results. The optical and unveiling diagrammatic representation of mathematical structures activates *direct perceptual operations* (for example identifying how a real function appears in its points and/or to infinity; how to really reach its limits).

In this paper, we will apply particular diagrams and their properties mentioned above to a mathematical situation, which concerns the tangent line of a real function. This notion is traditionally based on the classic standard ε, δ concept of limit. However, there are at least two valid rigorous alternative theories to this method, both based on infinitesimal numbers (even if they use two different concepts of them). First, the non-standard analysis invented by Abraham Robinson [4]; second, the smooth infinitesimal analysis constructed on the work of Kock-Lawvere [5].

The usual limit concept is intrinsically difficult to represent and not immediately assimilable by a beginner (see [3]). On the contrary, the infinitesimal methods help to avoid some troubles by introducing a pictorial device that allows the visualization of small details in the graph of a curve $y = f(x)$. We will use a "static" visualization of the concept of tangent line, because these infinitesimal methods do not involve the "dynamic" classical concept of limit. The diagrams we will present were invented by Stroyan [6], and improved by Tall [7, 8]: our intention is to continue to improve Tall's work by applying them

---

[3] The epistemic and cognitive role of mirror and unveiling diagrams in the discovery of non-Euclidean geometry is illustrated in [2].

to many other different situations, in order to focus upon their explanatory and heuristic role.

2 Non-Standard Analysis

2.1 The Hyperreal Line and Microscopes

In this section we will work on the hyperreal number system $\mathbb{R}^*$ and will assume the non-standard analysis given by Abraham Robinson [4][4]. In the following pages, we will briefly describe some of the properties of the hyperreal numbers.

$\mathbb{R}^*$ is a non-archimedean ordered field in which the standard real line $\mathbb{R}$ is embedded. Consequently, $\mathbb{R}^*$ contains infinite and infinitesimal numbers defined as follows:

Definition 1. *Let $s \in \mathbb{R}^*$. Then*

(i) *s is infinitesimal if $|s| < 1/n$ for every natural number n;*
(ii) *s is infinite if $|s| > n$ for every natural number n;*
(iii) *s is finite if it is not infinite.*

From this definition, 0 is an infinitesimal number, and by the archimedean axiom, it is the unique infinitesimal real number. However, for our purposes, it is important to observe that $\mathbb{R}^*$ contains at least (in fact, infinitely many) one infinitesimal number unequal to 0. Its reciprocal is an infinite number. For a nice presentation of a model of the hyperreal line see [11].

Definition 2. *Let $x, y \in \mathbb{R}^*$.*

(i) *x and y are infinitely close if $x - y$ is infinitesimal. In this case, we write $x \simeq y$.*
(ii) *x and y are finitely close if $x - y$ is finite. In this case, we write $x \sim y$.*

It is easy to check that $\simeq$ is an equivalence relation. The equivalence classes are called *monads*.

Theorem 1. *Every finite hyperreal number is infinitely close to a unique real number.*

If x is a finite hyperreal number, the unique real number infinitely close to x is called the *standard part* of x and denoted by $\mathrm{st}(x)$. In particular, a number ε is infinitesimal if and only if $\varepsilon \simeq 0$ (and then $\mathrm{st}(\varepsilon) = 0$).

Theorem 2.

(i) *Sums, differences and products of infinitesimals are infinitesimal.*
(ii) *The product of an infinitesimal and a finite number is infinitesimal.*

---
[4] For an easy introduction to non-standard calculus see Keisler [9, 10].

Theorem 3 (Standard Part Map Properties).

(i) $\operatorname{st}(x \pm y) = \operatorname{st}(x) \pm \operatorname{st}(y)$,
(ii) $\operatorname{st}(xy) = \operatorname{st}(x)\operatorname{st}(y)$,
(iii) $\operatorname{st}(x/y) = \operatorname{st}(x)/\operatorname{st}(y)$ if $\operatorname{st}(y) \neq 0$,
(iv) $\operatorname{st}(x) \leq \operatorname{st}(y)$ if $x \leq y$.

The power of the hyperreal world is the possibility to extend every function or relation defined on $\mathbb{R}$ to a function or relation defined on $\mathbb{R}^*$, in a very natural way. More precisely, it is possible to transfer every first order logic sentence referring to real numbers to the same sentence referring to hyperreal numbers (for details, see [11]). Consequently, if a real-valued function is defined by a formula (or a system of formulas), its extension can be obtained by applying the same formula to the hyperreal system. For example, the *natural extension* of $f(x) = \sqrt{x}, \{x \in \mathbb{R} : x \geq 0\}$ is the hyperreal function $f^*(x) = \sqrt{x}, \{x \in \mathbb{R}^* : x \geq 0\}$ (in the rest of this paper, we will omit to write the asterisk for f: the context will clarify any possible confusion). This makes the definition of derivative possible in the following way.

Definition 3. *Let f be a real-valued function defined in a neighborhood of x. The derivative of f in x is S if*

$$S = \operatorname{st}\left(\frac{f(x+\Delta x) - f(x)}{\Delta x}\right)$$

for every non-zero infinitesimal number Δx.

If we denote the quantity $f(x + \Delta x) - f(x)$ as Δy, the derivative is the real number infinitely close to the incremental ratio $\Delta y/\Delta x$ when Δx is a non-zero infinitesimal. The following theorem is useful.

Theorem 4 (Taylor's Formula). *Let f be a real-valued function defined and differentiable in a neighborhood of x, and Δx a non-zero infinitesimal. Then*

$$f(x + \Delta x) = f'(x)\Delta x + f(x) + \varepsilon \Delta x$$

for some infinitesimal ε.

If f is differentiable n times in a neighborhood of x. Then

$$f(x + \Delta x) = \sum_{k=0}^{n} \frac{f^{(k)}(x)}{k!}\Delta x^k + \varepsilon \Delta x^n$$

for some infinitesimal ε (where $f^{(k)}$ denotes the k-th derivative).

For the proof of this theorem, see [12] or [13]. As usual, the quantity $f'(x)\Delta x$ (depending on both x and Δx) is denoted by dy and called the *differential of f in x*.

Given two infinitesimals ε, δ, we say that ε is of *higher order* than δ, *same order* as δ, or *lower order* than δ if ε/δ is, respectively, infinitesimal, finite but not infinitesimal or infinite. It follows from this definition that, if ε is of higher order than δ, ε is an infinitesimal "smaller" than δ. Note that the difference between Δy and dy is an infinitesimal of higher order than Δx.

We report that each theorem of elementary calculus has its non-standard version. Our aim is to present a way of visualizing the hyperreal structure and functions, in order to make a good "mental" representation possible and help the intuition of the entities of the calculus. In the present and in the following section we will explain this method and the classification proposed by Tall (see [7] and [8]). In the subsequent sections, we will introduce new types of diagrams called microscopes "within" microscopes. Then, we will explain how to *discover* a property of a real function through these diagrams.

By zooming the difference between the numbers a and $a+\varepsilon$ (where $a \in \mathbb{R}$ and ε is a positive infinitesimal), we can use a very natural way, similar to the case of real numbers. We introduce the map $\mu : \mathbb{R}^* \to \mathbb{R}^*$ given by

$$\mu(x) = \frac{x-a}{\varepsilon}.$$

Thus $\mu(a) = 0$ and $\mu(a+\varepsilon) = 1$, that is, μ maps a and $a+\varepsilon$, two infinitely close points, onto clearly distinct points 0 and 1. We may also identify, through μ, a point a with its corresponding $\mu(a)$ (see Fig. 1).

In general, for all $\alpha, \delta \in \mathbb{R}^*$, the function $\mu : \mathbb{R}^* \to \mathbb{R}^*$ given by

$$\mu(x) = \frac{x-\alpha}{\delta} \quad (\delta \neq 0)$$

is called *δ-lens pointed at α*. What can we see through a lens? What kind of details can it reveal? We define *field of view* of μ the set of $x \in \mathbb{R}^*$ such that $\mu(x)$ is finite. Given a δ-lens μ, proceeding by taking the standard part of μ, we obtain a function from the field of view in $\mathbb{R}$, called the *optical δ-lens pointed in α*. The optical lenses are actually what we need to visualize infinitesimal quantities. In fact, our eyes are able to clearly distinguish only

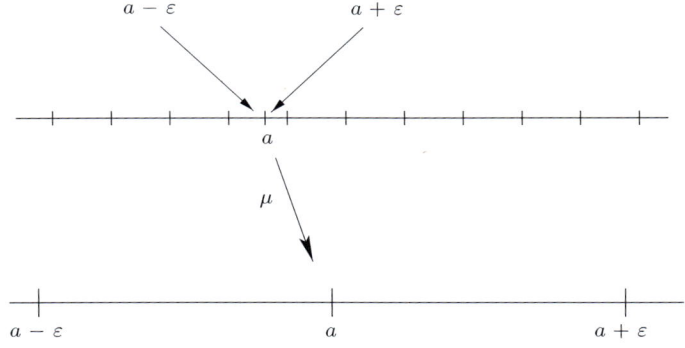

Fig. 1. The hyperreal line and the map μ.

images on the real plane $\mathbb{R}^2$. As such, the optical δ-lens translate on the $\mathbb{R}^2$ plane, in favor of our eyes, everything that differs from α *in the same order as* δ. Higher order details are "too small" to see and lower order details are "too far" to capture within the field of view. Two points in the field of view that differ by a quantity of higher order than δ appear the same through the optical δ-lens.

This method also works in two coordinates (and, in general, in n coordinates) by the application of a lens to every coordinate. The map

$$\mu : \mathbb{R}^{*2} \to \mathbb{R}^{*2}, \quad \mu(x,y) = \left(\frac{x-\alpha}{\delta}, \frac{y-\beta}{\delta} \right)$$

is called δ-*lens pointed in* (α, β). By considering the standard parts of every coordinate, we obtain an optical δ-lens in two dimensions, defined from the field of view of μ in $\mathbb{R}^2$. If δ is infinitesimal, the lens is called a *microscope*. Through an optical microscope, a differentiable function looks like a straight line, as we will see in the next section.

2.2 Microscopes and Differentiable Functions

Now we can easily generalize Tall's example [7] and [8] about the role of microscopes. An infinitesimal increment Δx of a differentiable function f from its point x can be written as follows

$$f(x + \Delta x) = f'(x)\Delta x + f(x) + \varepsilon \Delta x \tag{1}$$

where ε is infinitesimal. Thus, we can fix $(a, f(a))$ on the graph of f and point on it an optical Δx-lens to magnify infinitesimal details that are too small to see with the naked eye. We have

$$\mu(x,y) = \left(\frac{x-a}{\Delta x}, \frac{y-f(a)}{\Delta x} \right).$$

An infinitely close point $(a + \lambda, f(a + \lambda))$, when viewed through μ, becomes

$$\mu(a + \lambda, f(a + \lambda)) = \left(\frac{\lambda}{\Delta x}, \frac{f'(a)\lambda + \lambda \varepsilon}{\Delta x} \right).$$

Suppose that λ is of the same order as Δx, i.e. $\lambda/\Delta x$ is finite. This means that $\lambda \varepsilon / \Delta x$ is infinitesimal. By taking the standard parts, we have

$$\left(\operatorname{st}\left(\frac{\lambda}{\Delta x}\right), \operatorname{st}\left(\frac{f'(a)\lambda}{\Delta x} + \frac{\lambda \varepsilon}{\Delta x}\right) \right) = \left(\operatorname{st}\left(\frac{\lambda}{\Delta x}\right), f'(a) \operatorname{st}\left(\frac{\lambda}{\Delta x}\right) \right).$$

If a is fixed, putting $\operatorname{st}(\lambda/\Delta x) = t$, we see that the points on the graph in the field of view are mapped on the straight line $(t, f'(a)t)$, where t varies (see Figure 2). Note that the slope of the line is, in effect, the derivative of f

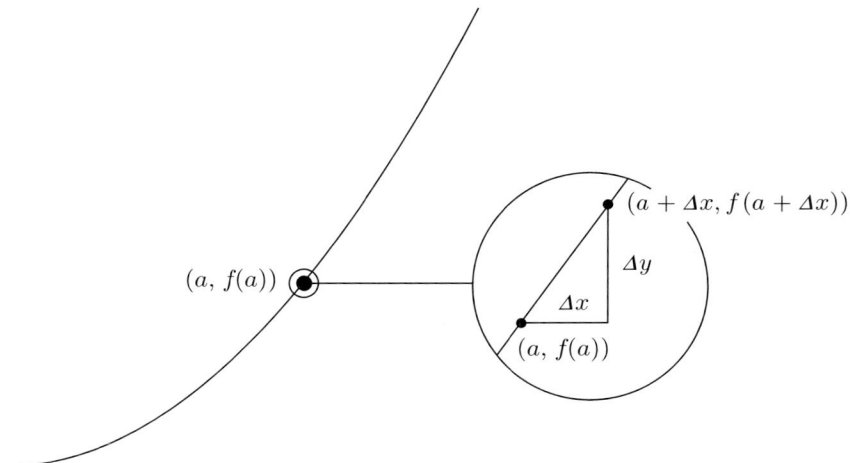

Fig. 2. A graph of a differentiable function through an optical Δx-lens.

at point a and the function is really indistinguishable from its tangent in an infinitesimal neighborhood of a.

In the following sections we will describe some interesting new mathematical situations in which such lenses can be used to construct a suitable mental representation.

2.3 Microscopes "within" Microscopes

This type of diagram was originally suggested and used by Keisler (see [9] and [10]), but not formalized by the construction of optical lenses.

Let f be a real function with continuous second derivative ($f \in C^2$). If we magnify an infinitesimal neighborhood by a more powerful tool than an optical Δx-lens, we can see other interesting properties of the curve. This is what we call a microscope "within" a microscope pointed at $(a+\Delta x, f(a+\Delta x))$ in the non-optical Δx-lens (because the optical lenses lose all infinitesimal detail). By an optical Δx-lens pointed in $(a, f(a))$, both the curve $y = f(x)$ and the tangent $y = f'(a)(x - a) + f(a)$ are mapped on the line $(t, f'(a)t)$, where $t = \mathrm{st}(\lambda/\Delta x)$ and λ is an infinitesimal of the same order as Δx. Now we can put $\lambda = \Delta x$ and point a Δx^2-lens in $(a+\Delta x, f(a+\Delta x))$. In order to visualize more details, we need to have more information about the function: our idea is to use Taylor's non-standard second order formula for f, i.e.

$$f(a + \Delta x) = f(a) + f'(a)\Delta x + \frac{1}{2}f''(a)\Delta x^2 + \varepsilon_1 \Delta x^2 \qquad (2)$$

where ε_1 is infinitesimal.

Thus the Δx^2-lens maps as follows

$$(x, y) \mapsto \left(\frac{x - (a + \Delta x)}{\Delta x^2}, \frac{y - f(a + \Delta x)}{\Delta x^2} \right)$$

and the point $(a + \Delta x, f(a + \Delta x))$ is mapped onto $(0, 0)$. Let λ be an infinitesimal of the same order as Δx^2. Taylor's second order formula gives

$$f(a + \Delta x + \lambda) = f(a) + f'(a)(\Delta x + \lambda) + \frac{1}{2} f''(a)(\Delta x + \lambda)^2 + \varepsilon_2 (\Delta x + \lambda)^2.$$

Therefore, we have

$$(a + \Delta x + \lambda, f(a + \Delta x + \lambda)) \mapsto \left(\frac{\lambda}{\Delta x^2}, \frac{f(a + \Delta x + \lambda) - f(a + \Delta x)}{\Delta x^2} \right) =$$
$$\left(\frac{\lambda}{\Delta x^2}, \frac{f'(a)\lambda + \frac{1}{2} f''(a)\lambda^2 + f''(a)\Delta x \lambda + \varepsilon_2 \Delta x^2 + \varepsilon_2 \lambda^2 + 2\varepsilon_2 \Delta x \lambda - \varepsilon_1 \Delta x^2}{\Delta x^2} \right)$$

and by taking the standard parts

$$\left(\mathrm{st} \left(\frac{\lambda}{\Delta x^2} \right), f'(a) \, \mathrm{st} \left(\frac{\lambda}{\Delta x^2} \right) \right)$$

as the other terms are all infinitesimals.

The point $(a + \Delta x + \lambda, f'(a)(\Delta x + \lambda) + f(a))$ on the graph of the tangent line is mapped on the point

$$\left(\frac{\lambda}{\Delta x^2}, \frac{f'(a)(\Delta x + \lambda) - f'(a)\Delta x - \frac{1}{2} f''(a)\Delta x^2 - \varepsilon_1 \Delta x^2}{\Delta x^2} \right) =$$
$$\left(\frac{\lambda}{\Delta x^2}, \frac{\lambda f'(a) - \frac{1}{2} f''(a)\Delta x^2 - \varepsilon_1 \Delta x^2}{\Delta x^2} \right) = \left(\frac{\lambda}{\Delta x^2}, f'(a) \frac{\lambda}{\Delta x^2} - \frac{1}{2} f''(a) - \varepsilon_1 \right)$$

and then the optical lens gives

$$\left(\mathrm{st} \left(\frac{\lambda}{\Delta x^2} \right), f'(a) \, \mathrm{st} \left(\frac{\lambda}{\Delta x^2} \right) - \frac{1}{2} f''(a) \right).$$

This suggests nice, new, (and mathematically justified) mental representations of the concept of tangent line: through the optical Δx^2-lens, the tangent line can be seen as the line $(t, f'(a)t - \frac{1}{2} f''(a))$ which means that the graph of the function and the graph of the tangent are distinct, straight, and parallel lines in a Δx^2-neighborhood of $(a + \Delta x, f(a + \Delta x))$. The fact that one line is either below or above the other, depends on the sign of $f''(a)$, in accordance with the standard real theory: if $f''(x)$ is positive (or negative) in a neighborhood, then f is convex (or concave) here and the tangent line is below (or above) the graph of the function (see Fig. 3).

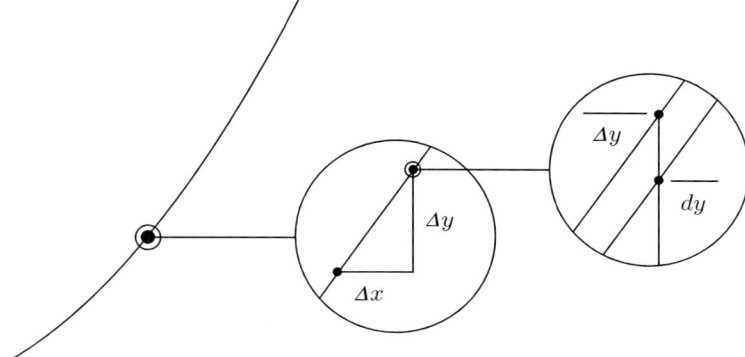

Fig. 3. A microscope "within" a microscope.

Note that from the equation
$$f(x + \Delta x) - f(x) = f'(x)\Delta x + \varepsilon \Delta x$$
that we can write as follows
$$\Delta y = dy + \varepsilon \Delta x$$
we see that the difference between the tangent and the curve at point $a + \Delta x$ has to be exactly $\varepsilon \Delta x$. However, through the second microscope we saw that the difference is $(1/2)f''(a)$. In fact, if we consider the distance $\varepsilon \Delta x$ through the optical Δx^2-lens, we obtain, by comparing the two equations (1) and (2)
$$\varepsilon \Delta x = \frac{1}{2}f''(a)\Delta x^2 + \varepsilon_1 \Delta x^2.$$
Thus
$$\text{st}\left(\frac{\varepsilon \Delta x}{\Delta x^2}\right) = \text{st}\left(\frac{\frac{1}{2}f''(a)\Delta x^2 + \varepsilon_1 \Delta x^2}{\Delta x^2}\right) = \frac{1}{2}f''(a)$$
is the expected result (see Figure 4).

2.4 A Cognitive Application of Microscopes within Microscopes

In this section we will show how a diagram easily allows the construction of a mathematical concept.

We saw that through a microscope within a microscope the curve and its tangent are respectively
$$y(t) = f'(a)t \quad \text{and} \quad y(t) = f'(a)t - \frac{1}{2}f''(a).$$

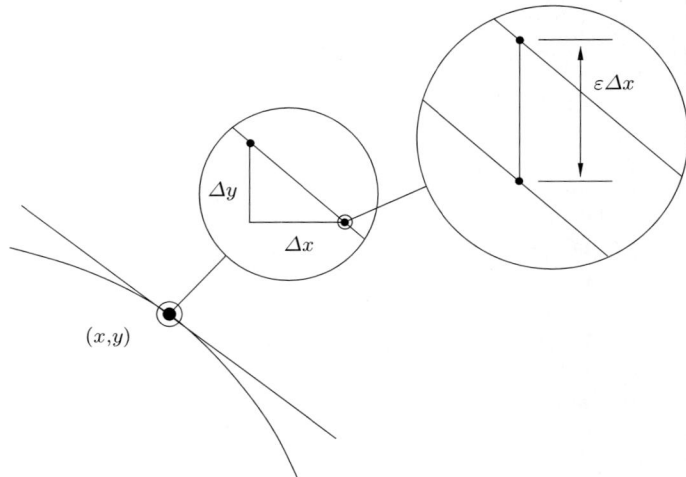

Fig. 4. The difference $\varepsilon \Delta x$ between the curve and the tangent line.

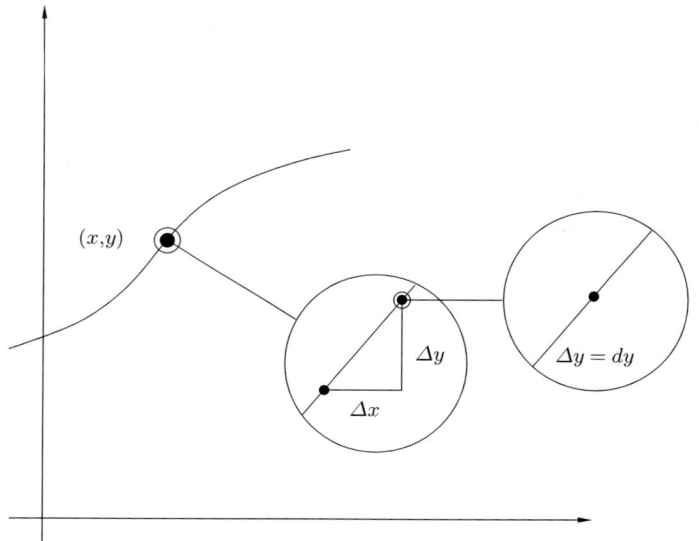

Fig. 5. An inflection point through a microscope within microscope.

Then, what happens when $f \in C^2$ is such that $f''(a) = 0$, for example when a is an inflection point for f? In this case the second microscope would still show the tangent line indistinguishable from the curve (see Figure 5). What does this mean? We can simply deduce that at an inflection point a curve that is twice differentiable has a particular behavior: here it is very slightly curved and much more similar to a straight line (its tangent). An expert mathematician would say that it has a small *curvature*. In fact, the curvature of a function at a point t of its domain is the quantity defined by

$$\frac{|f''(t)|}{(1+(f')^2(t))^{3/2}}$$

and it is a value of how much the curve locally differs from the tangent line. For example, a straight line has null curvature and a circle has constant curvature.

At an inflection point, a function $f \in C^2$ has curvature equal to 0. In other words, at this point the graph is much more than simply indistinguishable from its tangent, it has a more marked "straight local trend". In order to discover this property in standard calculus, the concept of curvature is necessary. On the contrary, the simpler idea of a microscope within a microscope allows the discovery of the same property immediately, easily and without the concept of curvature.

3 Smooth Infinitesimal Analysis

3.1 Introduction

Another traditional concept of infinitesimal quantity is the one corresponding to a "nilpotent" element, that is, a number so small that it has a null power. Smooth infinitesimal analysis is a good approach to the calculus based on rigorous nilpotent elements. This method, different from the non-standard one, came about due to the works of Lawvere and Kock on category theory (see [5]). As we did in the case of non-standard analysis, we will explain it, starting from some informal and intuitive ideas. However, as with every rigorous theory, the consistency of the smooth infinitesimal analysis is provided by the construction of several models of it, as shown in [14] and [15].

The main intuitive idea is that dx is so small in order to satisfy the following conditions

$$dx \neq 0 \quad \text{and} \quad dx \cdot dx = dx^2 = 0.$$

Of course, it is impossible to find such a number in both real or hyperreal systems. Anyway, try thinking the number 0.0000001. It is certainly small, but not equal to zero. Might its square (the number 0.00000000000001) be considered "the same" as zero? In particular practical situations, it would not be so absurd to consider 0.00000000000001 as zero. Let us try doing so in the following argumentation.

Our aim is to develop a method in order to find the slope of the tangent of a curve $y = f(x)$. Let $f(x) = x^2$ and let x and $x+dx$ be two points on the graph with this intuitive "infinitesimal" distance dx. Putting $dy = f(x+dx) - f(x)$, we obtain

$$dy = (x+dx)^2 - x^2 = x^2 + 2xdx + dx^2 - x^2 = 2xdx + \underbrace{dx^2}_{=0} = 2xdx$$

and therefore
$$\text{slope in } x = \frac{dy}{dx} = \frac{2x\,dx}{dx} = 2x$$
the correct result. In the following pages, we will go through the suggestive presentation proposed by Bell [14], which allows reasoning to be carried out in a rigorous way maintaining the same idea.

Let $\mathbb{S}$ be a model of the smooth infinitesimal analysis, containing all the geometric entities, the real line, Euclidean spaces, maps and transformations between them. We will refer to it as a *smooth world*. In order to assure the consistency of our theory with nilpotent infinitesimals, we are compelled to adopt a different logical perspective: we have to reject the classical *law of excluded middle* ($A \vee \neg A$) and its logical equivalent *law of double negation* ($\neg\neg A \to A$). This kind of logic is called *constructive* or *intuitionistic logic*, and, for practical purposes, it can be identified with the Classical Logic without the Aristotelian law of *tertium non datur* (excluded middle – classical logic may be thought of as a particular case of intuitionistic logic in which that law is postulated)[5].

Constructive logic allows us to make the presence of infinitesimals possible in the following way. If we call two points x, y on the real line *distinct* when $x \neq y$ (i.e. $\neg x = y$), and *indistinguishable* when $\neg x \neq y$, then two indistinguishable points will *not* necessarily be equal, as the excluded-middle law would impose. Therefore, if we define a point as indistinguishable from 0 *infinitesimal*, and the "infinitesimal neighborhood of 0" as the set of all points indistinguishable from 0, it is not a contradiction to affirm that it does not reduce to $\{0\}$.

At the same time, we cannot assert the existence of an infinitesimal which is $\neq 0$, either because this would imply that such an element would possess the property of being both distinguishable and indistinguishable from 0, which is clearly impossible (also in intuitionistic logic we have to assume the law of non-contradiction). Therefore, we cannot think about our new infinitesimals in an *actual sense* like the non-standard ones (i.e., as distinct elements of the hyperreal line), but only in this so-called *potential sense*[6]. Nevertheless, this kind of existence is enough to develop our Smooth Infinitesimal Analysis in a very simple manner, by using calculations with nilpotent infinitesimals rather than the concept of limit (and paying attention to only exploit rules from Intuitionistic Logic).

3.2 A Geometric View of $\mathbb{S}$

The general geometrical idea in the smooth infinitesimal analysis is expressed in the following principle:

---

[5] For more details about the intuitionism as foundation of mathematics see [16–18] and [19]. An axiomatic explanation of intuitionistic propositional calculus can be found in [14] and [20].

[6] See, for details, [14].

Principle of Local Straightness of Curves 1 *For any curve C and any point on it, there is a non-degenerate segment of C around the point which is straight.*

This principle allows a very different concept of continuum regarding the real numbers, because in the case of standard (and also non-standard) analysis, this continuum is constructed as a collection of discrete points.

Hence, in an infinitesimal neighborhood of a point, the tangent of a curve in that point exactly coincides with the curve itself: this will allow us a natural application of infinitesimal microscopes[7].

An immediate consequence of the Principle of Local Straightness is the existence of nilpotent infinitesimals. In order to prove this, consider the curve C with equation $y = x^2$. Let Δ be the straight portion of the curve around the origin. So, if Δ is the intersection of the curve with its tangent (the x-axis), it is the set of points x such that $x^2 = 0$. Since Δ has to be non-degenerate, it follows the existence of a nilpotent element not coincident with 0.

In the following section, we will provide a rigorous treatment of these geometrical concepts.

3.3 An Axiomatic System for $\mathbb{S}$

The main object in the world $\mathbb{S}$ is the straight line R, called the *smooth real line*. Its algebraic and order structures are defined by the following axioms

Axiom 1 *R has two points 0 and 1 and maps $- : R \to R$, $+ : R \times R \to R$ and $\cdot : R \times R \to R$ that make it into a non-trivial field, that is, for every x, y, z in R,*

i) $0 + x = x + 0 = x, \quad x + (-x) = (-x) + x = 0, \quad x + y = y + x,$
ii) $1 \cdot x = x \cdot 1 = x, \quad x \cdot y = y \cdot x,$
iii) $(x + y) + z = x + (y + z), \quad (x \cdot y) \cdot z = x \cdot (y \cdot z),$
iv) $x \cdot (y + z) = (x \cdot y) + (x \cdot z)$
v) $\neg(0 = 1),$
vi) $\neg(x = 0) \to \exists y(x \cdot y = 1).$

Axiom 2 *There is a relation $<$ on R which makes it into an ordered field in which square roots of positive elements can be extracted. That is, for every x, y, z in R,*

i) $(x < y \land y < z) \to x < z, \quad \neg(x < x),$
ii) $x < y \to x + z < y + z, \quad x < y \land 0 < z \to x \cdot z < y \cdot z,$
iii) $0 < 1, \quad 0 < x \lor x < 1,$
iv) $0 < x \to \exists y(x = y^2),$
v) $x \neq y \to y < x \lor x < y.$

---

[7] See section 3.6.

Further, every function $f: R \to R$ (also by considering its restriction to Δ) is characterized by the following two axioms.

Axiom 3 (Principle of Microaffineness) *Let $\Delta = \{x \in R : x^2 = 0\}$. For every $f : \Delta \to R$, there exists one and only one $b \in R$ such that for every $\varepsilon \in \Delta$ we have*
$$f(\varepsilon) = f(0) + b \cdot \varepsilon.$$

Axiom 4 (Principle of Costancy) *Let $f : R \to R$ such that $f(x + \varepsilon) = f(x)$ for every $x \in R$ and every $\varepsilon \in \Delta$. Then,*
$$\forall x, y \in R \quad f(x) = f(y).$$

The elements of Δ are called nilpotent (or nilsquare) infinitesimals.

The consistency of these four axioms is guaranteed by the fact that mathematical structures (models) satisfying them have been "constructed", using the rules of intuitionistic logic. For further details see Bell [14], Lavendhomme [20], and Moerdijk and Reyes [15].

The Principle of Microaffineness says that the graph of f is straight in a small neighborhood of $(0, f(0))$ with slope b. That principle has two immediate consequences:

1. *Δ does not consist of 0 alone.* Let $f : \Delta \to R$ be the function defined by $f(\varepsilon) = \varepsilon^2$ and suppose 0 as the only element of Δ. Then $f(\varepsilon) = f(0) + b \cdot \varepsilon$ for any $b \in R$, but this is impossible since b is unique.
2. *Principle of Microcancellation: for any $a, b \in R$, if $\varepsilon \cdot a = \varepsilon \cdot b$ for all $\varepsilon \in \Delta$, then $a = b$.* Consider the function $f : \Delta \to R$ defined by $f(\varepsilon) = \varepsilon \cdot a$. Then, $f(\varepsilon) = \varepsilon \cdot a = \varepsilon \cdot b$, and $a = b$ follows from the uniqueness of b. In particular, if $\varepsilon a = 0$ for all $\varepsilon \in \Delta$, then $a = 0$.

As in the case of non-standard analysis, we can say that two points $a, b \in R$ are *infinitely close* if they differ by an infinitesimal, i.e., if $a = b + \varepsilon$ where ε is infinitesimal. In this case we will write $a \simeq b$.

The natural translation of the concept of continuity in smooth infinitesimal analysis is given in the following way, in terms of infinitesimal distances:

Definition 4. *The function $f : R \to R$ is continuous if and only if $f(x) \simeq f(y)$ whenever $x \simeq y$.*

The following theorem fixes some interesting properties of Δ.

Theorem 5. *In the smooth world $\mathbb{S}$*

(i) *Δ is included in the closed interval $[0, 0]$;*
(ii) *every element of Δ is indistinguishable from 0;*
(iii) *it is false that, for all $\varepsilon \in R$, either $\varepsilon = 0$ or $\varepsilon \neq 0$.*

We shall say that a subset I of R is *microstable* if $a + \varepsilon \in I$ whenever $a \in I$ and $\varepsilon \in \Delta$. The following theorem shows that every closed interval is microstable.

Theorem 6. *For any $a, b \in R$ and all $\varepsilon, \delta \in \Delta$*

$$[a, b] = [a + \varepsilon, b + \delta].$$

In order to see that an interval $[a, b]$ is microstable, consider an element x in it and note that $[a, x] = [a, x + \varepsilon]$, that is, also $x + \varepsilon$ is in $[a, b]$. The proofs of these theorems can be found in [14].

3.4 Differential Calculus in the Smooth World

At this point, we are able to reconstruct in a very natural manner the "smooth" differential calculus. We will start by defining the derivative of an arbitrary function, by considering it defined in a microstable part I of R (in particular, in a closed interval).

Let $f : I \to R$ be a function in $\mathbb{S}$, where I is a microstable part of R. Given $x \in I$, consider the map $g_x : \Delta \to R$ defined by

$$g_x(\varepsilon) = f(x + \varepsilon).$$

By Microaffineness, there is a unique $b_x \in R$ such that for all $\varepsilon \in \Delta$

$$g_x(\varepsilon) = g_x(0) + b_x \cdot \varepsilon$$

or, equivalently, such that for all $\varepsilon \in \Delta$

$$f(x + \varepsilon) = f(x) + b_x \cdot \varepsilon. \tag{3}$$

Definition 5. *The derivative of a function $f : I \to R$ is the function $f' : I \to R$ defined by*

$$f'(x) = b_x$$

where b_x is the number provided by (3). Then, for arbitrary $x \in R$ and $\varepsilon \in \Delta$, we can write

$$f(x + \varepsilon) = f(x) + \varepsilon f'(x).$$

Note that this process can be iterated to obtain derivatives of every order, that is, any smooth function is infinitely differentiable.

This definition of derivative allow us to derive all the standard formulas of the differential calculus in a very easy way. For example, the product rule can be derived as follows:

$$\begin{aligned}(fg)(x + \varepsilon) &= (fg)(x) + \varepsilon(fg)'(x) \\ &= f(x)g(x) + \varepsilon(fg)'(x)\end{aligned} \tag{4}$$

and

$$\begin{aligned}f(x + \varepsilon)g(x + \varepsilon) &= [f(x) + \varepsilon f'(x)][g(x) + \varepsilon g'(x)] \\ &= f(x)g(x) + \varepsilon[f'(x)g(x) + f(x)g'(x)] + \varepsilon^2 f'(x)g'(x).\end{aligned} \tag{5}$$

By comparing (4) and (5), we obtain

$$\varepsilon(fg)'(x) = \varepsilon[f'(x)g(x) + f(x)g'(x)]$$

as $\varepsilon^2 = 0$. Hence, since this works for every $\varepsilon \in \Delta$, by Microcancellation we have

$$(fg)'(x) = f'(x)g(x) + f(x)g'(x).$$

3.5 Higher-Order Infinitesimals and the Principle of Micropolynomiality

Let ε, η be two infinitesimals. The product $\varepsilon \cdot \eta$ is not necessarily equal to 0: for example, if $\varepsilon \cdot \eta = 0$ for all $\varepsilon \in \Delta$, by Microcancellation we would get $\eta = 0$. What about the element $\delta = \varepsilon + \eta$? Is it a nilsquare infinitesimal? Let us try calculating

$$\delta^2 = (\varepsilon + \eta)^2 = \varepsilon^2 + \eta^2 + 2\varepsilon\eta = 2\varepsilon\eta$$

and even if it is indistinguishable from 0, we know that it may be not coincident with 0. However, if we calculate the third power of δ

$$\delta^3 = 2\varepsilon\eta \cdot (\varepsilon + \eta) = 2\varepsilon^2\eta + 2\varepsilon\eta^2 = 0.$$

Then, $\delta = \varepsilon + \eta$ is still a nilpotent element (even if not nilsquare). In general, it can be shown that for any $\varepsilon_1, \ldots, \varepsilon_2 \in \Delta$ we have $(\varepsilon_1 + \ldots + \varepsilon_n)^{n+1} = 0$. This suggests the introduction of the concept of *nth-order (nilpotent) infinitesimal*, that is, an element $x \in R$ such that $x^{n+1} = 0$ $(n \geq 1)$. The set of all nth-order infinitesimals is indicated with Δ_n. Note that $\Delta_1 = \Delta$ and $\Delta_i \subseteq \Delta_j$ for $i \leq j$.

Any property we obtained for the nilsquare infinitesimals is easily generalizable to the generic nilpotent infinitesimals. In particular, they are still indistinguishable from 0, and the set of all nilpotent infinitesimals is included in the interval $[0,0]$. Now we shall say that a subset I of R is *microstable* if, for every n, we have $a + \delta \in I$ whenever $a \in I$ and $\delta \in \Delta_n$. Again, every closed interval is microstable in this sense.

At this point, we can generalize the Axiom 3 given for our smooth world $\mathbb{S}$, by substituting it with a more powerful axiom, in order to get the idea that a smooth function locally behaves exactly as a polynomial of degree $n \geq 1$. Of course, this new axiom is still compatible with the construction of a model $\mathbb{S}$ (see [15]).

Axiom 3 (Principle of Micropolynomiality) *Let $\Delta_n = \{x \in R : x^{n+1} = 0\}$. For every $n \geq 1$ and every $f : \Delta_n \to R$, there exist unique $b_1, \ldots, b_n \in R$ such that for all $\delta \in \Delta_n$ we have*

$$f(\delta) = f(0) + \sum_{k=1}^{n} b_k \delta^k.$$

Directly from that principle, the following important theorem can be proved[8].

Theorem 7 (Taylor's Theorem). *Let $f : I \to R$ be a function in $\mathbb{S}$, where I is a microstable part of R. For any $n \geq 1$, any $x \in I$ and any $\delta \in \Delta_n$ we have*

$$f(x+\delta) = f(x) + \sum_{k=1}^{n} \frac{f^{(k)}(x)}{k!} \delta^k.$$

3.6 Microscopes in the Smooth World

The Principle of Micropolynomiality and Taylor's Theorem claim that the fact that every smooth function locally looks like a polynomial is an intrinsic property. This implies a natural application of the infinitesimal microscopes in the smooth world $\mathbb{S}$. The definition of this concept is natural and easy.

Definition 6. *Let $f : I \to R$ be a function in $\mathbb{S}$, where I is a microstable part of R. Given $x \in I$, the first-order microscope pointed in $(x, f(x))$ is the function $Mf_x^1 : R \to R$ defined by*

$$Mf_x^1 : dx \mapsto f'(x)dx.$$

The first-order optical microscope pointed in $(x, f(x))$ is the graph of the function Mf_x^1, that is, the set of all pairs $(dx, Mf_x^1(dx))$ where $dx \in R$.

We can generalize this definition in the following way.

Definition 7. *Let $f : I \to R$ be a function in $\mathbb{S}$, where I is a microstable part of R. Given $x \in I$, the nth-order microcomponent of f in $(x, f(x))$ is the function mf_x^n defined by*

$$mf_x^n : dx \mapsto \frac{f^{(n)}(x)}{n!} dx^n.$$

The nth-order microscope pointed in $(x, f(x))$ is the function $Mf_x^n : R \to R$ defined by

$$Mf_x^n = \sum_{k=1}^{n} mf_x^n$$

$$Mf_x^n : dx \mapsto \sum_{k=1}^{n} \frac{f^{(k)}(x)}{k!} dx^k.$$

The nth-order optical microscope pointed in $(x, f(x))$ is the graph of the function Mf_x^n, that is, the set of all pairs $(dx, Mf_x^n(dx))$ where $dx \in R$.

---

[8] For the proof of this theorem see [14].

In the Figures 6 and 7 we can find an example of a smooth function viewed through a first-order and a second-order optical microscope respectively. Remember that a second-order infinitesimal is an element δ such that $\delta^3 = 0$. Even if we cannot compare, for example, a second-order infinitesimal and a first-order (nilsquare) one, we can intuitively think that a first-order infinitesimal is "smaller" than a second-order infinitesimal. Let us go back to informal and intuitive argumentations. We can still compare a nilsquare infinitesimal to the number 0.0000001 and a second-order infinitesimal to the number 0.0001. The square of the former is 0.00000000000001 and the

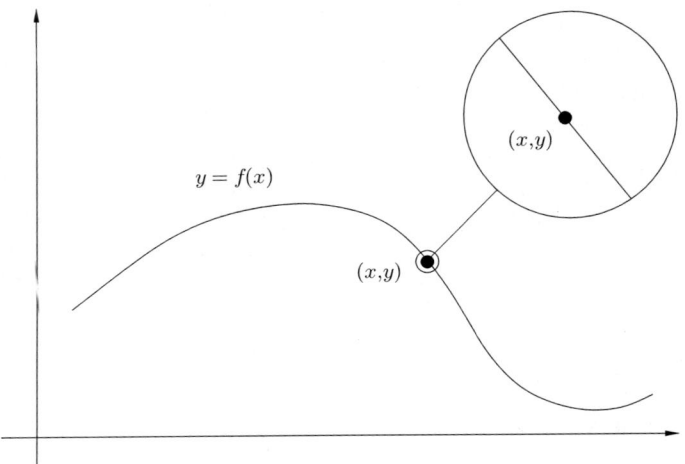

Fig. 6. A smooth function through a first-order optical microscope.

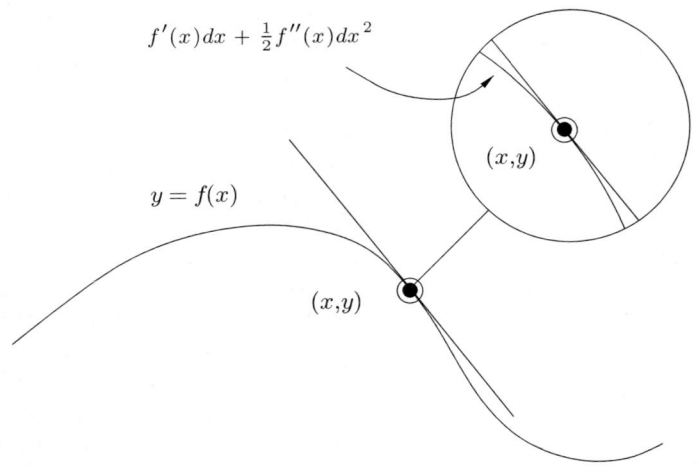

Fig. 7. A smooth function through a second-order optical microscope.

square of the latter is 0.00000001. If we are allowed to consider the former the "same" as zero, we cannot do the same with 0.00000001, because it needs a further power to reach for example the number 0.000000000001, that is sufficiently small to be considered as zero.

The role of the optical microscopes in the smooth world is to formalize these intuitive ideas related to the comparison between infinitesimals of different orders and real numbers. In fact, this could not be done only with the axiomatization given for $\mathbb{S}$: we cannot establish an order relation between infinitesimals, because from $\varepsilon < \delta$ we would obtain the result that at least one of them is distinguishable from zero. For this, consider that from $\varepsilon - \delta < 0$ would follow $\varepsilon - \delta \neq 0$, but this is impossible since $\varepsilon - \delta$ is a nilpotent element.

So, if we look at the curve through a second-order optical microscope, we will see that the curve looks like a piece of parabola. This can be generalized in the following way: *through a nth-order optical microscope a curve looks like a polynomial of degree n*.

We can now introduce the concept of *microscope within microscope* in the smooth world. Figure 8 shows a first-order optical microscope within a second-order optical microscope. The latter is less powerful, because it is based on second-order infinitesimals ("smaller" than the first-order ones). Therefore, this kind of microscope is not able to look so deeply to show that the curve mingles with the tangent line. However, it is powerful enough to show that in a second-order neighborhood the curve looks bent, exactly like a second degree polynomial. In order to look at the curve more in depth, we can point within the second-order microscope a more powerful first-order microscope. In that case, we will see that the curve and the tangent line look like the same thing. Of course, such a procedure can be iterated: we can always point a nth-order microscope within a kth-order microscope whenever $n < k$.

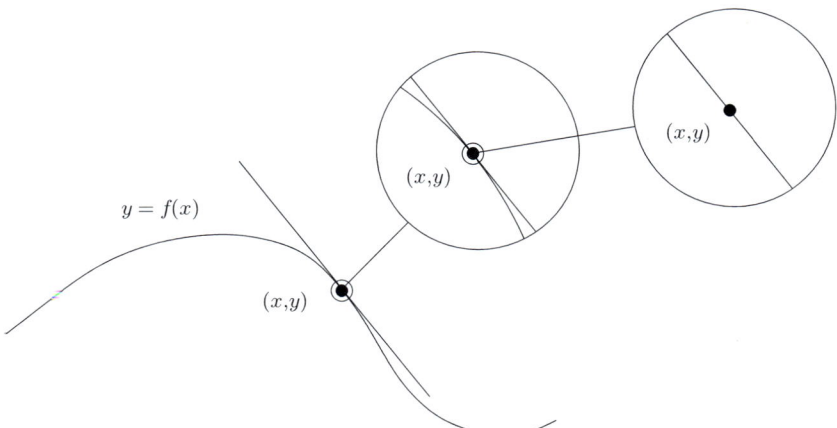

Fig. 8. A smooth function through a second-order optical microscope and a first-order microscope within the former.

As we have shown, the concepts of microscope within the non-standard and smooth analysis look quite different. In the non-standard world, what we can see through an infinitesimal microscope is either a straight line or a system of straight lines. Otherwise, in the smooth world we can see a straight line only through a first-order microscope: through a higher-order microscope we will see a bent curve. More precisely, in a smooth world, microscopes emphasize the polynomial nature of a neighborhood of a point in the graph of a curve. In this case, the role of microscopes has been essentially explanatory (and they may also be used as a didactic tool). This allows us to get two different kinds of mental representations of the infinitesimal world, both rigorously justified. The non-standard approach focuses the behavior of the curve related to the tangent line: in that case, the exploration with the infinitesimal microscopes shows that in an infinitesimal neighborhood we will find only straight pieces of the curve parallel to the tangent line. Instead, the smooth approach is strictly related to Taylor's Theorem, and not only to the tangent line: by exploring further through an infinitesimal microscope, we will see an exact correspondence of the curve and a polynomial of degree n (the degree depends on the order of the microscope).

4 Conclusion

The diagrams we have described provide explanations which allow a better understanding of calculus. In fact, we think they improve and complete non-standard and smooth infinitesimal methods, but also they may be used in the teaching of the standard calculus: they are useful tools from the psychological (didactic) and the epistemological point of view, because they propose a good – and mathematically justified – mental representation of the behavior of a real function in many "critical" situation (at small neighborhoods by looking at infinitesimally small details...). Further, we are convinced that these kind of diagrams can be exploited in other everyday non-mathematical applications (finding routes, road signs, buildings maps, for example), in connection to various zooming effects of spatial reasoning. We think the activity of magnification of optical diagrams can be studied in other areas of model-based reasoning, such as the ones involving creative, analogical, and spatial inferences.

References

1. Magnani, L.: *Abduction, Reason, and Science. Processes of Discovery and Explanation*. Kluwer Academic/Plenum Publishers, New York (2001)
2. Magnani, L.: Epistemic mediators and model-based discovery in science. In Magnani, L., Nersessian, N., eds.: *Model-Based Reasoning: Science, Technology, Values*, New York, Kluwer Academic/Plenum Publishers (2002) 305–329

3. Sullivan, K.A.: The teaching of elementary calculus using the non-standard approach. *American Mathematical Monthly* (1976) 370–375
4. Robinson, A.: *Non-Standard Analysis.* North Holland, Amsterdam (1966)
5. Kock, A.: *Synthetic Differential Geometry.* LMS Lecture Notes Series 51. Cambridge University Press, Cambridge (1981)
6. Stroyan, K.D.: Uniform continuity and rates of growth of meromorphic functions. In Luxemburg, W.J., Robinson, A., eds.: *Contributions to Non-Standard Analysis*, Amsterdam, North-Holland (1972) 47–64
7. Tall, D.: Elementary axioms and pictures for infinitesimal calculus. *Bulletin of the IMA* **18** (1982) 43–48
8. Tall, D.: Natural and formal infinities. *Educational Studies in Mathematics* **48** (2001) 199–238
9. Keisler, H.J.: *Elementary Calculus.* Prindle, Weber, and Schmidt, Boston (1976)
10. Keisler, H.J.: *Foundations of Infinitesimal Calculus.* Prindle, Weber, and Schmidt, Boston (1976)
11. Hurd, A.E., Loeb, P.A.: *An Introduction to Nonstandard Real Analysis.* Academic Press, Orlando (1985)
12. Goldblatt, R.: *Lectures on the Hyperreals: An Introduction to Nonstandard Analysis.* Springer-Verlag, New York (1998)
13. Stroyan, K.D., Luxemburg, W.A.J.: *Introduction to the Theory of Infinitesimals.* Academic Press, New York (1976)
14. Bell, J.L.: *A Primer of Infinitesimal Analysis.* Cambridge University Press, Cambridge (1998)
15. Moerdijk, I., Reyes, G.E.: *Models for Smooth Infinitesimal Analysis.* Springer-Verlag, New York (1991)
16. Browuer, L.E.J.: Intuitionism and formalism. In Benacerraf, P., Putnam, H., eds.: *Philosophy of Mathematics. Selected Readings.* Prentice-Hall, Inc., Englewood Cliffs, New Jersey (1964) 66–77
17. Dalen, D.V., ed.: *Browuer's Cambridge Lectures on Intuitionism.* Cambridge University Press, Cambridge (1981)
18. Heyting, A.: Disputation. In Benacerraf, P., Putnam, H., eds.: *Philosophy of Mathematics. Selected Readings.* Prentice-Hall, Inc., Englewood Cliffs, New Jersey (1964) 55–65
19. Heyting, A.: The intuitionist foundations of mathematics. In Benacerraf, P., Putnam, H., eds.: *Philosophy of Mathematics. Selected Readings.* Prentice-Hall, Inc., Englewood Cliffs, New Jersey (1964) 42–54
20. Lavendhomme, R.: *Basic Concepts of Synthetic Differential Geometry.* Kluwer Academic Publisher, Dordrecht (1996)

Part II

Models, Mental Models, Representations, and Medical Reasoning

Cognition, Environment and the Collapse of Civilizations

Michael E. Gorman

School of Engineering and Applied Science, University of Virginia, Charlottesville, VA, USA
meg3c@virginia.edu

Summary. Jared Diamond, in his provocative book *Collapse*, describes multiple cases where civilizations went through periods of collapse, e.g., the Mayans, the Anasazi and the natives of Easter Island. These collapses were caused by changes in the local system brought about by a combination of natural and human activity. So, for example, the Mayan and Anasazi civilizations developed agricultural technologies during periods when the climate was favorable. The result was an expanded population which could not be supported when extended droughts occurred. But collapse was not inevitable; all of these cultures made choice; indeed, some like the Anasazi survived by changing practices. Our global civilizations have now created the anthropocene, in which human, natural and technological systems are inextricably intertwined. Human activity can, in a very short time, create holes in the ozone layer and change the climate across the entire planet. According to Brad Allenby, our species has a responsibility to manage the global ecosystem. Indeed, it the key to the survival of civilization. This presentation will explore the cognitive capabilities needed to undertake this task, incorporating the latest results from a workshop on trading zones and interactional expertise, to be held at Arizona State University from May 21 to 25. Modeling will be one of the tools discussed in the talk, including results on the economic and health impacts of air pollution in China.

According to Jared Diamond, there are five reasons why civilizations collapse [1].

1. Human environmental impacts – Civilizations that grow beyond their base of resources and that poison their own local environment cannot be sustained.
2. Climate change – Civilizations that cannot deal with local climate changes collapse. Often one and two are connected. A drought that could be weathered by a moderate-sized civilization with ample stores will overwhelm one that has expanded and used up local resources. Human environmental impacts now extend to the global system, further connecting one and two.

3. Dependence on trading relationships – Most civilizations depend on trading relationships with others. If these trading relationships are disrupted, and no alternative trading partners are available, civilizations can collapse.
4. External enemies – An alternative to trading is to simply take over a trade partner, thereby owning all of its resources and enslaving or removing its people.
5. Society's response – Some civilizations can adapt and grow in the face of disruptions that would doom others.
   The Pueblo civilization of the American Southwest has survived for thousands of years despite drought, Spanish invasion and other disruptions in part by moving and in part by preserving a core set of beliefs that remain mostly secret today [2]. Chinese civilization survived the Mongol invaders by absorbing them after conquest – turning them into another dynasty [3].

As an example, let us consider how Diamond's five factors apply to China.

1. China has a serious and growing pollution problem, fueled in part by increased energy demands, which leads to heavier use of coal. Older power plants and indoor coal pollution are especial problems [4].
2. As of 1994, China was the second-largest source of greenhouse gases – still well behind the United States, but growing.
3. China is dependent on trading partners like the U.S., which buys Chinese exports, and Sudan, which sells oil to China.
4. China has no external enemies at present, though it has a long-standing conflict with Taiwan that could lead to war.
5. Past civilizations have collapsed because governments focused more on maintaining power than coping with change. The Chinese government wants to maintain political power while relaxing the old Communist economic constraints. Chinese citizens can now get rich as long as they do not threaten the political status-quo. In the long term, this combination of economic freedom and political control may not be sustainable. Amartya Sen reminds us that there are no famines in true democracies [5]. When climate change or pollution cause suffering, a democratic government has to pay attention.

Global problems like climate change, war, hunger and disease require collaboration across national borders. Here even democracies may not feel the pressure for a solution: countries that suffer disproportionately from environmental damage cannot threaten politicians from another country with losing an election. The problem is complicated by differences in world views that may prevent distinct cultures from understanding one another, even if they wanted to do so.

1 The Problem of Incommensurabilty

The philosopher and historian of science Thomas Kuhn noticed a similar problem when scientists tried to communicate across paradigms. Barry Marshall and Robin Warren discovered in 1983 a new kind of bacteria in the stomachs of people who had ulcers. Their proposal differed from the current paradigm, which held that the stomach was a sterile environment, in which bacteria could not exist, and traced the cause of ulcers to excess peptic acid, exacerbated by stress [6]. The initial reception for the work by Marshall and Warren was hostile; one eminent infectious disease research described Marshall as a "madman" and another described his idea as "crazy".

Kuhn referred to this gap in understanding as the problem of incommensurability; those in the old paradigm (ulcers caused by stress) though the ideas from the new paradigm (ulcers caused by bacteria) were crazy. In the ulcer case, as in many others, incommensurability was eventually overcome: both of the researchers who thought the idea was "mad" and "crazy" eventually did significant work on the bacterial hypothesis. Marshall and Warren received the Nobel Prize in 2005 for this discovery, and now antibiotics are routinely used as part of a treatment program for ulcers. So incommensurability is not impermeable; it can be overcome when scientists eventually agree on methods and data. But this agreement can take a long time, and some scientists may never understand the new paradigm.

Peter Galison proposed that science and engineers create trading zones to communicate across different paradigms when designing systems like radar and particle accelerators [7]. Trading zones allow them to coordinate activities without completely overcoming incommensurability[1] Participants in these trading zones develop an interlanguage that allows them to coordinate activities despite disciplinary differences[2]. They progress from shared jargon to pidgins to creoles; the creole becomes a language that is taught to newcomers[3] to the trading zone, which may morph into a new discipline like bio-medical engineering.

Persons in a role similar to trade agents can substitute for, or complement, a creole. Early in the development of MRI, surgeons interpreted an artifact as

---

[1] Monique Lambert has observed that Jet Propulsion Laboratory engineers refer to their negotiations over where to land a rover as trades [8].
[2] Explored in a workshop on trading zones & interactional expertise, May 22–24, 2006 http://bart.tcc.virginia.edu/Tradzoneworkshop/index.htm, supported by the NSF (SES-0526096), the *Boston Consulting Group* and the *Center for Nanotechnology & Society*. Results from the workshop are used at various points in this article.
[3] Creoles are often developed by children who grow up in a trading zone where adults are using a pidgin [9].

a lesion [10]. This problem was solved by someone with background in both physics and surgery, who in effect facilitated a knowledge "trade" between designers of the device and its primary users.

This kind of "trade agent" must possess interactional expertise, or mastery of enough of the language of another discipline to speak fluently with experts [11]. The term "T-shaped" is often used to refer to this kind of ability, where the long stem of the T is disciplinary depth in one area and the crossing line at the top refers to interactional facility in one or more other areas.

Contributory expertise is used by Collins and Evans to refer to experts in one domain who learn enough about another to make an original research contribution. Walter and Luis Alvarez used their expertise in geology and physics, respectively, to come up with the comet theory of dinosaur extinction and identify and locate the evidence necessary to support the theory [12].

Interactional expertise, in contrast, involves fluency in the language, concepts and even "tribal knowledge" of other domains without being able to do the research. Here tribal knowledge refers to understanding the relative merits of the players in the field. Actual conduct of research involves significant tacit knowledge of procedures and practices [13].

2 Three States in Trading Zones

As a first step towards a taxonomy of different types and stages in trading zones, Mehalik and Gorman proposed three states, on a continuum from a situation where the trading zone is dominated by an elite group or individual to one where participants develop a shared mental model and move from trading to collaborating at a deep level [14]. In between these extremes are a variety of trading zones, including many where parties are more or less equal and free to leave and others where various kinds of inequality make trades unfair. Smaller trading zones can be nested within larger ones, and the nature of the trading zone can change, depending on where one draws the boundaries. As a start, let us take the three categories above and explore their implications for the survival of civilizations.

State 1, Imperialism: An elite has the overall problem representation and black boxes other parties in the trading zone into specific roles whose purpose those persons do not need to understand. An example is centralized agricultural schemes that work well for control, but poorly for producing food [15]. The problem with elites, as noted earlier in this paper, is that their obsession is with maintaining power, not looking out for the interests of the whole society. Therefore, they tend not to adapt well to climate changes and other social disruptions. Elites can maintain control via force, via ideology and even via imposing certain kinds of technology that restrict choices and make it easier to keep track of citizens. So, for example, the Chinese government can maintain control partly by placing restrictions on the use of the internet.

State 2, *Relatively* equal trading zones: In this category, there are a wide range of possibilities. Two obvious sub-types[4]:

A: Trading zones mediated primarily by an interlanguage.
B: Trading zones mediated primarily by interactional experts.

In this second category, consider how AIDs activists mastered the language of medical science in order to change the research protocol so they would not have to be in the placebo group [16]; they wanted to try any treatment that offered even a sliver of hope. In order to create this trading zone, activists had to acquire sufficient interactional expertise to speak the language of the scientists. The activists did not establish a creole; instead, they spent hours learning scientific terminology so they could form a trading zone.

An example of a new specialty that combines both trading zones and interactional expertise is Service Science Management and Engineering (SSME) [17]. This new kind of expertise will involve training service specialists who can co-evolve solutions with clients like companies and governments. This co-evolution will require interactional expertise, because the service scientist will not only need depth in an area of expertise but also the ability to interact with the client's culture. An example is designing computational solutions that really fit the needs of a client and can be implemented in their culture.

State 3: Shared mental model: On the opposite end of the continuum from imperialism is another case where there really is no trading zone, but only because participants have a shared mental model of their mission, of their goal. This mental model is dynamic, and every member of the team has an impact on modifying it. Examples are cutting-edge, mission-focused project teams like the one that created the sidewinder [18] and the groups that worked together to invent the ARPANET [19].

When there is a shared mental model, trades are replaced by collaboration, and there is no incommensurablity, in terms of understanding the problem that needs to be solved and why it is important. Group members flow to whatever work is most vital to the mission, jointly contributing to a new area of expertise. Participants develop a shared set of "routines, words, tools, ways of doing things, stories, gestures, symbols, genres, actions or concepts" [20, p. 88].

Differing mental models of what state a trading zone is in can lead to problems. Consider the recent collaboration between Toshiba and a small company named Lexar. Toshiba wanted Lexar's expertise on incorporating controllers into nand flash memory cards, in order to penetrate the US market. Lexar wanted to take advantage of Toshiba's ability to manufacture nand flash memory.

But the relationship was based on different understandings of the trading zone. Lexar thought it was Toshiba's strategic partner, working in a trading zone that would include a shared mental model and a relationship based

---

[4] These sub-types were first proposed to me by Harry Collins and Rob Evans, with whom I am collaborating on interactional expertise and trading zones.

on trust. Toshiba, in contrast, saw the relationship as one of several trading zones the company was forming in this area. So Toshiba engineers and managers who worked with Lexar felt comfortable interacting with companies pursuing controller designs that competed with Lexar. From an evolutionary standpoint, it makes sense for a big company like Toshiba to pursue multiple controller design trading zones, but only by using a "clean room" strategy, where Toshiba employees working with one company on its technology do not interact with anyone pursuing an alternate design, either within or outside of Toshiba. Lexar successfully sued Toshiba over this breach of trust [21].

3 Three States, Emerging Technologies and the Environment

Imperialistic control of the environment is exemplified by polluters that pass the cost of environmental clean-up on to the rest of society, without any mechanism in place to penalize the polluters. Imperialism can also come on the environmental end, in the form of organizations like Greenpeace that attack the fields of farmers who are producing GMOs. In both cases, there is no dialogue with other stakeholders.

A more equal trading zones is exemplified by the Governor's Commission for a Sustainable South Florida, which met from 1994 to 1999 to try to play a role in the development of a comprehensive plan for the Everglades. Boyd Fuller at MIT[5] studied how this group formed a trading zone, developing an interlanguage that included shared meanings for terms like "sustainability", "aquifer storage and recovery", and "hydroperiod". Furthermore, participants in the trading zone agreed on maxims for conduct like, "Keep your pet pigs at home. If you're going to have your pet pig, we'll have a pet pig festival. So for a couple of hours everybody will get out their pet pigs and parade them around – then we'll put them away and get back to the job of restoring the Everglades". This maxim is a way of avoiding the tendency for stakeholders to announce their incommensurable perspectives and "parade them around" without really moving beyond them to the point where negotiations could occur.

Ideally, the eventual result of such a trading zone should be a shared mental model, with respect to the goals of the system, but this will be very difficult to achieve, given that the Everglades serve as a reservoir for the Miami area, a wildlife refuge particularly sensitive to changes in flow and also an agricultural area where wastes from sugar farming and livestock flow into the water.

Trading zones offer a way in which civilizations can deal with human environmental impacts. When these impacts are extended to Diamond's second factor, global climate change, these trading zones have to be global in nature – like the Kyoto protocols.

---

[5] http://www-personal.umich.edu/~bwfuller/
Trading_Zone_Paper--Boyd_Fuller--Distribution--Jan_1-05.pdf

None of the current agreements over the Everglades incorporates the disruptive possibility of global warming. The global environment is a complex system, which means that small changes in one part of the system could lead to a shift in system state that is not entirely predictable. To deal with the global ecosystem, Brad Allenby calls for a new kind of expertise in Earth Systems Engineering and Management (ESEM) that can keep overall system impacts in front of participants in trading zones [22]. ESEM management will require adaptive management, made possible by:

1. Continuous and fine-grained monitoring of environmental conditions world-wide, so that changes in systems states can be detected.
2. Reversible technologies that permit alteration of technological systems as monitoring reveals their effects are not what was intended.
3. Collaboration across disciplines and stakeholder interests.
4. Explicit consideration of values.

Modeling will play a critical role in adaptively managing the ecosystem, and therefore in facilitating society's response to environmental changes. Climate models will play a significant role. A good example is Wang & McElroy's model of atmospheric pollution in China, which incorporates data from a new atmospheric station near Beijing jointly managed by Harvard and Tsinghua University [13]. Models can suggest trends and highlight improved metrics that can tell us what future scenario we are entering.

One problem is that there are so many models of parts of systems, often based on incommensurable assumptions and differing in:

- programming languages & algorithms
- mathematical techniques, and
- levels of resolution.

Therefore, a modeling creole is necessary. Consider DOME, a kind of modeling interlanguage, which allows product designers and environmental consultants to collaborate by linking their different models [24]. DOME sets requirements for input, which is similar to requiring that participants in an exchange at least agree on shared definitions of key terms. DOME is designed so that, "A change in any part of the system model will propagate through the system, so that the distributed models together form a concurrent system model" (p. 41). DOME therefore facilitates the kinds of trading zones involved in designing environmental technologies.

4 Modeling Principles

Those who do modeling of global environmental systems would be well to keep the following five principles in mind:

1. Make assumptions transparent.
2. Be aware how sensitive your model is to changes in initial conditions or in key variables over time.

3. Iterate and improve, both data and models.
4. Strive for comparability or even integration with other models (creole).
5. Humility: a model is always a representation of reality, and it is probably better to have several models based on different assumptions, when one is looking into the future.

5 Cultural Incommensurability

One of the central problems facing civilization at this point is that problems like managing the environment do not honor national borders. The Mayans and the natives of Easter Island could not manage their local environment sufficiently to sustain their civilization [1]. Consider how much harder it is to manage climate change, or the availability of food and water, across national boundaries. Here deep differences in values and ideology create incommensurabilities that exceed those usually experienced in scientific and engineering situations. One culture may not even see another as fully human, and therefore see nothing wrong with taking their resources, polluting their air, etc.

Overcoming values incommensurabilities requires the exercise of moral imagination. According to Mark Johnson, we learn practical ethics from stories, which become mental models for virtuous behavior [25]. These mental models can become unquestioned assumptions, which become confused with reality by those who hold them.

Moral imagination consists of seeing that these "realities" are mental models, and that alternative models, e.g., those of other stakeholders, are worth understanding [26].

Note that moral imagination is not the same as relativism. There are moral belief systems that are wrong, e.g., systems based on imperialistic control – like slavery. There are certainly moral maxims like Kant's "never use people merely as a means to an end" that can apply universally.

Properly applied, moral imagination will lead to improved models for how to progress ethically, as a civilization. Moral imagination is much like the advice for modelers, above: do not confuse your moral models with reality, be humble, seek additional data and be receptive to alternative models that suggest the value of taking a different perspective. In the case of Kant's maxim, for example, what would happen if we added "ecosystem" to human? We would need to imagine and evaluate the consequences of this shift, which might transform civilization – or, improperly applied, make it impossible.

6 Technology and the Future of Civilization

Technology promises to extend human capabilities to realms reserved for Gods in traditional stories, giving human beings the ability to:

1. Control evolution.
2. Change "human nature" by altering our own genetic code and making ourselves into cyborgs [27].
3. Manage the global ecosystem.

Civilizations that race ahead with these developments and seek to protect their gains will achieve a temporary advantage, but the spread of nuclear weapons and the rise of terrorism show the limits to this strategy. All civilizations are not only part of a single ecosystem, they are also connected by distributed, high-speed communications and are part of an increasingly global economy. Not everyone has the opportunity to participate in this new interconnected world, and those who cannot will resent and resist it as another device by which the powerful and rich get stronger at the expense of others – a global imperialistic trading zone.

To constitute progress, technological development should enhance freedom and opportunity worldwide. This kind of progress will require us to engage in moral imagination, developing new stories for a global civilization that is involved in constant transformation and self-examination.

Of Diamond's five factors that determine the fate of civilizations, the last one – civilization's response – is the most important one. Human beings can see the system of which they are a part, can look critically at their own actions, collective and individual. Modeling is an important tool for imagining the consequences of present actions, or failures to act. If civilization survives, it will be radically altered by the accelerating pace of technological development. Part of this development will be extraordinary new tools for modeling, that will help us manage the future we create.

Such modeling tools could even facilitate moral imagination. Arizona State University's Decision Theater (http://dt.asu.edu) is an example. This environment allows stakeholders to envision the consequences of different development plans for Phoenix. This system could evolve into one that allows stakeholders to visualize their imagine desirable futures, share them with others, and modify them on the fly, based on discussion. Modeling tools alone cannot span incommensurable ideological divides. But they can facilitate the development of trading zones over the challenges and opportunities that face our increasingly global civilization.

References

1. Diamond, J.: *Collapse: How Societies Choose to Fail or Succeed.* Viking, New York (2005)
2. Roberts, D.: *The Pueblo Revolt: The Secret Rebellion that Drove the Spaniards out of the Southwest.* Simon & Schuster, New York (2004)
3. Shouyi, B.: *An Outline History of China.* Foreign Languages Press, Beijing (2002)

4. McElroy, M.B., Nielsen, C.R., Lydon, P., eds.: *Energizing China: Reconciling Environmental Protection and Economic Growth*. Cambridge University Press, Cambridge, MA (1998)
5. Sen, A.: *Development as Freedom*. Random House, New York (1999)
6. Thagard, P.: Ulcers and bacteria I: Discovery and acceptance. *Studies in History and Philosophy of Science. Part C: Studies in History and Philosophy of Biology and Biomedical Sciences* **29** (1998) 107–136
7. Galison, P.: *Image & Logic: A Material Culture of Microphysics*. The University of Chicago Press, Chicago (1997)
8. Lambert, M.H., Shaw, B.: *Transactive Memory and Exception Handling in High-Performance Project Teams* (2002) No. CIFE Technical Report 137.
9. Pinker, S.: *The Language Instinct*. W. Morrow and Co, New York, NY (1994)
10. Baird, D., Cohen, M.: Why trade? *Perspectives on Science* **7**(2) (1999) 231–254
11. Collins, H.M., Evans, R.: The third wave of science studies. *Social Studies of Science* **32**(2) (2002) 235–296
12. Alvarez, W.: *T.Rex and the Crater of Doom*. Princeton University Press, Princeton, NJ (1997)
13. Collins, H.: Tacit knowledge and scientific networks. In Barnes, B., Edge, D., eds.: *Science in Context*. MIT Press (1982) 44–64
14. Gorman, M.E., Mehalik, M.M.: Turning good into gold: A comparative study of two environmental invention networks. *Science, Technology & Human Values* **27**(4) (2002) 499–529
15. Scott, J.C.: *Seeing Like a State: How Certain Schemes to Improve the Human Condition Have Failed*. Yale University Press, New Haven (1998)
16. Epstein, S.: The construction of lay expertise: Aids activism and the forging of credibility in the reform of clinical trials. *Science, Technology & Human Values* **20**(4) (1995) 408–437
17. Spohrer, J.C., McDavid, D., Maglio, P.P., Cortada, J.W.: Nbic convergence and technology-business coevolution: Towards a services science to increase productivity capacity. In Bainbridge, B., Roco, M.C., eds.: *Managing Nano-Bio-Info-Cogno Innovations: Converging Technologies in Society*. Springer (2006) 227–253
18. Westrum, R., Wilcox, H.A.: Sidewinder. *Invention & Technology* (Fall) (2004) 57–63
19. Hughes, T.: *Rescuing Prometheus*. Pantheon books, New York (1998)
20. Wenger, E.: *Communities of Practice: Learning, Meaning, and Identity*. Cambridge University Press, Cambridge (1998)
21. Gorman, M.E.: STS, ethics, and knowledge transfer in the courtroom. *Social Studies of Science* **31**(7) (2006)
22. Allenby, B.: Technology at the global scale: Integrative cognitivism and earth systems engineering management. In Gorman, M.E., Tweney, R.D., Gooding, D.C., Kincannon, A., eds.: *Scientific and Technological Thinking*. Lawrence Erlbaum Associates (2005) 303–344
23. Wang, Y.X., McElroy, M.B., Wong, T., Palmer, P.I.: Asian emissions of co and nox: Constraints from aircraft and chinese station data. *Journal of Geophysical Research* **109** (2004) 1–26
24. Borland, N., Wallace, D.: A collaborative Internet-based modeling approach. *Journal of Industrial Ecology* **3**(2–3) (2000) 33–46

Cognitive Aspects of Tacit Knowledge and Cultural Diversity*

Riccardo Viale and Andrea Pozzali

Fondazione Rosselli, Turin and Università degli Studi di Milano-Bicocca, Milan, Italy
riccardo.viale@fondazionerosselli.it, andrea.pozzali@unimib.it

Summary. Tacit knowledge is pervasive in many aspects of human life. In the past it was analyzed mainly as behavioral skill and know-how in practical knowledge such as craftsmanship. Afterwhile it was also applied to more intellectual skills such as piano playing and science. But this interpretation of tacit knowledge lacks to include fundamental cognitive dimensions such as background knowledge and implicit cognitive rules. The first deals with cultural values and principles that drive our interpretation of the reality. The second deals with the inferential rules that drive our reasoning and decision making processes. In the last ten years Cognitive Anthropology has collected a great amount of data showing deep differences between westerner (mainly American) and easterner (mainly Chinese) way of thinking. In our opinion these differences are based on a different tacit background knowledge that causes different implicit cognitive rules.

1 Introduction: Different Types of Tacit Knowledge

The concept of "tacit knowledge", introduced in modern epistemological literature thanks to the seminal work of the scientist and philosopher of science Michael Polanyi [1, 2], has experienced over the years an ever widening application in a growing number of disparate disciplines, that range from psychology to mathematics, from econometrics to religious thought, from aesthetics to evolutive economy.

The need to develop more detailed taxonomies of the characteristics that can be attributed to tacit knowledge has been unanimously recognized in the literature.

One distinction that has for a long time contributed to orient the debate, and that goes back to the work of Ryle [3], is the one between *know how* and

---

* The first two paragraphs are based on Pozzali, A., Viale, R.: Cognition, Types of 'Tacit Knowledge' and Technology Transfer. In Topol, R., Walliser, B., eds.: Cognitive Economics: New Trends. Elsevier, Oxford (2007)

know that, or in almost equivalent terms between procedural and declarative knowledge [1]. This distinction is relevant here as for long time tacit knowledge has been in a certain way confined to the domain of *know how*, as a component of skills and physical abilities. More recent contributions have tried to come to more refined classifications, like, for example, in the case of Gorman [5], who identifies four categories: *information (know what)*, *skills (know how)*, *judgment (know when)* and *wisdom (know why)*. Four are also the categories identified by Johnson et al. [6], who, however, substitute *know when* with *know who*:

Know what - should indicate knowledge regarding "facts", assimilable to so-called "information". This type of knowledge is easily codified and communicated, also thanks to its decomposability into many elementary components or "raw data";

Know why - should refer to knowledge related to principles and to general laws present in nature, in society and in the human mind;

Know how - indicates *skills*, understood, however, not in the limited sense of mere physical type abilities, but in a general sense as "the capacity to do something", that can present also theoretical and abstract elements:

> Even finding the solution to complex mathematical problems is based on intuition and on skills related to pattern recognition that are rooted in experience-based learning rather than on the carrying out of a series of distinct logical operations [7, pp. 101–102] [6, p. 250];

Know who – encloses all the knowledge related to "who knows what", that is, the capacity to individuate within the whole available knowledge base the most appropriate expertise to solve determined problems:

> The general trend towards a more composite knowledge base, with new products typically combining many technologies, each rooted in several different scientific disciplines, makes access to many different sources of knowledge more essential. [6, p. 251]

All these classifications share a common method which consists in individuating a series of types of knowledge and subsequently indicating to what extent the single types can be considered more or less codifiable. In this sense, the classic distinction between *know how* and *know that* represented a sort of alternative formulation (or, if you prefer, of specification) of the tacit knowledge/explicit knowledge dichotomy. In fact, *know how* ended up being identified as the only field where it was possible to track down forms of tacit knowledge, while *know that* was considered almost totally explicit. The subsequent classifications by Gorman and by Johnson, Lorenz and Lundvall represent a notable advance in the debate as they both recognize that forms of tacit knowledge, far from being confined exclusively in the context of *know how*, can also be traced in other types of knowledge. None of these classifications, however, has tried to analyze the possibility that "tacit knowledge", far from representing a concept that defines a perfectly homogeneous series of phenomena, can take on internal distinctions, or said more simply that different types of tacit knowledge can exist[1].

2 A New Tripartition of Tacit Knowledge

Distinguishing these different types is important for two reasons: in the first place because this classification is a prerequisite for the conduction of more detailed empirical analyses, and in second place because different types of tacit knowledge can be learned, and consequently transmitted, with different mechanisms. In order to carry out a similar analysis it is opportune to refer directly to the classic tripartition between forms of knowledge in use in the epistemological literature [8, 9], which distinguishes *competential knowledge* (ability), *direct knowledge* (knowledge as familiarity) and *propositional knowledge* (or *justified true belief* or knowledge as "correct information").

In a similar way, tacit knowledge can be classified in the following three categories:

Tacit knowledge as competence: this class includes all the forms of physical abilities and skills that refer to the capacity of a subject to know how to perform certain activities without being able to describe the knowledge he used to do the task. This type of tacit knowledge can have an automatic and unreflected character (for example, in the case of knowing how to breathe) or it can be the fruit of a conscious learning or training process (for example, in the case of knowing how to play the piano). This kind of tacit knowledge operates in particular in physical-like abilities such as swimming or riding a bicycle: in all these *skilful performances*, the activity is carried out by following a set of rules that are not explicitly known by the person following them. In other words, usually a person is able to ride a bicycle or to swim even if he does not know how he is able to do it. The same holds also for more complicated and less common abilities, that are at the base of the development of craftsmanship (for example, the ability to make a violin) and of technological innovations (such as nuclear weapons, cf. [10] or aircrafts, cf. [11]. In all these cases the actual practice, that is the ability to carry on the given activity, can not be described correctly in all its details; even when a description can be formulated, this is always incomplete and is not enough to allow for knowledge transfer[2]:

---

[1] German's classification admits the possibility that tacit knowledge may be present as a constitutive element of a series of different types of knowledge, such as, for example, heuristics, mental patterns, physical abilities, moral imagination and so on, but it does not specify concretely the modalities with which this can take place.

[2] By the way, this explain why in our times, with all the modern technology we can dispose of, we are still not able to recreate or emulate Stradivari's mastery in making violins!

> Rules of art can be useful, but they do not determine the practice of an art; they are maxims, which can serve as a guide to an art only if they can be integrated into the practical knowledge of the art. They cannot replace this knowledge. [1, p. 50]

This type of manual abilities is defined by Polanyi as the capacity to physically carry out a predefined series of actions in order to complete a complex activity. The classic example, used also to introduce an important distinction between subsidiary awareness and focal awareness, starts from the observation of an apparently simple operation, like hitting a nail with a hammer:

> When we use a hammer to drive in a nail, we attend to both the nail and hammer, *but in a different way*. We *watch* the effect of our strokes on the nail and try to wield the hammer so as to hit the nail most effectively. When we bring down the hammer we do not feel that its handle has struck our palm but that its head has struck the nail. Yet in a sense we are certainly alert to the feelings in our palm and the fingers that hold the hammer. They guide us in handling it effectively, and the degree of attention that we give to the nail is given to the same extent but in a different way to these feelings. The difference may be stated by saying that the later are not, like the nail, objects of our attention, but instruments of it. They are not watched in themselves; we watch something else while keeping intensely aware of them. I have a *subsidiary awareness* of the feeling in the palm of my hand which is merged into my *focal awareness* of my driving in the nail. [1, p. 55]

The two forms of awareness are mutually exclusive. Shifting our focal awareness from the general nature of a determined action to the single details that the action is composed of produces in us a sort of "self-consciousness" that can act as an impediment, making it impossible for us to go on doing the action we have undertaken. This is what happens, for example, to a pianist when he shifts his focal awareness from the piece he is playing to the details of the movements of his hands: it is likely that at this point he will become confused to the point that he has to interrupt his performance. What is destroyed, in these cases, is the sense of context.

In the performance of complex tasks, therefore, we have a focused awareness only of some central details regarding the different operations being performed, while the rest of the details are left to subsidiary awareness. It is precisely the interaction between different forms of awareness that enables us to perform our various activities, which could not, for their nature, be performed in a fully "self-conscious" manner.

Tacit knowledge as Tacit Background Knowledge (TBK) (or as familiarity): in this class we find all those forms of interiorized regulations, of codes of conduct, of values and widespread knowledge that a determined subject knows thanks to his direct experience. This knowledge cannot be articulated or formalized because of its extremely dispersed nature, which makes it difficult to access to it by aware consciousness. This type of tacit knowledge has more than one affinity with the notion of *background*, introduced by Searle to find a solution to the problem of retrieving a stable foundation for the process of interpretation of rules and of representations, or in more precise terms, to prevent this process from turning into an infinite regression [12, 13]. *Background* is defined as that set of biological and cultural capacities, of assumptions, of presuppositions and of pre-theoretic convictions that are the preconditions of any form of theoretical knowledge. Even if background is a very complex structure, that has been the object of many reinterpretations and redefinitions, even by Searle himself, it is possible, in any case, to find between it and the concept of "knowledge as familiarity" some significant overlapping, especially if we consider those components of the "background" whose acquisition is mediated by processes of socialization and acculturation (and therefore in final analysis of experience understood in a broad sense).

On the other hand, this type of tacit knowledge shows many elements of contact also with "pre-theoretical" knowledge on which the analysis of sociologists of knowledge like Berger and Luckmann concentrate [14]. Every modern society is characterized by a huge amount of this kind of tacit knowledge, dispersed among every individual member of the society and transmitted from one generation to another through an endless and continuous process of socialization. It appears evident how the analysis of Berger and Luckman is in many respects less shareable than that of Searle, especially where they speak of "objective structures of the social world" and define pre-theoretic knowledge as the pure and simple "total sum of what everyone knows". If we want the analysis of the role of tacit background knowledge to gain an effective explicatory role and not to remain a pure and simple descriptive concept, we should always try to lower our focus to the individual level, analyzing how the cognitive capacities of the single individual filter and recombine the set of pre-existing social knowledge. The work of Searle on the "construction of social reality" offers some interesting methodological cues in this context, where it shows how the "objective structures of the social world", that Berger and Luckmann speak of, can in fact be analyzed and described as the fruit of thought and language processes that take place in individual minds.

It seems likely to assume that tacit background knowledge can act as a reference point and as an inevitable filter between the individual and the social level. If we want to find, in the economic literature, a sort of correspondence for this type of knowledge, we may look for example at the concept of "social capital" [15]. More in general, we can also think that tacit background knowledge can be one of the constitutional elements of all those forms of knowledge that are embedded in a specific social, political and/or geographical context [16–18] and that are used in many cases as explanatory variables in the analysis of the different competitiveness performance at the local level.

However, we should not be misled by the so-called social dimension of background knowledge. It is not a question of divesting the individual and cognitive dimension of possessing TBK. Every set of principles and values varies from one individual to another in its declination, accentuation and precision. Individuals may be immersed in the same cultural context, but the genesis of their TBK follows different paths, even solely at an infinitesimal level, generating a different baggage of principles and values. In fact, the individual learning of TBK depends on how the person assimilates principles and values from the context in which he or she lives. This is achieved through various methods: direct and explicit teaching by parents, teachers and "masters"; imitation of customs, behavior and lifestyles from which reference guiding principles are assimilated; reading and learning from texts and documents; information from old and new media. It is obvious that everyone has his own specific individual history of learning (because the teachers, media, books, friends, etc. with whom he or she interacts are certainly different, or at least not identical) and this generates different TBK. The unique qualities and idiosyncrasy of individual baggages of TBK is similar, in terms of the learning dynamics, to concentric circles: starting from one's own family nucleus and moving out to the global dimension, the shared content of TBK gradually diminishes.

Tacit knowledge as Implicit Cognitive Rules (ICRs): following the epistemological classification we have proposed as a reference point, we now come to the problem of finding a kind of tacit knowledge that can be considered as an analogous of "knowledge as justified true belief" or as "correct information". Under a certain point of view, this can be considered as an impossible task: how can we conceive, in fact, of an individual possessing a "tacit propositional knowledge"? How can we ascertain that the knowledge one subject has can be considered as a "justified true belief", if this knowledge is tacit, that is the subject is not able to express and formulate it? How can a person holds "tacit beliefs"? These are just some of the questions that immediately raise when one starts to conceive of the possibility to envision a type of tacit knowledge that is not merely a physical abilities or a social background knowledge. As a matter of fact, the possibility of considering tacit knowledge as having also a cognitive dimension was for many years substantially ruled out in epistemology and in cognitive sciences. The only way of considering tacit knowledge was limited to admitting that it could have a role in skill-like abilities. Other forms of tacit knowledge seem to represent no more than a logical absurdum.

In the last few years this kind of veto toward a form of "tacit cognition" is beginning to vacillate, thanks in particular to the empirical and theoretic evidences coming from cognitive psychology and from neurosciences. The first and perhaps the most significant example of a form of tacit knowledge that cannot be considered either a physical-type skill, or a form of "social capital", is linguistic knowledge [19, pp. 263-273]. This form of knowledge does not represent, in a strict sense, a form of skill, but must be considered as an actual cognitive system, defined in terms of mental states and structures that cannot be articulated in words nor described in a complete formal language. The completely tacit nature of this linguistic knowledge is such that a language, in fact, cannot be "taught", but must be more properly "learned" by subjects [20].

Moreover, not only the acquisition, but also the utilization of linguistic knowledge does not seem to imply a reference to the formalized rules of language, but rather an automatic and mostly unconsciously reference to the acquired abilities: *"the knowledge of grammatical structures [...] is not present in a conscious way in most of the cases where we use the language effectively and perfectly correctly"* [21, p. 357][3].

Other examples of cognitive forms, not *skill-like* nor *background-like*, of tacit knowledge come from the substantial number of studies on implicit learning processes [22–24], in particular those relating to experiments in artificial grammar and probabilistic sequence learning[4]. The typical experiment of artificial grammar learning consists in giving subjects a series of alphanumeric strings, some of which generated from a hidden grammatical structure, others completely casual. After completing this phase, subjects are given other alphanumeric strings and they are asked to distinguish between the grammatical and the non-grammatical ones. The results show that the subjects are able to successfully perform this recognition task, though they are unable to explain in an articulated form the type of logical path that led them to these results, nor can they describe the characteristics of the hidden grammatical structure. Even more interesting experiments, with a similar structure, are those related to the control of complex systems, in which a subject is asked to maximize an unknown function selecting the values to be attributed to specified variables [28]. On the whole, it is possible to say that research on implicit learning shows how subjects are able to make use of the hidden structural characteristics that make up the essence of a given phenomenon, though they are not able to come to the complete and explicit knowledge of these same characteristics.

The knowledge that enables the subjects of implicit learning experiments to obtain this type of results can be considered, together with linguistic knowledge, as a type of tacit knowledge that is neither a purely physical "skill", nor a form of "familiarity" or "background" knowledge. Obviously, we can not in any case consider it is a type of "justified true belief", or as a "propositional knowledge", for the reasons already explained. How can we try then to define it?

We propose to define this kind of tacit knowledge as *implicit cognitive rules* that can guide the actions and decisions of a subject while at the same time remaining confined to the tacit domain. As we know that admitting the possibility that a cognitive rule can be implicitly held can represent a highly controversial point a clarification is here needed. The problem seems to lie in

---

[3] Even if in certain cases it is possible to admit that, in the case of language, we can reach the formulation of an explicit rule, the fact remains that the total formalization and codification of linguistic knowledge has not yet been reached, in spite of the considerable research efforts expended over the years.

[4] To remain in the field of neurosciences, further empirical evidence supporting the role of tacit knowledge in individual cognitive processes comes also from research on implicit memory and perception phenomena, cf. [25–27].

the fact that the representational theory of mind, that can be considered the mainstream in cognitive science, in a certain way requires that in order to be causally efficacious representations have to be tokened in a conscious way. The evidences coming from implicit learning research, but also from recent studies on phenomena of implicit memory and subliminal perception, should make us consider more in depth the possibility that not all knowledge need to be tokened in order to play a causal role, as Cleeremans and Jimenez clearly state:

> We suggest to eliminate the "knowledge box" as a requirement for the definition of knowledge, and to assume that representations can simultaneously constitute knowledge and be causally efficacious without ever being tokened in any way. For instance, observing that "butter" has been perceived in a subliminal perception experiment because it exerts detectable effects on performance does not imply that the property of "butter" has been somehow represented in the subject's knowledge box [...]. It simply means that the relevant neural pathways were activated sufficiently to bias further processing in the relevant direction when the stem completion or lexical decision task is actually performed. The knowledge embedded in such pathways is knowledge that is simultaneously causally efficacious and fully implicit. [29, p. 771]

The type of tacit knowledge subjects seem able to develop in implicit learning experiments is knowledge that can not be expressed and at the same time surely has a direct causal impact on subjects' decisions and performances. We can consider it as a kind of tacit analogous of other well known cognitive mechanisms such as pragmatic schemes, heuristics, mental models and so on. As it is knowledge able to influence the decisions made by the subject, it is a real cognitive rule, that is held in an implicit way. For this reason we propose to categorize it as implicit cognitive rules.

Even if empirical research on this type of tacit knowledge is still in great part lacking, we suspect that it may be considered as an important element in the development of heuristics, rules of thumb and case-based expertise that are commonly used in decision-making processes [30]. In economic literature, we could maybe find this type of tacit knowledge as being one of the component of "expert knowledge" and of "organizational routines" [31, 32]. We believe the clarification of these elements to be one of the main future topics for the advancement of tacit knowledge research in cognitive science and in economics both.

The distinction between different types of tacit knowledge is a useful heuristic instrument to develop deeper and more accurate empirical analyses. Compared to alternative distinctions, like for example the one by Collins [33], the one we are proposing has the advantage of dividing tacit knowledge into three distinct forms, each of which can be easily detected in an empirical way and characterized on the basis of its specific mechanisms of acquisition, codification and transfer.

As for the mechanisms with which the different forms of tacit knowledge can be acquired and transmitted, we can indicate the following points (which could be aspects worth further empirical analysis):

tacit knowledge as competence (skills, know-how) can be learned and transmitted fundamentally through processes of *imitation and apprenticeship* based on *face-to-face* interaction and *on the job learning by doing/learning by using* [34, 35]; for a description of the neurological processes that seem to be involved in the acquisition of skill like abilities and other similar physical competences, see [36, 37];

tacit knowledge as tacit background knowledge is acquired, as we have seen, mainly through processes of socialization (to which we can also add mechanisms of implicit learning in some cases); the same mechanisms are at the base of the circulation and transmission of this type of tacit knowledge within a determined social, economic and institutional context;

tacit knowledge as implicit cognitive rules is acquired through processes of implicit learning like the ones remembered above [22, 38–41]. The mechanisms that allow the transmission of this type of knowledge have not yet been analyzed in a thorough manner. One of the first objectives of current research on tacit knowledge should be precisely the study of this particular field of analysis.

3 Cultural Diversity of Implicit Cognitive Rules: the Role of Tacit Background Knowledge

The tripartition described above does not mean that there are no connections and blurred boundaries between the three types of tacit knowledge. In particular, the relationship between TBK and ICRs appears to be one of strong cognitive integration. The close relationship between TBK and ICRs is highlighted in the results of numerous studies on developmental psychology and cognitive anthropology. Our inferential and heuristic skills appear to be based on typical components of TBK. Moreover, our reasoning, judgment and decision-making processes seem to rely on principles that are genetically inherited from our parents.

As described by Viale [42] infants are endowed with an innate set of principles that allows them to begin to interact with the world. Among these principles, one of the most important allows a causal attribution to relations between physical events. At around the age of 6 months, the infant is able to apply the principle of cohesion – a moving object maintains its connectedness and boundaries – the principle of continuity – a moving object traces exactly one connected path over space and time – and the principle of contact – objects move together if and only if they touch [43]. Moreover, there is the theory of biology and the theory of psychology. These theories show that infants individuate some theory-specific causal mechanisms to explain interactions among the entities in a domain. A child has an intuition of what characterizes a living being from an artefact or an object. Between the ages of 2 and 5, the child assumes that external states of affairs may cause mental states and that there is a causal chain from perception to beliefs to intentions and to actions [44].

What are the features of these principles? Data from developmental studies and a certain universality of causal perception in crosscultural studies seem to support the hypothesis that we are endowed with earlydeveloped cognitive structures corresponding to maturational properties of the mind-brain. They orient the subject's attention towards certain types of clues, but they also constitute definite presumptions about the existence of various ontological categories, as well as what can be expected from objects belonging to those different categories. Moreover, they provide subjects with "modes of construal" [45], different ways of recognizing similarities in the environment and making inferences from them.

The previous Piagetian notion of formally defined stages, characterized by principles which apply across conceptual domains, has been replaced by a series of domain-specific developmental schedules, constrained by corresponding domain-specific principles. These principles constitute a core of probably innate "intuitive theories" which are implicit and constrain the later development of the explicit representations of the various domains. As Gelman highlights, "different sets of principles guide the generation of different plans of action as well as the assimilation and structuring of experiences" [46, p. 80]. They establish the boundaries for each domain which single out stimuli that are relevant to the conceptual development of the domain.

Data reported by developmental psychologists show how the capacity for reasoning and decision-making is built on a foundation of implicit principles, of innate origin, contained in the child's tacit background knowledge. In addition to the universal principles described earlier, the child also assimilates cultural-based schemes and principles that determine the development of cognitive styles valid only at local level [42]. These take the form of principles, values, and theories of a metaphysical, ontological and epistemological nature that vary depending on cultural context and which generate different implicit cognitive rules. These different rules provide a unique characterization of the way of perceiving and representing external reality, the way of using empirical data inductively, of using deductive methods of reasoning, of categorizing phenomena, of making probability judgments, etc. This cultural and acquired aspect of TBK gives rise to profound differences between various cultural areas in terms of the cognitive style of ICRs [47]. A case in point is provided in the studies of the cognitive and perceptive differences among Asians and Americans reported by Nisbett et al. [48–50]. They rely on an impressive number of cognitive tests that try to compare the way of reasoning of North Americans, mainly university students, and East Asians – Korean, Chinese and Japanese – mainly university students. The East Asians and the

Americans respond in qualitatively different ways to the same stimulus situation in many different tests. For example, American participants showed large primacy effects in judgements about covariation, whereas Chinese participants showed none. "Control illusion" increased the degree of covariation seen and the reported accuracy of Americans but tended to have the opposite effects on Chinese. Koreans were greatly influenced in their causal attribution by the sort of situational information that has no effect for Americans. Koreans showed great hindsight bias effects under conditions where Americans showed none. Finally, Americans responded to contradiction by polarizing their beliefs, whereas Chinese responded by moderating their beliefs.

We can summarize the results as follows.

The *American vs. East Asian style of thinking* [49].

1. *Explanation*: East Asians tend to explain events, both social and physical, more with respect to the field and Americans tend to explain events more with respect to a target object and its properties.
2. *Prediction* and *"postdiction"*: East Asians tend to make predictions with reference to a wider variety of factors than Americans do. Consequently, they are less surprised by any given outcome and they are more prone to "hindsight bias", or the tendency to regard events as having been inevitable in retrospect.
3. *Attention*: since East Asians locate causality in the field instead of the object, they tend to be more accurate at "covariation detection", that is the perception of relationship within the field.
4. *Control*: Americans are more subject to the "illusion of control", that is, a greater expectation of success when the individual is involved in interaction with the object – even when that interaction could not logically have an effect on the outcome.
5. *Relationships and similarities* vs. *rules and categories*: East Asians tend to group objects and events on the basis of their relationships to one another, for example, "A is a part of B". Americans would be expected to group them more on the basis of category membership, for example, "A and B are both Xs". Americans are inclined to learn rulebased categories more readily than East Asians and to rely on categories more for purposes of inductive and deductive inference.
6. *Logic* vs. *experiential knowledge*: East Asians are more influenced by prior beliefs in judging the soundness of a formal argument. Americans are more able at setting aside prior beliefs in favor of reasoning based on logical rules.
7. *Dialectics* vs. *the law of noncontradiction*: East Asians are inclined to seek compromise solutions to problems ("Middle Way") and to reconcile contradictory propositions. Americans tend to seek solutions to problems in which a given principle drives out all but one competing solution, to prefer arguments based on logic, and to reject one or both of two propositions that could be construed as contradicting one another.

The crucial thesis of Nisbett et al. [48, 49] is that the different ways of reasoning, that is the different ICRs, are not a contingent and superficial feature, but they are rooted in two completely different systems of thinking, that is, in different metaphysical and epistemological principles contained in the TBK, that shape the American and East Asian cognition differently. These two different systems of thinking originated causally from two different sociocultural environments: the old Greek trading society and classical philosophy on one hand and the old Chinese agricultural society and Confucian philosophy on the other. In fact, according to them, social organization and economic structure are the major determinants of the causal chain metaphysics-epistemology-cognition. Different socioeconomic configurations generate fixed irreversible different causal chains. Different social and economic variables gave birth to different styles of thought that we can summarize under the heading of "holistic" and "analytic" thought. Nowadays, these different styles of thought continue to be effective in differentiating the reasoning processes of contemporary Americans and East Asians.

Norenzayan [51] also confirms, experimentally, the results of Nisbett et al. [48–50]. The cultural differences between Western and Asiatic populations are examined in a variety of cognitive tasks that involve formal and intuitive reasoning. "Formal reasoning is rulebased, emphasizes logical inference, represents concepts by necessary and sufficient features, and overlooks sense experience when it conflicts with rules of logic. Intuitive reasoning is experience-based, resists decontextualizing or separating form from content, relies on sense experience and concrete instances, and overlooks rules and logic when they are at odds with intuition. The reasoning of European American, Asian American, and East Asian university students was compared under conditions where a cognitive conflict was activated between formal and intuitive strategies of thinking. The test showed that European Americans were more willing to set aside intuition and follow rules than East Asians".

Norenzayan [51] agrees with the previous consideration about the relationships between TBK and ICRs. The human mind is equipped with basic cognitive primitives and possesses cognitive processes that carry out many tasks, such as exemplar-based categorization, deductive reasoning, causal attribution, and so on. However, this basic endowment does not rule out differentiated development in response to cultural and environmental stimuli. These differences are manifested in various ways. Firstly, different cultural practices can make a given cognitive process, which is universally available in principle, accessible in a differentiated way. Asians appear to have a greater propensity than Westerners for exemplar-based categorization, and a lesser propensity to decontextualize deductive arguments and more to explain behavior by referring to the situational context. Secondly, through discoveries and inventions, societies often introduce artificial and complex new ways of thinking which differentiate one culture from another. One need only think of the statistic and probabilistic revolution in the 17th century and its impact on Western rationality and decision-making models. Or the development and influence of the ancient Taoist notion of yin and yang in the contemporary Chinese way of reasoning in relation to modal concepts like change, moderation and relativism.

In conclusion, the cultural diversities of TBK lead to different ICRs. This diversity at the level of TBK is often an underlying factor for difficulties involving social coordination and the communication and transmission of knowledge. This can often be seen in the relationship between individuals belonging to radically different cultures, for example from Eastern and Western cultures:

> There are very dramatic social-psychological differences between East Asians as a group and people of European culture as a group. East Asians live in an interdependent world in which the self is part of a large whole; Westerners live in a world in which the self is a unitary free agent. Easterners value success and achievement in good part because they reflect well on the groups they belong to; Westerners value these things because they are badges of personal merit. Easterners value fitting in and engage in self-criticism to make sure that they do so; Westerners value individuality and strive to make themselves look good. Easterners are highly attuned to the feelings of others and strive for interpersonal harmony; Westerners are more concerned with knowing themselves and are prepared to sacrifice harmony for fairness. Easterners are accepting of hierarchy and group control; Westerners are more likely to prefer equality and scope for personal action. Asians avoid controversy and debate; Westerners have faith in the rhetoric of argumentation in arenas from the law to politics to science. [50, pp. 76–78]

The different composition of TBK in terms of its principles and values generates profound differences between various aspects of everyday life and social organization. In particular, as is highlighted by Nisbett [50, pp. 193–201], there are dramatic differences in the way in which medicine, science, law, contracts, conflicts, rhetorics, political relations, human rights and religion are developed and perceived. These differences emerge as the result of contextual diversity in the causal relationship between TBK and ICRs. Such diversity is also found in more homogeneous cultural settings. For example, the difficulty of establishing relations and transferring knowledge between academic research laboratories and businesses appears to be caused precisely by contextual diversity in the relationship between TBK and ICRs[5].

---

[5] In a study of the sociocognitive difference between academic and industrial research, we have hypothesized that the difficulties of collaboration and transferring knowledge are based on the presence of different values in TBK, such as a different evaluation of time, different importance given to money and increased importance attributed to scientific reputation, which generate different decision-making ICRs in terms of risk assessment, treatment of sunk costs, and the falsification or confirmation of hypotheses [52].

Acknowledgement. This paper was developed within the scope of a research project financed by the Italian Ministry of University and Research (FIRB project 2003 - Prot. RBNE033K2R "A multidimensional approach to technology transfer for more efficient organizational models"). Acknowledgements are made to the Ministry, to the general coordinator of the FIRB project and to all the partners.

References

1. Polanyi, M.: *Personal Knowledge: Towards a Post-critical Philosophy.* Routledge & Kegan Paul, Oxford (1958)
2. Polanyi, M.: *The Tacit Dimension.* Doubleday, New York (1967)
3. Ryle, G.: *The Concept of Mind.* Chicago University Press, Chicago (1949)
4. Anderson, J.: *The Architecture of Cognition.* Harvard University Press, Cambridge (1983)
5. Gorman, M.: Types of knowledge and their roles in technology transfer. *Journal of Technology Transfer* **27** (2002) 219–231
6. Johnson, B., Lorenz, E., Lundvall, B.: Why all this fuss about codified and tacit knowledge? *Industrial and Corporate Change* **11** (2002) 245–262
7. Ziman, J.: *Reliable Knowledge.* Cambridge University Press, Cambridge (1979)
8. Lehrer, K.: *Theory of Knowledge.* Routledge, Oxford (1990)
9. Dancy, J., Sosa, E., eds.: *TA Companion to Epistemology.* Basil Blackwell, Oxford (1992)
10. MacKenzie, D., Spinardi, G.: Tacit knowledge, weapons design and the uninvention of nuclear weapons. *American Journal of Sociology* **101** (1995) 44–99
11. Vincenti, W.: *What Engineers Know and How They Know It.* Johns Hopkins University Press, Baltimore (1990)
12. Searle, J.: *The Rediscovery of the Mind.* MIT Press, Cambridge (1992)
13. Searle, J.: *The Construction of Social Reality.* Free Press, New York (1995)
14. Berger, P., Luckmann, T.: *The Social Construction of Reality. A Treatise in the Sociology of Knowledge.* Doubleday, New York (1966)
15. Woolcock, M.: Social capital and economic development: Toward a theoretical synthesis and policy framework. *Theory and Society* **27(2)** (1998) 151–208
16. Granovetter, M.: Economic action and social structure: the problem of embeddedness. *American Journal of Sociology* **49** (1985) 323–334
17. Saxenian, A.: *Regional Advantage: Culture and Competition in Silicon Valley and Route 128.* Harvard University Press, Cambridge (1994)
18. Lawson, C., Lorenz, E.: Collective learning, tacit knowledge and regional innovative capacity. *Regional Studies* **33** (1999) 305–317
19. Chomsky, N.: *Knowledge of Language.* Praeger, New York (1986)
20. Chomsky, N.: *Reflections on Language.* Fontana, Glasgow (1975)
21. Damasio, A.: *The Feeling of What Happens: Body and Emotion in the Making of Consciousness.* William Heinemann, London (1999)
22. Reber, A.: *Implicit Learning and Tacit Knowledge. An Essay on the Cognitive Unconscious.* Oxford University Press, Oxford (1993)
23. Cleeremans, A.: Implicit learning in the presence of multiple cues. In: *Proceedings of the 17th Annual Conference of the Cognitive Science Society.* (1995)
24. Cleeremans, A., Destrebecqz, A., Boyer, M.: Implicit learning: News from the front. *Trends in Cognitive Science* **2** (1998) 406–416

25. Atkinson, A., Thomas, M., Cleeremans, A.: Consciousness: Mapping the theoretical landscape. *Trends in Cognitive Science* **4** (2000) 372–382
26. Raichle, M.: The neural correlates of consciousness: An analysis of cognitive skill learning. *Philosophical Transactions: Biological Science* **353** (1998) 1889–1901
27. Zeman, A.: Consciousness [invited review]. *Brain* **124** (2001) 1263–1289
28. Broadbent, D., Fitzgerald, P., Broadbent, M.: Implicit and explicit knowledge in the control of complex systems. *British Journal of Psychology* **77** (1986) 33–50
29. Cleeremans, A., Jiménez, L.: Fishing with the wrong nets: How the implicit slips through the representational theory of mind. *Behavioral and Brain Sciences* **22** (1999) 771
30. Gigerenzer, G.: *Adaptive Thinking. Rationality in the Real World.* Oxford University Press, Oxford (2000)
31. Nonaka, I., Takeuchi, H.: *The Knowledge-Creating Company.* Oxford University Press, New York (1995)
32. Cohen, M., Burkhart, R., Dosi, G., Egidi, M., Marengo, L., Warglien, M., Winter, S.: Routines and other recurrent action patterns of organizations: Contemporary research issues. *Industrial and Corporate Change* **5** (1996) 653–698
33. Collins, H.: Tacit knowledge, trust, and the Q of sapphire. *Social Studies of Science* **31** (2001) 71–85
34. Nelson, R., Winter, S.: *An Evolutionary Theory of Economic Change.* Harvard University Press, Cambridge (1982)
35. Anderson, J.: Skill acquisition: Compilation of weak-method problem solutions. *Psychological Review* **94** (1987) 192–210
36. Passingham, R.: Functional organisation of the motor system. In Frackowiak, R., Friston, K., Frith, C., Dolan, R., Mazziotta, J., eds.: *Human Brain Function.* Academic Press (2006)
37. Petersen, S., van Mier, H., Fiez, J., Raichle, M.: The effects of practice on the functional anatomy of task performance. In: *Proceedings of the National Academy of Science USA.* (1998) 853–60
38. Berry, D.: The problem of implicit knowledge. *Expert Systems: The International Journal of Knowledge Engineering* **4** (1987) 144–151
39. Berry, D., Broadbent, D.: Interactive tasks and the implicit-explicit distinction. *British Journal of Psychology* **79** (1998) 251–272
40. Berry, D., Dienes, Z.: The relationship between implicit memory and implicit learning. *British Journal of Psychology* **82** (1991) 359–373
41. Dienes, Z., Berry, D.: Implicit learning: Below the subjective threshold. *Psychonomic Bulletin Review* **4** (1997) 3–23
42. Viale, R.: Introduction: Local or universal principles of reasoning? In Viale, R., Andler, D., Hirschfeld, L., eds.: *Biological and Cultural Bases of Human Inference.* Lawrence Erlbaum Associates Inc. (2006)
43. Spelke, E., Phillips, A., Woodward, A.: Infants' knowledge of object motion and human action. In Sperber, D., Premack, D., Premack, A., eds.: *Causal Cognition.* Oxford University Press (forthcoming)
44. Sperber, D., Premack, D., Premack, A.: *Causal cognition.* Oxford University Press, Oxford (1995)
45. Keil, F.: The growth of causal understandings of natural kinds. In Sperber, D.e.a., ed.: *Causal cognition.* Oxford (1995)

46. Gelman, R.: First principles organize attention to and learning about relevant data: Number and the animate-inanimate distinction as examples. *Cognitive Sciences* **14** (1990) 79–106
47. Viale, R., Pozzali, A.: *Tacit Knowledge and Cognition* (forthcoming)
48. Nisbett, R., Masuda, T.: Culture and point of view. In Viale, R., Andler, D., Hirschfeld, L., eds.: *Biological and Cultural Bases of Human Inference*. Lawrence Erlbaum Associates Inc. (2006)
49. Nisbett, R., Peng, K., Choi, I., Norenzayan, A.: Culture and systems of thought: Holistic vs. analytic cognition. *Psychological review* **108** (2001) 291–310
50. Nisbett, R.: *The Geography of Thought*. The Free Press, New York (2003)
51. Norenzayan, A.: Cultural variation in reasoning. In Viale, R., Andler, D., Hirschfeld, L., eds.: *Biological and Cultural Bases of Human Inference*. Lawrence Erlbaum Associates Inc. (2006)
52. Viale, R., Pozzali, A., Passerini, G.: *Sociocognitive Models in Academic and Industrial Research* (forthcoming) Fondazione Rosselli, www.fondazionerosselli.it.

The Functional-Analogical Explanation in Chinese Science and Technology
A Case Study of the Theory of Yin-Yang and Five Elements

Huaxia Zhang and Zhilin Zhang

China's major Research Base of Philosophy of Science and Technology, Shanxi University, Taiyuan, China, and Department of Philosophy, Sun Yat-sen University, Guangzhou, P.R. China
{ssszhx,hsszhang}@sysu.edu.cn

Summary. Philosophy offers the culture of science and technology a certain research tradition composed of two elements, ontological assumptions and epistemological-methodological principles, which restricts specific explanation model. Through the case study of the theory of Yin-Yang and Five Elements, this paper states the ontological assumptions and epistemological-methodological principles of the research tradition in ancient China, analyzes the functional-analogical explanation model's positive and negative influences on Chinese culture of science and technology, tries to answer the famous "Needham's Problem", and criticizes some trends in the culture of science and technology in contemporary China.

The development of science and technology is dependent on cultural backgrounds. In the view of philosophy, cultural backgrounds offer science and technology ontological assumptions and epistemological-methodological principles which form the research tradition in culture of science and technology [6]. With different ontological assumptions and epistemological-methodological principles, different scientific communities raise different explanation models for understanding experiential phenomena.

The theory of Yin-Yang and Five elements is a fundamental research tradition and explanation model in Chinese culture of science and technology. The authors in this paper hold that the model is one of functional-analogical explanations based on intuition, and that it has profound influences on Chinese culture of science and technology.

1 The Theory of Yin-Yang and Five Elements in the View of the Research Tradition

The term "research tradition" in this paper is from Laudan, who states: "A research tradition is a set of general assumptions about the entities and processes in a domain of study, and about the appropriate methods to the used for investigating the problems and constructing the theories in that domain" [6, p. 81]. According to this definition, a research tradition contains two key elements, ontological assumptions and epistemological-methodological principles.

We hold that the research tradition dominating ancient Chinese culture of science and technology can be featured by three key concepts, Qi, Yin-Yang, and Five Elements. Historically, the three concepts have different origins. Once they merged into one, however, it gradually became the research tradition ruling the whole Chinese culture of science and technology. Moreover, we can say that it is a kind of "perpetual philosophy", in Joseph Needham's word. The ontological assumptions of this research tradition include the following:

(A) Qi is the noumenon (or first principle) which produces and forms every thing in the world. Chuang Tru states: "It is Qi that unifies the whole word" (*Chuang Tru*, Chapter 22, "Zhibei Travels"). Qi is not like air, one of Four Elements, in Aristotle's philosophy. It is continuous and permeates the cosmos, and it is invisible. Therefore, Qi is not like the atoms in the eyes of philosophers in ancient Greece, nor is it like their concept of void, for the universe is permeated with Qi, as Wang Fuzhi states: "The Void contains Qi, and Qi is full of the Void. There is no not-being"(*Comment on Zhang Zai's Correcting Youthfel Ignorance*).

(B) From the point of view of a functional analogy, Qi is classified into two kinds, Yin's and Yang's, and all the things in the world can be classified in Yin and Yang. Yang originally refers to the part of a mountain facing the Sun, while Yin refers to the part upon which the Sun does not shines. However, using character-imagination and function-analogy, ancient Chinese people infinitely extended the original meanings of Yin and Yang, and thought that every thing can belong to Yin or Yang. For example, the Sun belongs to Yang while the Moon to Yin; brightness to Yang while darkness to Yin; the heaven to Yang while the earth to Yin; South to Yang while North to Ying; Heat to Yang while Cold to Yin; Hardness to Yang while Softness to Yin; movement to Yang while still to Yin; man to Yang while woman to Yin; the back of the human body to Yang while the abdomen to Ying; the five internal organs (heart, liver, spleen, lung, and kidney) to Yang while the six hollow organs (gallbladder, stomach, large intestine, small intestine, bladder, and "saujiao") to Yin, and so on. Evidently, Yin and Yang are not like the concepts such as the South Pole and the North Pole, positive charge and negative charge, positive particle and negative particle, because the former is based on intuitive imagination and analogy, while the latter is based on logical analysis and experiment. Generally speaking, the moving, the high, the warm,

Thinking of the quotations, and considering the common translation in the long history of combination between Chinese culture and the Western one, we have chosen the translation of "Five Elements". At the same time, we do not neglect the meanings of "movement", "change", "process", and "function". These meanings fully show Chinese intuitive-analogical-thinking perspective, by which everything in the world – and its natures – belongs to one of Five Element. According to this perspective, the classification of all things can be made on the basis of Five Elements (see Table 1).

There are two basic interactions, "Sheng" and "Ke", among the five. "Sheng" means that one thing can promote, produce, cause, and push another thing, while "Ke" means that one thing can restrain, win, and conquer another thing. The sequence of "Sheng" is that wood promotes fire, which promotes earth, which promotes metal which promotes water, which promotes wood. It can be expressed as shown in Figure 1.

The sequence of "Ke" is that wood restrains earth, which restrains water, which restrains fire, which restrains metal, which restrains wood. Cf. Figure 2:

These two sequences can also be expressed by Figure 3. Tong Zhongshu explains: "Broken lines represent 'Ke', while unbroken lines represent 'Sheng' ".

We have described above the ontological assumptions of Chinese research tradition. Let us now turn to its epistemological-methodological principles:

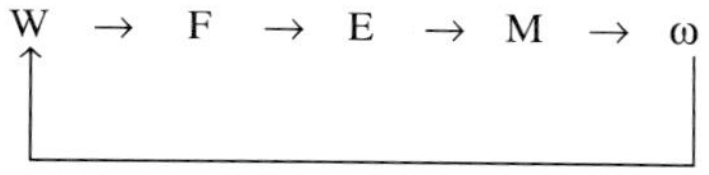

$$W \rightarrow F \rightarrow E \rightarrow M \rightarrow \omega$$

Fig. 1. The sequence of promotion among Five Elements.

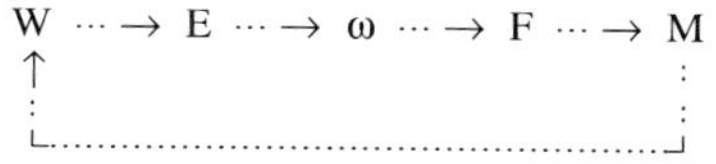

$$W \cdots \rightarrow E \cdots \rightarrow \omega \cdots \rightarrow F \cdots \rightarrow M$$

Fig. 2. The sequence of restraint among Five Elements.

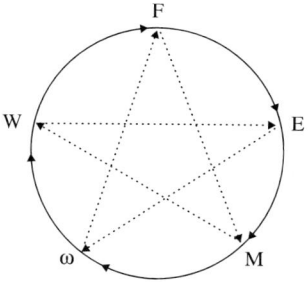

Fig. 3. The broken line represent restraints while the unbroken line represent promotion.

(a) Everything in the world is composed of Qi which has two kinds, Yin's and Yang's. The principle which make us able to study all problems and to explain all phenomena is that we have to recognize whether Yin's and Yang's are in the state of dynamic equilibrium and harmony or not. This principle is manifested in the sequences of "Sheng" and "Ke" among Five Elements.
(b) Therefore, the intuitive relationship and functional analogy between the Five Elements and all things in the world is the principle that make us able to explain all experiential phenomena.
(c) "Sheng", among the Five Elements, shows a positive relationship. Looking at the Figure 1, the move strongly Wo acts on F, the more strongly F acts on E; the more weakly Wo acts on F, the more weakly F acts on E. The other positive relationships are the same. As a result, "Sheng" among Five Elements forms a circle of positive feedbacks.
(d) "Ke", among the Five Elements, shows a negative relationship. Looking at the Figure 2, the more strongly Wo acts on E, the less weakly E acts on Wa; the less weakly Wo acts on E, the more strongly E acts on Wa. The other negative relationships are the same. Therefore, "Ke", among the Five Elements, forms a circle of negative feedbacks.
(e) The relationship between "Sheng" and "Ke" is negative. Looking at the Figure 3, the more strongly Wo acts on F, the less weakly Wa acts on F; the less weakly Wo acts on F, the more strongly Wa acts on F. The other relationships between "Sheng" and "Ke" are the same. Therefore, both the circle of positive feedbacks and the circle of negative feedbacks form a system of dynamic equilibrium.

These epistemological-methodological principles have been showed at work in medicine, astronomy, alchemy, etc. We will only describe some example coming from medicine in ancient China. In *Yellow Emperor's Inner Classics*, there are many interesting statements: "Qi is the substance of life", "man's essence of life, energy, saliva, fluid, blood, and pulse all can be reduced into Qi", "If various kinds of Qi fit each other, human body is in a harmonious state; if not, man is ill". In the perspective of Chinese medicine, health is dependent on harmony of Qi, while illness is dependent on disharmony of it. Therefore, The essence of therapy is the adjustment of Qi. The adjustment actually consists in adjusting the relationship between Yin and Yang. For "excess of Yin or Yang means illness", "excess of Yang leads to Yang's illness, and excess of Yin leads to Yin's illness", "when a good diagnostician observes patient's face and feels the pulse, he should make a judgement on the classification of patient's illness at first (Does it belong to Yin's illness, or to Yang's one?)" "Those who know Yang should know Yin, and those who know Yin should know Yang".

The coordination of Yin and Yang is reflected in the dynamic equilibrium of the Five Elements' "Sheng" and "Ke". The classification of five viscera into Five Elements is based on the intuitive imagination of functions and analogy (cf. Table 1). Therefore, the mythological principles of Five Elements "Sheng" and "Ke" are useful in Chinese medicine. For example, if a patient coughs, has soft phlegm, and has no appetite, his illness is diagnosed as energys deficiency in spleen, which leads to that saliva and fluid cannot circulate efficiently, thus causing Sheldon drop of vital energy in lung, and causing cough. In this case, the cause of the illness is the obstruction of the movement of spleen's energy to lung (obstruction of the wood promotion of metal). The therapeutical method is to enrich the energy of speed and lung (to aid earth and then metal).

2 The Theory of Yin-Yang and Five Elements and the Functional-Analogical Explanation

What is an explanation? Nagel says: "Explanation is to answer the 'why'-questions" [7, p. 15]. In other words, an explanation is a way of saving the phenomena, to provide reasons for the statements which we want to understand, and to answer our question, why would matters be like what the statements state?

How does the theory of Yin-Yang and Five Elements answer the question of "why"? Let us look at the following examples.

> Question 1: Why has Zhou Dynasty as a unified dynasty been displaced by Qin Dynasty?

Scholars in ancient China answer the question on the basis of the classification of Dynasties in accordance to the classification of Five Elements. According to the Figure 2, they make the following conclusion:

> Qin Dynasty (Water) – Zhou Dynasty (Fire) – Shang Dynasty (Metal)
> – Xia Dynasty (Wood) – Yellow Emperor (Earth)

> In Lu's *Spring and Autumn* it is stated: "When a new emperor is going to appear, there must be auspicious signs. At the time of the Yellow Emperor, there appeared the biggest earthworm and the mole cricket. The emperor said that Earth dominates the world. Because of that, Yellow was regarded as the best color, and Earth as the most important. At the time of Yu (the emperor of Xia Dynasty), trees and grasses did not wither in fall and winter as usual. Yu said that Wood dominates. Because of that, green was regarded as the best color, and Wood as the most important. In time of Tang (the emperor of Shang Dynasty), Sword jumped from water. Tang said that Metal dominates. Because of that, White was regarded as the best color, and Metal as the most important. In time of Wenwang (the emperor of Zhou Dynasty), birds keeping red books on their mouth gathered at Zhou's altar. Wenwang said that Fire dominates. Because of that, Red was regarded as the best color, and Fire as the most important. The replacement of Fire would be Water, and Water would dominate. Because of that, Black would be regarded as the best color,

and Water as the most important. And the next turn would be for Earth."
We can see that the statement of "Water conquers Fire" is applied to answer
the question 1; and we shall note that it is predicted that the replacement of
Qin Dynasty must belong to Earth. It is equivalent to Boyang Fu's prediction
of Zhou Dynasty's perishing on the basis of the disorder of Yin and Yang.
The logical structure of the explanation is the following:

Explanans:

1. The classes of functions: Water, Fire, Metal, Wood, Earth.
2. The relations between classes: Water conquers Fire.
3. What class the events belong to: Qin Dynasty belongs to Water, and Zhou Dynasty belongs to Fire.

Explanandum:

4. Qin Dynasty replaced Zhou Dynasty and became the ruler of China.

According to the covering-law model of scientific explanation [4], if we take the statement, "Water conquers Fire" as a universal law, the explanation described above obeys to the D-N model (deductive-nomological model); if we take the statement as a statistical law, the explanation obeys to the I-S model (the inductive-statistical model). However, it is questionable to consider a statement like "Water conquers Fire" in the theory of Yin-Yang and Five Elements a "law". If you state that all the things belonging to Water have the function to conquer Fire only on the basis of the observation that Water can extinguish Fire, without analysis or tests through experiments, it will be difficult to say that your statement is qualified to be considered a universal law, needless to say, a universal law in terms of a relationship between cause and effect. Furthermore, in both D-N explanation and I-S explanation, the explanans should contain statements about the experiential conditions under which the laws holds.

However, in the above structure of explanation, the explanans does not contemplates a "law-like" statement like "Water conquers Fire". On the basis of "intuitive imagination and functional analogy", the man who makes the explanation holds: (1) everything in the world can be classified into five kinds (Five Elements); (2) there is a relationship between Water and Fire: Water can conquer Fire; and (3) Qin Dynasty belongs to Water, while Zhou Dynasty to Fire. Then he assumes these statements as the explanans for the explanandum: (4) Qin Dynasty replaced Zhou Dynasty. There is neither analysis of causation on Zhou Dynasty's perishing and Qin Dynasty's rise, nor explanation of purpose. What he offers is the functional-analogical explanation which has two characteristics: 1) explanans include no laws in strict sense, but functional classes and their relationship; and 2) the relationship between explanans and explanandum is not deductive or inductive, but analogical. It presents a low probability. In other words, the quality of the explanation is very low. In addition, we should note that it is an *ad hoc* explanation. In order to explain Qin Dynasty's rise, the Dynasty must be classified as Water.

Question 2: Why do I often have hectic fever in the afternoon, feel upset, and have a red face, emaciation, night sweat, dryness in my mouth and throat, red tongue with little coating, and weak pulse?

On the basis of the theory of Yin-Yang and Five Elements, Chinese medicine offers an explanation: it is because of deficiency of Yin in kidney. It is a functional-analogical explanation which is different from that given in Western medicine. The latter uses tests (of blood, urine, X-ray, etc.) to detect the cause of the disease (for example, the effects of bacteria, virus, etc.), and to ascertain the focus for example of an infection. In the light of certain medical causal laws (e.g., under certain condition, certain kind of bacteria would certainly causes certain kind of disease), certain boundary conditions, and the particular statements from the tests about the patient, western doctors can logically explain the whole symptoms in Question 2: we hypothesize a Chronic nephritis, or a pulmonary tuberculosis, or a diabetes, etc. On the other hand, doctors in Chinese medicine probe neither substantial causal relationship, nor the focus of disease (infection, for example). They only make functional classification, beginning with classifying human body into five internal organs (liver, heart, spleen, lung, kidney) on the basis of the Five Elements.

It is very important to note that the organs, in the perspective of Chinese medicine, are different from those in the perspective of Western medicine: the former is functional while the latter is "substantial" and anatomical [15]. In order to make a more plentiful and complex classification of diseases, doctors in Chinese medicine add the Eight Concepts of Yin-Yang, exterior-interior, chills-fever, and void-solid, taking Yin-Yang as the key. Matching five internal organs with the Eight Concepts, there can be totally 80 basic functional kinds of diseases. Considering the degrees of every kind of diseases and their combination, there are surely more than one million logically possible kinds of diseases. Chinese medicine's explanation of symptoms actually is related to the classification into different functional sets. Of course, the task of making the classification is not a simple job. It needs technical knowledge and expertise also related to the suggestion of the therapy. Again, we can see that the model of functional-analogical explanation is applied to explain the symptoms in questions. The process can be summarize in the following way:

Explanans:

1. The classes of functions: a system of classification of diseases is related to the pathological changes of man's five internal organs and six hollow organs and the Eight Concepts mentioned above.
2. The characters of the class: the disease of deficiency of Yin in kidney has symptoms stated in question 2.

3. The characters of the individual in question: I have most of the symptoms stated in question 2.

Explanandum: Therefore, I have the disease of deficiency of Yin in my kidney.

If I ask the Chinese doctor a question, "What on earth is my disease?" I will get no answer. Because he does not have empirical tests and universal "causal" laws, and he consequently cannot make D-N or I-S explanation for my symptoms. In the end, the doctor writes out a prescription of "Liuwei dihuang Wan", and tells me the daily dosage.

Question 3: Why can the medicine of "Liuwei dihuany Wan" cure my disease of deficiency of Yin in my kidney?

The explanation is based on the theory of Yin-Yang and Five Elements. According to the theory, Chinese doctor classifies medicinal herbs into two kinds (Yin's on Yang's), five kinds (Five Elements), and four kinds (the ascending, the descending, the floating, the sinking) and so on. Then, after having ascertained the classification of the medicine as "Linwei dihuang wan", he states that the medicine has the effect to nourish Yin and to enrich the energy in kidney. Because of "Sheng" and "Ke" among the Five Elements, the medicine which I have taken can help my body recover dynamic equilibrium.

Putting a phenomenon into a certain class, and then using the universal characters of the class to explain or predict characters of the phenomenon, was a common way to understand the world in early human history. And it also existed later on in other historical times. The value of this kind of explanation depends on whether the classifications are proper, and whether the supposed universal characters of a class are applicable to all the individuals. If they not, it will be dependent on the degree of the applicability, and on whether the phenomenon to be explained belongs to the class. The key problem is how to make classifications, and what is the role of the functional classification in various classifications.

Let us consider the following three kinds:

1. Natural Kind/Class. It is a set of elements which have common characters, $P_1, P_2, \ldots, P_n$, checked through analysis and experiment. Separated, each of the characters is necessary for the set, while united, the whole of them is sufficient for the set and having all of them is a sufficient condition for the elements to belong to the set. Therefore, the common characters can be used to define the class. Certain atoms, certain molecules, or certain species can all be taken as natural kinds. For example, all molecules of water (H_2O) have the same basic properties under the same conditions. The statement, "All water freezes at $0\,°C$", can explain the phenomenon, "Today water in this lake freezes". For natural kinds, classification explanation is actually a form of the D-N model. The explanation is deductive.

2. Class through Family Resemblances. The concept of "family resemblances" is due to Wittgenstein [14]. According to him, the members in a "family" have no common characters, but there is a complex net of overlap of similarities. For example, mixture, earth, furniture, ship, and the most typical one "game" are all concepts belonging to such classes. Zhang Zhilin, one of the authors in this paper, finds out something similar to "essences" (it is better to call them "quasi-essences") from Wittgenstein's concept of family resemblances. They are common characters of most members in the same "family", or characters in the core of the net of family resemblances [16, p. 46]. Because of the quasi-essence in the class with family resemblances, when someone gives a family resemblance explanation to the characters of certain phenomenon, he cannot deduce the characters of the phenomenon from characters of the class. However, he can use induction with high probability to support his explanation. For example, when he explains somebody's death, he states that because the person had a serious disease he died. Such a serious disease is a class with family resemblances, and most members in the class have a common character of death. Many of I-S explanations belong to the explanation of class with family resemblances.
3. Functional Class through Intuition. Because the class is based on intuition, analogy, and conjecture, it is not at the same level as the natural kinds. In this class, there are no common characters for all of its members, or for most of them. As a result, the ability of explanation based on functional lass is not high. In comparison with the above two explanations, this one only has supports from weak induction and reasoning from analogy. This is why using the theory of Yin-Yang and Five Elements to explain Qin Dynasty's victory is farfetched. However, before people can make D-N and I-S explanations, the explanation based on functional class through intuition is a primary alternative of explanation after all.

3 The Explanation Based on Yin-Yang and Five Elements

How to value the rule of functional-analogical explanation based on Yin-Yang and Five Elements in the development of Chinese science and technology is a big problem. In this paper we try to answer Joseph Needham's famous difficult question, and criticize some trends in the Chinese culture of science and technology.

1. The model of functional-analogical explanation has both positive and negative effects on Chinese science and technology. Chinese people have accumulated vast knowledge about nature and technology in the history of civilization. At some stage of the development of knowledge, theoretical explanations have to be formed. As a system of analogy, classification and arrangement of phenomena, the research tradition of Yin-Yang and Five Elements was

and is very helpful for Chinese people to systematize their experiences and knowledge, and to accumulate and develop them. For example, when people applied the theory of Yin-Yang and Five elements to make pills of immortality, their actions unintentionally lead to the invention of powder. The theory is also essential for Chinese medicine. It seems that up to now, no alternative can be found to replace the classification system of Yin-Yang and Five Elements which can classify both diseases and traditional Chinese medicine to make appropriate and effective correspondence between the two and produce the effect that traditional medicine can not go beyond. Furthermore, the research tradition of Yin-Yang and Five Elements has the advantage to offer a philosophy of nature in terms of wholeness and of dynamic equilibrium. This holistic perspective is helpful from many respects.

However, the research tradition, on the whole, is out of date, for it is intuitive and not experimental, synthetical and not analytical, conjectural and not logical. The concepts of Yin, Yang, and Five Elements all are intuitive, they have no clear definition, and even they cannot be defined. For example, the concept of "Metal" refers to metals, gold, autumn, and dryness, etc. It has so many different meanings that you cannot define it. How to use logic with these ambiguous concepts? Reasoning in this tradition depends on conjectures and imagination, and it is neither inductive nor deductive. Why does liver belong to Wood? How can you infer that "liver produces veins" from "Wood's ability to grow"? Why do we say it belongs to Wood? Is not there wood which is sweet or bitter? The theory of Yin-Yang and Five Elements cannot answer these questions. Because of the lack of conformity to logical requirements, it may more easily lead to absurdities such as seeking human body's immortality and practicing divination. Moreover, as we already said, all the basic theories of Chinese medicine based on the ideas of Yin-Yang and Five Elements are not set up by experimental methods. Because Chinese medicine cannot explain how moisture reaches bladder through intestines, it has to fabricate the organ of "Sanjiao" as a passage. And it states that there are "Jing" and "Qi" breathed in and out the human body, and that some part of them circulates through all internal organs and the other part circulates in the whole body through fictitious channels. All these have not been verified by experiments.

We can conclude that the lack of analytic spirit and the lack of experimentation is the fatal disadvantage of the research tradition and the explanation model of Yin-Yang and Five Elements. "Lack of analytic spirit" means lack of methods of analysis and induction to discover true laws in terms of cause and effect. "Lack of experimental spirit" does not mean that Chinese philosophers of nature and technicians do not have anything to do with experimentation and empirical results. Indeed, it is well known that in order to ascertain the healing properties of various herbs, Shen Nong ate them and was poisoned by toxic herbs for seventy times in a day. In Tang Dynasty, alchemists who tried to make pills of immortality did a plenty of experiments, and having blown up many people and buildings, they unintentionally invented powder.

Shen Nong and these alchemists are all good experimentalists. However, the non-experimental and anti-experimental spirit represented by the theory of Yin-Yang and Five Elements is: (i) There is no such systematic experimental approach to reveal causation, by which scientific experiments can be designed and done; (ii) theories are not designed to be testable or falsifiable; and (iii) experimentation is not taken as the criterion of truth even if propositions and theories can withstand empirical tests. In this sense, we hold that the research tradition lacks the experimental spirit. What they do at most belongs to the context of discovery, but not to the context of justification.

2. Joseph Needham's difficult question may be answered by checking Chinese research tradition. In his study of the history of Chinese science and technology, Needham in 1938 raised a question which afterward was expressed in the following way: "Since Chinese people made so many achievements of science and technology in early time, why did not they develop modern science?" [10]. We hold that the Chinese research tradition for a long time was too indifferent to scientific experiments, to analytical and inductive methods, and to strict logical reasoning. In China there is no sufficient changes in scientific methodology able to create and promote a full interest in experimental method, in analysis, and in mathematical and logical tools. It is well known that modern science in Europe began with a "revolution" in which the experimental method, the practice of induction and deduction, and the exploitation of mathematical tools, became central. Galileo, Bacon, and Descartes violently discussed the cognitive value of the Aristotelian tradition and the kinds of "explanations" related to it.

On the other hand, the Chinese research traditions is specifically stubborn. Although some thinkers, such as monists in pre-Qin Dynasty and Wang Chong in Han Dynasty, were skeptical about it, no new research traditions had been set up. Intuitive-functional-analogical method and explanation model may be helpful for science research, but if this method and this model stay at the core of the research tradition, there is no room for modern science. Therefore, we think that the following Einstein's statement is the correct answer to Needham's difficult question, "The development of Western Science is based on two great achievements, the system of formal logic invented by ancient Greek philosophers (in Euclid's geometry) and the possibility to find out relationships between cause and effect through systematic experiments (on the period of the Renaissance). In my point of view, it is not surprising that Chinese sages do not step into the two" [2, p. 574]. It shall be noted that the reason why the sages cannot do that task is that because their minds are dominated by the research tradition of Yin-yang and Five Elements.

The Needham's question presents another aspect. Since modern science only developed in European civilization, but not in Chinese culture, why was Chinese civilization much more effective in obtaining knowledge of nature and applying it to human practical needs in the period from 100 B.C. to 1400 A.D.? [9]. Since we are not historian of science, we cannot use many historical facts to answer the question. But we can suggest some remarks. When it is said that Chinese science and technology surpassed Western counterparts, this mainly refers to technology, such as in the case of the invention powder, compass, and printing. Although there is a close relationship between scientific and technological knowledge, they are two different ones at all, they often had and independent life. Knowing how without knowing why is certainly possible. Chinese traditional focus of science and technology on practical results is related to disregard the theoretical aspects of science. This is the reason why modern science did not appear in China. However, it is not surprising that the focus on practical aspects generated those many results of Chinese technology that surpassed Western ones before the fifteenth century.

3. We do not think we have to give up the functional-analogical explanation model of Yin-Yang and Five Elements. What we have stated above does not mean that the model has to be eliminated. What we mean is that the model is a primitive or auxiliary way to known the world, and that it cannot occupy the key position in a modern system of science. We strongly maintain that the research tradition of Yin-Yang and Five Elements should have space to continue to develop freely. It is still helpful to diagnose and to choose and make therapies in Chinese medicine. We think that three kinds of medicine, Western (modern), Chinese (traditional), and the combination of the two should be developed at the same time. In the combination of the two traditions, the Chinese one can be a great treasure.

However, we should distinguish between the medical technical knowledge on hers, acupuncture, qigong etc. and its philosophical basis, the explanation model of Yin-Yang and Five Elements. Furthermore, we disagree with the obscure statement that Chinese medicine is holist while Western one reductive. Finally, we certainly also disagree with Qian (1996), who thinks it is necessary to abandon the science of life based on modern Western medicine, and focus on a Chinese theory of medicine enhanced by Marxist philosophy in a core position. He thinks this can reestablish a correct science of human body, which he calls "Somatic Science". Furthermore, he also maintains that the research of Qigong and extrasensory perception is the key to open the door of science of human body, and that this research can lead to a new scientific revolution. Of course we are not opposed to the freedom to make research on this "science of human body". It is a pity that we could not discover the publications in the field.

Acknowledgement. The research work of this paper is supported by China's major Research Base of Philosophy of Science and Technology, Shanxi University, Taiyuan, China (No. 04JZD0004).

References

1. Dupré, J.: Natural Kinds. In W.H. Newton-Smith, ed., *A Companion to the philosophy of Science*. Blackwell Publishers (2000)

2. Einstein, A.: *Einstein's Works*, Vol. 1. Business Press, Beijing (1976)
3. Feng, Y.: *New History of Chinese Philosophy*, Vol. 1. People's Press, Beijing (1983)
4. Hempel, C.G.: *Aspects of Scientific Explanation and Other Essays in the Philosophy of Science.* The Free Press, New York (1965)
5. Hou, W., Zhao, J., Du, G.: *History of Chinese Thought.* People's Press, Beijing (1957)
6. Laudan, L.: *Progress and Its Problems: Towards a Theory of Scientific Growth.* University of California Press, Berkeley (1977)
7. Nagel, E.: *The Structure of Science.* Hackett Pub. Co., Indianapolis (1979)
8. Needham, J.: *Science and Civilization in China*, Vol. II. Caves Books. LTD. (1985)
9. Needham, J.: Science and society in the East and the West. *Magazine of Nature* **12** (1990)
10. Needham, J.: Preface. In G. Wang, *Needham and China.* Shanghai's Popular Science Press, Shanghai (1992)
11. Qian, X.: *Free Talk about Science of Human Body and Development of Modern Science and Technology.* People's Press, Beijing (1996)
12. Ren, Y., Liu, C., eds.: *Collected Essays on Inner Classics.* Hubei People's Press, Wuhan (1982)
13. Wang, G.: *Needham and China.* Shanghai's Popular Science Press, Shanghai (1992)
14. Wittgenstein, L.: *Philosophical Investigations.* Translated by G.E.M. Anscombe, Basil Blackwell, Oxford (1967) 3rd Ed.
15. Yin, H., ed.: *The Fundamental Theory of Chinese Medicine.* The Science and Technology's Press of Shanghai, Shanghai (2000)
16. Zhang, Z., Chen, S.: *Anti-essentialism and The Problems of Knowledge.* Guangdong People's Press, Guangzhou (1995)

Model-Based Reasoning and Diagnosis in Traditional Chinese Medicine (TCM)

Zhikang Wang

Department of Social Sciences and Education, Sun Yat-sen University, Guangzhou, P.R. China
zdwangzk@tom.com

Summary. It is common knowledge that there is an essential methodological distinction in dealing with diagnosis in Traditional Chinese Medicine(TCM) and in Modern Medicine(MM). For a long time, understanding the diagnosis in TCM has been quite disregarded. The concept of *model-based reasoning* can help us to get a new and clearer understanding of the cognitive process involved with TCM. In the first part of this paper I will present the most common models coming from the theory of TCM. In the second I will describe how these models are applied. Finally, I will discuss from a methodological point of view the significance of model-based reasoning in the diagnosis of TCM.

1 Introduction

There are three aspects in the research about how people understand objects: (1) the metaphysical consideration of introspective experience, which leads to the controversy between empiricism and rationalism; (2) the scientific research of behavior and symbolic logic, which leads to the controversy between environmental determination and apriorism; and (3) the discussion based on the analogy with computers, which leads to the argument on the contract between computationalism and naturalism.

The core of modern cognitive psychology is an attempt to generalize "cognition" taking advantage of computational analogies [7, p. 4]. Cognition is considered a process that can be analyzed through logical models and procedures.

We can hardly discover the origin of creativity and some related thinking processes with the help of the above approach, because the logic and calculating rules which we learn are only some procedures, just like those in computers. This implies it is difficult for us to grasp the characters of many instinctive thinking processes based on those methodological tools coming from modern science. The creativity of human beings is fundamentally based on a particular natural process. Hence, "returning to nature" becomes an important strategy of research on cognition.

Recently, rising silently in international academic community, there is a trend which focuses on the "returning to nature" in the research about the processes of cognition and thinking. Namely, this trend shows the passage from the approach which takes computers as analogical subjects and pure mathematical logic and algorithms as the central tools, to the approach which studies and imitates the actual human and animal cognitive processes.

Let us come back to the central problem of this paper. As it is well-known, practice shows that both the diagnostic methods of traditional Chinese medicine and of modern Western medicine are fruitful, but there are great differences in their methodologies. One of the differences is that modern Western medicine is built on the basis of logical analysis. Western thinking inclines more toward use of logic and of reasoning methods which have been confirmed by past experience. Chinese medicine originated before modern experimental science was established. At that time logical deductive rules, and important mathematical laws, were still not found and made clear. Hence, TCM diagnoses may be closer to the actual ways of thinking of humans. Let us see in the following the consistency of this assumption.

2 TCM: Theories or Models?

Some people claim TCM does not fulfil good epistemological requirements and so it is not science, but a kind of philosophy; some people believe that TCM is another kind of science, different from modern Western science. There are distinctive explanations of TCM. However, we must admit that many concepts describing and inferring the physical situation of a human body in TCM do not originate experimentally, for example through dissection and testing, and consequently it is hard to analyze TCM in the framework of modern science and logic. Nevertheless, I contend we can still justify TCM on the practical effectiveness of the actual diagnoses and therapies.

2.1 The Difference Between Theories and Models

In modern scientific methodology, there is an obvious difference between theories and models. People usually believe a model is the intermediary or a bridge from reality to theory. A scientific theory is also established on the basis of mental models and so constructed with the help of results made available by creative hypotheses. Of course the theory also take advantage of a final "logical" assessment. Therefore, in modern scientific methodology, the model is a transitional form or tool in theoretical reorganization. The model can be discarded when the theory is established and made standard. So, the understanding of science is based on theoretical knowledge, rather than on models.

Obviously, theories are the fruit of a logical organization, while models just serve as intermediate tools. Models are always considered as important methods to form theories, rather than the basis of reliable scientific judgments. Theories are the achievements of scientific research, while models are related to mental prototypes and to the intuitive imagination based on these prototypes: they can just promote scientific discovery.

In the framework of that movement of "returning to nature" I have quoted above, the classical views on theories and models just mentioned have been changed. During the process of thinking logic and models are intertwined. Scientists know that it is difficult to make research completely depending on strict logical requirements. In the analysis of actual problems, both models and logical (and mathematical) tools are fruitful.

2.2 Theories and Models in TCM

The conclusion above is obvious in TCM. In Chinese traditional medicine the distinction between theories and models was extremely obscure, and it is almost impossible to be described.

The so-called theories of TCM, in fact, can be seen as piles of inter-relative models. One can regard TCM as theories or model systems. But when you name TCM as a theory, you must add that the word is not used in the modern sense. The relationship between concepts or categories in modern scientific theory is deduced by logic, while the basis of TCM is ancient Chinese Philosophy. Moreover, there is a difference between "models" in TCM and in modern science. TCM pays more attention to specific visual objects, rather than to abstract objects (such as, mathematical ones), and though some models of TCM are reshaped through philosophical ideas, they remain considerably concrete.

In complex systems such as TCM, various models interact or even overlap, they possess a layered structure. Nevertheless, the relations among the various models are very clear. The diagnostic reasoning in TCM completely relies on such a model system. I think that the new area of research of the so-called "model-based reasoning" enables us to better understand TCM and its concept of diagnosis. I am convinced that by exploring the TCM diagnostic processes new research and new logical methods on modeling, intuition, emotions, and beliefs, can be promoted. TCM can be a good source of new ideas and perspectives.

The transcendent idea of "the association between man and his universe and the symbolic relation of parts of the body to the universe" – directly transplanted from Chinese philosophy – is the essence of TCM. Such an idea indicates that the human body is a miniature of the universe and of the society. Each part of a human body is regarded as similar to a part of the universe or of the society and it is related to the corresponding component of nature or society. Likewise, the internal mechanisms of the human body resemble the mechanisms of the universe or of the society and exhibit a similar predictability, and similar cause/effect relationships. Thus, a sound basis for reasonable judgments in all deductions of TCM is established. Based on this premise, TCM will take various objects in nature or society as reflective models of the human body and will judge the internal situation of the body according to direct observation, prior experience, anticipation, and intuition. The diversity of nature and society and their complex relations lead to the analogous disorder of TCM. Hence, it is an "art" for doctors engaged in TCM to understand how to handle the relations among different models, in order to make an accurate assessment of a pathological situation of the human body.

Concepts used in TCM are derived from expressions used to describe objects and phenomena in society and nature. They do not form a consistent conceptual system and are not linked through strictly logical relations. The so-called theories of TCM basically are coherent visual models based on objects in nature and society.

2.3 Systems of Models in TCM

From its principal idea of "the unification of heaven and human beings", TCM constructs a set of complete systems of models about what is happening inside the human body using the results from direct observation and imagination of objects and their changes in nature and society [14]. In TCM model systems, there are three fundamental models: "Yin-Yang" "Wu-Hsing" (Five Elements) and "Ch'i "(Gas).

(1) The model of "Yin-Yang" balance

"Yin" and "Yang" come from the Great Absolute. They are the fundamental operational forces of the universe and everything that is in it. Yin means the shady side, and Yang the sunny side of a given location, event, matter, etc. Hence, Yin can be earth, moon, night, female, negative, death, destroying, cold, etc; Yang can stand for heaven, sun, day, male, life-creating, hot and so forth.

In the human body, the skin or external parts are Yang and the internal parts are Yin. The dorsal side is Yang and the ventral side is Yin. The hollow organs called "Fu" are Yang, and the solid organs called "Tsang" are Yin, in ancient Chinese terms. In the case of the organs in the body, there are six Yin organs: the lungs, spleen, heart, kidneys, liver, and the envelope of the heart. The last one is considered to be a functional entity rather than the anatomical pericardium. The Yang organs are: the large intestine, stomach, gallbladder, urinary bladder, small intestine, and the triple-burner, or the "three reaction chambers". The triple-burner is again a conceptual or functional entity, which is divided into three portions, the superior burner for the reaction of air and blood, the middle burner for the reaction of food stuffs, and the inferior burner for the production of waste and its elimination [4].

The model of Yin-Yang balance shows that the human body is a contradictory community, composed of the interactions and mutual restrictions of two forces "Yin-Yang" [3, p. 18].

(2) The model of "Wu-Hsing" or Five Elements (which exhibit mutual production and restriction).

The Five Elements, or Wu-Hsing in Chinese ("Wu" means five, and "Hsing" means walking), consist of Water, Fire, Wood, Metal, and Earth. They denote not exactly the members of a series of five categories or types of fundamental matters, but rather five kinds of fundamental processes.

The Five Elements are arranged in a sequential order to express the mutual influences, as follows:

The order of generation: Water → Wood → Fire → Earth → Metal → (Water)

The order of subjugation: Water → Fire → Metal → Wood → Earth → (Water)

The "Five Elements" are applied to all matters and events, as well as seasons, i.e. everything that is foreseeable and related to human life. According to the conviction of "the unification of heaven and human beings", parts or organs in the human body match the Five Elements distinctively, and their changes follow the changes of the Five Elements.

(3) The model of "Ch'i" or Gas Movement

The Chinese word "Ch'i" literally means gas or air. But Ch'i does not mean gas or air in Chinese philosophy and medicine. Ch'i means life or energy, acting like gas or air. Changes in gas are quite visual, but the results are hard to grasp. So the visual sense about the movement of gas or air is regarded as the circulating template of the original energy of life in TCM.

The Ch'i circulates in the body, beginning daily from the lungs and following the progression as listed below:

Lung → Large Intestine → Stomach → Spleen → Heart → Small Intestine → Urinary Bladder → Kidneys → Envelope of the Heart → Triple-burner → Gallbladder → Liver → Lungs

These twelve organs are active in subsequent stages beginning from the lungs in the early morning and ending with the liver after midnight. The circulation of the Ch'i must never stop and does not have to be obstructed, and it should not be excessive. Both obstruction ("deficiency") and outpouring ("excess") in the Ch'i, at any location of the body, will cause disease.

From the analysis of the three fundamental TCM models above, we can see that the TCM cognitive perspective is based on its wholeness, representing "the unification of heaven and human beings". TCM sets many models about the human body from different perspectives in order to compare and check them reciprocally.

In addition to the integrated models of the human body, there are models related to the parts (and part changes) of the human body. However, these models are not fixed, and they may be visual objects stored in the minds, which vary from person to person, or "thinking" objects established by imagination. For example, the models of "Eight Gang" have different forms and their prototypes in nature are concrete objects. "Gang" in TCM means outline. "Eight Gang" are eight aspects of an outline – or basic profile – of the human body. They are: outside, inside, cold, hot, hollow, solid, shady and sunny. "Eight Gang" are considered the bases of diagnosis because they can explain many phenomena of the human body[1].

There are other models which specify detailed phenomena about the human body. With the help of those models we could give explanations and judgments about the phenomena that can be learnt through intuition or observation, such as breath, blood, sweat, changes of pulse, ache of various parts, chills, fever, and so on, on the basis of real objects of the same kind outside the human body. A kind of "simulation" is performed.

3 How Does TCM Diagnose with Models?

Diagnosis in TCM is the process of finding or constructing a model which describes the situation of the patient, followed by the process of inferring the causes at play, and so the choice of the suitable therapy. Nearly all models are ready-made, already experienced and put in memory, or visual objects. The only thing the doctors engaged in TCM have to do is to make choices. They understand that their knowledge about the inner parts of the human body mainly comes from the analogy to visual objects in nature or society, rather than from experimental knowledge derived from dissections, experiments, or statistics. Furthermore, they apply the method of trial and error to avoid potential mistakes.

It is very important to select the model that correctly expresses the patient's condition. However, the bases for choosing are very limited, and accurate answers cannot be obtained only with the help of information gained by looking, smelling, asking, and feeling pulse. By investigating and interviewing, we find out that doctors tend to turn their attention to – so to say – the upper level, and to get answers based on objects of this upper level, when there are various choices. The direction and commitment of the doctors' attention develop until it reaches the uppermost thinking level of "the unification of heaven and human beings". "From models to models" is the method of inferring of the TCM diagnosis. One model infers another one, in a framework of reciprocal limitations. All models appeal to mental objects mainly visual.

After having reached a judgment related to the patient's physical condition, the TCM doctors's "thinking" further addresses those models that illustrate the physical condition, temporarily setting aside the observation of patient's body. With the help of this reflection upon models, doctors acknowledge the causes of the actual patient's condition. There is only direct association of ideas. In TCM, especially ancient TCM, the internal structure and mechanisms of the human body are basically considered as a black box.

---

[1] On the models of "Yin-Yang" "Five Elements" "Gas Movement" "Eight Gang", and the idea of "the unification between man and universe", cf. The Basic Theory of TCM" [2], TCM Diagnostics [10], and "the Chinese medical system" [4].

Changes in the human body can be interpreted with the only help of a mere phenomenological analysis together with the analogy to the objects of nature or society.

We are dealing of a kind of visual judgment from one phenomenon to another. In some cases there are diagnostic situations that make doctors feel puzzled. Doctors tend to be tolerant to accept all possible results and delay their judgment after treatment. Doctors know that the simple elimination or change of causes cannot consequentially change the results that those causes have generated in the patient's body. Now let us make a comparison between modern cybernetics theory and TCM diagnosis (see Figs. 1 and 2).

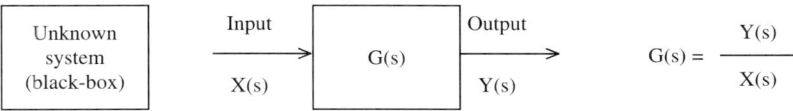

Fig. 1. The recognition and operation on an unknown system in modern cybernetics.

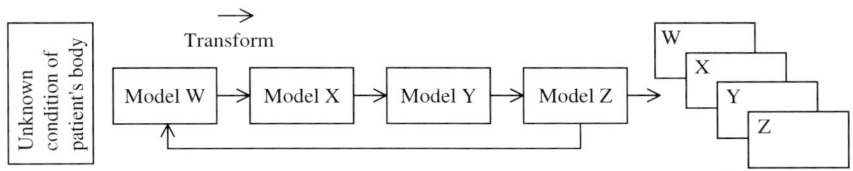

Fig. 2. The recognition and judgment of the patient's body in diagnosis of TCM.

$X(s)$ indicates the driving function, $Y(s)$ the response function, $G(s)$ the transfer function. Their relationship is shown in the formula. The transfer function completely symbolizes the characterization of the system, and so long as we know the transfer function, the response function can be drawn from the driving function. The transfer function can be found if the driving function and response function are clear. The transfer function is a phenomenological description of the system. It has nothing to do with the importation and it does not indicate the physical structure of the system; moreover, lots of systems with different physical properties may have the same transfer function. In this way, based on the whole system, the behavior of it is abstracted into the process of the information transfer, setting aside the physical nature and the movement states.

The models W, X, Y, Z are the visual objects with a certain relationship in four inter-related groups. W means the situation of the objects that have functions and effects on the human body; X means the situation of the inner objects of a normal (healthy) body; Y means the situation of the inner objects of a sick body; and Z means the situation of external objects of a sick body. From this perspective we observe there is a reciprocal relationship among the models, all based on the visual objects. Thanks to this comparison, it is

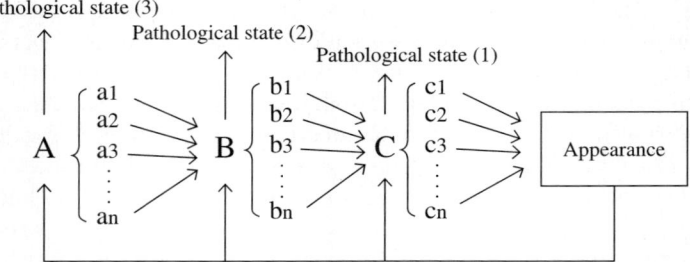

Fig. 3. The process of model-based reasoning in the diagnosis of TCM. Here, C, B, A indicate the visual mental models from large comprehensive objects. c_n, b_n, a_n indicate the mental models from smaller concrete objects. Once c_n, b_n, a_n are locked by C, B, A, c_n will change to C; b_n will change to B; a_n will change to A.

easy to see that modern cybernetics furnished a good model of some function of TCM. However, there are obvious distinctions between modern cybernetics and TCM. In sum, the reasoning processes of diagnosis in TCM can be illustrated like indicated in Figure 3.

A process of tracing causes is needed. A series of real world objects interacting one another are always directly used as prototypes of the models to describe the changes of different levels in patients' body appearance. Through them the causes are selected and established. This is a process of visual reasoning made by successive comparisons and transformations.

The third task of diagnosis in TCM is related to the decision about the treatment according to the situation of patients. It is a quite simple task for modern Western medicine, since it can be done immediately after the diagnosis by establishing the so-called "therapy". If the doctors can find out the cause, they usually know what to prescribe for the disease. However, it is still a difficult task for TCM to give a prescription, because the results of TCM diagnoses are metaphors of things or the vivid descriptions of visual models, while the effects of medicines depend on their physical and chemical attribution. In general TCM regulates the features and the effects of various medicines by visual mental models. A certain medicine may have a "wood-attribute" or "gold-attribute", "Yin" or "Yang", getting rid of "cold" or "damp", supplement "blood" or "gas", and so on. So, also medicines are model-based, and the disease of patients first of all has to be cured in the doctors' minds.

4 The Methodological Meaning of Model-based Reasoning in TCM Diagnosis

Compared with deductive and inductive reasoning, the model-based reasoning through visualizations, that acts in TCM diagnosis, is closer to the every thinking of humans, and seems closer to the physiologic mechanisms of human mind. We have to further clarify this belief from an epistemological perspective. There are four questions to be addressed. (1) "Can the method of the TCM diagnosis be used to explain scientific discovery?" (2) "Do scientists "discover" by exploiting visual mental models?" (3) "What are the characters of such visual mental models[2], and where are they stored?" (4) "What are the final bases of the rationality of human thinking?"

4.1 Abductive Reasoning

The principle that "one cause causes a result, and a result certainly is caused by just one known cause" is not respected even in the case of reasoning about simple objects. New knowledge can never be obtained if it just relies on old knowledge. The rules of logic and of algorithms would only be all-powerful if all objects in the world were locked by chains of essential causal relations, and if the qualities of all objects were provided by the permutation and combination of their components. Even if there were all-powerful rules of logic and of algorithms, man would have to know and grasp these rules before he can use them consciously. It is known that the world is not simple but complicated, that the objects have not only certain causes but also uncertain causes. Human progress and scientific development show that the causes of objects are unceasingly revealed. The related problem of abduction (reasoning to hypotheses, possibly causal), is still a problem that continuously puzzles the philosophers of science.

Already studied by Aristotle, Peirce considered abduction a third form of reasoning besides deduction and induction. For Peirce, this third form of reasoning is "hypothesis", "retroduction", or "inference to hypothesis", that accomplishes the task of "inferring the cause from a result" [9]. Peirce noticed that the first figure of categorical syllogism called Barbara can be modified in two ways. One is induction, the other is abduction/hypothesis. After Peirce, many pursuers have studied both of two modified forms of deductive *modus ponens*, especially in order to study the various forms of the abductive inference to hypotheses. Magnani [5, 6] depicted different kind of abduction, theoretical and manipulative aspects, selective and creative cases, model-based and sentential levels. Today most of researchers agree with Peirce's view that the so-called "abductive reasoning" is the process of looking for an interpretative hypothesis, that is, it is an "inferential" creative process of generating a new hypothesis [6].

Because of "the fallacy of affirming the consequent", this kind of reasoning lacks soundness in the perspective of classical logic. It seems that tracing causes is a complex process, which can hardly be reduced into one or several kinds of logical reasoning forms, and that "abduction" includes various forms of reasoning that support each other. Presently, most researches are interested

---

[2] The question may be asked in another form "what is the format of mental model?" [12].

to study the role in abduction of intuition, insight, and subconscious. I urge that the problem of abduction may be further enhanced also taking advantage of the study of diagnosis of TCM above.

An effective diagnosis is a kind of complicated "selective" [6] abduction. The problem here is first of all related to the selection of a series of hypotheses applicable to the objects, and to the definition of the most plausible one. In the TCM diagnosis, the whole process of inference is model-based. First, many mental models deriving from the audiovisual objects of nature, society, and daily life have to be formed, by means of induction, deduction, analogy, metaphor, imagination, etc. (any method may be used in this step). Second, it is necessary to define the models being consistent with the explanation of results, and produce various hypotheses to be selected in the following step, by further searching and transforming the mental models. Given the fact that usually there are many hypotheses which possess the same power of explanation, the expert reasoning ends when one hypothesis is locked. The selected hypotheses are locked step by step until the satisfactory result appears. The rationality of such reasoning is attributed in TCM to the transformation of visual models. The clarification that we have got from examining the diagnosis of TCM is that abduction can be also seen as a process of locking hypotheses by models.

4.2 The Layers of Consciousness and the Natural Manner of Human Cognition

Our investigation on the role of model-based reasoning in TCM diagnosis can shed new light on the understanding of the layered structure of human consciousness. Human thinking is a complex hierarchical system composed by consciousness at different, so to say, administrative-levels (consciousness-layers). The consciousness patterns that we know are aligned – direct perception / indirect perception / rational faculty / world outlook / consciousness / subconscious / top-consciousness[3] – from the lower level to the upper level, and from concrete to abstract in the mind. They are comparatively independent, inter-containing, inter-restraining, and have inter-contact, forming complex interacting relationships. During the process of actual thinking, many consciousness-layers work together, and support each other. This process is creative and the creativity comes from the interacting relation, which possesses qualities that contain and are contained by each other, and which is irreducible, striding across layers and non-linearly among layers. [11].

When one pattern of consciousness changes, not only the consciousness in the same layer but also the consciousness in other layers – even the framework of all consciousness – will change. If there is contradiction among

---

[3] The existence of top-consciousness is especially emphasized, which is the consciousness related to the physical mechanisms, and which is still the unclear and unconfirmed part of consciousness [12].

the consciousness-layers we might feel uncomfortable. And we might keep thinking spontaneously and continuously until the relationship of different consciousness-layers returns to harmony. I believe it is the best explanation of the processes "to rationality" [13].

The carriers of consciousness in different layers may be something like sentences but also non linguistic representations [1]. I suppose that in most cases, when the objects cannot be described in terms of language, the carriers of consciousness in different layers are for example mental things from visual objects of the real world. Hence, the most direct thinking tools are the objects that can be grasped by sense organs – the intuitive mental model: the most direct way of thinking in humans seem to be the visual one.

The analysis above is supported by the model-based reasoning aspects of TCM diagnosis, in which the reasoning tools mostly come from visual mental models deriving from real objects. The models of TCM seem confuse. Yet they can be categorized in different consciousness layers: some belong to direct perceptive forms, such as the models of "out", "in", "cold" and "hot"; some tend to the rational form, such as "gas" and "energy-channel" models; some belong to the philosophical forms, they are "Yin-Yang" and "five elements" models; some are the mixture of intuition and subconscious, e.g. the model of "the unification between man and universe". These visual mental models of different kinds are stored in corresponding consciousness levels, whose mutual function requests them to be consistent and rational.

Relating the systems of models of TCM to the consciousness hierarchical-structure of humans, we can finally understand why the human mind can accomplish judgments about unknown objects (and so can also "discover"), merely with the help of limited information coming from direct perception, unaided by the additional devices derived from the systematic observation and experiment.

Now we may outline a "natural" mechanism of human cognition as follow: various prototypes appear in different consciousness-layers, while simultaneously the complicated relations among the consciousness-layers make the prototypes change in different layers, becoming updated and selected progressively. These new prototypes are the supports for the rationality of a real mental model which can leads to the existence of a new mental model indicating a discovery of something new. Hence, the human cognition, not only is based on logic but also on imagination, intuition, introspection, emotion, and various levels of consciousness. All the elements in consciousness hierarchical structure may contribute to judgment.

References

1. Bermúdez, J.L.: *Thinking without Words*. Oxford University Press, New York (2003)
2. Yin, H., Zhang, B., eds.: *The Basic Theory of TCM*. Shianghai Science and Technology Publishing House, Shianghai (1984)

3. Xiong, J.: *The Pith of the Theories of Nei-Ching (The Yellow Emperor's Canon of medicine)*. Hunan People Publishing House, Changsha, China (1993)
4. Kao, F.F.: China, Chinese medicine, and the Chinese medical system. In F.F. Kao and J.J. Kao, eds., *Recent Advances in Acupuncture Research*. Institute for Advanced Research In Asian Science and Medicine Publishing, (1979) 1–39
5. Magnani, L.: Model-based creative abduction. In L. Magnani, N.J. Nersessian, and P. Thagard, eds., *Model-Based Reasoning in Science Discoery*. Kluwer Academic/Plenum Publishers, New York (1999)
6. Magnani, Lorenzo, and Nancy Nersessian (eds.) (2001), *Model-Based Reasoning: Science, Technology, Values*. Dordrecht: Kluwer
7. Neisser, U.: *Cognitive Psychology*. Appleton-Century-Crofts, New York (1967)
8. Nersessian, N.J.: Model-based reasoning in conceptual change. In L. Magnani, N.J. Nersessian, and P. Thagard, eds., *Model-Based Reasoning Science Discovery*. Kluwer Academic/Plenum Publishers, New York (1999)
9. Peirce, C.S.: *Collected Papers, Vol. 1–6*, C. Hartsshorne and P. Weiss, eds.; *Vol. 7–8*, A. Burks, ed. Harvard University Press, Cambridge, MA (1931–35, 1958)
10. Deng, T., Guo, Z., eds.: *TCM Diagnostics*. Shanghai Science and Technology Publishing House Shianghai, China (1984)
11. Wang, Z.: The concept of complexity: Its source, definition, characteristic and function. *Philosophical Research* **3** (1990)
12. Wang, Z.: On the administrative level structure and complexity of thinking system. *Studies in Dialectics of Nature* **10** (2003)
13. Wang, Z.: Complexity of human thinking systems and the principle of future intelligent machines. In H.W. Chu, M. Savoie, Y. Hkvarko, and F. Ramos, eds., *Proceedings of the International Conference on Computing, Communications and Control Technologies, Volume V*. International Institute of Informatics and Systemics Press, USA (2004)
14. Fang, Z.: Preface. In Qu. F., ed., *Study on Clinically Theoretical Thinking of TCM*. Chinese Curatorial Science and Technology Publishing House, Beijing (1992)

Model-Based Reasoning in Cognitive Science

Yi-dong Wei

Research Center for Philosophy of Science and Technology in Shanxi University,
Taiyuan, P.R. China
weiyidong@sxu.edu.cn

Summary. This paper addresses the different models and their functions in cognitive science. First, this paper discusses the various uses and meanings of various models in science and two kinds of functions of each model such as idealizations and representations of the real world. In cognitive science, cognitive architectures were used as cognitive models. Second, this paper discusses Neil Stillings' global cognitive architecture as well as an example. In addition, this paper focuses on several forms of cognitive models including the von Neumann model, symbol system model and production system model. Further, it was argued that the connectionist model was a better approach to understanding the mechanisms of human cognition through the use of simulated networks of simple, neuron-like processing units. Finally, four models in neuroscience were addressed, being: different models of sensory processing, Marshall-Newcombe's symbolic model of reading, model of memory system, and Mishkin-Appenzeller's model of visual memory functions. These models are approaches to physical implementation, not a computational approach to cognition.

Cognitive science is an interdisciplinary field that has developed from the convergence of a common set of questions discussed in psychology, linguistics, computer science, philosophy and neuroscience. In these disciplines, the method of model building is often applied when studying the nature of the mind. The goal of this paper is to address the various models used in cognitive science.

1 Models and their Roles

One might ask by what processes can scientists discover which cannot be observed? Thus, it is the art of model making. But what are the models used in the different fields of science? What are their meanings and roles? One might say in general that a model is a tool for thinking and explaining, of which we make representations of some subject matter making it easier to think about.

1.1 Variety of Uses and Meanings of Models

The word "model" has different uses and meanings in different areas of science. In mathematics and logic, abstract systems of signs are developed for various purposes. Using systems of objects can be associated with these signs to give them a meaning. An example is "systems of objects" that may be called "models". Mathematical models have been used to study human perception, learning judgments and choices. "Statistical models have become the primary tool for expressing relationships between variables" [7, p. 5] and so on. Therefore, according to Harré, the concept of model "covers systems of objects that are used for meaning making, that is, for interpreting, as well as for representing" [9, p. 43]. An example being, a toy car is a model of a real car, Bohr's atom model is an analog of the solar system, and highly-branched tree represents history of biology.

In biology, model organisms are used to study processes that cannot be easily measured in humans. In engineering, models of physical structures are tested in wind tunnels. In Physics, data, and both experimental and theoretical models are often used for explaining or representing the physical system of the world [8].

In Psychology, computer simulation models have been created for cognitive phenomena. Regarding cognitive Psychology, "mental modeling has been investigated in a wide range of phenomena, including: reasoning about causality in physical systems; the role of representations of knowledge in reasoning; analogical reasoning; deductive and inductive inferring; and comprehending narratives" [19, p. 10]. An example being, in the psychology of remembering, store models are used to represent the features of memory. For example, Baddeley [5] used a model which expressed what we know of the phenomena of remembering. Rolls [28] treated the hippocampus as if it were a neural net, or, as if it were indeed a connectionist device. The model regarding remembering as storage and the idea of remembering as storage is attributed to different theories as to the nature of the stored material. According to Atkinson and Shiffrin [4], each perceptual mode, being sight, hearing and touch, was the source of its own kind of memory items. Therefore each sensory modality needed its own independent memory "store" for its own type of items.

In cognitive science, models are often used as architectures of the mind. There are two main types of theories of cognitive architectures. The first type is motivated by the digital computer, it is often called von Neumann architecture model which includes production system, the information processing system or physical symbolic system. The second type of model is based on an associative architecture, which often called the connectionist architecture, or parallel distributed processing (PDP) model, or neural network. Models of the von Neumann type assume that processing involves serial, rule-governed operations on symbolic representations. The associative model of memory explains how remembering part of an event can cue retrieval of the rest of the event by claiming that an association between the two parts was constructed when the event was first encoded. Activation from the representation of the first part of the event flows to the representation of the second part through an associative connection.

Thus, it is thought that another type of cognitive architecture which attempts to integrate the von Neumann and associative or connectionist architecture models. This type of architecture may be called the "hybrid architecture" model which includes elements of both the von Neumann-type and associative components. Such alliances of processing systems appear necessary on both the theoretical and empirical levels [31], because only the von Neumann components appear capable of manipulating variables in a manner matching human competence, although associative components appear better able to capture the context-specificity of human judgment and performance as well as people's ability to deal with and integrate many types of information simultaneously. An example being, ACT is an important hybrid architecture [1] which includes both the production system and an associative network. It posits three memories: a production, a declarative, and a working memory, as well as processes which interrelate them. This was an early attempt to build an architecture that took advantage of both the von Neumann and associative principles. However integrating these very different attitudes in a productive way continues to be an ongoing challenge.

1.2 Two Types of Basic Models and their Roles

Though the term "model" has various uses and meanings, it has common features and functions. In sciences, models which are based on the identity of subject and source are extremely common. They serve to bring out salient features of the systems which are under investigation. An example being, an anatomical model of the brain is based on the discernible attributes of the brain. This kind of construction is an analytical model, which represents the result of an analysis and ranking of the attributes of some natural systems which are both the source and the subject of the model. Regarding the realist program in cognitive science, there are models which have subjects systems and structures that are as yet unobserved. How does one know which attributes to assign to a model of anything that cannot be perceived? The goal in which to develop a technique is to both abstract from and idealize a plausible source.

One example being, no one has ever been able directly to observe the real constituents of a gas. The molecular model represents those unknown constituents. The concept of a molecule is arrived at by abstraction from and idealization of the properties of perceptible material things. Models of this kind must play a predominant role in the building up of scientific explanations. They are the key to realism, since they are the main devices by which the disciplined imagination of scientists moves beyond the boundaries of what is perceptible. This kind of construction is an explanatory model. Regarding the distinction between subject and source, the difference between the two basic kinds of models is easily expressed. According to Harré, "for analytical models, source and subject are the same, while for explanatory models they are generally different" [9, p. 44]. In light of the two types of models, Harré described two main roles of models in scientific research: the analytical role and the explanatory role.

Models as Idealizations of their Subjects: the Analytical Role

Some people feel that a model, which is a version of some complex natural entity, is created by abstraction, that is, ignoring some of its aspects, and by idealization, that is, smoothing out and simplifying others. For example, "natural history museums sometimes have models showing a cross-section of the local landscape, displaying the geological strata below the surface, separated by nice smooth edges with each stratum uniformly colored. Taken together, abstraction (not every detail in a stratum needs to be reproduced in the geological model) and idealization (not every kink and break in the strata boundaries needs to be reproduced in the model) lead to a simplification of the natural state of affairs in the model representing it" [9, p. 45].

In the above case, the geological strata below a landscape are the source and the subject of the model. Such a model is a useful representation of the known, it may play a role in explanations of the character of the landscape as we observe it. Some analytical models are constructed in which their source and subject are the same. However, it is also the case that sometimes a powerful analytical model can be developed by drawing on a source different from its subject.

Models as Representations of the Unknown Things: the Explanatory Role

People have questioned how could scientists ever create a representation of the entities to which we have no access by means of observations and direct experimentation of the entities themselves, and cannot observe the molecular motions, even with the most powerful microscopes. At best, scientists can observe the random Brownian motion of visible particles suspended in a liquid. This phenomenon is most convincingly explained as the "effect" on the visible particles of being struck by invisible moving particles. One then asks how can a scientist build a model of something unknown? The possibility of this challenge follows from the method that a model of an unknown subject can be constructed by drawing on some source other than that subject. For example, Benjamin Franklin did not know how electricity is propagated in a conductor, but he knew that water passes through pipes. Thus, he invented a model of the propagation of electricity imagined as a fluid. He devised a conception of the electric fluid, not by abstracting it from electrical phenomena, but by drawing on the flow of water in a pipe as a process analogous to the flow of electricity in a conductor. This leap of the imagination was expressed by the use of the metaphor "Electricity is a fluid". This is at the very heart of scientific creativity. Harré demonstrated that "the working process is something like this:(1) (observed) unknown process P produces a certain kind of observable phenomenon O; (2)(imagined) an iconic model, M, of P "produces" a certain kind of "observable phenomenon", "O"; (3) If "O" is a good likeness of O, and M is ontologically plausible as a possible existent if it were realized in the location of P, we can say that M more or less faithfully represents P" [9, p. 46].

Others have asked how models have been used as the core of theories? Taking Darwin's theory of natural selection as an example, it can be looked upon as both a prescription of a model for understanding the history of living things, and also as a hypothesis regarding the main process by which that history was brought about. Darwin described how farmers and gardeners produced new breeds of plants and animals. They had used selective breeding, so that only those specimens that exhibited the attributes the stockbreeder wanted were allowed to reproduce. In that way, new animal and plant forms were produced. That was artificial selection by domestication. What happens in nature? Just as there are variations in each generation on the farm and in the garden that are exploited by the stockbreeder, so too there are variations in nature. Thus, the model for nature is the farm. Better-adapted animals and plants breed more freely, and more of their offspring survive. This mechanism matches the farmer's way of producing new breeds by controlling the reproduction of organisms. Thus, by building a model we have managed to create a picture of a process that could never be observed in a human lifetime.

2 The Global Cognitive Architecture of Mind

In this section, cognitive architectures as cognitive models are discussed. A cognitive architecture refers to the design and organization of the mind, or, "refers to information-processing capacities and mechanisms of a system that are built in" [34, p. 16]. Theories of cognitive architecture provided can be contrasted with other types of cognitive theories in providing a set of principles for constructing cognitive models, rather than a set of hypotheses to be empirically tested. Cognitive scientists often use the method of model building to study cognitive phenomena.

Neil Stillings [34] described a global cognitive architecture of the mind in Figure 1. One can see that the architecture consists of three systems which involve sensory system, central system including language, and the motor system. It is known that human beings can receive information through their senses, think and take physical action through voluntary muscular movements. Thus, one can initially conceive of the human information processor as consisting of a central thinking system that receives information regarding the world from a sensory system and issues movement commands to the motor system. In addition, Language plays a large and biologically unique role in human

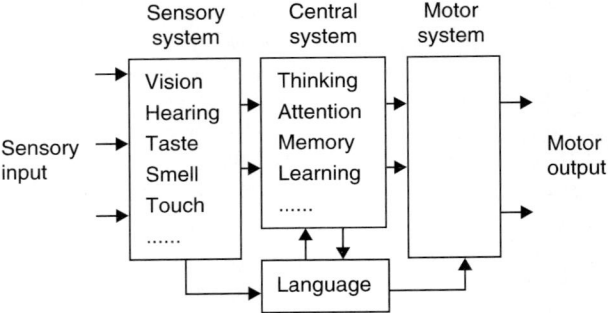

Fig. 1. Neil Stillings' global cognitive architecture.

cognition, and is a specialized subsystem of the central cognitive processor. Neil Stillings claimed that senses (vision, hearing etc) could be represented as separate subsystems, raising questions about whether they intercommunicate directly, or only by the central system. In contrast, he emphasized that we must recognize that "the boundaries between modules and the channels of communication may not be so clear-cut" [34, p. 19]. Stillings provided an example that the separation between central processes and the sensory and motor systems is not as clear as the data indicates. Similarly, the boundaries between the linguistic system and sensory, motor, and other central systems are not completely clear too. Stillings noted that it is necessary to recognize that any simple partitioning of the cognitive architecture is partly an intellectual convenience that obscures some of the complexities of human information processing.

The models of central system include language and the motor system which will be discussed in later sections. This author will provide an example of the application of the global cognitive architecture in sensory system. Taking the model of vision motion for example, Wang and Mathur and Kock [35] have proposed a connectionist and neural network model of visual motion computation (see Figure 2). They argued that the activity of direction-selective MT neurons represents the velocity, or optical flow field. In Figure 2, the sustained channel carries the outputs of retinal units that respond to local spatial variations in an intensity that remains relatively stable over time. In addition, the transient channel carries signals from units that have the same spatial profiles but that respond only to rapid temporal changes in local contrast. Thus, the orientation-selective units correspond to the cortical simple cells. This model combines the signals in the sustained and transient channels to feed direction-selective cortical simple cells that are sensitive to oriented edges that are moving in a direction perpendicular to the edge. Thus, both types of units are common in the primary visual cortex. An optical flow unit represents a particular spatial location and direction of image motion. For each point in the image there are flow units that represent a number of different

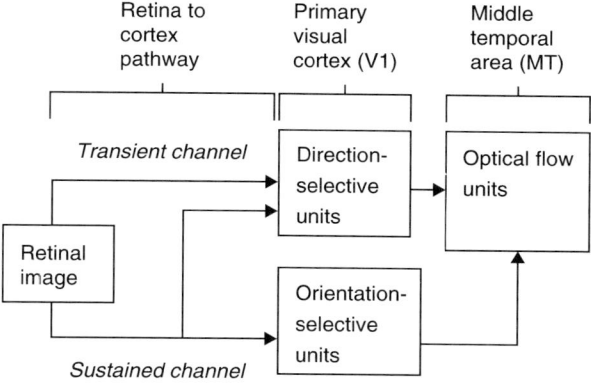

Fig. 2. A block diagram of optical flow model of vision motion.

directions of motion. The coding of direction is relatively coarse, and the optical flow at each point in the image is represented by the population of responses of all the flow units for that point. Each flow unit receives input from all orientation-selective and direction-selective units at its spatial location. These units appear to force flow cells to respond in terms of the local motion data they carry. In addition, each flow unit is also connected to other flow units that encode other directions of motion and neighboring spatial locations. These lateral interactions adjust the responses of the flow units so that they yield a smooth flow field in which motion directions in nearby regions are similar.

3 The Cognitivist Architecture: Computer as Brain

The cognitivist architecture is often called the information processing system, or physical symbol system. It is motivated by the digital computer in which the currency is information in the form of symbols. The most common digital computer model is called von Neumann architecture that consists of a central processing unit (CPU), a memory unit, and input and output units (see Figure 3). Information is input, stored, and transformed algorithmically to derive an output. The framework plays an important role in helping make the computational theory of mind. It has spawned three classes of theories of cognitive architectures.

The first architecture of this type is the production system, which claims that the mind consists of a working memory, a large set of production rules, and a set of precedence rules which determines the order of firing of production rules. An example being, General Problem Solver (GPS), which was proposed by Newell, Simon, and Shaw. This was the first theory of this type. Newell and Simon [22] described the production models in Human Problem

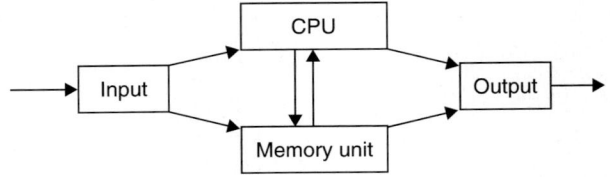

Fig. 3. The model of von Neumann architecture.

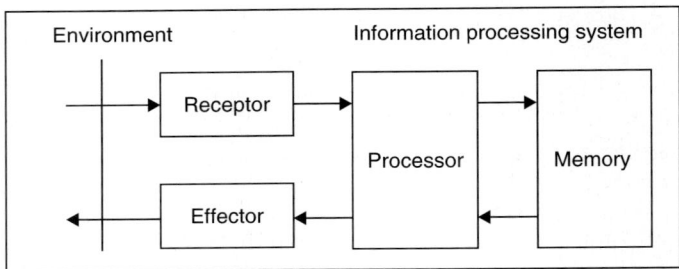

Fig. 4. The general structure of information processing.

Solving that discusses a rigorous theory of human problem solving from an information-processing perspective. They pointed out, "such a representation is no metaphor, but a precise symbolic model on the basis of which pertinent specific aspects of human problem solving behavior can be calculated" [22, p. 5].

Further, a production system is a set of operators that manipulate symbols stored in a working memory (see Figure 4). Each operator can be thought of as a condition-action pair, or as an "if-then" rule. This means "if certain conditions are satisfied, then the specified actions take place". In general, all of the operators in a production system scan the working memory for the presence of their condition. When a particular operator finds its condition, it seizes control and prevents the other operators from working. It then performs its action, which usually involves rewriting some of the information in the working memory. The control is broadcasted by the working memory, and is seized by one of the set of productions. After the production performs its action, control is released, and is again broadcasted by the memory.

The conditions of a production are propositions that state properties of, or relations among, the components of the system being modeled, in its current state. In implementing production systems these conditions are usually stored in a working memory, which may represent short-term memory or current sensory information, or an activated portion of semantic memory [3]. To activate a production, all of the conditions specified in its "if" clause must be satisfied by one or more elements in working memory. The actions that are then initiated may include actions on the system's environment, or actions the change its memories, including erasing and creating working memory elements. There are many models of production system including SOAR [20], CAPS [10], ACT-R [2], and EPIC [15]. These models have been applied to a set of phenomena in cognitive psychology. An example being, Anderson has demonstrated that a production rule analysis of cognitive skills, along with the learning mechanisms posited in the ACT model, provides detailed and explanatory accounts of a range of regularities in cognitive skill acquisition in complex domains such as learning to program LISP.

Here, the author will discuss the architecture of LEX (Learn by Experimentation) in learning system.

The LEX model was developed by Mitchel, Utgoff and Banerji (1883). It attempts to integrate symbolic mathematical expressions. The LEX is related to an early Program called SAINT (Symbolic Automatic INTegrator) which used heuristic rules to solve integration problems. The LEX's task is to learn the kind of knowledge that was built into SAINT to make it a powerful problem solver. The LEX program contains four modules: Problem solver (the LEX's performance element), Critic, Generalizer (the LEX's learning element), and Problem Generator (see Figure 5). The LEX's knowledge base consists primarily of two sorts of domain-specific knowledge: a collection of if-then rules representing integration techniques, and a hierarchy of classes of mathematical functions and objects. Its environment consists of integration problems with integrands that are instances of the mathematical functions and expressions it knows about. The functions of the four modules are as follows:

(1) The Problem solver tries to solve the problem at hand with its available store of operators, which includes the current status of its heuristics.

(2) The Critic module analyzes the trace of a successful solution to gain positive and negative instances. A positive instance is a state on a path that leads to a successful solution; a negative instance is a problem state on a path that leads away from the solution.

(3) The Generalizer rewrites its knowledge of heuristics on the basis of what the Critic tells it: in addition, it narrows the most general statement of

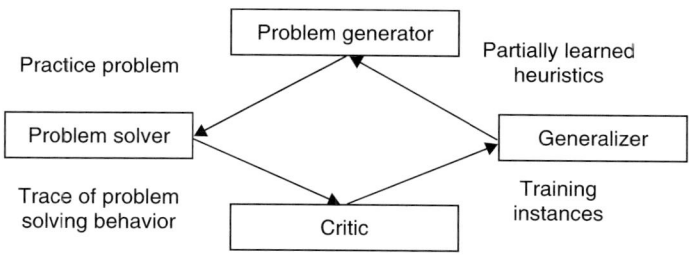

Fig. 5. The architecture of LEX system.

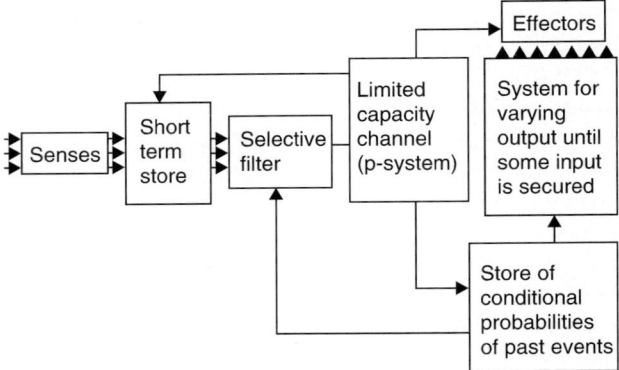

Fig. 6. Broadbent's filter model.

the heuristic on the basis of negative instances and generalizes from the most specific on the basis of positive instances.

(4) The Problem generator poses new problems to solve that will help to further refine knowledge of the heuristics.

A second class of von Neumann-inspired cognitive architecture is the information processing theory. This theory posits a sequence of processing stages from input through encoding, memory storage and retrieval, to output, different from the production system, which posits a particular language of symbolic transformation. An example being, Broadbent's [6] filter model which is an instance of this kind of architecture (see Figure 6). This model asserts that there exists a limited capacity stage of perception (p-system), and that this stage is preceded by parallel analysis of simple stimulus features, and that access to the p-system is controlled by a selective filter. In addition, Short-term and long-term (store of conditional probabilities of past events) memory systems were postulated and integrated into the information processing system.

The third class of cognitive architectures which was inspired by the digital computer emphasizes veridical representation of the structure of human knowledge. This computer model distinguishes program from data. Thus, the computer modeler has the option of putting most of the structure to be represented in the computer program. The representational model has the option of putting it in the data that the program operates on. The two models use sophisticated data structures to model organized knowledge and posit two memory stores: a working memory and a memory for structured data. Currently, there are various kinds of structured data formats, such as Minsky's frame [16], Rumelhart and Ortony's schemata [26], Schank and Abelson's scripts [30], which both specialize in the representations of different aspects of the world.

4 The Connectionist Architecture: Neural Net as Model

A connectionist architecture is frequently called parallel distributed processing (PDP) model, or neural network which made up of the interconnected artificial neurons. This model is based on an associative architecture in which the currency is activation that flows through a network of associative links. It is based on the assumption that natural cognition takes place through the interactions of large numbers of simple processing units. The connectionist cognitive modeling is an approach to understanding the mechanisms of human cognition through the use of simulated networks of simple, neuron-like processing units. In the connectionist systems, an active mental representation, such as a precept, is a pattern of activation over the set of processing units in the model. This processing takes place via the propagation of activation among the units, via weighted connections. An example being, the associative model of memory explains how remembering part of an event can cue retrieval of the rest of the event by claiming that an association between the two parts was constructed when the event was first encoded. Activation from the representation of the first part of the event flows to the representation of the second part through an associative connection.

Further, a connectionist network is also a system of interconnected, simple processing units that can be used to classify patterns presented to it. Such a network usually consists of three kinds of processing units: input units, hidden units and output units (see Figure 7). The input units encode the stimulus or activity pattern that the network will eventually classify; the hidden units detect features or regularities in the input patterns, which can be used to mediate classification; and the output units represent the network's response to mediate classification on the basis of features or regularities that have been detected by the hidden units. Thus, the processing units communicate by sending numerical signals through weighted connections.

Next, the artificial neural net is a two-dimensional interconnected set of artificial neurons (see Figure 7). The simplest artificial net is one in which all neurons are connected to all others. These connections link the nodes into a net in such a way that an output of one neuron is an input to another. The connection is any link between one artificial neuron and another. Thus, it is obvious that many neurons are processing inputs simultaneously, in parallel. Such a net has edges. "The choice of the edge in which to input 'information'

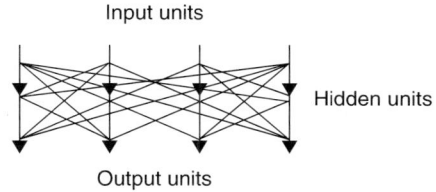

Fig. 7. The structure of connectionist model.

and the choice of which will display the final output of the net is important for the setting up of the system" [9, p. 195].

Often, a processing unit carries out three central functions: first, a processor computes the whole signal that it receives from other units. The net input function is used to carry out this calculation. Once having determined its net input, the processing unit transforms it into an internal level of activity, which typically ranges between 0 and 1. Second, the internal activity level is calculated by means of an activation function. In addition, "the processing unit uses an output function to convert its internal activity into a signal to be sent to other units" [7, p. 48].

This signal is sent by one processor to another, and is transmitted through a weighted connection, which is typically described as being analogous to a synapse. The connection itself is only a communication channel. The weight associated with the connection defines its nature and strength. For example, inhibitory connections are defined with negative weights, and excitatory connections are defined with positive weights. A strong connection has a weight with a large absolute value, while a weak connection has a weight with a near-zero absolute value. The pattern of connection in a PDP network defines the clausal relations between the processors and is therefore analogous to a program in a connectional computer [32].

Further, the connectionist models are most often applied to what might be called natural cognitive tasks, which include perceiving the world of objects and events and interpreting it for the purpose of organized behavior, retrieving contextually appropriate information from memory, perceiving and understanding language. When attempting to explain perception, the connectionists claim that perception is a highly context-dependent process. The early connectionist model, which captured the joint role of stimulus and context information, was the interactive activation model [13]. This model contained units for familiar words, for letters in each position within the words, and for features of letters in each position. Thus, mutually consistent units had mutually excitatory connections. Mutually inconsistent units had mutually inhibitory connections. Therefore, simulations of perception as occurring through the excitatory and inhibitory interactions among these units have led to a detailed account of a large body of psychological evidence regarding the role of context in letter perception.

A second example is NETtalk, which was developed by Sejnowski and Rosenberg [29]. The goal of NETtalk was to learn the very difficult mapping from English spelling to pronunciation. NETtalk is a feed-forward network consisting of layers of input, hidden, and output units (see Figure 8). The input layer contains 203 units, are divided into 7 groups, each containing 29 units. Each group can encode a single character using a local code. The input to the network consists of a string of seven characters, which can be letters, spaces(#), or punctuation marks. These seven characters are encoded by groups of input units. The input layer is fully connected to the hidden layer

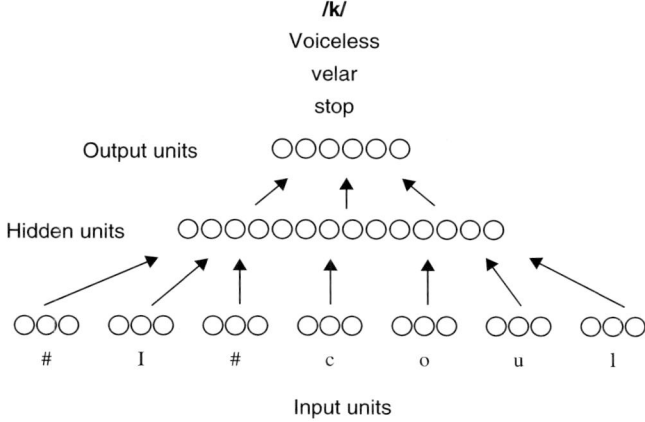

Fig. 8. The architecture of NETtalk.

of 80 units. The hidden layer is in turn fully connected to the output layer of 26 units. Each output unit represents a feature of English pronunciation. The goal of this network is to activate the output units that correspond to the correct pronunciation of the center character of input string. For example, given the "c" in the figure, the network should activate the units for voiceless, velar, and stop, which represent the correct phoneme, /k/.

One fundamental assumption of connectionist models of memory is that memory is inherently a constructive process, which takes place through the interactions of simple processing units. An early model of memory retrieval [12] demonstrated how multiple items in memory can become partially activated, thereby filling in missing information, when memory is probed. The partial activation is based on similarity of the item in memory to the probe and to the information initially retrieved in response to the probe.

In addition, the connectionist models have also been applied to aspects of concept learning, prototype formation, the acquisition of conceptual representation of concepts, language comprehension, reasoning and problem solving. For example, an early connectionist model of Rumelhart and McClelland [27] demonstrated that a network model that learned connection weights to generate the past tense of the word from its present tense was capable of capturing a number of aspects of the acquisition of the past tense.

However, the PDP ideas drew attention to the fact that the steps in the simulation of a cognitive process are not taken sequentially as they would be in a von Neumann machine. They are combined by virtue of the structure of the connections in the net. This has an important theoretical and practical consequence. Items of information are not stored as such at specific sites. The very expression "items of information" is misleading when applied to PDP. When "Taken strictly, the concept of 'representation' has no application in connectionism" [9, p. 191]. This term is used only as a metaphor, and should probably be avoided, since the implications are quite misleading. There are no representations of items in the net. One could say that a body of knowledge is represented in the whole net, though it hardly seems helpful. If the computing mechanism is an artificial neural net, it must run by PDP, several artificial neurons processing inputs at the same time. If the mechanism runs by PDP, it must take the form of a network of interconnected neurons, each processing the inputs it receives from others and activating connections with others.

5 Models in Neuroscience

In the above sections, this author has taken a computational approach to cognition, which analyzes information processes as computations over abstractly defined representations without paying attention to physical implementation, with the exception of the optical flow model of vision motion (see Figure 2). However, human cognition is physically realized in the nervous system, so in this section, this author will discuss a few models in neuroscience, which is the field that contains the levels of analysis that are required for the study of physical processes and structures in the nervous system.

5.1 Different Models of Sensory Processing

Pinel [25] in Biopsychology described various models of sensory processing (see Figure 9). In the figure, information is passed from the primary cortex to secondary sensory regions and then to the association cortex, where information from different sensory modalities is integrated. In the simplest model, a single stream of information follows a sequential progression, moving from receptors to higher brain regions, gradually being transformed from raw sensory data into perceptual representations. Each level of the system analyzes the information available at that stage and then passes its analysis on to the next level.

5.2 Marshall-Newcombe's Symbolic Model of Reading

Marshall and Newcombe [11] proposed a symbolic model of reading, which contains two parallel information-processing routes for the reading of individual words (see Figure 10). One route is based on sight vocabulary and is known as the lexical or direct route. The other route is called the phonological route, and relies on regularities in the correspondence between spelling and sound and on the morphological and phonological structure of spoken language. This route can "sound out" a word in the absence of a direct processing route. In Figure 10, each box in the chart represents a computational module, or component, that takes one or more representations as input and maps that input

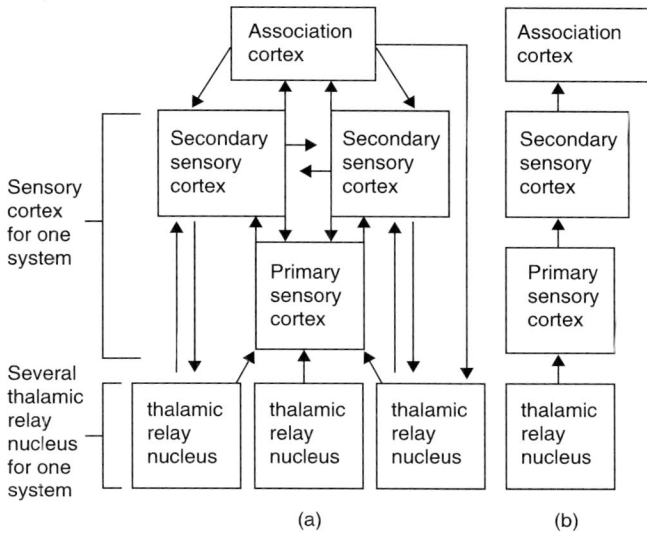

Fig. 9. Models of sensory processing:(a) a parallel, multi-representation, hierarchical model of a sensory system;(b) a serial, hierarchical, model of a sensory system.

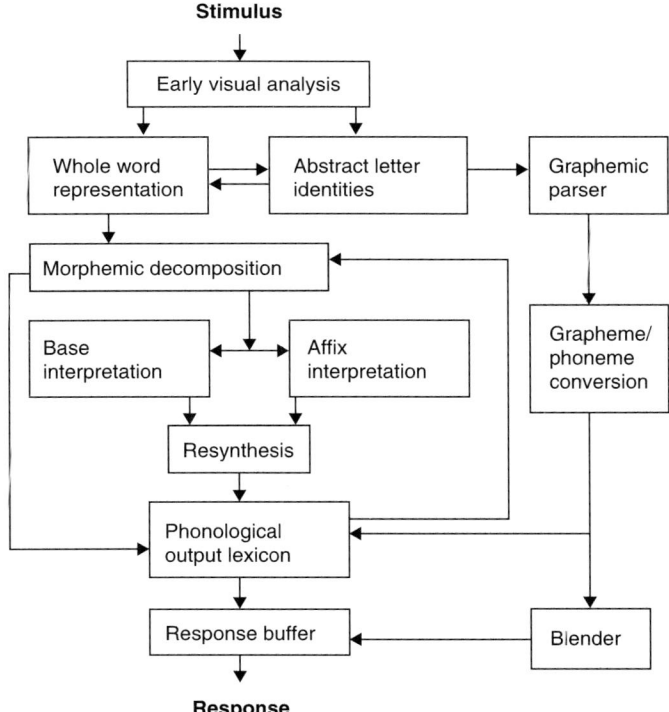

Fig. 10. A symbolic model of reading.

onto an output representation. The label in a box describes the output representation that is computed by that module. The arrows in the chart represent the flow of information among the modules.

5.3 Squire's Model of Memory System

Memory is a subsystem of the global cognitive architecture. Squire [33] described a classification scheme for types of memory (see Figure 11).

Declarative memory covers two subspecies, being episodic and semantic memory. The episodic memory is remembering incidents from the past, and semantic memory is remembering the meaning of words. The ability to remember the meanings of symbols does not require recall of the context in which the meaning was first learned. Procedural memory comprehends the maintenance of abilities and skills that have been learned at some time in the past. Typically, procedural remembering is evident in remembering how to perform certain tasks, such as proving the Pythagoras theorem.

5.4 Mishkin-Appenzeller's Model of Visual Memory Functions

One may ask how does the visual memory of human beings work? Mishkin and Appenzeller [17] have proposed a model in which the medial temporal (Hippocampus and amygdala) and diencephalic (mamillary bodies and thalamus) regions function together as a memory system. This model incorporates information not only about these structures but also about brain regions that are connected to them, such as the prefrontal cortex, basal forebrain, and temporal and parietal structures (see Figure 12). In this model, each region serves different roles in the memory process. Sensory information moves from the primary sensory cortical regions to the associative cortex, where long-term memories are stabilized by a concurrent feedback activity from the hippocampus and amygdala. This feedback circuit has two loops: an indirect route which passes from the diencephalic structures through the prefrontal cortex and then the basal forebrain and another route which connects more directly through the basal forebrain. The prefrontal cortex has been hypothesized to organize and associate behavioral responses based on the current sensory input.

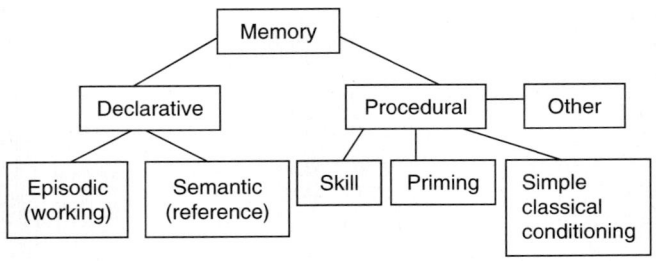

Fig. 11. Squire's model of memory system.

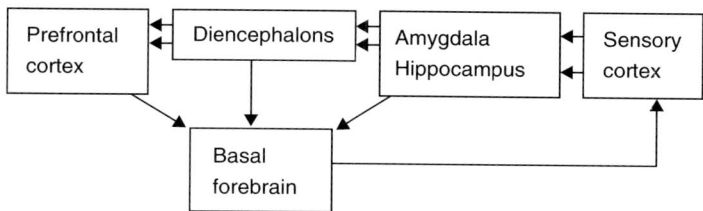

Fig. 12. Mishkin-Appenzeller's model of visual memory functions.

The hippocampus and amygdala are thought to function as a working memory system, to help consolidate new perception from short-term to long-term memory and also to assist in memory recall by associating the many features of an episodic memory across different sensory modalities.

6 Conclusion

As has been discussed above, models have different uses and meanings in different sciences. In general, something can be represented differently in two ways. The first is word-representation, by which a symbol can be given an established use to represent something; the second is iconic-representation, by which a scientist can represent what he or she wants. A model of something is an analog, representing its subject because of the balance of similarities and differences between the model and its subject. The roles of models may be (1) as tools for thinking; (2) as representations of the real world; (2) as explanations of theories; (4) as cognitive architectures of mind; (5) as simulations of sources or prototypes.

There are both models of representations and of idealizations. Both of these use the word "model" for real or imagined things, which are either analogs of something else or idealized forms of some types of things, and can be found in cognitive science. Scientific thinking and experimenting as model building and model using attempt to resolve the seemingly intractable problem of justifying claims to have reliable, though in principle revisable knowledge regarding regions of the world we cannot observe.

In cognitive science, this author feels that the cognitive architectures are used as models. There appear to be two main types of cognitive architectures. The first one is motivated by the digital computer in which the currency is information in the form of symbols. The second is based on an associative architecture in which the currency is activation that flows through a network of associative links. Both of these are applied to explain the working mechanism of human mind.

An artificial neuron is a good application of the connectionist model. It can be transformed into a scientific explanation of thinking just as artificial neurons can be related to real nerve cells and their patterned arrangement in the human brains in humans. It is necessary and useful for cognitive scientists to use model building and thus model-based reasoning to study the mental phenomena and the mechanism and functions of the human brain.

References

1. Anderson, J.R.: *The Architecture of Cognition.* Harvard University Press, Cambridge, MA (1983)
2. Anderson, J.R.: *The Adaptive Character of Thought.* Erlbaum, Hillsdale, NJ (1993a)
3. Anderson, J.R.: *Rules of Mind.* Erlbaum, Hillsdale, NJ (1993b)
4. Atkinson, R.C., Shiffrin, R.M.: Human memory: a proposed system and its control processes. In K.W. Spence and J.T. Spence, eds., *Psychology of Learning and Motivation II.* Academic Press, New York (1968)
5. Baddeley, A.: *Human Memory: Theory and Practice.* Allyn & Bacon, Boston MA and London (1998)
6. Broadbent, D.E.: *Perception and Communication.* Pergamon, Oxford (1958)
7. Dawson, M.R.W.: *Minds and Machines: Connectionism and Psychological Modeling.* Blackwell Publishing Ltd (2004)
8. Giere, R.N.: Using models to represent reality. In L. Magnani, N.J. Nersessian, and P. Thagard, eds., *Model-Based Reasoning in Scientific Discovery.* Kluwer Academic/Plenum Publishers, New York (1999) 41–58
9. Harré, R.: *Cognitive Science: a Philosophical Introduction.* SAGE, London (2002)
10. Just, M.A., Carpenter, P.A.: A capacity theory of comprehension: individual differences in working memory. *Psychological Review* **99** (1992) 122–149
11. Marshall, J.C., Newcombe, F.: Lexical access: a perspective from pathology. *Cognition* **10** (1981) 209–214
12. McClelland, J.L.: Retrieving general and specific information from stored knowledge of specifics. *Proceedings of the Third Annual Conference of the Cognitive Science Society,* Berkeley, CA (1981) 170–172
13. McClelland, J.L., Rumellhart, D.E.: An interactive activation model of context effects in letter perception: part1: an account of basic findings. *Psychological Review* **88** (1981) 375–407
14. McLeod, P., Plunkett, K., Rolls, E.T.: *Introduction to Connectionist Modeling of Cognitive Processes.* Oxford University Press, Oxford (1998)
15. Meyer, D.E., Kieras, D.E.: A computational theory of executive cognitive processes and multiple-task performance: Part1: basic mechanism. *Psychological Review* **104** (1997) 3–65
16. Minsky, M.A.: Framework for representing knowledge. In: P. Winston, ed., *The Psychology of Computer Vision.* McGraw-Hill, New York (1975)
17. Mishkin, M., Appenzeller, T.: The anatomy of memory. *Scientific American* **256** (1987) 80–89
18. Mitchell, T.M., Utgoff, P.E., and Banerji, R.: Learning by experimentation: Acquiring and refining problem-solving heuristics. In Michalski, R.S., Carbonell, J., and Mitchell, T.M. eds., *Machine Learning: An Artificial Intelligence Approach. vol1.* Tioga, Palo Alto, Calif (1983)

19. Nersessian, N.J.: Model-based reasoning in conceptual change. In L. Magnani, N.J. Nersessian, P. Thagard, eds., *Model-Based Reasoning in Scientific Discovery*. Kluwer Academic/Plenum Publishers, New York (1999), pp5–22
20. Newell, A.: *Unified Theories of Cognition*. Harvard University Press, Cambridge, MA (1990)
21. Newell, A., Simon, H.A.: The simulation of human thought. *Current Trends in Psychological Theory*. Pittsburgh University Press, Pittsburgh PA (1961)
22. Newell, A., Simon, H.A.: *Human Problem Solving*. Prentice Hall, Englewood Cliffs, NF (1972)
23. Newell, A., Simon, H.A.: Computer science as empirical inquiry: symbols and search. *Communications of the ACM* **19** (1976) 113–126
24. Newell, A., Simon, H.A., Shaw, J.C.: Elements of a theory of human problem solving. *Psychological Review* **65** (1958) 151–166
25. Pinel, J.P.J.: *Biopsychology*. Allyn and Bacon, Boston (1990)
26. Rumelhart, D.E., Ortony, A.: The representation of knowledge in memory. In R.C. Anderson et al., eds., *Schooling and the Acquisition of Knowledge*. Erlbaum, Hillsdale, NJ (1976)
27. Rumelhart, D.E., McClelland, J.L.: On learning the past tenses of English verbs. In J.L. McClelland, D.E. Rumelhart, and the PDP Research Group, eds., *Parallel Distributed Processing: explorations in the microstructure of cognition, vol.2*. MIT Press, Cambridge, MA (1986) 216–271
28. Rolls, E.C.: The representation and storage of information in neural networks in the primate cerebral cortex and hippocampus. In R. Durbin, C. Miall, and G. Mitchison, eds., *The Computing Neuron*. Addison-Wesley, Reading MA (1989)
29. Sejnowski, T.J., Rosenberg, C.R.: Parallel networks that learn to pronounce English text. *Complex Systems* **1** (1987) 145–168
30. Schank, R.C., Abelson, R.: *Scripts, Plans, Goals and Understanding*. Erlbaum, Hillsdale, NJ (1977)
31. Sloman, S.A.: The empirical case for two systems of reasoning. *Psychological Bulletin* **119** (1996) 3–22
32. Smolensky, P.: On the proper treatment of connectionism. *Behavioral and Brain Science* **11** (1988) 1–74
33. Squire, L.R.: *Memory and Brain*. Oxford University Press, New York (1987)
34. Stillings, N.A. et al.: *Cognitive Science: an introduction*, second edition. MIT Press, Cambridge, Mass (1995)
35. Wang, H.T., Mathur, B., Kock, K.: I thought I saw it move: computing optical flow in the primate visual system. In M.A. Gluck and D.E. Rumelhart, eds., *Neuroscience and Connectionist Theory*. Erlbaum, Hillsdale, NJ (1990)

An Examination of Model-Based Reasoning in Science and Medicine in India

Sundari Krishnamurthy

Stella Maris College (Autonomous), Chennai, India
ksundari03@yahoo.com

Summary. Indian Philosophy can be traced to the Vedic literature. However the six Schools of Indian Philosophy - Nyaya, Vaisesika, Sankhya, Yoga, Purva and Uttara Mimamsa are found much later, some of them contemporary to Buddhism. Model-based reasoning can be found in the sutra literature of Indian philosophical writers of the six Schools. Comparison can be made with western philosophical theories of model-based reasoning. The impact of Indian Philosophy on sciences-physics, chemistry, biology and metallurgy together with ayurveda, the indigenous system of medicine is also assessed.

1 Indian Philosophy and Model-Based Reasoning

Sanskrit was the language in which much of ancient Indian literature mathematics, astronomy, physics, medicine, architecture, fine arts and performing arts, was recorded. Therefore, it is appropriate to search for the roots of "model", "modeling" and "model-based reasoning" in Sanskrit etymology and grammar. *Adarsah* is the first word denoting a model [6, p. 294]. This word together with *Upama* and *Pratima* connote "pattern for imitation" or in a more earthy sense, as "reflection of reality". The second denotation of "model" in Sanskrit, is *pratirupam, pratikritih, pratimurtih* [74, p. 510], and means a reflection of a form or structure or representation. This is further enlarged in the third definition of model, as *pramanam prama, pramata* which indicate the method of obtaining knowledge. This knowledge could be through a mathematical or algebraic equation or a geometric diagram. Hence the model *prama* is a knowledge construct and need not necessarily be of real-world objects. The fourth word for model is *samskarah, samsthanam* and *akarah* and indicate "model" to mean "a model of propriety" or something or someone worthwhile of emulating, thus pointing, to a value-laden entity.

All of the above definitions and usages of models, in ancient India, bear a great deal of consonance to contemporary thinking. *The Stanford Encyclopedia* of *Philosophy*[1] echoes the *adarsah* meaning of model by stating that "to model a phenomenon is to construct a formal theory that describes it and explains it". In a closely related sense, it can be said that a system or structure is planned to be built, by writing a description of it. This writing is often done by using a formal language devised for it. The Universal Modeling Language (UML) is a formal language designed solely for this purpose. This is the essential meaning of *pratirupam: rupam* is form or structure and *prati* is description or facsimile. *Pratimurtih* conveys the meaning of "mould" which is parallel to Latin "modellus" and English "mould, module, model". A mould or moduli is an object in hand that expresses the design of some other objects in the world. For example, the architect's model on a palm leaf carries the form that serves to guide the masons, carpenters and builders, to construct the arches, columns and facades of an inspiring building.

From the etymological to the historic is the next step to take, in order to examine models and "modeling", in science and medicine, in ancient India. The Indus Valley Civilization can be traced to 3300–1700 BCE, and archeological remains, point to a very well-planned advanced city civilization with good water supply, drainage and sanitation systems. These point to the existence of knowledge systems in mathematics, calculus, astronomy and civil engineering. These can be examined for their contribution to model-based reasoning. Jewelry, pottery and other material artifacts indicate that these ancient Indian city dwellers of Mohenjadaro and Harappa (1700–1300 BCE) practiced mining, smelting and refining of ores and created wondrous art objects. The harbor at Lothal in the Kutch coast indicate that ship building, fishing and overseas maritime trades were briskly practiced.

Literary evidence from the four Vedas, *Rg-Veda, Sama-Veda, Yajur-Veda and Atharva-Veda* testify to the existence of the following six *Vedangas* or support literatures of the Vedas and these included: *Siksa* or the science of phonetics, *Nirukta* or etymology, *Chandas* or metrics, *Vyakarana* or grammar, *Kalpa* or rituals and *Jyotisha* or astronomy[2]. Language, meaning and their communication, can be examined for their theories in modeling. In this era of ancient Indian history, Vedic civilization is placed from 1500–500 BCE.

Cold logic and reasoning, precise and calculative mathematics and astronomy, rational and empirical advances in Ayurveda-the indigenous Indian herbal medical system, paced through India till the period of the Mauryan Empire (321–184 BCE). A world religion Buddhism and an eclectic Jainism were born in this period. The philosophies of these two religions also reveal the clear models, on which their world vision of a universal brotherhood was based.

Great heights were reached by the intellectuals of the period and this led to the golden period of the Guptas (240 BCE–550 CE), which was the period

---

[1] www.plato.stanford.edu/entries/model-theory.
[2] [50, p. 477–486].

of Renaissance in India. The artists and artisans of this period have produced yet unrivalled works of art, architecture, sculpture, painting, dance and music. Knowledge-systems in astronomy to zoology, mathematics to religion and philosophy were enriched.

The Chola Empire spanned from Afghanistan in the north to Ceylon in the south and Pakistan in the west to Burma in the east. The Chola emperors worked on establishing an effective administrative system to keep their huge empire well knit. They opened up traffic between people by constructing good roads and had a good judicial system to mete out justice to the people. Overseas seafaring trade brought rich revenues to the coffers of the State and the Chola emperors lavished it on the people, by constructing temples like the Big temple in Tanjore and other monuments which are still standing, defying time and the elements[3]. Chola bronze making and Tanjore art paintings are still made today in India in the villages of Tanjore Districts and sought after by art collectors.

The sheer muslin cloth and silks, exotic spices, sandalwood, musk and perfumes, gold and silver, wondrous monuments and fertile river valleys not only invited scholars and travelers like Hieun Tsang and Fa-Hien but also attracted conquerors. Alexander from Macedonia, the Grecians, Persians, Scythians Huns and Muslims all invaded different parts of India at different times and ages. The fortunes, social fabric and thought processes of the people of India were also swept by the whirlwinds of all these socio-economic and political changes. Yet in the *ashrams* (hermitages), in the forests and in the caves used by sages and saints for meditation, knowledge systems were being protected and sustained, with new thought and transmitted to the next generation.

2 Indian Philosophical Literature: *Sutras*

In keeping knowledge systems alive the ancient Indians did not give much importance to embossing in gold, silver or copper or carving in marble or stone. Rather they used a fragile medium of communication-a bubble of air. From the mind of the teacher-ideas, theories and practices-were passed to the student. Oral recitation, repetition and mnemonic storage devices were used to communicate record and store knowledge in the human memory. For this a new type of literature-*Sutra* literature evolved. A *Sutra* is a short aphorism, which consists of a few words (*alpaksaram*). Economy does not compromise on clarity (*asamdigdam*). This *sutra* or thread wove thoughts or ideas and connected teacher and student and served to keep knowledge alive.

Knowledge for centuries in the sutra literature was unblemished, untarnished and unaffected by the socio-economic and political changes sweeping through India. Be it Bharata's *Natya Sastra* on dance or *Caraka's* and

---

[3] A sixty ton single capstone which tops the temple tower is still an architectural master piece of this era.

Sushruta's treatises on medicine (*Caraka-sutra* and *Sushruta-Sutra*) or the various *Sutras* in philosophy: *Nyaya Sutra*, *Vaisesika Sutra*, *Sankhya-Sutra*, *Yoga-Sutra*, *Mimamsa-Sutra* and *Brahma Sutra*, not to mention the *Kama-Sutra* (on erotics) – all these *Sutras* are still available for us to read and reflect. It is this *Sutra* literature which is going to be mined for an understanding of model-based reasoning in Science and Medicine in ancient India.

3 *Nyaya Sutras*: Nyaya School of Philosophy

The *Nyaya Sutras* are attributed to Gautama, the founder of the *Nyaya* School. It is a moot question: as to whether he was the author of the sutras or merely the editor or compiler; and whether *Gautama* was one person or several people, who took it as an honorific title as "teacher" in this School. Without getting bogged down by these historic questions, the *Sutra* literature of all systems – its thoughts, ideas and practices – alone are going to be analyzed for their contributions to cognitive science.

At the very outset in the first *Sutra* itself Gautama declared "Supreme felicity is attained by the knowledge about the true nature of the sixteen categories viz *pramana*-means of right knowledge; *prameya*-object of right knowledge; *samsaya*-doubt; *prayojana*-purpose; *drstanta*-familiar instance; *siddhanta*-established tenet; *avayava*-members of a syllogism; *tarka*-confutation; *nirnaya*-ascertainment; *vada*-discussion; *jalpa*-wrangling; *vitanda*-cavil; *hetvabhasa*-fallacy; *chala*-quibble; *jati*-futility; *nigrahasthana*-occasion for rebuke"[4].

i. *Pramana* is the means of getting true knowledge. If an examination is made, of the sources or methods of knowledge, this is the only true path to salvation. The *Nyaya Sutras* thus established epistemology on a very firm footing. Every other philosopher, who followed in later centuries, would merely state that "we follow the *Nyaya* (in epistemology)"[5].

ii. *Prameya* is the object of true knowledge and according to the *Nyaya* are: (a) the *atma* (soul); (b) the *sarira* (body) which is the seat of organic activities and hence of pleasure and pain; (c) *indriyas* (senses) of smell, taste, sight, touch and hearing; (d) *artha* the objects of the senses; (e) *buddhi* cognition is the same as knowledge (*jnana*) and apprehension (*upalabdhi*); (f) *manas* or mind is an internal sense as it perceives everything within; (g) *pravritti* or activity may be good or bad and is of three kinds : vocal, mental and bodily; (h) *dosa* or mental defects such as *raga* (attachment) *dvesa* (hatred) and *moha* (infatuation) which are at the root of our activities; (i) *pretyabhava* or rebirth after death is brought about by our good or bad actions; (j) *phala* or experiences of pleasure

---

[4] [73, book I, chapter 1, sutra 1].
[5] Vidyabhusana [72] describes at great length the contributions of the Nyaya School to epistemology.

and pain which result from activities due to mental defects; (k) *duhkha* or suffering which is experienced by people in life; (l) *apavarga* or liberation from all pain and suffering[6].

iii. The third of the sixteen categories of *Nyaya* is *Samsaya* or doubt and is not a mere state of uncertainty. It is a "positive state of cognition of mutually exclusive characters in the same thing at the same time"[7].

iv. *Prayojana* or an end-in-view is obtained by obtaining desirable objects and to get rid of undesirable ones.

v. *Drstanta* or an example is an undisputed fact which illustrates a general rule. It is a very useful and necessary tool, which often resolves disputes and settles differences of opinions[8].

vi. *Siddhanta* is a doctrine that is taught and accepted. For example, according to *Nyaya*, the soul is a substance of which consciousness is a separable attribute.

vii. *Avayava* is a member of the syllogism in any one of the five members of a *Nyaya* syllogism. Inference or syllogism is an important source of knowledge.

viii. *Tarka* or hypothetical argument is an indirect way of justifying a conclusion or exposing the absurdity of its contradictory.

ix. *Nirnaya* is certain knowledge about anything and any legitimate method of knowledge can produce this as an end product.

x. *Vada* is a discussion, which is inducted with the help of *pramanas, tarka* and inference. The Nyaya logician attempted to state a proposition *(vadi)* and oppose it *(prativadi)* and in both the processes arrive at truth. This dialectical method was henceforth used by all subsequent philosophers[9].

xi. *Jalpa* is mere wrangling in which people aim only to confuse the other and obtain victory and not aim at truth. Lawyers, mass media analysts and politicians seem to be good at this method!

xii. *Vitanda* is a kind of debate in which the opponent's point of view is refuted and one's own position is not established.

xiii. *Hetvabhasa* is the *hetu* or reason which helps to determine the fallacies of reasoning.

xiv. *Chala* is an unfair play on words or is a method to contradict a statement by using different meanings of words.

xv. *Jati* technically means an unfair reply based on false analogy. Since analogy is useful in determining a hypothesis, which is the first step in scientific investigation, this is a useful tool of refutation.

---

[6] Kuppuswami Sastri [60] provided from Tarka-Sangraha all the classic details of Nyaya.

[7] Nyaya Sutra. I, 1, 9–22 [73].

[8] Chatterjee [18] is a standard text for teaching Nyaya and details all the sixteen categories, especially *pramana*.

[9] The indologist Keith [46] has appreciated the dialectical contributions of the Nyaya School.

xvi. *Nigrahasthana* literally means a ground of defeat in debate. Primarily, misunderstanding or wrong understanding or want of understanding causes defeat in a debate.

3.1 *Nyaya* School of Philosophy and Model-Based Reasoning

A superficial look at all the above sixteen categories listed by *Nyaya*, seems like a medley of disparate ideas, appliquéd into a crazy quilt. But seasoned scholars like Bhaduri [11][10] and Suguira and Singer [65][11] have described and evaluated the contributions of this School to metaphysics, logic and epistemology. Subsequent philosophers became "self conscious", that is, they became critical about their methods of knowledge, analyzed their metaphysical stand points and even critiqued their ethics.

Thus, Gautama's *Nyaya Sutra*, Vatsyayana's *Nyaya bhasya* Uddyotakara's *Nyaya-Vartika*, Vacaspati's *Nyaya-Vartika-tatparya tika*, Udayana's *Nyaya-Vartika-tatparya-parisuddhi and Kusumanjali* and Jayanta's *Nyaya manjari* are a rich treasure house of the contributions of the Nyaya School of philosophy from 400 BC to the Navya – Nyaya School of Bengal in the twelfth century[12].

Contemporary Western philosophers have stated that formal logic is concerned with sets of axioms and their deductive consequences and also with the interpretations of these axioms and theorems in "models" – that is, sets of entities that satisfy the axioms[13]. Any set of entities that constitutes an interpretation of all the axioms and theorems of a system and in which those axioms and theorems hold true is called a "model".

Writers like Max Black in *Models and Metaphors* [14, chapters 3 and 13] have denied that a model need not have any causal or explanatory force. However, the *Nyaya* School has created a tremendous impact on subsequent cognitive development in the areas of perception, inference, language, metaphor, verbal testimony, theistic proofs and a realistic analysis of the world, its environment and people.

4 *Vaisesika* School of Philosophy

The *Vaisesika* School of philosophy is supposed to have existed prior to the *Nyaya* School and its founder was Kanada, the author of the *Vaisesika Sutra*. The ultimate purpose of life is to be free from finiteness and pain. Taking a line from the *Svetasvatara Upanisad*, vi, 20, "when men shall roll up the sky

---

[10] Bhaduri has evaluated the metaphysical contributions of the Nyaya School, especially its anlaysis of the triad of entities: God, Man and World.
[11] Suguira and Singer have described the influence of the Nyaya School and its presence in China and Japan.
[12] English translations are available in Ramakrishna Publications, Madras and Calcutta.
[13] Hesse [37] discussed as to how the term "model" has become fashionable in the philosophy of science.

as a piece of leather, then shall there be an end of pain without the knowledge of *Siva*". According to *Kanada*, the passage of life should be through *dharma* or duty. We can infer this, as he sets out in the first sutra, "Now, therefore we shall explain dharma (righteousness)"[14]. This line of thinking seems to follow the popular dictum *"dharanad-iti-dharma"* – "that what holds together is *dharma*". Physical order in the universe through *rta* and moral order through *karma or adrsta* would maintain *dharma* and hence did not need the presence of an entity called "God".

This system takes its name from particularity (*visesa*) for it emphasizes the significance of particulars or individuals and is definitely pluralistic. The *Vaisesika* is definitely a system of physics and metaphysics. It adopts a six-fold classification of objects of experience (*padarthas*): (1) substance (*dravya*); (2) quality (*guna*); (3) activity (*karma*); (4) generality (*samanya*); (5) particularity (*visesa*); (6) inherence (*samavaya*) and adds a seventh called (7) non-existence (*abhava*)[15].

A *dravya* (substance) is a substratum of qualities and actions. It is the material cause (s*amavayikarana)* of other composite things produced from it. There are nine kinds of substances: (1) *prthvi* (earth); (2) *jala* (water), (3) *tejas* (light); (4) *vayu* (air); (5) *akasa* (ether); (6) *kala* (time); (7) *dik* (space); (8) *atma* (soul) and (9) *manas* (mind). The first five substances are the physical elements and each possesses a unique *(visesa)* property. For example, earth has the particular quality of smell, water has taste, light has vision, air has sensation (or touch) and ether that of hearing (or sound). These five specific qualities are "sensed" by the five external senses, which are said to be constituted by the five respective physical elements.

The substances of earth, water, light and air, are of two kinds: (i) eternal (*nitya*) and (ii) non-eternal (*anitya*)[16]. The atoms (*paramanu*) of earth water, light and air are eternal because the atom is part less, and can neither be produced nor destroyed. But all combinations of the above four elements producing composite objects are non-eternal. Atoms are not perceivable but are inferred. For example, a composite object like a chair can be separated with its individual parts, each part can be sub-divided into yet smaller parts. Finally the minutest parts which are indivisible are called *paramanus or atoms*. [32, VI, 85–86].

Akasa (ether), the fifth substance is one and eternal. It is not made up of parts and does not depend on any other part for its existence. It is all pervading, has an unlimited dimension and its quality sound is perceived everywhere. Space *(dik)* and time *(kala)* are imperceptible substances and are each one, eternal and all pervading.

---

[14] Sinha [63]. trans. *Vaisesika-Sutras* [63, book I, chapter 1, verse 1].
[15] Ibid. *Vaisesika Sutra* [63, book I, chapter 1, verse 4].
[16] *Vaisesika sutras* [63, book II, chapter 1, verses 1–19]. Earth posses colour, taste, smell and touch as does water which in addition is fluid and viscid. Fire possess colour and touch. Air posses touch. None of these characteristics are in ether.

The eighth substance is soul (*atma*) and there are two types: *jiva*, (the individual soul) and *paramatma* (supreme soul). The ninth substance is *manas* (mind) which cannot be perceived but can only be inferred. It is an internal sense (*antarindriya*); it is atomic and perceives the pleasure and pain of the individual.

The second category recognized by the Vaisesika is quality or *guna*, which exists in a substance but has no quality or activity in itself[17]. There are altogether twenty-four kinds of qualities and these are *rupa* (color), *rasa* (taste), *gandha* (smell), *sparsa* (touch), *sabda* (sound), *sankhya* (number), *parimana* (magnitude), *prthaktva* (distinctness), *samyoga* (conjunction), *vibhaga* (disjunction), *paratva* (remoteness), *aparatva* (nearness), *buddhi* (cognition), *sukha* (pleasure), *duhkha* (pain), *iccha* (desire), *dvesa* (aversion), *prayatna* (effort), *gurutva* (nearness), *dravatva* (fluidity), *sneha* (viscosity), *samskara* (tendency), *dharma* (merit) and *adharma* (demerit). Each of these qualities may have sub-divisions as for example: color can be white, yellow, red etc; smell may be pleasant or unpleasant, sound is articulate or inarticulate and so on.

The third category of the Vaiseska, *karma* or action, is physical movement[18]. It has no quality but belongs to a substance and does not belong to *akasa* (ether), space, time and soul. There are five kinds of action *utksepana* or throwing upward, *avaksepana* or throwing downward, *akuncana* or contraction, *prasarana* or expansion and *gamana* or locomotion. The actions or movements of perceptible substances like earth, fire, water and light can be perceived by the senses of sight and touch. But the action of the mind (*manas*) which is an imperceptible substance does not admit of ordinary perception.

The fourth category described by the Vaisesika is generality (*samanya*)[19] which the western philosopher would term as "universal". Objects of a certain class bear a common value because they posses a common nature. Universals are eternal (*nitya*) entities which are distinct from, but inhere in, many individuals (*anekanugata*). There is the same (*eka*) universal in all the individuals of a class. Thus, *samanya* or the universal is a real entity which corresponds to a general idea or class concept in our mind. Universals may be distinguished into *para* (highest and all pervading) *apara* (or lowest) and *parapara* (or inter-mediate).

The fifth category is *visesa* (particularity) and the School derives its name from this, as a lot of emphasis has been placed on this category[20]. Particularity (*visesa*) is the opposite of the universal and connotes the unique individuality of substances which have no parts and are therefore eternal (*nitya*).

There are innumerable particularities since the individuals in whom they subsist are innumerable, while each individual is distinguished from the other

---

[17] Ibid. *Padarthasamgraha*, [32, chapter 7, verses 46–56].
[18] Ibid. *Padarthasamgraha,* [32, chapter 6, section 2, verse 138].
[19] Ibid. *Padarthasamgraha* [32, chapter 7, verse 154].
[20] Ibid. *Padarthasamgraha* [32, chapter 8, verse 156].

by *visesa* (particularity). They are distinguished by themselves (*svatah*) and are the ultimate in the analysis and explanation of the difference of things. Like atoms, *visesas*, are supersensible entities.

Samavaya or inherence is the sixth category described in the Vaisesika School[21]. It is a permanent or eternal relation between two entities, of which one inheres in the other. The whole is in its parts, a quality or an action is in a substance, or the universal is in the individuals and particularity is in some simple eternal substance. Inherence is an eternal relation between any two entities, which cannot exist without each other.

Non-existence or *abhava* is the only negative category, all the other six categories being positive according to the Vaisesika School[22]. *Abhava* is of two kinds: *samsargabhava* and *anyonyabhava*; *samsargabhava* means the absence of something in something else and can be expressed in a judgment "S is not in P". *Anyonyabhava* means the fact that one thing is not another thing and is expressed in a judgment "S is not P". *Samsargabhava* is of three kinds: *pragabhava* (antecedent non-existence), *dhvamsabhava* (non-existence of a thing by its destruction) and *atyantabhava* (absolute non- existence).

4.1 *Vaisesika* School and Model-Based Reasoning

The seven categories with all their sub-divisions have been used by the *Vaisesika* School to build up a model of the world, its existence and functioning. This is reminiscent of "simplifying models"; that is, systems deliberately simplify (and even falsify) the empirical situation under investigation for purposes of convenience in research or application. Hesse [37] describes idealization of ideal gases come into this category, as do such simplifying statistical approximations as "smoothed out universes" in cosmology[23]. Archaic models, which have been developed in now falsified theories but which still, have some use as convenient approximation in applied rather than pure science. Examples are the model of heat as a fluid, or of faculty psychology, in which man is seen as a nexus of interacting faculties of reason, will and emotion. Insofar as they are still useful, these models must retain sufficient positive analogy, in other respects to enable some correct conclusion to be drawn from the comparison of system and model.

Indian physics, Indian atomism, principle of relativity and theories of light are based on the writings of Kanada and others of the *Vaisesika* School of philosophy. It is interesting to note, that all these concepts of physics were based on logic and philosophy, and lacked any empirical bases for want of commensurate technology like the electron microscope.

Similarly, the principle of relativity (not Einstein's) was available in an embryonic form, in the ancient Indian philosophical concept of *sapekshavad*

---

[21] Ibid. Padarthasamgraha [32, chapter 9, verse 157].
[22] This category was not stated by Kanada but is described only by later writers of the Vaisesika School.
[23] Hesse [38, pp. 198–214].

literally, theory of relativity, propagated in the sixty century BC. Several Indian texts in *Vaisesika*, speak of the relativity of time and space and have stated that "just as a man in a boat sees the trees on the bank move in the opposite direction, so an observer on the Equator sees the stationary stars as moving precisely towards the west". These theories attracted the attention of indologists and as Basham has stated, "They were brilliant imaginative explanations of the physical structure of the world, and in a large measure, agreed with the discoveries of modern physics"[24].

5 *Nyaya* and *Vaisesika* Schools of Philosophy: Impact on Indian Science and Medicine

Both the *Nyaya* and *Vaisesika* Schools of philosophy have encouraged scientific thought processes in physics and chemistry. Empirical investigations and applications in society are evident from the material evidence discovered in archeological excavations. Metallurgy has been central to all stages of civilization in India. Coins, necklaces, rings, bangles, statues, arrow tips, spears and swords are the artifacts found in different parts of India and are dated to different periods of history.

In India, certain objects which are still present bear mute testimony to the high level of the knowledge-system in metallurgy established in ancient India. Dated to the Gupta period, around the fifth century, is an iron pillar, which stands today by the side of the Qutab Minar World Heritage site. It is 73.2 meters tall with a diameter of 15 meters at the base, tapering to 2.5 meters at the top. It is estimated to weigh 6 tonnes and stands exposed to all the elements of nature, for the past several centuries. Yet it has not rusted over the years. Several such metal objects with high polish and finish and which is agelessly rust proof are found all over India. Mercury was used popularly for treatment of a variety of diseases and an alchemist, Nagarjuna had written a treatise called *"Rasaratnakara"* which dealt with the preparation of *rasa* (mercury) compound. This book also contains a survey on the status of metallurgy and alchemy in the land. Also mentioned in the treatise are the extraction of metals such as silver, gold, tin and copper from their ores and their purification.

Cave paintings in Ajanta and Ellora, World Heritage sites and on the roof of temples like the Brihadesvara at Tanjore, bear mute testimony to ancient India's advanced chemical science. Other examples of the practical applications of chemistry are seen in the distillation of perfumes and fragrant ointments. Ancient Indian literature[25] describes the use of these by men and women and other herbal beauty products which were produced by

---

[24] Basham was an Australian Indologist and has written, "The Wonder that was India".

[25] For example, the *Kama-sutra* of Vatsyayana.

the ayurvedic medical practitioners of the time; many of these are still available as products in India. Manufacturing of dyes and chemicals, preparations of pigments and colors were widely used in the textile and paint industries. Homes, palaces and other buildings used mica, glass and highly polished mirrors for ornamentation. Hakuja Ui [71] has a Chinese text on the Vaisesika philosophy and its contributions to empirical science[26].

6 *Sankhya* School of Philosophy

Kapila, the founder of the *Sankhya* system is said to have lived in the seventh century BC and is the author of the *Sankhya pravacana sutra*. However, it is the *Sankhya-Karika* of Isvarakrishna of the third century A.D. which is the most widely used text[27]. The *Sankhya* School is widely known for its theory of evolution which has had a tremendous impact on the sciences: biology, psychology and medicine. As a School of philosophy it is dualistic and recognizes the existence of *Purusa* (spirit) and *prakriti* (matter). "Primal nature (*prakriti*) is not an evolute; the seven[28] beginning with the great one (*Mahat*) are both evolvents and evolutes; the sixteen[29] are only evolutes; the spirit (*purusa*) is neither evolvent nor evolute."[30]

The entire evolutionary theory of Sankhya is based on its theory of causation termed as *sat-karya vada* which essentially states that the effect (*karya*) is related to its material cause and exists in it even before it is produced. There is an invariable relation between a material cause and its effect. Only certain effects can be produced from certain causes. The fact that only a potent cause can produce a desired effect goes to show that the effect must be potentially contained in the cause. It can also be said that the effect is a transformation of the cause and hence is not different but is essentially identical with the material cause.

This theory of causation accepted by the *Sankhya* School of philosophy had led to the acceptance of *Purusa* (spirit) and *Prakriti* (matter) as being the primordial cause of the Universe. We perceive that plants, animals, human beings (sentient beings) and other insentient things like mountains, lakes, rivers and oceans exist in the world. Since these are products or effects which are non-eternal and also exhibit both inert, insentient nature as also intelligent thought processes, they must be produced from material causes which are both material and mental, respectively *purusa* and *prakriti* exist as causes. *Prakriti*

---

[26] Hakuji Ui, trans by Thomas is a Chinese text with Introduction, Translation and Notes.
[27] Sastri [66] trans. *Sankhya Karika*.
[28] Buddhi or intellect, ahankara or individuation and five subtle elements or tanmatras.
[29] The five organs of sense (jnanendriyas), the five of action (karmendriyas), the mind and the five gross elements (panca bhutas).
[30] Jha [33, III, p. 31] *Sankhya Karika*.

is an unintelligent or unconscious principle, which is uncaused, eternal and all pervading, very fine and always ready to produce all the world of objects.

Prakriti is the first type of ultimate reality of *Sankhya* and is constituted of three *gunas*: *sattva, rajas* and *tamas*. These are like three strands in a rope. *Guna* thus means a constituent element or component and not an attribute or quality. These three *gunas* are inferred from the existence of *sattva, rajas* and *tamas* in all the objects of the world and there is a *tadatmya* (identity) between the effect and the cause.

Sattva is buoyant (*laghu*) and illuminating (*prakasaka*)[31]. *Sattva* is found in all things buoyant: the blazing up of fire, the upward course of vapor and air and any upward motion. Similarly, pleasure in all its forms such as satisfaction, joy, happiness, bliss and contentment are all produced in our minds, because of the presence of *sattva* in the mind. Also the conscious manifestation of the senses, mind and intellect, the luminosity of light and the power of reflection in the mirror and crystal are due to *sattva*.

Rajas is the principle of activity in all things. *Rajas* is exciting or stimulating (*upastambhaka*) and mobile (*cala*). *Sattva* and *tamas* are immobile and inactive and *rajas* help them perform their functions[32]. Fire spreads, the wind blows, the senses follow their objects and the mind becomes restless because of *rajas*. Being of the nature of pain (*duhkha*) *rajas* causes painful experiences.

Tamas is sluggish (*guru*) and enveloping (*varanaka*) as it obstructs the manifestation of objects. It always resists *rajas* and hence restrains (*niyama*) the motion of things. It produces ignorance and creates confusion and bewilderment (*moha*) as *rajas* counteracts the power of manifestation of the mind. Since it opposes *rajas* or activity, *tamas* causes sleepiness, drowsiness and laziness. It produces *visada* or the state of apathy and indifference.

Sattva, rajas and *tamas* have been compared to whiteness, redness and blackness; they are in constant conflict as well as cooperation. Just as the oil, wick and flame, which are relatively opposed to one another, cooperate to produce the light of a lamp, so do the *gunas* cooperate to produce the objects of the world. Therefore all the three *gunas* are present in everything of the world, great or small, fine or gross. But each of the *gunas* tries to suppress and dominate the others. The nature of each individual thing is determined by the predominant *guna*, while others are there in a subordinate position.

An important trait of the three is that they are constantly changing every moment. They undergo two kinds of transformation: during *pralaya* or dissolution of the world the *gunas* transform within themselves (*svarupa-parinama*) and in this stage cannot create anything. When one of the *gunas* dominates over the other we have *virupa parinanama* or change into the heterogeneous and this is the starting point of the world's evolution.

The second type of ultimate reality accepted by the *Sankhya* is *Purusa* or Self. This is again inferred from the perception of intelligence or sentience

---

[31] [33, XIII].
[32] [33, XIII].

in the world. It is different from the body, senses, mind and the intellect (*buddhi*). It is different from all the objects of the world as it is the subject of all knowledge. Consciousness is its very essence and is not a quality of it. It is uncaused, eternal and free from all attachment and unaffected by all objects.

The evolution of the world has as its starting point the *samyoga* (contact) between the *purusa* (self) evolve as it is inactive and *prakriti* cannot by itself evolve as it is unintelligent, just as the lame and the blind[33]. The equilibrium of the three *gunas* in *prakriti* is disturbed and *rajas* which is naturally active is disturbed first and through it the other two vibrate. The first evolute is *mahat*, in its cosmic aspect and *buddhi*, in the psychological aspect of the individual. Its special functions are ascertainment and decision. In its pure *sattvika* condition, it has such attributes as virtue (*dharma*), knowledge (*jnana*), detachment (*vairagya*) and excellence (*aisvarya*). But when mixed with *tamas*, *buddhi* has such contrary attributes as vice (*adharma*), ignorance (*ajnana*), attachment (*sakti*) and imperfection (*anaisvarya*). Buddhi stands closes to the self and hence reflects its consciousness; the mind and the senses function for *buddhi*.

Ahankara or the ego is the second product of *prakriti* and this creates a feeling of "I" and "mine" (*abhimana*). Ahankara is said to be of three kinds, depending on which of the three gunas dominate. When *sattva* dominates, it is called *vaikarika* or *sattvika*; *taijjasa* or *rajasa*, when *rajas* dominates. From the first (*vaikarika*) arise the eleven organs namely: the five organs of perception (*jnanendriya*), the five senses of action (*karmendriya*) and the mind (*manas*). From the third (*tamasa-ahankara*) are derived the five subtle elements (*tanmatras*). The second (*taijasa*) supplies the energy to the first and third.

The five organs of perception (*buddhindriya*) are the senses of sight, hearing, smell, taste and touch and they perceive color, sound, smell, taste and touch and *ahankara* enjoys them. The organs of action (*karmendriya*) are hands, legs, mouth, anus and sex organs which perform the function of apprehension, movement, speech, excretion and reproduction. All organs are powered by *sakti* energy. The mind (*manas*) is the central organ, which partakes of the organs of both knowledge and action. The mind, ego and intellect (*manas, ahankara* and *buddhi*) are internal organs (*antahkarana*) while the other ten organs are external (*bahyakaranas*). The mind interprets the indeterminate sense-data supplied by the external organs into determinate perceptions.

The three internal and ten external organs are called the thirteen *karanas* or organs of *Sankhya* philosophy. The five *tanmatras* (potential elements) are subtle and cannot be perceived. The five gross elements which arise from the subtle elements as follows: (i) from the essence of sound (*sabda tanmatra*) is produced *akasa* with the quality of sound, which is perceived by the air; (ii) from the essence of touch (*sparsa tanmatra*) combined with that of sound,

---

[33] [33, XXI] – just as a lame man and a blind man can cooperate and help each other to be mobile.

arisen air, with the attributes of sound and touch; (iii) out of the essence of color (*rupa tanmatra*) as mixed with those of touch and sound, there arises light or fire, with the properties of sound, touch and color; (iv) from the essence of taste (*rasa tanmatra*) combined with those of sound, touch and color is produced the element of water, with the qualities of sound, touch, color and taste; (v) finally, the essence of smell (*gandhatanmatra*) combined with the other four, gives rise to earth, which has all the five qualities of sound, touch, color, taste and smell. Thus the five physical elements of *akasa*, air, light, water and earth have respectively, the specific properties of sound, touch, color, taste and smell.

The whole course of evolution from *prakriti* to the gross physical elements is distinguished into two stages, namely the psychical (*pratyaya sarga* or *buddhi sarga*) and the physical (*tanmatra sarga or bhautika sarga*). The development of *prakriti* into *buddhi, ahankara and eleven* sense – motor organs in the first stage and the evolution of the five subtle physical essences (*tanmatras*) the gross elements (*panca mahabhutas*) and their products, occurs in the second cycle.

6.1 *Sankhya* School of Philosophy: Impact on Indian Sciences and Ayurveda

The *Sankhya* theory of evolution of the universe tries to describe it, not as a dance of blind atoms, or the push and pull of mechanical forces, which produce a world to no purpose, but as a stage for the fulfillment of a better spiritual life. The ultimate end of life is *mukti* (liberation of self) and the world is a training place for it.

Pain or suffering is of three kinds: *adhyatmika, adhibhautika and adhidaivika*. The first is due to intra-organic causes like bodily disorders and mental afflictions. The second is produced by nature like earthquake, floods, and fire, animals attacking and so on. The third is produced by supernatural forces. All human beings want to get rid of pain and enjoy pleasure. *Apavarga or purusartha* (the ultimate end) of the Sankhya is to get rid of pain and this is possible through the right knowledge of reality by getting rid of ignorance (*ajnana*); obtaining discrimination (*viveka*) between self and not-self. This twin-edged approach leads one to *mukti*, which is of two types *jivanmukti* (liberation while alive) and *videhamukti* (liberation after death). Liberation is thus achieved only by the three-fold destruction of misery (*duhkha-trayabhighata*).

The Sankhya theory of evolution had a tremendous impact on the sciences of biology, pharmacology and medicine in India. The models produced above formed the basis and framework for these sciences. *Ayurveda* is the oldest indigenous system of medicine[34] in India. It traces its roots to the Vedic

---

[34] "Veda" is science and *Ayus* means *ayush* or heatlh. Unlike allopathic medicine which is heavily based on the curative aspect, ayurveda is a way of life and emphasizes the promotive aspect of health.

period to the *Aswins*; twin brothers who were the Gods of Wind are said to be its founder. Whatever be its mythological roots, *ayurveda* uses medicines prepared from plants, herbs and other natural products. This is because it has modeled itself on the *Sankhya* theory of the evolution of the universe. Since all of creation is evolved from *prakriti* and *purusa*, one has to be in harmony with the rest of the universe to be healthy.

Ayurveda hypothesizes that each human being is made up of *sattva, rajas, and tamas*. That person in whom sattva predominates exhibits qualities of cleanliness (*saucam*), knowledge (*jnana*), purity (*suddhi*) and intellect (*mati*). If *rajas* predominate in a person, then he/she talks a lot (*bahu-bhasitam*), has ego (*mana*), is proud (*dama*) and is competitive (*valsara*). That person in whom *tamas* dominates, exhibits fear (*bhaya*), ignorance (*ajnana*) is often full of sleep (*nidra*) and laziness (*alasyam*) and is grief stricken (*visadam*)[35].

Each individual is said to have three humor or *tridosas: vata, pitha and kapha*. Literally, *vata* means air and is responsible for movement, *Pitha* means "bile" but also connotes "heat" and like fire is involved in transformation, *kapha* means "phlegm" but means "water" more as it is the binding element in the body. All these three – *vata, pitha and kapha* – should be in a state of equilibrium. If one or the other dominates then a person's health is disturbed and he/she becomes unhealthy.

Diagnosis of ailments is done by analyzing the states of *vata, pitha and kapha* in the person. *Vata* is exhibited through dryness (*ruksa*) and coldness (*sita*), *pitha* through heat (*usna*) and uncoutousness (*snighdha*), while *kapha* displays coldness (*shitala*) and unctuousness (*snighdha*). For example, if a person has a skin problem, then the *vaidya*[36], analyses if the skin of the patient is dry, cold, hot and so on and accordingly prescribes application of a warm or hot or cold oils or decoctions of herbal medicines. Any method of empirical treatment is judged by its efficacy and effectiveness and *Ayurveda* practioners and patients who follow its regimen, affirm to the end-results-a state of physical, mental well being! In fact, in India, *Ayurveda* has staged a comeback with the opening of many clinics in the cities of India, by individuals and multi-national corporations.

Ayurveda is said to aid a human being in the pursuit of the four goals of life: *dharma, artha, kama and moksa* and truly a healthy person can pursue these, said *Caraka*, the ancient ayurvedic physician[37]. Only a healthy person can work and accumulate material goods (*artha*) like a car, house and wealth. A person, who is not healthy, cannot pursue and obtain sensual fulfillment (*kama*), a person who has diabetes or cholesterol cannot eat all types of food and obtain fulfillment of the desires of taste. Fulfilling all of one's

---

[35] The writings of ancient *ayurvedic* practitioners Caraka and Susruta even today have a record of all these ideas.

[36] *Ayurvedic* practitioner or healer

[37] Caraka sutra, I.1, (i)
"*dharma-artha-kama-moksanam-arogyam-mulam-uttamam*".

duties (*dharma*) through the different roles each individual plays in life is possible only if one is fit and fine. The final goal of *mukti* or liberation is again something which can be actively pursued by a healthy person. Health is verily wealth for everyone.

Oils, massage, water and heat therapy and internal dosage through herbal decoctions, powders and gels are followed in *Ayurveda*. This is because it is modeled on the Sankhya philosophy that all of creation is an evolute of *Purusa* and prakriti and hence product of one (nature) can be used to treat the other (human beings). Thus, *Sankhya* model lead us to examine the western theories of modeling. The relation between model and thing modeled can be said to be a relation of analogy. Two kinds of analogy can be distinguished in connection with models in sciences [54, chapter 7]. First in the case of a logical model of a formal system, there is analogy of structure or isomorphism between model and system, deriving from the fact that the same formal axiomatic and deductive relations connect individuals and predicates of both the system and its model. This isomorphism consists of the correspondence between individuals and predicates of the system and the terms that are their interpretations in the model. Secondly, we must consider in a replica model, the material similarities between the parent system and its replica – for example, the wings of an aircraft and its replica, may have similar shape and hardness and may be made of the same material, although they differ in at least one respect size [21, p. 224].

Extending the above two to *Sankhya* modeling, one can deduce that the formal system of evolution was compared analogically by *ayurveda* for its understanding of the human body and its functioning. Analogy to the world of nature led *Ayurvedic* practitioners to pound, powder, grind, boil and heat to produce herbal oils, decoctions, pastes, powders and gels for treatment. It is possible to conceptualize a model and apply the modeling into practical and empirical usages in everyday life.

7 *Yoga* School of Philosophy

Another system which is interested in the practical aspect of disciplined activity is the *Yoga* School of philosophy, propounded by *Patanjali* (second century BC). Author of the *Yoga sutras*, *Patanjali* accepted the *Sankhya* psychology and metaphysics. Since each individual is a composite of the contradictory forces of matter (*prakriti*) and spirit (*purusa*) both these disparate elements have to be yoked (cognate with Sanskrit "*yoga*") together by a rigor of bodily, mental and ethical discipline. It is thus a methodical effort to attain perfection, through the control of the different elements of human nature, physical and psychical.

Atha-citta-vrtti-nirodhah is the first sutra in Patanjali's *Yoga Sutra*[38]. The modifications of *citta* (cognitive mental states) are numerous but can

---

[38] [56, p. 4].

be classified under five heads, viz, *pramana* or true cognition, *viparyaya* or false cognition, *vikalpa* or merely verbal cognition, *nidra* or sleep and *smrti* or memory. Whenever *citta* is modified the self is reflected in it and is apt to appropriate it as a state of itself, though it is above all these changes. The self becomes subject to five afflictions; *avidya* or wrong knowledge, *asmita* or false notion, *raga* or desire, *dvesa* or aversion and *abhinivesa* or instinctive fear of death. So long as the self is reflected in the *citta*, it identifies itself with the activities of the body and hence is subject to bondage. When the waves of empirical consciousness (*karya-citta*) die down and leave the *citta* in a state of calm (*karana-citta*), the self realizes its true nature, which is self shining intelligence. Hence *Patanjali* has formulated a system for reaching this state of bliss.

There are five levels or conditions of the mental life (*cittabhumi*), as the mind is composed of *sattva, rajas* and *tamas* and these are *ksipta* (restless), *mudha* (torpid), *viksipta* (distracted), *ekagra* (concentrated) and *niruddha* (restrained)[39]. The first three levels do not lend themselves for meditation, while the last two do so. Contemplation and meditation lead to self realization. There are two kinds of *Yoga* or *samadhi* namely, *samprajnata* and *asamprajnata*. Depending on the objects which are contemplated there are four kinds of *samprajnata* samadhi and these are: (1) *savitarka* when the mind is concentrated on gross physical objects; (2) *savicara* samadhi is when the mind is concentrating in subtle objects; (3) *sanandha* samadhi which is attained when the mind focuses on extremely subtle objects or abstractions and (4) *sasmita*, state when *ahankara* is contemplated.

When the mind paces itself through meditation it realizes that it is different from the body and senses. It becomes completely free and becomes liberated from pain and suffering and thus reaches *asampragnata* samadhi[40]. To reach this goal is neither easy nor simple. Physical exercise, control of breath, mental regimen emotional control and ethical purity are all required to yoke together the different aspects of the complex human personality. Hence Patanjali prescribes an eight-fold path (*astanga yoga*).

The first discipline or *sadhana* (method) is *yama* or restraint and consist of: (i) *ahimsa* (non-injury); (ii) *satya* or truthfulness; (iii) *asteya* (non-stealing); (iv) *brahmacarya* (continence) and (v) *aparigraha* (non-possession). It is a psychological law that a sound mind rests in a sound body and neither can be sound unless a person is ethical and can control his emotions.

The second discipline is *niyama* or culture. This stage emphasizes the cultivation of positive habits, emotions and training of the will and involves the following (i) *sauca* or purification of the body by bathing, and eating proper food and purification of the mind by cultivating good emotions such as friendliness, kindness, cheerfulness and indifference to vices; (ii) *santosa* or contentment, (iii) *tapas* or penance by cultivating endurance; (iv) *svadhyaya* or regular study of religious books and (v) *Isvarapranidhana* or submission to God and meditation on Him.

---

[39] *Yoga-bhasya* [64, chapters 1 and 2].
[40] Ibid. chapter I, verses 3–21.

The third discipline is *asana* or bodily postures; by the adoption of steady and comfortable posture, learnt from an expert teacher, one can condition the body, control the breath and preserve the vital energy. Maintaining the health of the body makes it a vehicle for concentrated thought.

The fourth discipline is *pranayama* or regulation of breath by paying attention to inhalation (*puraka*) exhalation (*recaka*) and retention (*kumbhaka*). Regular practice of these breathing techniques can strengthen the heart, lungs and nervous system and prepares the mind for prolonged concentration. These techniques should be learnt from a master *yogi*. The above four steps together with the fifth form the preliminary stage of external yoga practice and prepares the individual to launch into the more advanced stages of meditation. The sixth to the eighth stages marks the more advanced stages of *Yoga* practices as it is internal (*antaranga sadhana*).

The fifth stage is termed *pratyahara* or withdrawal of the senses from the respective sense objects, for in so doing, the mind follows itself and not the senses or their objects. The mind is like a monkey and so this stage requires an iron will and long practice to gain mastery.

The sixth stage of *Yoga* is *dharana* or attention. Concentrating on one point, like a distant object or the midpoint of the eyebrows or the navel, without bodily movement, mental distraction or emotional disturbance, calms and stills the person and prepares one for the next higher stage of *Yoga*.

Dhyana or meditation is the next step. This enables the practitioner to steadfastly contemplate any one thing and understand it clearly and completely. This stage sounds simple but is extremely difficult as there should be no break or disturbance in the thought process.

Samadhi or concentration is the final step in the practice of *Yoga*. In this stage, the mind is so deeply absorbed in the object of contemplation that it loses itself in the object and has no contemplation of itself.

A *yogin* who has mastered all the eight steps develops extraordinary powers like being able to see through closed doors, disappear from sight and appear elsewhere, appear at different places at the same time and so on. But a genuine *yogi* will not attempt to gain only these powers. The final aim of *Yoga* is the attainment of liberation and the *yogi* should not be lured by yogic powers[41]. The aim of *Yoga* is to explore the region of genuine super-physical experience and to reveal the reality of man and the world. According to Sri Aurobindo, man can become a superman, through yoga and other practices.

7.1 Yoga School of Philosophy and Model-Based Reasoning

Taking a leaf from the philosophy of science in the West one can evaluate the psychological modeling attempted by the *Yoga* School of philosophy. Models should have a heuristic function in relation to the theory that they are embedded in. A model should give an intellectually satisfying explanation

---

[41] The *Brhadaranyaka Upanisad* states that "he who says he is a *yogi* is an *ayogi* (rascal)".

of empirical data. More importantly, because of the dynamic nature of theories, models should be able to issue empirical predications over a wide domain of phenomena. Judged by the twin parameters of function and prediction, the psychological modeling found in the *Yoga* School of philosophy, has through empirical scientific and medical investigations revealed that they are scientifically valid. Individual practitioners of *Yoga*, tested before-after *Yoga* practices and have also given deductive proof of the benefits obtained from *Yoga* practices.

8 Indian Philosophical Schools and Model-Based Reasoning

Indian Schools of philosophy, *Nyaya*, *Vaisesika*, *Sankhya* and *Yoga* have been examined for their contributions to model-based reasoning. While *Nyaya* and *Vaisesika* have had an impact on Physics, Chemistry and other aspects of physical science, *Sankhya* and *Yoga* have largely influenced Biology, Psychology and Indigenous medicine. However, one must guard against blindly applying models and modeling techniques in the cognitive sciences [4, p. 213].

Firstly, one must guard against identifying or confusing model with theory, because the model may have implications that turn out to be untrue of the theory. Secondly, it is held that models are used in situations where deliberate simplification and distortion are intended and that therefore they cannot be identified with the theory of which they are imperfect interpretations. Thirdly, it can be stated that while models are accepted as essential ingredients of theories, there is no evidence other than their functions in relation to prediction and meaning, for endowing them with reality.

The above three arguments, to guard against the blind adoption of models and modeling is illustrated in the *Rasesvara darsana* or Mercurial system advocated and practiced in ancient India [23, pp. 137–144]. The virtues of mercury or quicksilver are extolled and it is stated that the gods-*Mahesa, Sukra, Kapila* and other acquired extra ordinary powers through the use of mercury as a *rasa* [23, p. 139]. It is further stated that "for ordinary people, mercury and air, swooning carry off diseases, dead they restore to life. Bound they give the power of flying about". Many used mercury, then and now, and suffered mercury poisoning. Modern medical research has shown that mercury *rasa* is not to be used for any kind of healing. Thus showing up that any cognitive device like model-based reasoning even if embedded in a very good theoretical system, should be well examined and tested before being practically utilized. In this Kalidasa's advice to test everything in the crucible of one's mind, seems to be very appropriate:

Puranam ityeva na-sadhu sarvam
na capi – kavyam – navamiti – vadhyam
santah – parikshyaya – antarat – bhajante
mudah – parah – pratyaya neyaya – buddih.

References

1. Abhedananda, S.: *Vedanta Philosophy*. Vedanta Society, New York (1899)
2. Abhedananda, Swami.: *A Study of Vedanta*. University of Calcutta, Calcutta (1934)
3. Abraham, R. and Shaw, C.: *Dynamics: The Geometry of Behavior*. Addison Wesley, Redwood City (1992)
4. Achenstein. P.: Variety and Analogy in Confirmation Theory. *Philosophy of Science* 30 (1963) 207–221
5. Allween, G., Barwise, J., eds.: *Logical Reasoning with Diagrams*. Oxford University Press, Oxford (1996)
6. Apte, V.S.: *The Student's English Sanskrit Dictionary*. Motilal Banarsidass Pub. Pvt. Ltd., New Delhi, 14th ed. (1993)
7. Banerjee, S.C.: *The Sankhya Philosophy: Sankhyakarika with Gaudapada's Scolia and Narayana's Gloss*. University of Calcutta, Calcutta (1909)
8. Barnett, L.D.: *Brahma-Knowledge: an Outline of the Philosophy of the Vedanta as set forth by the Upanishads and by Sankara*. John Murray, London (1907)
9. Behanan, K.T.: *Yoga, A Scientific Evaluation*. Macmillan Co., New York (1937)
10. Besant, Annie.: *An Introduction to Yoga*. Theosophical Publishing House, Madras (1920)
11. Bhaduri, S.: *Studies in Nyaya-Vaisesika Metaphysics*. Bhandarkar Oriental Research Institute, Poona (1947)
12. Bhattacharyya, K.C.: *Studies in Vedantism*. University of Calcutta, Calcutta (1909)
13. Bhisagaratna, Kaviraj Kunja Lal., ed.: *The Sushruta Samhita*, Eng trans. In 3 vols. Calcutta: Pub. by editor (1907)
14. Black, M.: *Models and Metaphors*. NY, Ithaca (1962)
15. Boerger, E. and Staerk, R.: *Abstract State Machines: A Method for High Level System Design And Analysis*. Springer-Verlag, Berlin (2003)
16. Boole, G.: *The Mathematical Analysis of Logic*. Macmillan, Barclay and Macmillan, Cambridge (1847)
17. Brodou, V.: *Indian Philosophy in Modern Times*. Progress Publishers, Moscow (1993)
18. Chatterjee, S.: *The Nyaya Theory of Knowledge*. University of Calcutta, Calcutta, 2nd ed. (1950)
19. Chatterjee, S.: *An Introduction to Philosophy*. University of Calcutta, Ballygunge (1975)
20. Chattopadhyaya, B.D.: *What is Living and What is Dead in Indian Philosophy*. Peoples Publishing House, New Delhi (1976)
21. Braithwaite, R.B.: Models in the Empirical Sciences. In *Proceeding of the congress of the International Union for the Logic, Methodology and Philosophy of Science*, California (1960)
22. Coster, F.G.H.: *Yoga and Western Philosophy: a Comparison*. Oxford University Press, Oxford (1949)
23. Cowell, E.B. and Gough, A.E., trans.: *The Sarva Darsana Samgraha or Review of the Different Systems of Hindu Philosophy* by Madhava Acharya. Cosmo Publications, New Delhi (1970)
24. Dasgupta, S.: *History of Indian Philosophy Vol I–IV*. Motilal Banarsidass Pub., Delhi (1986)

25. Dasgupta, S.N.: *The Study of Patanjali*. University of Calcutta, Calcutta (1920)
26. Doets, K.: *Basic Model Theory*. CLSI Publications, Stanford (1996)
27. Etchemandy, J.: *The Concepts of Logical Consequence*. Harvard University Press, Cambridge, MA (1990)
28. Faddegon, B.: *The Vaicesika System*. J. Muller, Amsterdam (1918)
29. Franklin, E.: *Beginnings of Indian Philosophy. Selections from the Rid Veda, Arthava Veda, Upanisads and Mahabharata*. George Allen & Unwin Pub., London (1968)
30. Frege, G.: *On the Foundations of Geometry and Formal Theories of Arithmetic*. Translation by E. Kluge. Yale University Press, New Haven, Connecticut (1971)
31. Fowler, M.: *UML Distilled*. Addison-Wesley, Boston (2000)
32. Ganganatha, Jha, trans.: *The Padarthadharmasamgraha of Prasastapada with the Nyayakandali of Sridhara*. E.J. Lazarus & Co, Allahabad (1916)
33. Ganganatha, Jha, trans.: *The Tattva-Kaumudi on the Samkhyakarika*. The Oriental Book Agency, Poona (1934)
34. Ganganatha, Jha: *Gautama's Nyayasutras with Vatsyayana's Bhasya*. Oriental Book Agency, Poona (1939)
35. Garnham, A.: *Mental Models And Interpretation of Anaphora*. Taylor And Francis, Philadelphia, PA (2001)
36. Gentner, D. and Stevens, A., eds.: *Mental Models*. Lawrence Erlbaum, Hillsdale, NJ (1983)
37. Hesse, M.: Models and Analogy in Science. In P.Edwards, ed., *The Encyclopedia of Philosophy in four volumes*. Macmillan Publishing Co. & The Free Press, London & NY (1978)
38. Hesse, M.B.: Models in Physics. *British Journal for the Philosophy of Science* 4 (1953)
39. Hiriyanna, M.: *Essentials of Indian Philosophy*. George Allen & Unwin Pub, Bombay (1975)
40. Hiriyanna, M.: *Outlines of Indian Philosophy*. George Allen & Unwin Pub, Bombay (1975)
41. Hodges, W.: *A Shorter Model Theory*. Cambridge University Press, Cambridge (1997)
42. Johnson Laird, P.: *Mental Models: Towards A Cognitive Science Of Language, Inference And Consciousness*. Cambridge University Press, Cambridge (1983)
43. Joseph, G.G.: *The Crest of the Peacock: Non-European Roots of Mathematics*. Penguin Books, London (2000)
44. Kak, S.C.: Birth And Early Development of Indian Astronomy. In Selin, Helaine, *Across Cultures: The History of Non-Western Astronomy (303-340)*. Kluwer, Boston (2002)
45. Keith, A.B.: *The Samkhya System*. Oxford University Press, London (1918)
46. Keith, A.B.: *Indian Logic and Atomism: An Exposition of the Nyaya and Vaicesika Systems*. The Clarendon Press, Oxford (1921)
47. Mahadevan, T.M.P.: *Gaudapada: A Study in Early Vedanta*. University of Madras, Madras (1952)
48. Majundar, A.K. & Majundar, J.K.: *The Sankhya Conception of Personality; or a New Interpretation of the Sankhya Philosophy*. University of Calcutta, Calcutta (1930)
49. Manzano, M.: *Model Theory*. Oxford University Press, Oxford (1999)

50. Mehendale, M.A.: Language and Literature in the Upanishads. In Munshi, K.M. (ed).: *The History and Culture of the Indian People*. Vol 7. Bharatiya Vidya Bhavan, Mumbai (1996)
51. Morgan, M.S., Morrison, E., eds.: *Models as Mediators*. Cambridge University Press, Cambridge (1999)
52. Muller, Max, F., ed.: *The Sacred Books of the East, in fifty volumes*. Translated by various oriental scholars. Motilal Banarsidass, Varanasi (1977) 4th ed.
53. Munshi, K.M., ed.: *The History and Culture of the Indian People in Ten volumes*. Bharatiya Vidya Bhavan, Mumbai (1996) 7th ed.
54. Cohen, M.R., Nagel, E.: *An Introduction to Logic and Scientific Method*. Simon Publications, New York (1934)
55. Pickover, C.A.: *Computers, Pattern, Chaos and Beauty*. St. Martins Press, New York (1990)
56. Prasada, R.: *Yoga Sutras of Patanjali in Sacred Books of the Hindus, Vol 4*. The Panini Office, Allahabad (1924)
57. Radhakrishnan, S.: *Indian Philosophy in Vol I & II*. George Allen & Unwin Pub, London (1962)
58. Radhakrishnan, S.: *Contemporary Indian Philosophy*. George Allen & Unwin, London (1976)
59. Rothmaler, P.: *Introduction to Model Theory*. Gordan and Breach, Amsterdam (2000)
60. Sastri, Kuppuswami S.: *A Primer of Indian Logic according to Annambhatta's Tarkasamgraha*. P. Varadachary & Co., Madras (1932)
61. Sharma, C.: *Critical Survey of Indian Philosophy*. Motilal Banarsidass Publishers, Varanasi (1986)
62. Sengupta, N., Sengupta, B., eds.: *Caraka Samhita*. Eng trans. Chaukambha Orientalia, Varanasi (1991), 1st ed.
63. Sinha, N.: *The Vaisesika Sutras of Kanada in the Sacred Books of the Hindus*. The Panini Office, Allahabad (1923)
64. Swami, S.P.: *Aphorism of Yoga*. Faber & Faber, London (1938)
65. Sugiura, S., Singer, E.A., ed.: *Hindu Logic as Preserved in China and Japan*. University of Pennsylvania, Philadelphia (1900)
66. Suryanarayana Sastri, S.S.: *The Sankhya Karika of Isvara Krsna*. University of Madras, Madras (1935)
67. Suppes, P.: *Studies in the Methodology And Foundations of Science*., N.J. Van Nostrand, Netherlands (1969)
68. Suppes, P.: *Theory of Definitions: Introduction to Logic*. Van Nostrand Reinhold, New York (1957)
69. Teresi, D.: *Lost Discoveries: The Ancient Roots of Modern Science from the Babylonians to the Maya*. Simon & Schuster, New York (2002)
70. Thurston, H.: *Early Astronomy*. Springer-Verlag, New York (1994)
71. Ui, Hakuja, trans. and Thomas F.W.: *The Vaisesika Philosophy according to the Dasopadartha-sastra: Chinese Text with Introduction, Translation, and Notes*. Royal Asiatic Society, London (1917)
72. Vidyabhusana, S.C.: *A History of Indian Logic*. University of Calcutta, Calcutta (1921)
73. Vidyabhusana, S.C.: Nyaya Sutras of Gotama. In *Sacred Books of the Hindus, Vol 8*. The Panini Office, Allahabad (1930)
74. Williams, Sir Monier. M.: *Dictionary: English and Sanskrit*. Motilal Banarsidass Pub, Varanasi (1956)

Ontology, Artefacts, and Models of Reasoning

Pasi Pohjola

University of Jyväkylä Department of Social Sciences and Philosophy, University of Jyväkylä, Finland
pasipoh@jyu.fi

Summary. In recent philosophical studies on technological artefacts, an idea of dual nature of artefacts has been emphasized. Although this idea of dual nature is not a novel one, recent studies have extensively developed ontological aspects of technological artefacts. According to various authors there are two constitutive elements of technological artefacts that can be described in terms of physical properties of objects and intentional action. In this paper, ontology of artefacts is connected to issues discussed by C.S. Peirce under topics of abductive reasoning and philosophy.

The intention in this paper is to develop an account of models for creating novel artefacts that includes the creative aspect in terms of abductive reasoning. In the model developed in this paper, the ontological aspect is discussed as preliminary conditions for reasoning involved in the process of creating artefacts. To give a thorough account of these preliminary conditions and, thus, to relate them to abductive reasoning, the discussion in this paper exploits ideas from Peirces own writings. Especially issues that Peirce discusses under his trichotomy of philosophical disciplines are applied for demonstrating how preliminary conditions that occupy thought relate to abductive reasoning and ontology of artefacts.

The structured tasks and arguments of this paper are: (i.) Describe the idea of preliminary conditions of thinking through Perices trichotomy of philosophy; (ii.) Discuss extensively how idea of preliminary conditions of (i.) relates to abductive reasoning; (iii.) Demonstrate the general idea of dual nature of artefacts; (iv.) Develop a model of novel artefacts by combining together the ideas of (i.), (ii.) and (iii.). The general argument here is that there are certain forms of description as preliminary conditions that present themselves in ontology of artefacts and act as constructing elements of creative reasoning (described in terms of abductive reasoning) in the process of creating novel artefacts.

1 Introduction

In recent philosophical studies on technological artefacts, it is proposed that artefacts have a dual nature [15]. Although this idea of dual nature is not a novel one, recent studies have extensively developed ontological aspects of technological artefacts from this point of view. According to various authors there are two constitutive elements of technological artefacts: Physical properties of objects and intentional action [11, 34]. This idea of dual nature captures also one relevant issue of technological innovations. On one hand, technical artefacts have physical structure and physical properties. On the other hand, artefacts have functions that are relative to intentional uses. It has been argued recently that one central issue of artefact design is related to reasoning from physical structure and properties to functions and uses, and vice versa [14, pp. 139].

The purpose of this paper is to elaborate recently developed ideas of ontology of artefacts in the context of innovative reasoning, and to propose a model of innovative reasoning in technology. The model is a conceptual model founded on ontological foundations of technical artefacts. As a model of reasoning it elaborates on features of intentionality and action and practical reasoning. The model can be labeled as action based and contextual – cognitive model of reasoning and it does not follow straight forwardly the cognitive assumptions of the traditional computational approach in Cognitive Science and AI. What comes to representations, the model intends to act as a framework for externalized representations rather then a representational model of consciousness. The purpose of developing the model is, as it is argued in the following, to propose a normative structuring of framework for reasoning and for the essential phenomena involved in the reasoning.

There are certain reasons and assumptions for approaching this issue form a perspective that differs from computational-representational view of reasoning, although this view might not be incompatible with it. First, there are logical and epistemological reasons for abandoning traditional computational assumption [32, pp. 139–153] [29] [7] and computational account of reasoning. F. ex. in research on abductive reasoning non-monotonic logics are applied to capture dynamic features of reasoning [21, p. 223] [19, p. 30]. Secondly, in theories of meaning and in recent epistemology a strong emphasis on context is given [27] [16, pp. 131–168], which suggests more dynamic and socially sensitive view of meaning than traditional computational semantics. The purpose of the model is to capture this contextual aspect of representations and contextually created knowledge, and to propose a conceptual framework for representations and reasoning. Also, the approach to reasoning here is action-based and in intentional or goal targeted action, the model intends to function more as a structuring normative framework of reasoning rather then a representation of some actual cognitive process of reasoning (for action-based reasoning, see f. ex. [20]).

Central issues in this work are, in addition to ontology of artefacts, abductive reasoning as a fundamental from of creative model-based reasoning and preliminary conditions of thought. The reason for considering preliminary conditions of thought is that in Charles Sanders Peirce's philosophy there is an interesting but rather rarely discussed inter-connection between fundamentals of thought and abductive reasoning. Although this issue cannot be discussed here extensively, some ideas are taken here to promote the relation between thought, concepts and reasoning. Similar kind of issue is discussed recently in context of linguistic meaning and creative reasoning in [27]. The reason for considering these issues from Peirce's philosophy is motivated by the fact that the model developed here is a general framework and therefore it has to take into account fundamental features of thought and reasoning.

The discussion of the paper divides into four parts. First part is about preliminary conditions of thought and reasoning, and how ontology and representations of knowledge relates to them. These discussions are influenced by Peirce's philosophy, although this text does not attempt to make an orthodox scholarly interpretation of Peirce's philosophy. Second part is a description of ontology of artefacts founded on the idea of dual nature. The purpose of this part is to discuss what the essential features of phenomena are that reasoning in technological innovation processes is about. The third part discusses intentionality and reasoning in designing and creating technical artefacts and illustrates the action based account of reasoning. Last part is about abductive and model-based reasoning and how the ontology can be extended into a conceptual model of reasoning in technology.

2 Thought, Reasoning and Knowledge Representation

There are two issues that the philosophy of Charles Sanders Peirce is typically known of: Abductive or retroductive reasoning and theory of signs or semiotics. Both of these are parts of his programme of pragmatism that he developed through out his career. In his later work he also provided a systematic philosophical background to his programme or system. One main purposes of his philosophy was the investigation of foundations of creative and knowledge producing reasoning, and symbolic representation of knowledge and thought (or consciousness). The tasks of this enterprise were to explicate the foundations for meaningful thoughts in experience and as products of reasoning. These foundations are, according to Peirce, investigated in three categories of philosophical disciplines, where each category of disciplines has it own part in the investigation (see [26, pp. 133–178]).

Peirce is famously known of his obsession to triadic explanations. In Peirce's trichotomy of philosophical disciplines the most fundamental category of philosophical study is Phaneroscopy, and the other two are Semeiotics and Metaphysics. In his ordering of philosophical disciplines Pheneroscopy is prior to Semeiotics and Semeiotics is prior to Metaphysics. According to Perice, Phaneroscopy is the description of the phaneron and he writes that

> by the phaneron I mean the collective total of all that is in any way or in any sense present to the mind, quite regardless of whether it corresponds to any real thing or not. [25, 1.284]

Phaneroscopy is something that provides theoretical and methodological principles for other philosophical disciplines, such as Semeiotics to which also abductive reasoning belongs to [23, 24].

In Perice's system of philosophical disciplines phaneroscopy is a study of most fundamental features of thought. In his discussion of this most fundamental discipline of philosophical study, he writes that:

> I invite you to consider, not everything in the phaneron, but only its indecomposable elements that is, those that are logically indecomposable or indecomposable to direct inspection. [25, 1.288]

According to Peirce, phaneroscopy as a discipline brings about the most fundamental categories of thought. These categories come as a trichotomy, typical to his philosophy, and he calls these categories as firstness, secondness and thirdness. In his Harvard lectures on the general method and programme of Pragmatism, he claims that firstness refers to Quality of Feeling, secondness to Reaction and thirdness to Representation [26, pp. 160–178]. In his previous Havard lecture dealing with phenomenology he claims that a proper theory of categories is such that it has its representative in the discipline that has to do with signs and logic [26, pp. 145–159].

To make sense what Peirce intends to argue with his three categories of thought one should reflect what kinds of preliminary conditions for meaningful thoughts there are, i.e. on what general grounds do we produce meaningful thoughts. For one thing, thoughts ought to be, at least in principle, such that they can be conceptually represented. Also, thoughts have to have qualitative content, which makes identification and recognition possible. This is something that is included into the first category called firstness. For example, if I have pain in my left foot, it has to have such qualitative content that it can be pain of something and not just pain. The category of secondness referring to reaction is the fundamental property that preoccupies every distinction and makes differentiation of experiences possible, such as having or not having pain. Thirdness as a category of representation refers to how qualitative content of firstness and foundations for categorization and differentiation of secondness become representations, such as I have pain in my left foot.

It can be said that the three fundamental categories stand as basic building blocks for meaning bearing thoughts. In Peirce's seven Harvard lectures on Pragmatism he also considers the question of meaning and what ought to be a theory of meaning. For one thing, the issue of meaning is something that relates the basic structures of thought with the fundamental logical and grammatical structures of symbolic representations. But this is not all that constitutes meaning. Also the content of thought has to be represented in symbols. Where Phaneroscopy is about general categories of thought and logic is about proper reasoning and forms of representations, meaning (of signs) deals with the content of representations expressed in signs and in certain forms defined by logic and grammar. According to Peirce, the very possibility of meaning, especially the meaning of propositions, is dependent on certain general categories of thought (thirdness including firstness and secondness) and its representative logical or grammatical forms. These issues of logic, grammar, meaning and reasoning are discussed in Peirce's second category of philosophical disciplines.

The issues of investigation belonging to disciplines under Peirce's conception of Semeiotics (semiotics) are provided with methodological and theoretical principles from Phaneroscopy. The disciplines belonging under the heading of Semeiotics are, according to Pape [24], speculative grammar, critic and methodeutic, and they can be described as a formal doctrine of signs (distinguished from concrete investigation of meanings of signs and sentences). Without going into discussions about Peirce's semiotics and the theory of signs, it has to be emphasized how the ideas of phaneroscopy, especially thirdness, are reflected and represented in logic and grammar. Peirce himself is explicit about the role of thirdness (including firstness and secondness) in having meaningful thoughts and constructing propositions. According to Peirce, thirdness is a general feature of conciousness that comes to represents itself in logical or grammatical forms of propositions. So the general forms of thought become represented in symbolic forms and the meaning is dependent on both structure and content of thought represented in symbols.

In his trichotomy, Peirce's third category of philosophical disciplines is Metaphysics and according to his classification, both Phaneroscopy and Semeiotics provide principles for Metaphysics. This is also the discipline where ontology of artefacts discussed below belongs to. By generalizing a great deal Peirce's work, and in a manner that probably doesn't do justice to his thinking, it can be said that for Metaphysics the prior philosophical disciplines provide preliminary conditions in terms of structure and form of thought and (a theory of) signs for representation of content (or knowledge). In other words, what is expressed in Metaphysics (or in ontology) is constituted by features of Phaneroscopy and Semeiotics, including also abductive reasoning. In following sections the constitution of these features are discussed in the context of ontology of artefacts, which clearly belongs under Perice's conception of Metaphysics.

The discussions above provide a short and probably in orthodox scholar view an unjustified description of Peirce's view of how thoughts become representations and systems of representations. For discussions of model-based reasoning the interesting question is how these issues relate to reasoning. It is a well known fact that for Peirce all thinking is in signs (see f. ex. [20, p. 13]). This idea has been influential for the proponents of representational theory in cognitive science [8, pp. 143–159]. Whether or not one is willing to accept the representational thesis, the relevant issue of Peirce's philosophy is the interrelation of these three disciplines. The way we construct conceptualizations and categorizations of the world in ontology and metaphysics is effected by the fundamental features of thoughts (or what kind of preliminary conditions for thoughts exist) and how these are represented in signs, grammar and logic (in a broad sense). As it has been previously argued, reasoning that can be labeled as creative is bound by certain forms or structures of reasoning such as analogy, models, logical forms, etc. [19, 27]. The proposal that is made here through discussing Peirce's philosophy is that these very general and fundamental issues discussed above have relevance for studying and constructing models of reasoning also in specified areas such as technological innovations.

3 Ontology of Artefacts

Recently there have been several attempts to develop and discuss the ontological status of artefacts [2, 11, 14, 22]. The developments in the account of ontology of technical artefacts described here follow some ideas these authors have suggested and on most parts what has been suggested recently by authors promoting the dual nature view of artefacts. The ontology discussed here is mainly founded on the idea of dual nature of artefacts. This idea of dual nature is here discussed through a conception of use plan, where this idea of dual nature is related to intentional (and collective) use of artefacts through the functions of artefacts. The idea of use plans is initially introduced by Vermaas and Houkes [34] and it is revised and enhanced in various respects in this paper.

In his study on the sciences of the artificial, Herbert Simon introduces a description of artefacts in somewhat similar way as dual nature of artefacts does. According to Simon, technical artefacts have an inner and an outer environment. He writes that artefact can be thought of as an interface

> between an "inner" environment, the substance and organization of the artefact itself, and an "outer" environment, the surroundings in which it operates. [29, p. 9]

The relevance of Simon's work for the theory of artefacts is that he emphasizes the interrelation between the inner and the outer environment, i.e. interaction between the natural and the human or the social world. In technical or naturalistic views of technical artefacts, the emphasis is merely on the inner environment of the artefact, i.e. the physical properties of artefacts. In social construction view dominant in social studies of technology, the emphasis is on the other hand is on the outer environment, social structures and mechanisms relating to the physical object. The dual nature view of artefacts intends to overcome this dualism and to argue that both aspects are essential to artefacts.

In his recent study of technological design, Peter Kroes has argued that technical artefacts have a dual nature. Kroes claims that technological artefacts have a physical nature and also an intentional nature. What Kroes means by claiming that technical artefacts have intentional nature is that they have properties that are something more then mere physical properties of the object. These properties, described as intentional by Kroes, are meaningful only in accordance to the use of the artefacts [12]. This idea, also existent in

Herbert A. Simon's book Sciences of the Artificial, emphasizes that certain natural and physical properties are (functionally) meaningful only when they are acknowledged as something functional in relation to uses.

According to Kroes, what distinguishes artefacts from physical (and natural) objects is the function attached to it and this function is distinguishable only in a certain context of intentional human action. Drawing attention to intentional human action as the relevant feature of existence of artefacts, Kroes comes to speak of artefacts as intentional objects [12]. Kroes is certainly right in emphasizing the ontological significance of this duality of artefacts and also the relevance of the context of human action for the existence of artefacts. Technical functions of artefacts cannot be (merely) intrinsic functions of the object. They are established in a context that is external to the object and its properties. The manifestation of existing artefacts is, then, dependent on two contexts, its physical structure and the context of use (as context of human action). It could be said following Vermaas and Houkes that technical functions forms a conceptual drawbridge between structural and intentional natures of technical artefacts [34, p. 6].

Recently there have been some suggestions that this dual nature of artefacts and especially its intentional nature are dependent on social-collective notions. The intended and established uses and relative statuses of artefacts differ from accidental uses. The statuses of artefacts, recognition of functions and ways of use are socially determined in action. [22] One way to demonstrate this idea of dual nature of artefacts is to apply a conception of constitutive rule developed by John Searle, although there has been some criticism concerning the applicability of Searle's social ontology to technical artefacts (see [13, 22]). In his social ontology John Searle has explained the creation and maintaining of institutions by his constitutive rule X counts as Y (in context C) [30, pp. 31–57]. This rule for declarative act of stating something as an institution (i.e. stating a certain kind of social fact) is used for assigning a status function to something X to function as Y. To an extent, this constitutive rule can be applied for explaining the institutional mechanism involved in creating artefacts.

Process of creating a novel artefact can be described as an assignment of a status function (making it an institutional fact) using constitutive rule X counts as Y (in context C). Certain physical and functional features count as the artefact in context of actual and potential users. The idea John Searle wants to emphasize with his constitutive rule is that creating an institution necessarily involves a performative act making it accessible and acceptable for others. The performative act is needed, for there cannot be private institutions. When an artefact is created, it involves performative act of declaration, stating that a physical object with its physical properties function as the intended artefact. At the same time this performative act is a creation of an initial institution.

For example, the intentional use of heart beat measurement and pulse meters for analyzing training effects in sports can be described by assignment of a status function: heart rate measurement counts as training effect in context of endurance sports. It should be noted that when using heart rate measurement, a natural property is used as the basis of an artefact. Although heart rate is very simple and straight forward natural property, in use of heart rate monitors it is dependent on social facts such as the function of the artefact and what it is used for. For example conceptions used in training and coaching, such as appropriate heart rate zones for effective aerobic capacity training, are socially constructed facts that are not directly measurable data. This relation between social institutions and the use of heart rate measurement in exercise is even more evident in contemporary pulse meters (sometimes also called as wrist computers because of their complex functionalities). The pulse meters can offer, based on your personal data, appropriate exercises and levels of exercise for fat burning etc.

The application of Searle's constitutive rule demonstrates how the status of an artefact with certain physical properties is related to social institutions. It is not, though, a fully fledged account of how (created) functions relate to intentions in use of artefacts. One suggestion about functions and function ascriptions of artefacts is made by Vermaas and Houkes. Their ICE-theory of function ascriptions to technical artefacts is an analysis of how functions of technical artefacts connect and separate the conceptual parts of dual nature of artefacts, i.e. intentional and structural natures of artefacts. Their idea is to present an action-theoretic account of artefacts that can contribute to theorizing use and design of technical artefacts. In their theory that intends to describe artefacts through its functions, one central concept is use plans that connects the function and structural properties to intentional use of technical artefacts. In similar way to the idea of constitution, Vermaas and Houkes intend to describe the connection between two natures of technical artefacts [34].

One central idea in the ICE-function theory of Houkes and Vermaas is that it is based on an action-theoretic account. According to them

> a technical function of an artefact can be roughly described as the role the artefact plays in a use plan for the artefact that is justified and communicated to prospective users. [34, p. 8]

The proper function of an artefact, to use this term in a vague sense, is the one that the object gets through the execution of use plan. The criteria for function ascriptions of an artefact are created by artefacts role in a use plan and through a process of communication the function becomes justified in the context of prospective users. Because the authors of ICE-theory emphasize the roles of use and other intentional properties, they claim that

> technical functions cannot easily be interpreted as intrinsic or essential properties of artefacts. [34, p. 8]

The ICE-function theory is, according to its authors, an abstract version of three general theories of functions: Intentionalist, causal and evolutionist theories. The intentionalist content of the theory states that certain capacity C is a function ascription of an artefact x in relation to use plan p, if an agent a has a belief that this capacity leads successfully to goals of the use plan p when executed. The causal role is related to the justification of the beliefs in the intentionalist content and the evolutionary content states that certain agents u have selected the artefact because of its capacity C to the use plan p and, thus, communicated this to other agents d. In other words, certain account A of an artefact x has a capacity C that can be executed to achieve a goal in a use plan p and certain user a has beliefs about this and these beliefs can be justified causally. Also this account A of an artefact a with its distinguished capacity C has been selected and probably manipulated for this use plan by some agents u and this has been communicated to other agents d [34, pp. 8–10].

For example, let's assume that certain agent a is hungry and wants to cut bread to make sandwiches. So the goal of a use plan p in this case is to cut bread. An account A of an artefact x (knife) here is a bread knife and it has a sharp blade property C. The account A here distinguishes the artefact x from other uses, such as the knife with its sharp blade being a perfectly suitable murder weapon. In order for C to be a function ascription in relation to p, the agent a has to have belief that the artefact x has the capacity C and a belief that x's capacity to C contributes to successful goal attainment when manipulated in the execution of p. So in this case, the agent a need to believe that the knife has a sharp blade and that the sharp blade is a reason for succeeding to cut bread. Also this property has to be causally justifiable, i.e. the knife as a bread knife actually cuts bread. In addition, a group of distinguished engineers have developed bread slicing p and have selected the knife x, because of its sharp blade property C, for this task and by naming it a bread knife they have communicated this use plan p to other agents.

Following from the discussions above, the ontology of artefacts to serve the purposes of following model creation can be summarized as follows. According to the dual nature of artefacts, an artefact has an institutional status attached to its physical properties that can be described with a constitutive rule X counts as Y in context C. But according to ICT-theory, in order to have an institutional status (such as being a mobile phone), the object has to possess functions relative to its status that are means for certain goal attaining use. In other words, an artefact has to have an institutional status of being an object with certain functions that can be used as means for certain goal attaining action of the use plan.

4 Intentionality and Reasoning in Design

Before going into the details of the model itself, some preliminary issues have to be discussed. In the dual nature of artefacts and in the ICT-theory, intentionality in relation to use of artefacts has a significant role. Intentionality plays part also in the action-based approach to reasoning, because the context where the reasoning in question takes place is intentional and goal attaining action. Intentionality as directedness of thought was also an issue considered by Peirce and actually intentionality has a role in his theory of conciousness or thought. In the model discussed here, intentionality and intentional action have central role in several respects and these are discussed in this section. One of them is directedness of action and thought that is present both in the use plan and the reasoning itself. Another is intentionality in reasoning in terms of plans which is more related to practical reasoning then theoretical reasoning in engineering design.

Terrence Love in his paper on philosophy of design has examined some contemporary attitudes towards theorizing technical design [18, 295-301]. According to Love, theoretical accounts of design have been to a great extent about natural facts involved in designing artefacts. Terrence Love labels this kind of approach to design, where interests are merely on the object of design, as design theory approach. In design theory approach, the theory of design is based on empirical knowledge of physical phenomena and includes, as limits, knowledge of applicability of laws of natural sciences to what is to be created in design processes [18, pp. 293–294]. Because artefact design is carried out mainly by humans, human design processes have also been under theoretical interest of psychologists, cognitive scientists and social scientists (mainly in organization theory and more recently in field of research called design science). The human approach, dealing to a large extent with personal cognitive capacities and social facts of design organization, is more targeted to what is called problem analysis, the ways of making decisions and ways of setting environments for decision making (design methodologies, design environment, best practice strategies, organizational structure, knowledge management etc.) [18, p. 293] [31, p. 132–140].

The problem in these theoretical approaches to technical design that Love highlights is that human approach and design theory approach have remained as independent areas of research. According to Love's suggestion, an enhanced approach to design should overcome the dichotomy between above mentioned approaches to engineering design. Also according to Kroes, one central problematic issue in engineering design is

> what kinds of inference patterns are involved in reasoning from statements about functions to statements about physical structure and vice versa? [14, p. 140]

By looking into reasoning and inferences in engineering design practice, it is possible to clarify how these two approaches can be integrated. Inferences and reasoning in design connects uses and functions to physical properties if the reasoning has a framework including something like use plans. The suggested perspective to answering Kroes' question here begins from intentionality in design action and reasoning in design.

Artefact creation and engineering design can be analysed as intentional goal attaining action, although some authors have suggested different kinds of approaches [9, 28]. In her account of intentionality, G.E.M. Anscombe characterizes intentional action by stating that descriptions of intentional action are descriptions of means to descriptions of ends [1]. This kind of means to ends analysis of design action is not entirely new in (methodology of) engineering sciences either. For example, H.A. Simon in his book The Sciences of the Artificial refers to this analysis as an alternative for describing the logic of design [31, pp. 141–142]. For a reason or another, what could be an interesting investigation of history of technology studies itself, intentionality has not been much investigated subject in technology studies until recent years. On the other hand, intentionality, especially in terms of collective intentionality (or shared intentions) has been one of the recent issues of investigation in the philosophy of collective action and cooperation [6, 33]. Also, in their paper on design, Houkes et al have stretched the importance of collective intentionality for understanding design [10].

One other intentionality-based issue relating to the reasoning in engineering design is practical reasoning, because artefacts are objects of practical use. In his central contribution to the philosophy of human sciences, G.H. von Wright connects Anscombe's means to ends analysis of intentional action to Aristotelian practical reasoning described in a practical syllogism, as a method for understanding human action. The kind of description of human action introduced by von Wright is intended as a method for understanding the reasons why some action has been done. In general, intentional action has some ends through some means and to understand the action, one has to know what ends the action is aiming to and by what means. In von Wright's model, intentional action is reconstructed as practical reasoning by referring to means and ends, and this practical reasoning can be described in a practical syllogism. If person p aims to achieve certain goal G and has she has belief B that it can be achieved by doing A, then p begins to do A. This practical reasoning can be written into a syllogism in this way:

> p intends to bring about G; p considers that he cannot bring about G unless he does A. Therefore p sets himself to do A. [35, 36]

By suggesting practical syllogism as a methodological device for understanding action and practical reasoning involved in the action [17, p. 328], von Wright intends to describe the necessary and sufficient conditions for intentional action. For some person p to do intentionally A, person p must have certain goals and relating beliefs how these goals are achieved. In order for p to bring about G, she must consider A to be necessary and sufficient means for bringing about G. These necessary and sufficient conditions can be thought of as such conditions that they have logical demand for practical reasoning that is described in a syllogism. If the conclusion in the practical inference is to be true, i.e. that person p sets to doing A, the premises in the syllogism ought to be true. The person reasoning from means to ends must think that these means are necessary and sufficient for bringing about G. Because humans do fail in the achievement of the goal, the minimal condition is that p needs to be convinced that doing A will bring about G [36, pp. 98–110].

The logical demands that are stated in the practical syllogism are not, though, conditions that are carried out in every course of action man makes. Even our common sense experiences are against this idea, because we sometimes fail to achieve the set goals and we do not actually make conscious practical reasoning for every action that we carry out. Recently Martin Kusch has remarked that practical syllogism as a description of practical reasoning is mostly post actu model for understanding why something was done, i.e. a description of the reasons of action which is carried out. In recent literature, though, for example Bratman and Brandom have argued that practical reasoning is also relevant for making plans and carrying out rational action, and not merely post actu method for understanding actions [3–5]. Houkes et al argue, through a reference to Bratman, that making plans by using practical reasoning is essential part of meaningful (and rational) design action [10, p. 304]. When considering logical deduction involved in the syllogism, the deductive reconstruction of reasoning in intentional action should be thought as an ideal case, which can also show flaws in reasoning relative to design action. Also the syllogism of practical reasoning can be thought of as a logical and conceptual tool for plan making. In a connection to practical reasoning, intentional design action should be seen, at least in an ideal case, as meaningful action including explicit goal setting and explication of ways to achieve the set goals [14, pp. 147–148].

Vermaas and Houkes suggested a conception of use plans as a conceptual tool for describing function ascription in design and manufacturing of technical artefacts. In a design plan, certain properties of an artefact are thought of as functions that are means for achieving certain assigned goal of the use plan. By application of this concept, the goals of intentional actions of designers and users manifest in function ascriptions of artefacts. The designers create functions as means for users to achieve a goal described in a use plan. The discussion of intentionality of design here propose more explicit role on practical reasoning of uses and functions in addition to theoretical reasoning of physical structure and functions. It also intends to promote the explicit role of plans in design including practical reasoning and intentional means ends analysis.

5 Reasoning, Ontology and Models

The basic conviction of reasoning adopted here is that creative reasoning is: (i) Non-monotonic and (ii) form and content bound. These types of reasoning described above are here denoted as abductive reasoning. Non-monotonicity means that learning new knowledge or incorporating new knowledge to a system can have effects on reasoning in terms of validity and truth. The non-monotonicity of model-based reasoning does not imply that reasoning is arbitrary in regard to existing knowledge base in the sense that any kind of knowledge incorporated can have effect on the reasoning. One of the main reasons for discussing Peirce's philosophy and the ontology of artifacts is to argue that reasoning is relative to the form and structure of representation of knowledge.

It has been proposed that one of the puzzling questions of (methodology of) engineering sciences and engineering design is reasoning form structure to functions of an artifact and vice versa [14]. The problem is how to reason from functions to proper physical structure or from certain physical structure to functions of artifact use. One of the reasons for this puzzle is that there isn't any straight forward deduction or causal law from certain physical properties and structures to technical functions of artifacts. The relation between these two natures of artifacts can be realized in various ways, i.e. constitution relation is not determined by the physical properties alone. One way of approaching this question is to consider it from the perspective of use plans.

When approach reasoning in engineering science form perspective of use plans, one is not confronted with merely theoretical or scientific reasoning (such as reasoning in scientific discoveries), but also with practical reasoning. Technical design and engineering practices are purposeful, intentional and goal attaining actions. In order to avoid unnecessary complexity here, practical reasoning is here considered only in relation to use plans. The intentional content of a designer or some other agent conducting the reasoning relates to the functionality of the artifact, i.e. for what purpose the artifact is used or for what goal attaining action the functions serve as means. The reasoning in technology is a combination of theoretical and practical reasoning concerning physical phenomena and intentionality of using artefacts.

Reasoning in engineering sciences from structure to functions and vice versa can be described here as abductive reasoning (for accounts of abductive reasoning see [19]). Especially the aspect of theoretical reasoning included in the process of creating new artefacts has obvious similarities to hypothesis generating reasoning in science. Because the constitution relation between physical facts and institutional facts (or physical properties and functions in uses) can become realized in various ways, initial suggestion of functions to serve as means to some goal attaining action is hypothetical. The difference in the acceptance of a hypothesis between engineering sciences and natural science is that natural science hypothesis has to fulfil only the causal (and evolutionary) criteria of ICT-theory. Hypothesis has to explain causal properties of phenomena and it has to be accepted in the scientific community. Technical artefacts have to fulfil also the intentional criteria. The artefact has to be usable for the intended purpose (causal), it has to be used for the intended purpose (intentional) and the usage has to be socially acknowledged and accepted by prospective artefact users.

In context of scientific reasoning and discoveries it has been recently argued that hypothesis generating reasoning is structured, guided and bound in several respects. Abductive reasoning is typically seen to relate to problem solving situations in science, such as when confronting anomalous phenomena that needs to be explained. It has been recently suggested that the possible hypotheses are generated from a limited amount of possibilities, and the limits of reasoning relate to possible descriptions of phenomena that needs to be explained [27, pp. 327–329]. This idea has an immediate relation to issues discussed above in context of Peirce's philosophy. The possible hypotheses are relative to how we can perceive and think of phenomena and how we can describe and conceptualize it.

In reasoning about technology the possible outcomes are bound also by the practical context of use in addition to the theoretical context of explaining physical phenomena. In the ICT-theory suggested by Vermaas and Houkes the manipulated physical properties manifest as functions only in relation to use. Therefore reasoning from structure and physical properties to functions and vice versa has to take explicitly into account the context of use and intentionality of users. It can be said that fully fledged account of technical artefacts has to take into account the physical-intentional context of functions and the intentional-social context of uses. On causal criteria the artefacts have to function as it is intended. On intentional criteria the artefacts have to be used as means for some goal attaining action. On evolutionist criteria the artefacts have to become socially accepted objects with certain functions as means for a goal attaining action.

Following Bratman's suggestion that practical reasoning is used also in planning of intentional action [5, pp. 28–49], a general model of technological reasoning is suggested here. Following Peirce's ideas of representing thoughts and the discussions about possibilities of phenomena in context of abductive reasoning, it is suggested that ontology of artefacts can provide a conceptual framework for the reasoning, because such ontology is about the essential phenomena involved in reasoning about functions and uses in relation to physical properties. The model should be developed from the perspective of engineering design and manufacturing as intentional action, meaning that actions are targeted to producing an artefact with certain functions and a relative use plan.

The basic structure of the model is the constitution relation between X, Y and C in the constitutive rule. It is the basic condition for technical artefacts stating that some non-institutional facts X are performatively assigned an institutional status Y in some context C. The X term refers to physical properties and structure of the artefact. The institutional status Y of the artefact is a term that integrates these physical properties to functions and uses. The term Y has three essential parts: intentional use, (technical and status) functions and social acceptance. First of all, in order for some object to have an institutional status, it has to have a socially recognized status and it has to be collectively used, so also its functions have to be collectively recognized.

In use plans this means that the evolutionist perspective has to be strongly emphasized, so the means-ends analysis of intentional use should be viewed as social behavior and not merely individual behavior. Also the relevance of the context has to be emphasized, because it is the environment including the social environment where the constitution relation becomes manifested. There are various examples of unsuccessful innovations where functional failure is not caused by physical properties but social unwillingness to use the artefact. So successful artefacts are dependent on consistent use plans in such way that it has causal properties that can be intentionally used for the purpose the artefact is created. Furthermore, this purposeful use have to be something that can be shared among it provisional users.

6 Conclusion

In this paper it is suggested that ontological investigation of technical artefacts can contribute to technological reasoning, and especially when approached from the perspective of model-based reasoning relating to creating technical artefacts. This suggestion here is supported by discussing the relation of reasoning and representations of knowledge and thought. The main idea here was to propose a general conceptual framework of technical reasoning that elaborates on ontology of artefacts. The promotion of this idea is based on conviction that study of fundamental features of phenomena, such as technical artefacts, can explicate the necessary features that reasoning has to take into account.

In addition to ontology of artefacts, technological reasoning was also considered from the point of view of design and production of artefacts. It was claimed that reasoning should be thought of as intentional action having its goal in producing an artefact with a successful use plan. Where in the context of users technical functions were means for achieving the goals of the use plan, the means of design is reasoning, including models such as the one proposed here. It is obvious that the model itself is still a very general and initial suggestion and an extensive amount of theoretical work has to be done to enhance the model. The purpose of this initial work is to propose a perspective and framework for understanding and investigating reasoning in technology, and to promote the relevance of investigating the fundamentals of phenomena that the reasoning is about.

References

1. Anscombe G.E.M.: *Intention*. Basil Blackwell, Oxford (1957)
2. Baker, L.R.: The ontology of artefacts. *Philosophical Explorations* 7 (2004) 99–111
3. Brandom, R.: *Making it Explicit. Reasoning, Representing and Discursive Commitment*. Harvard University Press, Cambridge MA (1994)

4. Brandom, R.: Action, norms and practical reasoning. *Philosophical Perspectives* 12 (1998) 127–139
5. Bratman, M.: *Intention, Plans, and Practical Reason.* Harvard University Press, Cambridge MA (1987)
6. Bratman, M.: *Faces of Intention: Selected Essays on Intention and Agency.* Cambridge University Press, Cambridge (1999)
7. Collins, H.: *Artificial Experts. Social Knowledge and Intelligent Machines.* MIT Press, Cambridge, MA (1999)
8. von Eckhard, B.: *What is Cognitive Science?* MIT Press, Cambridge, MA (1996)
9. Gero, J., Kannengiesser, U.: The situated function-behavior-structure framework. *Design Studies* 25 (2004) 373–391
10. Houkes, W. et al.: Design and use as plans: an action-theoretical account. *Design Studies* 23 (2002) 303–320
11. Houkes, W., Meijers, A.: The ontology of artefacts: the hard problem. *Studies in History and Philosophy of Science* 37 (2006) 118–131
12. Kroes, P.: Design methodology and the nature of technical artefacts. *Design Studies* 23 (2002) 287–302
13. Kroes, P.: Screwdriver philosophy: Searle's analysis of technical functions. *Techné* 6 (2003) 22–35
14. Kroes, P.: Coherence of structural and functional descriptions of technical artefacts. *Studies in History and Philosophy of Science* 37(2006) 137–151
15. Kroes, P., Meijers, A.: The dual nature of artefacts. *Studies in History and Philosophy of Science* 37 (2006) 1–4
16. Kusch, M.: *Knowledge by Agreement.* Clarendon Press, Oxford (2002)
17. Kusch, M.: Explanation and understanding: the debate over von Wright's philosophy of action revisited. *Poznan Studies in the Philosophy of the Sciences and the Humanities* 80 (2003) 327–353
18. Love, T.: Philosophy of design. a meta-theoretical structure for design theory. *Design Studies* 21 (2000) 293–313
19. Magnani, L.: (2001) *Abduction, Reason, and Science. Processes of Discovery and Explanation.* Kluwer Academic / Plenum Publishers, New York
20. Magnani, L.: Conjectures and manipulations: external representation in scientific reasoning. *Mind & Society* 3 (2002) 9–31
21. Meheus, J., Batens, D.: A formal logic for abductive reasoning. *Logic Journal of the IGPL* 14 (2006) 221–236
22. Miller, S.: Artefacts and collective intentionality. *Techné* 9 (2005) 52–67
23. Paavola, S.: Abduction through grammar, critic and methodeutic. *Transactions of the Charles Sanders Peirce Society* 40(2) (2004) 245–270
24. Pape, H.: Charles S. Peirce on objects of thought and representation. *Nôus* 24 (1990) 375–395
25. Peirce, C.S.: *Collected Papers of Charles Sanders Peirce.* Harvard University Press, Cambridge, MA, 1931–1958. vols. 1–6, Hartshorne, C. and Weiss, P., eds.; vols. 7–8, Burks, A. W., ed.
26. Peirce, C.S.: *The Essential Peirce. Selected Philosophical Writings* Vol 2 (1893-1913). The Peirce edition project (eds.). Indiana University Press, Bloomington (1998)
27. Pohjola, P.: Abductive reasoning and linguistic meaning. *Logic Journal of the IGPL* 14 (2006) 321–332
28. Rosenman, M., Gero, J.: Purpose and function in design: from the socio-cultural to the techno-physical. *Design Studies* 19 (1998) 161–186

29. Searle, J.: *The Rediscovery of Mind.* MIT Press, Cambridge MA (1992)
30. Searle, J.: *The Construction of Social Reality.* Penguin Books, London (1995)
31. Simon, H.A.: *The Sciences of the Artificial* (2nd ed.). MIT Press, London (1981)
32. Thagard, P.: *Mind. Introduction to Cognitive Science.* MIT Press, Cambridge MA (1996)
33. Tuomela, R.: *The Philosophy of Social Practices: A Collective Acceptance View.* Cambridge University Press, Cambridge (2002)
34. Vermaas, P., Houkes, W.: Technical functions: a drawbridge between the intentional and structural natures of technical artefacts. *Studies in History and Philosophy of Science* 37 (2006) 5–18
35. von Wright, G.H.: Practical inference. *The Philosophical Review* 72 (1963) 159–179
36. von Wright, G.H.: *Explanation and Understanding.* Routledge and Kegan Paul, London (1971)

The Wondering Angels of the Fractal Art

Viorel Guliciuc

Stefan cel Mare University, Suceava, Romania
viorel_guliciuc@yahoo.com

Summary. Fractal Art is still a disputable topic among those touched by the "computer phobia". But in the postmodern era, the power of the fragment to announce the whole could never be underestimated. In fact, it is a form of Digital Art, having its own techniques and its own aesthetics. Moreover, we could detect in Fractal Art some connections with the Artificial Intelligence, if we agree that the Intelligence itself is already presupposed in any Theory of/on Complexity and so in any set/system of algorithms describing shapes with the self-similarity property.

1 On Digital Art

Computers are used in nearly each art form in such an omnipresent way that we should better ask: what isn't Digital Art? Computer generated art means "art images produced through the means of graphic manipulation computer programs" that receives and applies manipulations to "outside imagery information" and/or "internally stored forms: lines, shapes, effects, etc" [1].

However, for those afraid by the so-called "mechanization of art" and naturally doubtful on the art character of the Digital Art the question is: "how does one make art that springs from the cold soul of the motherboard and yet carries the caress of a human hand and heart?" Because Fractal Art is a particular type of Digital Art, we need focus on that origin. While nobody really worries "that the word processor has made writing too easy" and when digitally working as an artist, you should however expect "some strange resistance and lack of external validation" [2], because despite their wild variety, fractals are still highly recognizable "formula based" images.

Any digital artist is operating a reduction of all sorts of input into a homogeneous data flow, as he is integrating divergent visual material into a single work, in a sort of "a fluid synthesis of all sorts and kinds of media, materials, processes and styles". Digital Art must demonstrate and show the same will for innovations as it is in the fine arts. "... This means the creation of imagery that is both excitingly new and strangely familiar". The newness of the digital imagery goes far beyond any "isms". "The art that is produced in this matter is wildly divergent and one artist's work shares very little with another. There is, therefore, no discernible emergent style" [2].

2 Digital Art's Divisions

John Charles Macpherson thinks there are two major divisions of Digital Art. The evolutionist digital art "builds from the past and either brings images into the computer, or uses paint programs to create digital images, or a combination of the two" [3]. Here, the artist has "natural media" software at his disposal (e.g. Adobe, Corel, Jasc, ACD Systems etc.), which share the possibility to digitally re-create almost all the techniques used in modern art, working with "an assortment of tools designed to make marks which simulate on the computer screen and in print nearly all traditional paint and draw tools". The liberty achieved by the visual artist is the largest he ever reached. However, "digital work never reaches that level of material preciousness..." The medium adopted by the digital artist "works as fast as one's imagination" and offers "constant opportunities to refine composition and fine tune color"; moreover, "spontaneous accidents and the effects of gravity do not come easy in digital media"; that is why "what can be achieved in... a stroke of paint must be rendered laboriously by the digital artist" [2]. The time saved using digital tools is re-invested in experimentation, exploration and decision making. Using digital graphic manipulations the artist obtains more or less familiar, realistic, abstract, decorative or symbolic images. Some great works have been digitally created and were accepted by the artistic community. Digital art works are exhibited in traditional or virtual museums or art galleries. Thus, the evolutionist trend is the prevailing tendency in the contemporary Digital Art [3].

On the other hand and at the same time, there is the revolutionary trend, "that takes very little from the past and creates images from pure mathematical algorithms" [3]. In the search for newness and explicit new identity, the promoters of the Chaotic Fractal Art Movement or Algorithmic Art Movement have announced a Manifesto of the Art and Complexity Group [3]. It is an art form that "crystallizes a materialization field for lattices, scale plays, proliferation, self-similarity, hybridation, recursivity, lossy structures, butterfly effect, strange attractors, infinitization", producing "maximalist" art works [4].

The tool is refined until becomes a set of complex numerical functions and variables. The artist's instruments do not reproduce (or develop the equivalent of) the techniques and tools used by the modern art, whichever would be these. Here "the computer is used as brush" (Marilyn D. Brown) and the Fractal Art is "a programmer's art using software as the brush" [5]. Here we are dealing with "a radically different approach to any current or past art movement" – including the contemporary art movement [3], "forbidding any not algorithmic intervention", or only "tolerating a post-processing through graphic software, or through photomontage techniques" [4, 6].

3 Introducing Fractals

"A fractal is a rough or fragmented geometric shape that can be subdivided in parts, each of which is (at least approximately) a reduced-size copy of the whole" [7].

We are using the term "fractal" starting from 1975, when Benoît Mandelbrot has created it, in his work *The Fractal Geometry of Nature*, using the idea of creation from/of irregular fragments [8, p. 4]. An entire spot of related ideas originates there. Let's short remember two of them: 1) "... Fractals graphically portray the notion of 'worlds within worlds' which has obsessed Western culture from its tenth-century beginnings" [9]. 2) "Every shape embodies a way of thinking... the shape represents a system of thinking, a philosophy of behaviour, a way for modeling" [4, 6, 10].

There are many fractal mathematical structures: e.g. Sierpinski triangle, Koch snowflake, Peano curve, Mandelbrot set, Lorenz attractor [7].

4 On Fractal Art

Fractal imagery is powerful and seductive and it is always a personal artistic synthesis: "Fractals are patently beautiful with breathtaking depth, sumptuous color, dynamic flowing lines that tickle and delight the eye..." [2]. That is why "fractal pictures now ask for a place in the Art of our time". This happens "at a time when Modern Art is rejected, or at least ignored, by most of the public which did not receive the required training - indoctrination? - for appreciating this Art"; this happens, too, when "the same layman who often dislikes modernist artworks is spontaneously fond of fractal pictures with no need to be taught a non-Euclidean vision or the mysteries of self-similarity" [4, 6]. That is why we must go beyond simple explanations.

In fact, the discussions and the debates on the artistic nature of fractal images are not finished. For some critics, the Fractal Art is not exactly "art". That is why, rhetorically asking the question: "are fractal images a form of art, or simply beautiful images obtained through an algorithmic and not creative way?" We could agree "this question is very controversial, and contradictory opinions have often been exposed". Even the artists have been using self-repeating designs since immemorial time, this new form of art "chooses fractal geometry functions and the computational capabilities of the computer as its tools for the creation of infinitely detailed or stunningly simple forms that reveal and enhance the fantastic vision that comes from the realm of numbers" [7].

We should agree that despite the fact that "today, fractal art is still a curiosity, enjoyed only by a recondite few", this will change "for these strangely haunting images are not merely another fad or school or movement" but "the very mold into which our Western consciousness has been poured, and now, after a thousand years, the whole planet quivers in that same awesome matrix" [9].

We have over passed the long discussions on the existence of a distinct art form named Fractal Art. Instead, we are interested in the specificity of this form of art. Let's remember some of the most illuminating contemporary descriptions of the Fractal Art, as they are presented by some famous fractal artists or/and modern art critics.

Fractal Art may be considered as a separate genre of Digital Art, within the category of Algorithmic Art. "The fact that a fractal generation program is the originator of the image sets Fractal Art images apart from the other major categories of Digital Art. A fractal image can only be produced by a particular mathematical formula, and can be produced as an image only with a program designed to handle this specific type of mathematics". In order to explore or create fractal images we need the assistance of high speed computers. "Fractals begin as an original mathematical formula, a pure mathematical entity. This is Fractal Art's relationship to what is now called Algorithmic Art" [1].

"Fractal techniques are often used in the background in order to give complexity to pictures, but other techniques are possible as well (such as the plotter images of Roman Verostko or the plane mappings of Escher)" [4, 6]. That means Fractal Art is not exclusively digital art. Let's observe (!) that Fractal Art has being accepted as a specific form of Art especially after the critics have discovered fractal art elements in the artworks of Jackson Pollock and Mark Tobey. This happened decades ago, before our nowadays computer made fractal art. Based on Richard Taylor studies, Brett Yviett remarks, in an interview, that the most pleasant for the (human) eyes of the 220 human subjects tested are the fractal works with a fractality degree of 1.3 The Pollock's works have a fractal complexity between 1.12 and 1.7 like (as it is *Number 8*, created in 1950). Moreover, it seems that 1.3 is the fractal number of the African savanna relief. That was the place were the first hominids emerged. The pleasure to look at a fractal image could be a sort of immerse memory from that time... [11].

Fractals are so common that it is always possible to make fractals without knowing it. Fractal art could be an exploration of the inner boundaries of our humanity, revealing an unexpected relationship of the aesthetic emotion with the magic of the numbers...

5 Not only a Digital Art

Fractal Art can be considered as digital algorithmic art. It is a subclass of the visual digital art, an art form produced using a computer (PC, Mac), fractal and graphical software as essential tools in the creative process.

"Fractal Art does not have to be generated by a computer, see Pollock, the modern fractalists painters, so the further division to Digital Fractal Art seems necessary". We could consider as sub categories of Fractal Art, Traditional and Representative Fractal Art: It is important to have them well defined "to help evaluate the relative merits of any fractal art image" [1].

"I consider the definition of Fractal Art to mean images that are originally produced with a computer program that is dedicated to fractal image generation". Obviously, "minimal post processing is applied to the final, presented image". Traditional fractal images "focus on magnification 1, or other typical fractal forms found in most fractal imagery (Magnification 1, Mandelbrots and Julias, etc., minibrots, elephants, seahorses, spirals, etc)." *Grosso modo*, "Traditional Fractal images consist of unique fractal forms. The degree of fractality is high in these images" [1].

Representative fractal images are created by "focusing the main imaginative work on the overall compositional elements of the image". That is why "the fractal nature of the Representative Fractal image may not be immediately obvious. Compositional considerations are paramount in these fractal images." Such images are often inspired by other art movements. "Fractal elements may or may not be present in the final image. The degree of fractality can be small or non-apparent in these images" [1].

After a phase, "when the goal was to produce pure self-similar images of mathematical objects using the best possible colors and some other visual effects", nowadays the tendency is "to use fractal programs to create abstract images whether they clearly show fractal structures or not." The current interest is mixing several layers of individual fractals to create a single complex image. "Many artists use some post-processing transformations on the fractal images... Other artists combine fractal motifs with photographic pictures or with other images created with advanced graphics' programs" [7].

6 Again about the Fractal Art

The unexpected nature of the instrument and of the concrete modality of creation in digital art has brought and still brings critical reactions, both vehement and blind. Obviously, the privileged target is the fractal art. "Most advocates of the contemporary art movement consider any other visual form of artistic expression as non-art" The trivial and classifier aesthetics specific to this type of reporting is always due to the simple idea that "In general, contemporary art has to "make a statement", has to have "content", has to "push the envelope", has to be "disturbing", etc." [3].

Fractal Art is "a new way of looking at space and form". A new vision of the world has opened up, through the computer and its associated peripherals, as if a new type of camera has been invented. "This new tool has infinite focusing abilities and multitudes of configurable variables. It is almost hard not to discover new images." Obviously "as always it is not the tool which creates art, but the artists and their viewers" [12]. That is why "the learning curve" for the appreciation of this kind of digital art "seems almost as steep as for the manipulation of the tools themselves" [2].

What is actually happening is the shock of discovering that the Shape, the Light and the Color are all the time *before and beyond* the Word. This "shock" is felt by the fractal artists themselves, who persist in making the error to name their works (not only to invite the onlooker to have the same type of reading as the creator, but also to emphasize a mimetic character, intrinsic to the fractal forms), and disregard their status of explorers of a rather evocative artistic expressivity (and that is why it is closer to music than to the traditional plastic/fine art) than of an imitative one. Moreover, the above mentioned "shock" is felt by critics, too – but it's less felt by those who are fond of modern music and sculpture, for in these fields, mimesis is likely to happens as an evocation. Therefore, the most suitable attitude of a fractalist towards his own creation could be, perhaps, that one of allocating numbers as titles: in this way, the onlooker is invited to explore and create together the significations of the artwork.

We also could say that Fractal Art engages in a new way the relations between the Creator and the Work, against the common opinions. The "beachcomber model for sitting back while the computer does all the work, then simply choosing the best shell on the beach, is a misconception born out of computer phobia" [13]. Because the final intentionality is not present, there is no room, here, for the Fine Art.

As a matter of fact, things are completely different. "As a painter, confesses an artist, I am always looking to make visible the inner world of dreams, meditation, emotion. Fractals come closer to representing that world than any of the other graphic arts I have explored. I take the view, however debatable, that nothing is art that has not been fashioned by the hand and mind of a sentient being. Computers are tools with which to create, but computers in themselves are not the artists. Similarly, generated fractals in themselves may be fascinating images; but to become art, in my view, they require an artist to take hand and mind to them. I take one or more generated images and cut, paste, distort, add, subtract – in short, I play with them – until, like my paintings, they wordlessly tell me that they are complete" [14].

Moreover, during the awkward process of exploration of the variants of a fractal form, the creation "confiscates" its creator for its own aims, in order to find the perfect equilibrium between evocation and mimesis, as it happens in other artistic creations. "... The computer makes thousands of computations to arrive at this final rendering. But, the computer is not doing this alone. The artist chooses the formulas, the layers, the colors, more formulas and continues until the computer renders a finished print that the artist is pleased with" [15].

It is worth to be mentioned is the fact that, among the fractal artists, there is a strong tendency to process and to post-process the primary fractal images, in order to emphasize the desired effects, when the possibilities to vary the shape and the colors are, however, limited by the reciprocal interdependency of diverse series of variables – which command the identity of a fractal art work. The fractal artist cannot do whatever he wants, by using only fractal graphic programs (he is conditioned by the limited lifetime and by the iron-made frame of the sets of formulas and of their intrinsic possibilities of variation); that is why the fractal artist has to process the primary images he got, in most cases. He "must direct the assembly of the calculation formulas, mappings, coloring schemes, palettes, and their requisite parameters. Each and every element can and will be tweaked, adjusted, aligned, and re-tweaked in the effort to find the right combination", because he is free to manipulate all these facets of a fractal image. That "brings with it the obligation to understand their use and their effects" [16].

We could note that Fractal Art is Experimental Art: "the artist is never fully certain what effect variables will have on the image." This simply fact connects, rather than separates, the fractal artist with the traditional paint artists. "The chaos that occurs when a painter touches the canvas with the brush can never be fully controlled. Indeed, most accepted masters of the painterly arts find methods to guide this fractal force, as a horseman can never possess the power of the horse but only direct it to some degree". There are some differences, too, because "when working with only a fractal program, with no other graphic manipulation used, the image dictates to the artist what direction to proceed, and not vice versa, as with traditional drawing or painting techniques. The dedicated fractal artist spends hours making small changes to various parameters, looking for just the right effect or color scheme" [12].

The fractal artists do not create but explores, does not manufacture but discovers. "Instead of being repelled by computers, ... thoroughly immersed in their machinery they become an integral part of it". Moreover, "fractal artists are first to know the synthesis of man and machine Western philosophers have long dreamed of but never understood" [9, 17].

There has been noticed that both the sets of formulas expressing a type of fractal forms and the sets responsible for theirs color combinations, have the Golden Proportion inside them. "Colourists find the computers capabilities particularly useful (but infinitely frustrating) due the speed and ease with which areas and even the whole color balances of entire image can be changed" [18].

Fractal Art engages both the artist and the onlooker in recognition and contemplation. Considering the recognition, Ken Keller wrote: "at exhibits of my images the most common experience among the audience is one of recognition. Subconsciously everyone can relate to the general effects found in fractal imagery, even though this type of image has never appeared before in all of art history. It is a more visceral experience than the psychedelic era poster art. Somehow the fractal image has validity in human consciousness that seems to be innate". We could note that "there is some recognition there, even if they have never seen that picture before" [19].

Considering the contemplation issue, he wrote: "another common experience while viewing fractal imagery is one of contemplation. Almost everyone who has been captured for a moment by a fractal art print is instantly transported to a contemplative state. A glaze comes over their eyes, as if they are contemplating cloud patterns. Depending on the abstract qualities of the image, every viewer sees the print in their own private context. Everyone sees something different in highly abstracted fractal images. Often that personal perception of the image is so individual that it can only be resolved by the viewer, no matter how much they explain what they see and point to where they see it. This is not an effect restricted to fractal art. Inkblots do the same thing. Yet the complexity and elegance of the fractal image viewer's subjective constructions always amaze me. This could be a window into the human unconscious that still remains unexplored. In addition to being quite possibly the most colorful images ever produced by man, fractal images have the ability to be explored as separate visual universes. All one needs to enter the universe of a particular fractal image is the original fractal parameter file and the program that can render it. This interactivity of the fractal image is one direction that this art form will could evolve" [12].

7 Fractal Art's Techniques

"Fractal art uses several effects that are not generally found in other art techniques". When skillfully used, these unique effects produce "images that are important advances in contemporary art". We could use such effects "to explore fresh areas of human perception" [20].

First of these effects is the "self-similarity". "Self-similarity is a well known property inherent in fractal imagery. What is perhaps not as generally appreciated is how this effect actually works in physical reality when viewing large fractal art prints" [20].

The second of these effects is the so-called "fractal perspective". "A very finely detailed and complex fractal when printed at a large print size may display many different image compositions, each image appearing as one moves closer to the print surface. Physically moving closer to the print is analogous to virtually zooming in on a fractal with a computer fractal program. The change depends on the scale of your viewpoint, on how far you are from the fractal print and the resolution of the print". Ken Keller name this effect "self-similarity" and consider its persistence over scale ranges as "Fractal Perspective". "This image is very different than when viewed at 30 then 20 then 10 then 3 feet away from the surface of the image. This effect, which I call the fractal perspective effect, is a new type of aesthetic experience... The fractal perspective effect occurs for an appropriately constructed fractal print on a gallery wall as it does for a fractal in the virtual reality of the computer zoom. If you are walking toward a print in a gallery or zooming in on a fractal with a computer the effect is essentially the same. Of course the physical print in a gallery will involve more of your senses and attention and depends on the size and resolution of the actual print. The effect is more visceral when seen in physical reality than in virtual reality and is a completely fascinating experience. The ideal situation to demonstrate the fractal perspective would be to display a fractal print large enough to exhibit large scale self-similarity.

The image you would see at the greatest viewing distance would be the same image you would see at the closest viewing distance" [20].

The third characteristic of the Fractal Art is the "fractality". "The degree that the work deviates from this ideal situation is the degree to which it is similar to a normal but very large and detailed geometric, non-fractal drawing. The degree of fractality involved in the different viewpoint images should also be considered." We could agree that "fractality is a measure of the fractal's dimensional (scale) complexity. It can be a precisely defined value. It is determined by imposing ever smaller grids onto the image. Fractality is another aspect of fractal art that will be discussed in the future" [20].

8 FA and Cognitive Aesthetics

The specifications above could be better understood if we infer that Fractak Art's implicit esthetics may be rather considered from a cognitive point of view. What happens here is similar to what happens while reading a philosophical work. To pan beyond the shock of this assertion, we could re-interpretate Ch. S. Peirce's work and make a distinction between three types of notions: "1) 'play', 'desert', 'good', 'sky', 'red', 'to wish'; 2) 'neutrino', 'hexaedron', 'heresy', 'diesis', 'metonimy', 'deflation'; 3) 'so/therefore', "signifies', 'object', 'true', 'theory', 'identic' ".

The terms in the first list "have the characteristic of being used before being defined. Their usage rules can be induced from their concrete applications, and it is possible to spend yours whole life using the correctly, without managing to find any rule... But it is important not to be wrong upon the viewpoint: this is not due to the fact that these are polysemantic terms... The notion they imply is of a complex nature. We use and name them without any difficulty, on the one hand, and we hardly manage to define them, on the other hand. This happens because they are learnt through the means of prototypical approximations, involving values "by definition" or by following 'the family airs'. For instance, it is known that play is something common among chess, bridge, football, etc." [21, pp. 42–43]. We can say that these notions make the pragmatic (not empirical) type of knowledge possible. As far as visual arts are concerned, there is the equivalent of this notion type, especially when the mimetic character of the plastic work is obvious and the language is a classical, traditional one. In such a case, the aesthetic sensibility is "pragmatic" as much as we know that anyone could recognize the artistic forms, starting from his own life experience.

"The terms on the second list need to be first defined and then used, and they can be understood only by their reporting to a theory which "forgess" or define them axiomatically, id est (that is) a priori" [21, p. 42]. Here, the achieved knowledge will be a "theoretical" one. As far as the plastic/fine arts are regarded the equivalent of this type of notions is the evocative character of modern art and of its type of language. Both the aesthetic sensibility and the perception are conditioned by a minimum aesthetic education and by a minimum knowledge of the art history evolution. The reading and the interpretation of a modern art work are, consequently, more bookish and more "theoretical".

The third lexical list contains philosophical notions, "which intervene in both pragmatic and theoretical enunciations, without being a part of any of the two. In any science we can find expressions such as those present in the third list, and the task for defining them is not incumbent on any of the sciences. The common language uses the equally and with the same acceptability as the theoretical language; but their learning is not achieved through pragmatic prototypes. It may be sustained that these terms are "transcendental" if this is not understood that they "are traveling". To a certain extent, these terms are the music network, allowing us to talk about other things, pragmatically and theoretically; things that configure something like laying rules of any possible language. Due to these expressions, the notions outline in architectures of thinking" [21, p. 44].

Let's consider Fractal Art as that form of art inside which the natural balance between "pragmatic" and "theoretic", between mimesis and evocation, between the wild, uneducated perception and the educated, refined perception, fully manifests itself. Only here the Form, the Light and the Color free themselves from a certain reading dependence. They are retrieving the primordial condition of pure "childish" media, under the sign of the Infinite exploration. The play is, in this case, one with our own limits. Here, the limits of the humannity are sublimating themselves, inside the limitations of our artistic sensibility. It is the only game in which our capital of essential and native perplexities is not only fruitful but also increasing.

Fractal Art is actually as non-permissive and elitist as any form of art, and only the gifted ones can say (referring to the deeper meaning of the difficult relationship: artist – creation), as W.B. Yeats does in his poem *Before the World was made*: "I'm looking for the face I had/Before the world was made". Therefore, when our Ideas as well as our Images are knocking at our doors or windows, the Child inside us – the only one endowed with artistic sensibility – will recognize his Wondering Wings of Angel.

9 Conclusions

Our first conclusion is that against all the prejudices, "the advancement of new markets, modes of display and distribution", will certainly revolutionize all the aspects of what we now call art. "Style will become a tool for expression, not oppression. Art will become, simultaneously more personal and more pervasive". All the signs tell us "we will have 'symbiotic art'", in which "the observer will become a functionary of the art itself and the designer will become a poet of the senses" [2].

"The art historians of the future will look back at this period as the first struggling attempts to assimilate this powerful new vision into our human experience" [12]. We are actually doomed to be only pretexts in the de-virtualization of some Worlds, inside of which the pure Intelligibles try to play with the pure Sensibles. As in any other form of creation, in Fractal Art the rule governing the reification of possible worlds is that suggestively uttered by the famous Romanian philosopher Constantin Noica: "You would not have searched for me if you hadn't (previously) found me".

This happens with the Idea as the Word and with the Shape/Forms, Light and Colors as well. Only when the Intelligible (that is always more and deeper than the rational) comes to identify itself with the Sensible, the result is Art). Thus, in a piece of fractal art work, we may say the above-mentioned equilibrium can be found only in a small number of variants, despite those who think the oneness of the art works is threatened by the Form, Light and Color's infinite possibilities of variations. "Art is not about the tools used to make it; but in the organization of color, line, form, composition, rhythm and the interplay of all these in support of the subject matter or intent of the work itself. These are the basic and well established tenants of visual art and as fundamental to digital art work as to the cave paintings of Lascaux" [2].

Our second conclusion: in the previous pages, a point of view has been exposed around the idea that Fractal Art is the result of the interaction between the Human Intelligence and the Digital Tools able to modeling some aspects of the Complexity of the Light, Form and Color. That is why, in the most unexpected manner, we consider to have good reasons to express a hypothesis on the relation between the Intelligence and Fractal Art. I strongly think that, much more than in other forms of Digital Art, in Fractal Art the Human Intelligence is really meeting the Artificial Intelligence. On The Forum for Artificial Intelligence, Harold Henry Chaput, President of the *Austin Museum of Digital Art*, announce his conference "The Artificial Artist" as it follows: "artists and art theorists have long been developing algorithms and formal theories of art. Some of these theories have even been somewhat generative, in that they allow for the production of a work given a set of rules and guidance from the artist. Notable examples include the music of J. S. Bach and Arnold Schönberg, and the visual art of Wassily Kandinsky. The trend of algorithmic art, though, has not been to increase control but to remove it, allowing the artist to establish a set of initial conditions and let the work generate itself." Nothing unusual! Ok, but let's be amazed by the following observation: "An essential component of Algorithmic Art (as it is understood today) is the "surprise", the ability of the algorithm to produce something that neither the viewer nor the artists expects."

In the sentence above a major Idea is watching us. If the Intelligence means the capacity to produce/to integrate/to assimilate the Unusual, then there is Intelligence in the Digital Art/Fractal Art. Intelligence seems related to the Complexity and especially to the generation of the Complexity. Because we will not find a fractal software fully protected to some random results, in the very creation process of a fractal image we will find a true interaction between the Human Intelligence and Sensibility and the Algorithmic Artificial Intelligence. Indeed, "the rules of Algorithmic Art do not define or constrain the artist so that he or she would act like a machine. They liberate the machine, they grant it autonomy to freely produce original works of art."

In fact, presupposing the Complexity is the key for the understanding (= intelliggere!) of Being, we have already the presupposition that Intelligence is one of the Complexity manifestations itself. Whatever that Intelligence could be.

References

1. Keller, K.: (Recent evolution in fractal art. Traditional and representative fractal art) http://fractalartgallery.com/fractal_art_essay_01.htm.
2. Jarvis, J.: (Toward a digital aesthetic)
   http://www.dpandi.com/essays/jarvis.html.
3. Macpherson, J.: (An open letter to the critics, curators or owners of art galleries and to the artists involved in the emergent movement of the digital art) http://www.rofag.usv.ro/an_open_letter.htm.
4. Vassallo, C.R.: (Fascination of fractals)
5. Shapiro, G.: (A programmer's art using software as the brush)
   http://www.glyphs.com/art/fractals/geoffrey.html.
6. Vassallo, C.R.: (Two forerunners)
   http://perso.orange.fr/charles.vassallo/en/art/fractalist.html.
7. Louvet, J.P.: (Fractal art faq : questions les plus fréquentes sur l'art fractal) http://fractals.iut.u-bordeaux1.fr/f-art-faq/faq03.html.
8. Mandelbrot, B.: *The Fractal Geometry of Nature.* W H Freeman & Co, New York (1982)
9. Beck, A.: (Images created by man & chaos. some personal observations on fractal art & artists) http://www.glyphs.com/art/fractals/frac_art.html.
10. Conde, S.: The fractal artist. *Leonardo* **34(1)** (2001) 3–4
11. Mugur, P.D.: Fractali, labirinte, simetrii. O introducere în estetica cognitiv - un dialog cu brett yviett de paul doru mugur [fractals, labyrinths, simetries. an introduction in the congnitive aesthetics - interview with brett yviett by paul doru mugur]. *Respiro* **15** (2004)
12. Keller, K.: (Fractal Art. exploring the boundary between Creation & Discovery) http://fractalartgallery.com/what_is_fractal_art.htm.
13. Jarvis, J.: (In art lover's guide to digital art)
    http://www.dpandi.com/essays/jarvis2.html.
14. Brown, M.D.: (The computer as brush. fractals and the human hand) http://www.glyphs.com/art/fractals/marilyn.html.
15. Schiffhouer, C.: (What are fractals?)
    http://www.digitallace.gq.nu/what_are_fractals.htm.
16. Mitchell, K.: (The Fractal Art Manifesto) http://www.fractalus.com/kerry/articles/manifesto/fa-manifesto.html.
17. Beck, A.: (What is a fractal? and who is this guy mandelbrot?) http://www.glyphs.com/art/fractals/what_is.html.

18. Davison, M.: (Impact of digital imaging on fine art teaching and practice) http://www.agocg.ac.uk/reports/graphics/26/node5.htm.
19. (Lloyd, K.) http://www.thefractalfarm.com.
20. Keller, K.: (The fractal perspective. Self-similarity in fractal art) http://fractalartgallery.com/fractal_art_essay_02.htm.
21. Almeida, I.: L'interprétation abductive et les règles du raisonable. sémiotique et philosophie. *Documents de Travail et Pré-publications* **197-198-199** (1990) Centro Internazionale di Semiotica e di Linguistica, Urbino.

Part III

Logical and Computational Aspects of Model-Based Reasoning

Polynomizing: Logic Inference in Polynomial Format and the Legacy of Boole

Walter Carnielli

Centre for Logic, Epistemology and the History of Science, UNICAMP, Campinas, Brazil; Department of Philosophy, UNICAMP, Campinas, Brazil; Security and Quantum Information Group–IT, Lisbon, Portugal
carniell@cle.unicamp.br

Summary. *Polynomizing* is a term that intends to describe the uses of polynomial-like representations as a reasoning strategy and as a tool for scientific heuristics. I show how proof-theory and semantics for classical and several non-classical logics can be approached from this perspective, and discuss the assessment of this prospect, in particular to recover certain ideas of George Boole in unifying logic, algebra and the differential calculus.

1 From Finite and Hard to Infinite and Smooth

One of the most fascinating episodes of the history of Mathematics, which is nowadays almost considered a triviality, is the discovery of the polynomial representation (by infinite series) of numerical functions.

One can situate this historical point in the western historiography, although variants of his methods were already known before in Europe and in China and India as well, around the English mathematician Brook Taylor (1685–1731) and his book *Methodus incrementorum directa et inversa*, of 1715, which led to the development of the Taylor's and MacLaurin's expansions. Surprisingly, however, the importance of Taylor's discovery remained unrecognized until 1772, when J.L. Lagrange realized its relevance and proclaimed it to be "the principal foundation of differential calculus".

Any infinitely differentiable function $f(x)$, under certain circumstances, can be rewritten as an infinite polynomial series in the neighborhood of a base point x_0:

$$f(x) = \alpha_0(x_0) + \alpha_1(x_0) \cdot (x - x_0) + \alpha_2(x_0) \cdot (x - x_0)^2 + \ldots \alpha_n(x_0) \cdot (x - x_0)^n + \ldots$$

for certain coefficients $\alpha_k(x_0)$. What inspires amazement is that such coefficients are the derivatives of $f(x)$ itself calculated in the base point x_0, and the idea of a *local* representation for the function (depending on the point x_0) emerges. Much deep mathematics originated from the questions on how to restore the global behavior of a function from its local behavior (as singularity theory), and how far we have to go in the series to gain substantially all information contained in the function (as Morse theory).

This amounts to transcendental functions being represented by algebraic, polynomial functions – at the cost, however, of accepting infinite expansions. Although the Greeks used the notion of infinite in geometry and arithmetic, as in the famous arguments of Euclid's proof of the infinity of primes, one cannot lose sight, however, of the problems that surrounded the concept of infinity since the hellenistic times.

Also, the notions of finite and infinite were not coincident in ancient Greek and Chinese thought for example (see [23]), which indicates that the notion of infinity was not (and perhaps is not) absolute; this may be seen as a measure of the boldness of users of the infinite much before George Cantor who attacked the problem of conferring meaning to the "unthinkable". Hermann Weyl in [33] claims that "mathematics is the science of the infinite", and compares mathematics with religion: "... the religious intuition of the infinite, the $\alpha\pi\varepsilon\iota\rho o\nu$ takes hold of the Greek soul...". I want to argue that what lies within this idea of expanding simple constructions to the infinite, if not religion, is a powerful method, still to be completely clarified, which I venture to call *polynomizing*: the idea that something "finite" and complex[1] can be reduced by considering (possibly infinite) polynomial-like representations. I will consider several instances of this idea, particularly in the rise of modern logic by the hands of Boole.

The discovery of power series is, in a sense, a generalization of polynomials taking into account the possibility of extending the sum to the infinite. Polynomials were basically the only functions which could be manipulated by hand to approximate trigonometric functions, for instance, which were almost beyond the capacity of 17th century calculation.

Prior to the full development of integral calculus, the discovery of the formula $\frac{\pi}{4} = 1 - \frac{1}{3} + \frac{1}{5} - \frac{1}{7}\ldots$, independently obtained by Gottfried Willhem Leibniz (1646-1716) and by mathematicians in South India in the fifteenth century, attributed to Nilakantha (in Sanskrit verses, cf. [27]) is a good example of the difference between just thinking in terms of infinite objects and thinking in polynomial terms with regard to infinite expansions. John Wallis in 1650 found the expression $\frac{\pi}{2} = \frac{2\cdot 2\cdot 4\cdot 4\cdot 6\cdot 6\ldots}{1\cdot 1\cdot 3\cdot 3\cdot 5\cdot 5\ldots}$ which converges slowly and is hard to generalize. On the other hand, the former expression is a particular case of the expansion $\arctan(x) = x - \frac{x^3}{3} + \frac{x^5}{5} - \frac{x^7}{7}\ldots$ (for $-1 \leq x \leq 1$) discovered by James Gregory (1638-1675), even if Gregory himself (cf. [18], chapter 4), failed to see that for $x = 1$ it gives the expression for $\frac{\pi}{4}$.

---

[1] Of course, I do not mean here complexity of computation, in the sense of mere efficiency: complexity is here meant in a wider sense, though not divorced from that restricted meaning.

Apparently, Nilakantha was aware of the impossibility of representing π by means of a finite series of rational numbers, so the idea of infinite was probably seen as a key to solve the problem of representing π.

However, it is illuminating to see cases where infinite sums and infinite products work together to produce new mathematical knowledge. An example of such cases was a remarkable result proved by Leonhard Euler (1707–1783) in [19] about equating an infinite sum with an infinite product, which gives an alternative proof of the infinity of prime numbers:

$$\frac{2 \cdot 3 \cdot 5 \cdot 7 \cdot 11 \ldots}{1 \cdot 2 \cdot 4 \cdot 10 \ldots} = 1 + \frac{1}{2} + \frac{1}{3} + \frac{1}{4} + \frac{1}{5} \ldots$$

which, in contemporary notation, can be written as:

$$\prod_p \frac{1}{1 - \frac{1}{p}} = \sum_n \frac{1}{n}$$

for all primes p and natural numbers $n \geq 1$.

This formula, which coincides with a particular case of the celebrated Riemann Zeta function at the value $s = 1$, gives an alternative proof of the fact (already known by Euclides) that there exists an infinite number of primes, by taking into account that the left-hand harmonic series is divergent.

Though I am more interested here in "infinite" methods emerging from algebra, there is of course a geometric side in the advent of the infinite expedient to produce finite calculations: as a precursor to integral calculus, Bonaventura Cavalieri (1598-1647) had completely developed a method of indivisibles, as a means of determining the size of geometric figures similar to the methods of integral calculus in his *Geometria Indivisibilibus Continuorum Nova Quadam Ratione Promota* ("A Certain Method for the Development of a New Geometry of Continuous Indivisibles"), published in Bologna in 1635.

According to [26], a method similar to Cavalieri's had already been used in China around the third century to find the volume of a sphere.

Intuition in this direction not only impacted algebra and geometry, but certainly influenced (directly of indirectly) Boole and other logicians. I want to suggest that it is possible to identify an ancient tradition of what I called "polynomizing", which has also deeply influenced logic, but this approach, although present in many aspects in Boole's work, was for some reason relegated. However, as I plan to show, it can be regained inside the methods of logic – and in several aspects, from classical propositional to many valued, from paraconsistent to modal and even to first-order logics, and it may be used as a reasoning model, helping understand certain aspects of Boole's methods.

George Boole is reputed one of the greatest logicians or philosophers of logic of all times: John Corcoran in [16] considers the *Prior Analytics* by Aristotle and the attempts by George Boole (1815–1864) to codify the laws of thought ([6]) as the two most important logical works from before the advent of modern logic.

However, Boole is often accused of fallacies and incoherences, and his logic calculations are sometimes considered close to ridiculous: as in [17] puts it, "Readers of Boole's logical writings will be unpleasantly surprised to discover... how ill-constructed his theory actually was and how confused his explanations of it", and even Corcoran, on p.285 of [16] dares say that "Aristotle seems superior to Boole and closer to contemporary thinking. My guess would be that Aristotle would have less trouble understanding Gödel's results than Boole".

Indeed, some of calculations Boole proposed may seem awkward and inept, but the critiques would lose impetus if one regards Boole's dream of algebrizing logic and his search for missing links between ordinary algebra and Aristotelian Logic from the point of view of attempts to polynomize: Boole was more interested in the algebraic aspects of logic, by means of solving equations expressed in polynomial form, than he was in the logic aspects of algebra.

The intentions of this paper are twofold: firstly, to raise some ideas on recovering the algebraic setting of logic in a broad sense, showing how this can be applied to the clarification of some criticisms in Boole's work; a second intention is to propose a wider algebraic stand to the contemporary view of classical and non-classical logics.

2 Algebraic Proof Systems

Boole, in [6], attached great importance to the "index law" $x^2 = x$, placing it in such a central position that, for him, "... a fundamental law of Metaphysics is but the consequence of a law of thought".

From the purely mathematical side, this has connections to another important work of Boole: the invention of the calculus of finite differences of 1860 [7], preceded by his better known treatise on differential equations.

Boole was one the first to perceive clearly that the symbols in operations could be treated directly as objects of calculation, separated from the idea of quantity. However, Leibniz already admitted equations with no explicit arithmetical content such as $x+x = x$, and even talked about "blind thinking" to refer to pure reasoning reduced to arithmetical calculation (cf. [28]). It is interesting to know how Leibniz, in his *Elementa Calculi* of 1679, assigned numbers to concepts in such a way as to obtain a complete representation for Aristotelian syllogistic and complete version of algebraic logic in Boolean terms, although apparently Boole did not know his work (see [10] for proofs of correctness and completeness of the two mentioned systems).

Besides the "index law" $x^2 = x$, Boole assumed, differently from Leibniz, that $x + x = 0$ implies $x = 0$. He did not assume multiplicative inverse, but just additive inverse; from contemporary viewpoint, many problems behind Boole's methods are explained by the fact that he was not working within a *field*: he accepted the generalization of the index law $x^n = x$ in [4] and [5], but this will be rejected in [6]; indeed, for him $x^3 = x$ would lead to $x^3 - x$ to have as factors $x+1$ and $x-1$. The first could not be accepted, as $1+x$ would correspond to adding x to the universe 1, and -1 is equally non-interpretable, since it does not satisfy the index law $(-1)^2 = (-1)$. However, analogous difficulties will arise in the same index law $x^2 = x$, since it is equivalent to $x^2 - x = 0$, which has $x + 1$ and $x - 1$ as factors as well (for details see [22]).

Of course Boole was opening a path to future developments that would only come after his achievements, such as working with rings of characteristic 2 with unity. This would make simple to accept for instance $1 = -1$ and solve many of his difficulties – in particular, for his case, a Boolean ring with unity would suffice (a concept that possibly would have never been invented were it not for his difficulties!).

Departing from the idea that if the intuition behind the index law had some importance for classical logic (even if exaggerated by Boole), it seemed obvious that this law could be easily generalized to the "higher-order laws" of the form $x^n = x$ (that Boole had to reject) by employing polynomials over Galois fields: intending to explore such laws for non-classical, [11] studied the question for some finite-valued logics. I later learned that some methods for Boolean reasoning were developed by the Russian logician Platon Sergeevich Poretski in the 19th century (cf. [31]), and that [34], in 1927, proposed a translation of propositions into polynomials in the realm of classical logics (with the initial intention of giving a method of proof for the propositions of *Principia Mathematica*). More recently, analogous ideas have been considered by [32] (with the purposes of automatic proof theory by means of rewriting systems) and by [15] and [3] (by computing Gröbner bases with the purpose of investigating proof complexity).

However, (as far as I know) polynomial rings over Galois fields were not extensively used, nor the method extended to all finite-valued logics, to non-finite valued logics or to first-order logic.

In particular, by means of Boolean rings over finite fields (using finite or infinite formal series with sums and products over convenient variables), one can obtain a sound and complete method where any finite-valued derivations and classical propositional derivations reduce to solving equations in polynomial form; what is more surprising, the method can be applied to non-finite-valued propositional logics (as far as they can be represented through the dyadic semantics studied in [9]) by means of introducing multivariable polynomials in appropriate rings (cf. [13]). In particular, the same method permits to represent and compute the so-called "non-deterministic" logics of A. Avron, as in [14].

The polynomial ring calculus can be successfully extended at least to the monadic fragment of first-order logic that expresses traditional syllogisms, giving a new approach to Boole's representation. So Boole's intent to unify the two sides of logic, the propositional and the quantificational, could also be seen as related to the idea of polynomizing. I wish to discuss the role of this approach as a reasoning model and suggest its role in scientific discovery.

3 The Strange Methods of George Boole

The idea that Boolean algebra can be regarded as abstract rings is a consequence of the sophisticated result of M. H. Stone of 1936 (cf. [30]), and the fact that any Boolean algebra can be represented by algebras of classes is seen by Stone as a precise analogue of the fact that any abstract group is represented by an isomorphic group of permutations.

Based on ideas introduced in [13], I briefly review here the intuitions of using polynomials instead of formulas for finite many-valued logics. Given a propositional logic **L**, a *polynomial interpretation* for **L** is a translation $\Omega : \mathbf{L} \mapsto \mathbf{F}[X]$ of the wffs of **L** into a convenient polynomial ring $\mathbf{F}[X]$. Then a wff $\alpha \in \mathbf{L}$ is *satisfiable* if its polynomial translation $\alpha^* \in \mathbf{F}[X]$ gets values within a certain set $D \subseteq \mathbf{F}$ of distinguished truth-values when evaluated in the field **F**.

It is convenient to show first that any finite function can be expressed by means of polynomials over finite fields, using a particular case of the well-known Lagrange interpolation[2] (a simple but important fact that almost certainly belongs to the mathematical folklore of combinatorics and coding theory).

Theorem 1. *(Representation of finite functions in $GF(p^k)$) Let A be any finite set and $f : A^m \mapsto A$ be any function with m variables on A. Let $GF(p^k)$ be a Galois field with cardinality greater than the cardinality of A. Then f can be represented as a polynomial function in $GF(p^k)[x_1, \ldots, x_m]$.*

Proof: The proof is just sketched for the case of binary functions. Suppose, without loss of generality, that the elements of A are $\{0, 1, \ldots, r-1\} \subset GF(p^k)$.

Define functions $\delta_{\langle m,n \rangle}(x, y)$ as:

$$\delta_{\langle m,n \rangle}(x, y) = \prod_{i \neq n, j \neq m} (x - i) \cdot (y - j) \cdot \prod_{i \neq n, j \neq m} (n - i)^{-1} \cdot (m - j)^{-1}$$

Clearly, $\delta_{\langle m,n \rangle}(x, y) = 1$ if $\langle x, y \rangle = \langle m, n \rangle$, and 0 otherwise.

Now, if $f : A^2 \mapsto A$ has values $f(i, j) \in GF(p^k)$, then:

$$p(x, y) = f(0, 0) \cdot \delta_{\langle 0,0 \rangle}(x, y) + \ldots + f(m-1, m-1) \delta_{\langle m-1, m-1 \rangle}(x, y)$$

is a polynomial in $GF(p^k)[x, y]$ which represents $f(x, y)$. Of course, a similar construction can be obtained for the general case. □

For our construction, it is essential to work within a Galois field $GF(p^k)$: for example, the binary function $f(x, y) = max\{x, y\}$ and the unary function $g(x) = 0$ if $x \neq 2$, and $g(2) = 3$ are both representable in $GF(2^2)[x, y]$, but not in $Z_4[x, y]$.

---

[2] I am indebted to Odilon Otávio Luciano from the the Department of Mathematics at the University of São Paulo (IME-USP) for this remark.

The method above gives a particularly expeditious way to compute polynomials over Z_3, since in this case the denominator $\prod_{i \neq n, j \neq m}(n-i) \cdot (m-j)$ of the functions $\delta_{\langle m,n \rangle}(x,y)$ is easily seen to be the unity: indeed, $\prod_{i \neq n}(n-i)$ and $\prod_{j \neq m}(m-j)$ are products of distinct non-zero factors, and can only be $1 \cdot 2$ or $2 \cdot 1$ in Z_3; hence the product $\prod_{i \neq n, j \neq m}(n-i) \cdot (m-j)$ is 1.

This is interesting since the vast majority of examples and usage of many-valued logics falls into the three-valued case. Thus, for the specific case of converting three-valued logics into polynomial form, the functions $\delta_{\langle m,n \rangle}(x,y)$ are:

$\delta_{\langle 0,0 \rangle}(x,y) = (x-1) \cdot (x-2) \cdot (y-1) \cdot (y-2)$
$\delta_{\langle 0,1 \rangle}(x,y) = (x-1) \cdot (x-2) \cdot y \cdot (y-2)$
$\delta_{\langle 0,2 \rangle}(x,y) = (x-1) \cdot (x-2) \cdot y \cdot (y-1)$
$\delta_{\langle 1,0 \rangle}(x,y) = x \cdot (x-2) \cdot (y-1) \cdot (y-2)$
$\delta_{\langle 1,1 \rangle}(x,y) = x \cdot (x-2) \cdot y \cdot (y-2)$
$\delta_{\langle 1,2 \rangle}(x,y) = x \cdot (x-2) \cdot y \cdot (y-2)$
$\delta_{\langle 2,0 \rangle}(x,y) = x \cdot (x-1) \cdot (y-1) \cdot (y-2)$
$\delta_{\langle 2,1 \rangle}(x,y) = x \cdot (x-1) \cdot y \cdot (y-2)$
$\delta_{\langle 2,2 \rangle}(x,y) = x \cdot (x-1) \cdot y \cdot (y-1)$

We suppose, then, that all calculations are done within a convenient field $GF(p^n)$; there are two basic sets of rules to manipulate polynomials:

a) Index rules

1. $p \cdot x \vdash_\approx 0$, where $p \cdot x$ means $x + x + \ldots + x$ p times
2. $x^i \cdot x^j \vdash_\approx x^k (mod\ q(x))$ where $q(x)$ is a convenient primitive polynomial that defines $GF(p^n)$, and $k = i + j (mod\ p^n - 1)$

b) Ring rules

1. $f + (g + h) \vdash_\approx (f + g) + h$
2. $(f + g) \vdash_\approx (g + f)$
3. $f + 0 \vdash_\approx f$
4. $f + (-f) \vdash_\approx 0$
5. $f \cdot (g \cdot h) \vdash_\approx (f \cdot g) \cdot h$
6. $f \cdot (g + h) \vdash_\approx (f \cdot g) + (f \cdot h)$

We also need some explicit metarules : For $f, g, h \in \mathbf{F}[X]$:

1. Uniform Substitution: $\dfrac{f \vdash_\approx g}{f[x:h] \vdash_\approx g[x:h]}$
2. Leibnitz Rule: $\dfrac{f \vdash_\approx g}{h[x:f] \vdash_\approx h[x:g]}$

The index rules are justified by taking into account that the Galois field $GF(p^n)$ has characteristic p, and that $GF(p^n)$ is defined as the quotient of $GF(p)[x]$ by its ideal (g). The ring rules come from the fact that $GF(p^n)[x_1, \ldots, x_n]$ is a polynomial ring. Uniform Substitution and Leibnitz Rule can be easily justified by induction on polynomial functions.

Now, for the definitions of deduction and proof in the polynomial ring calculus for the logic **L**, let $\Gamma \cup \{\alpha\}$ be wffs in **L** and Γ^*, α^* be their translations in polynomial form: the following general completeness of the method can be proven for any finite-valued logic: $\Gamma \vdash_\mathbf{L} \alpha$ iff $\Gamma^* \vdash_\approx \alpha^*$ where $\vdash_\approx$ denotes the derivation of $\alpha^* \in D$ (in the equational logic defined by the above rules) from the hypothesis $\Gamma^* \in D$ (see [13] for details).

Of course, when the the set D of distinguished values is a singleton, say $D = \{1\}$, then derivations $\Gamma^* \vdash_\approx \alpha^*$ reduce to proving $\alpha^* \approx 1$ from the hypothesis $\Gamma^* \approx 1$, and (when $\Gamma = \emptyset$) proofs reduce to showing directly that $\alpha^* \approx 1$ by high-school manipulation of polynomials.

As an illustration, consider the case of classical logic **PC**. Define in this case the translation $\Omega : \mathbf{PC} \mapsto Z_2[X]$ of **PC** into the Boolean ring $Z_2[X]$ as:

- $\Omega(p_i) = x_i$ for each atomic variable p_i
- $\Omega(\neg \alpha) = 1 + \Omega(\alpha)$
- $\Omega(\alpha \wedge \beta) = \Omega(\alpha) \cdot \Omega(\beta)$
- $\Omega(\alpha \vee \beta) = \Omega(\alpha) \cdot \Omega(\beta) + \Omega(\alpha) + \Omega(\beta)$
- $\Omega(\alpha \to \beta) = \Omega(\alpha) \cdot \Omega(\beta) + \Omega(\alpha) + 1$

Thus, for instance, having translated atomic variables p_i as fresh variables x_i, we have:

- $x^2 \approx x$
- $x + x \approx 0$
- $\neg \alpha \approx 1 + x$
- $\alpha \wedge \beta \approx x \cdot y$
- $\alpha \vee \beta \approx x \cdot y + x + y$
- $\alpha \to \beta \approx x \cdot y + x + 1$

Proving *reductio ad absurdum*, for example, amounts to:

- $\alpha \to \beta, \alpha \to \neg \beta \vdash_{PC} \neg \alpha$. Translating into polynomial form, we have to check that: $(x \cdot y + x + 1) \cdot (x \cdot (y+1) + x + 1) \cdot x \vdash_\approx 0$
- But easily: $(x \cdot y + x + 1) \cdot (x \cdot (y+1) + x + 1) \cdot x \approx (x \cdot y + x + 1) \cdot (x \cdot y + 1) \cdot x \approx (x^2 \cdot y^2 + x \cdot y + x^2 \cdot y + x + x \cdot y + 1) \cdot x \approx (\widehat{x \cdot y} + \widehat{x \cdot y} + \widehat{x \cdot y} + x + \widehat{x \cdot y} + 1) \cdot x \approx (x+1) \cdot x \approx x^2 + x \approx 0$ taking into account (as indicated) that here $x^2 \approx x$ and $x + x \approx 0$.

Now, for classical logic **PC** this seems to be an obvious usage of Boolean algebras and Boolean rings, but the same can be done for all finite-valued logics[3]; of special interest are the cases of the well-known (see e.g. [21]) three-valued logics of Lukasiewicz, Gödel, Kleene, Sette, and the four-valued logic of Belnap (see [13] for concrete examples of polynomial proofs in three-valued logics).

However, some authors (as for instance [16]) see solving equations as opposed to performing deductions. In that paper, J. Corcoran points to two fallacies of Boole: the first (p. 280 and 281) is that Boole overlooks indirect

---

[3] In a certain sense, this is the algebraic analogue of the universal method for provability in many-valued logics investigated in [11].

reasoning, or *reductio ad absurdum*, an important and productive form of inference: "This... is very likely part of why he missed indirect deduction (or *reductio* reasoning). There is no such thing as indirect equation-solving, of course."

This criticism cannot be taken literally, as we just have seen (as an example) an equational proof of *reductio ad absurdum*. The second fallacy, according to Corcoran, is the *Solutions Fallacy*, which involves confusing solutions to an equation with its consequences. For example, the equation $x = (x \cdot y)$ (in a Boolean algebra – an unavoidable anachronism!) does not imply the solution $x = 0$ (since x and y may both equal 1). However, $x = 0$ does imply $x = (x \cdot y)$. In the defense of Boole, it could be added, as S. Burris suggests in [8], that there exists a kind of "universal error" of Boole's interpreters: Boole used existential import in his Aristotelian arguments but this is usually not taken into account. For example, Boole used Aristotelian semantics, accepting arguments (as *Conversion by Limitation*) which only make sense if all classes are non-empty: "All A is B", therefore "Some B is A".

To make our meaning clearer, let us recall an example by Boole from his paper of 1848 [5, pp. 7–8], proving contraposition. The sentence "All Ys are Xs" is formalized in his algebra as $y = v \cdot x$ (meaning: Y is the intersection of X with some non-empty V) and he seeks the value of $1-x$ (i.e., the class not-X).

Boole uses $x \cdot y$ to denote intersection, $x + y$ to denote union (provided $x \cdot y = 0$) and $x - y$ to denote class difference (provided $y \subseteq x$); thus $1 - x$ denotes the complement of x; 0 denotes the empty class, and 1 is the universe of discourse.

A point which caused some confusion in Boole's intuition is that $x \cdot y$ can be interpreted as intersection or conjunction, but $+$ cannot be interpreted as union or disjunction: indeed, while $x \vee (y \wedge z) = (x \vee y) \wedge (x \vee z)$ holds in propositional logic, $x + (y \cdot z) = (x+y) \cdot (x+z)$ does not hold universally in what is today called a Boolean ring; it can be easily checked that:

$x + (y \cdot z) = (x + y) \cdot (x + z)$ iff $x \cdot (y + z) = 0$.

This is, of course, due to the fact that $+$ should be seen as "exclusive or", not usual disjunction.

First, Boole solves the equation $y = v \cdot (1-z)$ in the new variable z (putting $1 - x = z$): $z = v \cdot (1 - y) + \frac{1}{0}(1-v) \cdot y + \frac{0}{0}(1-v) \cdot (1-y)$.

He then considers $\frac{1}{0}$ as an "infinite coefficient" and thus the term $\frac{1}{0}(1-v) \cdot y$ vanishes, but $\frac{0}{0}$ is to be replaced by "an arbitrary elective symbol w".

Thus the equation becomes: $z = v \cdot (1 - y) + w \cdot (1 - v) \cdot (1 - y)$ or $1 - x = (v + w \cdot (1 - v)) \cdot (1 - y)$.

He then argues that $(v + w \cdot (1 - v))$ represents a class, since it satisfies the "index law"[4] $(v + w \cdot (1-v))^n = (v + w \cdot (1-v))$, and therefore can be

---

[4] It is truly remarkable that Boole referred in this case to his "index law" with exponent n (see [5]) where a square would be sufficient. He failed to see, however, the distinction between different exponents – and this distinction is the key to the approach developed here.

replaced by an "elective symbol" u: therefore $1 - x = u \cdot (1 - y)$, i.e., "All not-Xs are not-Ys".

This "elective symbol" was a source of problems and misunderstanding, and has been widely criticized since the beginning (see [28]). But the point is that Boole is actually trying to reason at the same time with algebra and with classes, so in a certain way anticipating the results that would only be clarified by M. Stone more than 80 years later. I think that Boole was much more innovative than logicians would suppose: he even mixed ideas of differential calculus to logic, algebra and probability, a blend that we are far from understanding in the general case of non-classical logics.

4 Infinite Polynomials and Aristotelian Logic

In [29], Ernst Schröder, in his reformulation of Boole's logic, already considered addition and multiplication as logical operations and stressed their dual character. He introduced the symbols $\prod$ and $\sum$ as arithmetic analogues of conjunction and disjunction; quantification could thus be seen as "indefinite" operations (indefinite logical addition for existential quantification, and indefinite logical multiplication for existential quantification).

However, the approach I am considering here is significantly distinct: Schröder was probably influenced by the ideas of Charles S. Peirce, and the way he chose to see logic as a model of absolute algebra does not seem to generalize to logics other than classical. What I suggest, instead, is a way to employ algebra to represent and calculate logical inference[5].

With all this as motivation, I now examine some preliminary ideas on expressing first-order logic (**FOL**) in polynomial form, but treating the monadic case only.

The translation rules for interpreting propositional logic in terms of polynomials over Z_2 can be extended to first-order logic by adding clauses defining a translation $\Omega : \textbf{FOL} \mapsto Z_2[X]$:

1. For each constant c_i (in a denumerable universe), $\Omega(A(c_i)) = x_i^A$ (i.e., a new variable in $Z_2[X]$)
2. $\Omega(\forall z A(z)) = \prod_{i=1}^{\infty} x_i^A$. This results in:
3. $\Omega(\exists z A(z)) = \Omega(\neg \forall z \neg A(z)) = 1 + \prod_{1=1}^{\infty}(1 + x_i^A)$

It is to be noted that now polynomials are infinite (i.e. formal series in $Z_2[X]$). To simplify notation, let $\Omega(\forall z A(z)) = \prod x_i$ and $\Omega(\exists z A(z)) = 1 + \prod(1 + x_i)$.

It is instructive to see some examples of proofs in **FOL**.

---

[5] Perhaps, if it contributes to a better understanding of the approach, one might call it an *algebra ratiocinator*.

- $\forall z A(z) \to \exists z A(z)$:
  $(\prod x_i) \cdot (1 + \prod(1+x_i)) + \prod x_i + 1 \approx (\prod x_i) \cdot (\prod(1+x_i)) + \prod x_i + \prod x_i + 1 \approx$
  $(\prod x_i \cdot (1+x_i)) + \prod x_i + \prod x_i + 1 \approx 1$ since $\prod x_i + \prod x_i \approx 0$ and $x_i \cdot (1+x_i) \approx 0$
  for each x_i

As another example, consider:

- $(\forall z A(z) \to \forall z B(z)) \to \forall z (A(z) \to B(z))$:
  1. Let $\alpha := (\forall z A(z) \to \forall z B(z))$: $\prod x \cdot \prod y + \prod x + 1 \approx \prod x \cdot y + \prod x + 1$
  2. Let $\beta := \forall z(A(z) \to B(z))$: $\prod(x \cdot y + x + 1)$
  3. $\alpha \to \beta$: $(\prod x \cdot y + \prod x + 1) \cdot \prod(x \cdot y + x + 1) + (\prod x \cdot y + \prod x + 1) + 1 \approx$
  4. $\prod(x \cdot y + x \cdot y + x \cdot y) + \prod(x \cdot y + x + x) + \prod(x \cdot y + x + 1) + (\prod x \cdot y + \prod x + 1) + 1 \approx$
  5. $\prod(x \cdot y) + \prod(x \cdot y) + \prod(x \cdot y + x + 1) + (\prod x \cdot y + \prod x) \approx$
  6. $\prod(x \cdot y + x + 1) + \prod x \cdot y + \prod x \not\approx 1$

This method, like other proof procedures (such as tableaux), can also be used to find counter-models: why is $\prod(x \cdot y + x + 1) + \prod x \cdot y + \prod x \not\approx 1$? Well, if there are x and x' such that $x = 0$ and $x' = 1$, and some y such that $y = 0$, then: $\prod(x' \cdot y + x' + 1) + \prod x \cdot y + \prod x = \prod(0 + 1 + 1) + \prod 0 + \prod 0 = 0$ which is precisely the intuitive counter-model: $x = 0$ corresponds to a false instance $A(a)$, $x' = 1$ to a true instance $A(b)$, and $y = 0$ to a false instance $B(c)$.

Boole's analysis of Syllogistic Logic can now be recovered in polynomial form. Recall the four Aristotelian categorical forms:

A – All A is B: $\forall z(A(z) \to B(z))$
I – Some A is B: $\exists z(A(z) \wedge B(z))$
E – No A is B: $\forall z(A(z) \to \neg B(z))$
O – Some A is not B: $\exists z(A(z) \wedge \neg B(z))$

where **A** and **I** are affirmative (respectively, universal and existential), **E** and **O** are negative (respectively, universal and existential), $\mathbf{A} = \neg\mathbf{O}$ and $\mathbf{I} = \neg\mathbf{E}$.

Recalling our simplified notation, the categorial propositions are expressed in polynomial form as follows:

1. $\Omega(\forall z A(z)) = \prod x_i$
2. $\Omega(\exists z A(z)) = 1 + \prod(1 + x_i)$
3. $\Omega(\neg \alpha) = 1 + x$
4. $\Omega(\alpha \wedge \beta) = x \cdot y$
5. $\Omega(\alpha \to \beta) = x \cdot y + x + 1$

Mnemonically:

- **A** (All A is B): $\prod(a \cdot b + a + 1)$
- **I** (Some A is B): $1 + \prod(1 + a \cdot b)$

It is now possible to recover Boole's interpretation: for the form **A**,

- **A** holds iff $\prod(a \cdot b + a + 1) = 1$ iff $a \cdot b + a + 1 = 1$ for every a, b iff $a \cdot b + a = 0$ for every a, b iff $a \cdot b = a$ for every a, b

which coincides with Boole's formalization of **A** as "$AB = A$" in his book [4] of 1847.

It is important to remark here that $a \cdot b = a$ implies $a = 0$ if $b = 0$, and that $a \cdot b = a$ holds "vacuously" if $a = 0$.

Similarly, for the form **I**:

- **I** holds iff $1 + \prod(1 + a \cdot b) = 1$ iff $\prod(1 + a \cdot b) = 0$ iff $1 + a_0 \cdot b_0 = 0$ for some a_0, b_0 iff $a_0 \cdot b_0 = 1$ for some a_0, b_0

which by its turn coincides with Boole's formalization of **I** as "$AB = V$" in his article [5] of 1848.

5 Proving Syllogisms in Polynomial Form

As an example let us show how to use this technique to prove the syllogism *Barbara* (mode **AAA** of the First Figure):

From		
A	All A is B	$a \cdot b = a$ for every a, b
A	All B is C	$b \cdot c = b$ for every b, c
conclude		
A	All A is C	$a \cdot c = a$ for every a, c

The proof runs as follows (recalling the mnemonic abbreviation above):

1. $a \cdot b = a$ hypothesis 1
2. $b \cdot c = b$ hypothesis 2
3. $a \cdot b \cdot c = a \cdot b$ from (2), multiplying by a
4. $a \cdot c = a$ from (3) and using (1), replacing $a \cdot b$ by a

As another example, it is instructive to prove the syllogism *Darii* (mode **AII** of the First Figure):

From		
A	All B is C	$b \cdot c = b$ for every b, c
I	Some A is B	$a_0 \cdot b_0 = 1$ for some a_0, b_0
conclude		
I	Some A is C	$a_0 \cdot c_0 = 1$ for some a_0, c_0

The proof is as follows:

1. $a_0 \cdot b_0 = 1$ hypothesis 2
2. $b_0 = b_0 \cdot c_0$ instance of hypothesis 1

3. $a_0 \cdot b_0 = a_0 \cdot b_0 \cdot c_0$ from (2)
4. $a_0 \cdot b_0 \cdot c_0 = c_0$ from (1)
5. $a_0 \cdot b_0 = c_0$ from (3) and (4)
6. $a_0 \cdot a_0 \cdot b_0 = a_0 \cdot c_0$ from (5),
7. $a_0 \cdot b_0 = a_0 \cdot c_0$ from (6), since $a_0 \cdot a_0 = a_0$
8. hence $1 = a_0 \cdot c_0$ from (1) and (7)

It is well known that from *Barbara* and *Darii* all the 19 valid syllogistic forms can be deduced, by means of the following rules:

- Conversion: Some A is B/Some B is A (in our notation: $a_0 \cdot b_0 = 1/b_0 \cdot a_0 = 1$);
- Conversion by limitation: All A is B/ Some A is B (in our notation: $a \cdot b = a/a_0 \cdot b_0 = 1$).

Conversion is an obviously valid rule in our setting. Conversion by limitation is more complicated, since our method takes into account the "contemporary" semantics, where classes can be empty, as remarked before. In order to adapt it to "Aristotelian" semantics that assumes existential import, we have to suppose that there exists a_0 such that $a_0 = 1$. Thus, since supposing "All A is B" implies $a \cdot b = a$ holds for all a, there must be b_0 such that $a_0 \cdot b_0 = 1$. If not, then $a_0 \cdot b = 0$ for any b, which implies $a_0 = 0$, a contradiction.

6 Conclusions

The methods described in this paper have a promising potentiality for any truth-functional multiple-valued logic; there is an exciting area of research in designing new proof theory techniques for such logics, simplifying applications to multiple-valued logics in decision tables and discovering patterns, and in several other fields (it is well known that multiple-valued logics find applications in artificial intelligence, database theory and data mining, modeling reasoning and model checking, for instance). It is important to emphasize that the method is also plainly applicable to non-finite valued logics, and also to represent binary semantics for many-valued logics[6] (cf. [13]) and even to quantum circuits and quantum gates (cf. [1]). The arguments advanced here try to conceptualize this approach, in particular when extended to quantification and non-finite valued logics, as heritage of an admirable tradition in mathematical thinking, which may have been disregarded by logicians. We should keep in mind that one of Boole's ideas was relating logic to probability

---

[6] In such cases, the binary semantics for a finite-valued logic may be simpler and more philosophically palatable than the multiple-valued one, and even permits a completely different approach to the logic, but at the cost of truth-functionality; see [9].

theory and to the fascinating method of finite differences. By exploring it conveniently, we would gain new tools in logic and in our patterns of reasoning, and assess Boole's work from a fresh perspective.

As B. Mates points out in [25], Boolean insights rehabilitated Stoic logic, rather than Stoicism supporting Boole. Starting from a historical background leading up to a modern perspective on algebraic logic, the excellent survey [2] accurately concludes (p. 511) that the ideas of Boole have not borne their full fruit yet.

We are suggesting here that Boolean insights also rehabilitated a method of looking at logic which boldly mixes logic with the roots of differential calculus. Is it possible to re-analyze some of Boole's deep intuitions, and return to a more closely algebraic approach, which is appropriate for several logics? Or to bring methods of differential calculus to logic, via polynomizing? It does not appear to be easy to extend this type of calculus to full **FOL** and to higher-order logic or even to modal logics – an especially interesting application would be to extend it to the finite variables fragment of **FOL** – but it seems to be a very rewarding challenge.

Acknowledgement. This research was supported by FAPESP Thematic Research Project Grant 2004/14107-2 (Brazil), by a CNPq Research Grant Level 1 (Brazil) and by EU-FEDER via CLC (Portugal).

References

1. Agudelo, J.C., Carnielli, W.A.: Quantum algorithms, paraconsistent computation and Deutsch's problem. In Bhanu Prasad et al., eds.: *Proceedings of the 2nd Indian International Conference on Artificial Intelligence*, Pune, India, December 20–22 (2005) IICAI 2005, 1609–1628. Pre-print available from *CLE e-Prints* vol. 5(10) (2005)
   ftp://logica.cle.unicamp.br/pub/e-prints/MTPs-CompQuant%28Ing%29.pdf
2. Ahmed, T.S.: Algebraic logic, where does it stand today? *Bull. Symbolic Logic* 11(4) (2005) 465–516
3. Beame, P., Impagliazzo, R., Krajicek, J., T. Pitassi, T., Pudlak, P.: Lower bounds on Hilbert's Nullstellensatz and propositional proofs. *Proceedings of the London Mathematical Society* 73 (1996) 1–26
4. Boole, G.: *The Mathematical Analysis of Logic, Being an Essay Towards a Calculus of Deductive Reasoning.* Macmillan, Barclay and Macmillan, London (1847) (Reprinted by Basil Blackwell, Oxford, 1965)
5. Boole, G.: The calculus of logic. *Cambridge and Dublin Math. Journal* 3 (1848) 183–198
6. Boole, G.: *An Investigation of the Laws of Thought, on Which are Founded the Mathematical Theories of Logic and Probabilities.* Walton and Maberley, London (1854) (Reprinted by Dover Books, New York, 1954)
7. Boole, G.: *Calculus of Finite Differences.* 5th Edition, Chelsea Publishing (originally published in 1860) (1970)

8. Burris, S.: The laws of Boole's thought. Unpublished (2002), Preprint at http://www.thoralf.uwaterloo.ca/htdocs/MYWORKS/PREPRINTS/aboole.pdf
9. Caleiro, C., Carnielli, W.A., Coniglio, M.E., Marcos, J.: Two's company: "The humbug of many logical values". In Béziau J.-Y., Birkhäuser, eds.: *Logica Universalis*. Verlag, Basel, Switzerland (2005) 169–189 Preprint available at http://wslc.math.ist.utl.pt/ftp/pub/CaleiroC/05-CCCM-dyadic.pdf.
10. Caicedo, X., Martín, A.: Completud de dos cálculos lógicos de Leibniz. *Theoria* 16(3) (2001) 539–558
11. Carnielli, W.A.: Systematization of the finite many-valued logics through the method of tableaux. *The Journal of Symbolic Logic* 52(2) (1987) 473–493
12. Carnielli, W.A.: A polynomial proof system for Lukasiewicz logics. *Second Principia International Symposium*. August 6-10, Florianpolis, SC, Brazil (2001)
13. Carnielli, W.A.: Polynomial ring calculus for many-valued logics. *Proceedings of the 35th International Symposium on Multiple-Valued Logic. IEEE Computer Society*. Calgary, Canad. IEEE Computer Society, pp. 20–25, 2005. Available from *CLE e-Prints* vol. 5(3) (2005) at http://www.cle.unicamp.br/e-prints/vol_5,n_3,2005.html
14. Carnielli, W.A., Coniglio, M.E.: Polynomial formulations of non-deterministic semantics for logics of formal inconsistency. Manuscript (2006)
15. Clegg, M., Edmonds, J., Impagliazzo, R.: Using the Gröbner bases algorithm to find proofs of unsatisfiability. *Proceedings of the 28th Annual ACM Symposium on Theory of Computing*. Philadelphia, Pennsylvania, USA (1996) 174–183
16. Corcoran, J.: Aristotle's Prior Analytics and Boole's Laws of Thought. *Hist. and Ph. of Logic* 24 (2003) 261–288
17. Dummett, M.: Review of "Studies in Logic and Probability by George Boole". Rhees, R., Open Court, 1952, *J. of Symb. Log.* 24 (1959) 203–209
18. Eves, H.: *An Introduction to the History of Mathematics*. (6th ed.) Saunders, New York (1990)
19. Euler, L.: Variae observations circa series infinitas, *Commentarii academiae scientiarum Petropolitanae* 9 (1737) 160–188. Reprinted in *Opera Omnia*, Series I volume 14, Birkhuser, 216–244. Available on line at www.EulerArchive.org.
20. Giusti, E.B.: *Cavalieri and the Theory of Indivisibles*. Cremonese, Roma (1980)
21. Gottwald, S.: *A Treatise on Many-Valued Logics, Studies in Logic and Computation*. Research Studies Press Ltd. Hertfordshire, England (2001)
22. Hailperin, T.: *Boole's Logic and Probability: A Critical Exposititon from the Standpoint of Contemporary Algebra, Logic, and Probability Theory*. North-Holland Studies In Logic and the Foundations of Mathematics (1986)
23. Lloyd, G.: Finite and infinite in Greece and China. *Chinese Science* 13 (1996) 11–34
24. MacHale, D.: *George Boole: His Life and Work*. Boole Press (1985)
25. Mates, B.: *Stoic Logic*. University of California Press, Berkeley, CA (1953)
26. Martzloff, J.-C.: *A History of Chinese Mathematics*. Springer-Verlag, Berlin (1997)
27. Roy, R.: The discovery of the series formula for π by Leibniz, Gregory and Nilakantha. *Mathematics Magazine* 63(5) (1990) 291–306
28. Schroeder, M.: A brief history of the notation of Boole's algebra. *Nordic Journal of Philosophical Logic* 2(1) (1997) 41–62
29. Schröder, E.: Vorlesungen über die Algebra der Logik (exakte Logik) 2 1, B.G. Teubner, Leipzig, 1891 (reprinted by Chelsea, New York, 1966).

30. Stone, M.H.: The theory of representations for boolean algebras. *Trans. of the Amer. Math. Soc* 40 (1936) 37–111
31. Styazhkin, M.I.: *History of Mathematical Logic from Leibniz to Peano*. The M.I.T. Press, Cambridge (1969)
32. Wu, J.-Z., Tan, H.-Y., Li, Y.: An algebraic method to decide the deduction problem in many-valued logics. *Journal of Applied Non-Classical Logics* 8(4) (1998) 353–60
33. Weyl, H.: God and the Universe: The Open World. Yale University Press (1932) Reprinted as "The Open World" by Ox Bow Press, 1989.
34. Zhegalkin, I.I.: O tekhnyke vychyslenyi predlozhenyi v symbolytscheskoi logykye (On a technique of evaluation of propositions in symbolic logic). *Matematicheskii Sbornik* 34(1) (1927) 9–28 (in Russian)

Abductive Inference and Iterated Conditionals

Claudio Pizzi

Dipartimento di Filosofia e Scienze Sociali, Università di Siena, Siena, Italy
pizzic@unisi.it

Summary. The first part of the paper aims to stressing the analogy between conditional inference and abductive inference, making evident that in both cases what is here called "reasonable" inference involves a choice between a finite set of incompatible conclusions, selecting the most information preserving-consequent in the case of standard conditionals and the most information-preserving antecedent in the case of abductive conditionals. The consequentialist view of conditionals which is endorsed in this perspective is then extended to cover the case of higher degree conditionals, introducing in the semantical analysis the notion of inferential agents reasoning about the activity of other inferential agents. It is then shown (i) that iterated conditionals are essential in the treatment of redundant causation (ii) that abductive conditionals are essential parts of iterated conditionals in the analysis of causal preemption (iii) that there is a widespread use of second-degree conditionals involving first degree abductive conditionals. The final section is devoted to remind that Peirce's original notion of abductive inference was actually defined in terms of second degree conditionals.

1 The Notion of ε-implication

There is no doubt that Stalnaker-Lewis conditional logics introduced an important change of paradigm in the study of conditional inference[1]. However, many features of the theorems involving Stalnaker-Lewis conditionals have been object of criticism inasmuch as the truth of such conditionals turns out to prescind from any kind of relevance or dependence nexus between the clauses. In front of such difficulty a natural move has been to go back to the tradition of the so-called "consequentialist" theory of conditionals originally proposed by Chisholm, Goodman and Reichenbach in the '40, when the tool of possible-worlds semantics was not yet developed. The present author has

---

[1] For a survey on conditional logics, see Nute [11], where Stalnaker-Lewis systems are termed "minimal change theories".

defended the idea that the consequentialist view, in the form of what here will be called *reasonable* inference, grants a unified treatment of counterfactual, inductive and abductive conditionals. However, even in this trend of investigations, little attention has been given to two points which I will try to focus in the following sections: a) Higher degree conditionals, i.e. conditionals having other conditionals as antecedents or consequents, are essential to the reconstruction of scientific reasoning, even when they do not appear in the superficial structure of the statements. b) The consequentialist view is able to provide a clear and straightforward semantical interpretation of higher degree conditionals and grants a deeper understanding of the relation between counterfactual and abductive conditionals[2].

The treatment will be semiformal. However, it is useful to presuppose a formal language which may provide a joint formal representation of both probabilistic and modal notions. Fattorosi-Barnaba's and Amati's [5], for example, offers the instance of a system for additive graded modalities which may be chosen as a reference formal system. If $\Diamond^n A$ is intended as *A is probable at degree* $> n$, $\Diamond A$ is coincident with $\Diamond^0 A$ and $\Box A$ with $\Box^0 A$. $Pr(A) = n$ may be then put equivalent to $\neg\Diamond^n A \wedge \neg\Diamond^{1-n}\neg A$ and $Pr(B|A)$ may then be defined in terms of B and A in a standard way, i.e. as:

$$\frac{Pr(A \wedge B)}{Pr(A)} \quad (Pr(A) \neq 0)$$

In this linguistic framework other operators may be introduced by definition. For instance, if ε is a negligible value ≥ 0, we may define what we may call *ε-implication* in this way:

(1) $A \;_\varepsilon\!\to B =_{Df} Pr(B|A) = 1 - \varepsilon$

The notion of information content in terms of possible world semantics is that A ε-implies B iff "almost all the accessible A-worlds are B-worlds". The notion of information content ($Cont$) for sake of simplicity will be introduced here by definition as $Cont(B|A) =_{Df} 1 - Pr(B|A)$[3].

Some obvious properties of this notion are:

1. $Pr(A) = 1 - Cont(A)$
2. $Cont(A \wedge B) = Cont(A) + Cont(B) - Cont(A \vee B)$
3. $Cont(A \wedge B) \geq Cont(A)$

What about the information content of a physical law L? According to an extensionalist view of laws, a law L is an infinitary conjunction of statements. So in normal contexts it happens that if A is any finitary truth-functional and non contradictory statement, $Cont(L) > Cont(A)$. Let us recall that in

---

[2] For a first outline of this theory see Pizzi [14].
[3] For this definition see for instance Hintikka [2].

Carnap's inductive logic any physical law receives probability value 0 – so it has information content 1 – while in Hintikka's inductive logics laws receive a probability value which is different from 0 but anyway low[4]. These results give substance to the idea that every law will be normally more informative than every finite combination of single facts. Something should be said about the controversial question of second or higher degree probabilities, so about the content of probabilistic statements[5]. We are especially interested in giving a value to $Cont(A \, _\varepsilon \rightarrow B)$, i.e. to the content of $Pr(B|A) = 1 - \varepsilon$. There is no problem in introducing a **S4**-style axiom for Pr. i.e.

(2) $Pr(q|p) = 1 - \varepsilon \supset Pr(Pr(q|p) = 1 - \varepsilon) = 1 - \varepsilon$

From (2) of course, if ε is 0 and p is a tautology $\top$, we have:

(3) $\Box q \supset \Box\Box q$

This minimal principle is however not of help in evaluating the content of a conditional. As we will say in the next section, in evaluating an argument from A to B, we should take care of the laws which are essentially used in the argument itself. This suggests that the information value of a conditional should be proportional to the information content of the laws essentially involved in the derivation, and such a content, as already said, is very high. This criterion marks a difference between what we shall call rational and reasonable inference.

2 Rational and Reasonable Inference

A basic idea that we intend to develop here is that every rational inference rests on the choice of the best consequent in a set of consequents or in the choice of the best antecedent among a class of possible antecedents for a given consequent. But which is the *best* consequent or the *best* antecedent? We could leave this notion sufficiently vague or we can make it depend on some variable parameter of evaluation[6], but here we prefer a non-neutral policy: the best inferential conclusions will be here defined as the ones which are more information-preserving with respect to some given set of background knowledge **K**. Let us call **CR** (Corpus Rationale) the infinite set of all true statements, including the laws $L_1 \ldots L_n$. **CR** is closed under logical rules and may be thought as the infinite set of sentences describing the actual world $w°$. Let **K** be a finite subset of **CR** containing a finite set of true statements and a theory T consisting of a finite subsets of the laws in **CR**. Let use K to denote the conjunction of the members of **K**. **K**(A) will be a subset of **K**

---

[4] See Carnap [1] and Hintikka [3, 4].
[5] For a recent approach to higher degree probabilities see [6].
[6] See for instance Rescher [19].

revised in dependence of A. Of course $\mathbf{K}(\top) = \mathbf{K}$ More specifically, $\mathbf{K}(A)$ is a subset of $\mathbf{K}$ which is obtained by selecting the most informative statements of $\mathbf{K}$ compatible with A. Let us consider, to begin with, the case in which A and B are truth-functional statements. The symbol $>_C$ in wffs whose form is $A >_C C_i$ will be used for standard conditionals, namely for conditionals which are *factual, afactual* or *counterfactual*.

Given a certain finite set $\mathbf{K}$ with respect to which the conditional is evaluated as true or false, there are two possibilities to be considered:

(a1) $\Diamond(K \wedge A)$ (a2) $\neg\Diamond(K \wedge A)$

In the first case the conditional is afactual or factual, in the second case it is counterfactual[7]. We will say that $A >_C C_i$ is true or false with respect to K only if there are at least two statements C_i and C_j such that:

b.1 $\neg\Diamond(K \wedge A \wedge C_i \wedge C_j)$
b.2 There are at least two $\mathbf{K}_i, \mathbf{K}_j \subseteq \mathbf{K}$ such that $\Diamond(A \wedge K_i)$ and $\Diamond(A \wedge K_j)$
b.3 $(A \wedge K_i \ \varepsilon - implies \ C_i)$ and $(A \wedge K_j \ \varepsilon - implies \ C_j)$
b.4 $\underline{Cont(A \wedge K_i) \succ Cont(A \wedge K_j)}$

The underlined clause $b.4$ makes it clear that C_i is the preferred conclusion due to the fact that $\mathbf{K_i}$ is a subset of $\mathbf{K}$ which ε-implies C_i and has higher information content than the rival set $\mathbf{K_j}$. An instance of rational inference is offered by the choice between the following two counterfactuals:

(4) If Socrates were a donkey, Socrates would be four-legged: $A >_C C_1$
(5) If Socrates were a donkey, Socrates would be a two-legged donkey: $A >_C C_2$

Here $\mathbf{K_1}$ is {Every donkey is four-legged}, $\mathbf{K_2}$ is {Socrates is two legged}, $\mathbf{K}$ is {Socrates is not a donkey} $\cup \ \mathbf{K}_1 \cup \mathbf{K}_2$. The theory T is the law in $\mathbf{K}_1$. The two conclusions C_1 and C_2 are incompatible, and a fortiori A, C_1, C_2 form an incompatible triad. The counterfactual (4) is "true" because its rival (5) relies on a set $\mathbf{K}_2$ which has a lower information content, due to the fact that this singular statement is less informative than the laws belonging to $\mathbf{K}_1$. The case of factual or afactual conditionals suggests that we have to extend our conditions by making the further assumption that $\mathbf{K}$ should be always contain inside T the metalaw known as principle of Uniformity of Nature (UN). As Goodman showed in the *grue-bleen* paradox, one could infer both C_i and C_j from the same premise A, but if we have in $\mathbf{K}$ also UN the conclusion, say, that emeralds will be blue after 3000 is incompatible with the consequence of UN stating that the properties of substances are spatio-temporal invariant[8].

---

[7] The factual conditionals, or since-conditionals, are conditionals such that A belongs to $\mathbf{K}$, while this is not required in afactual conditionals.
[8] In Rescher [19] a coherence theory of inductive reasoning is introduced. If 100 black ravens have been observed, this is compatible with the conclusion that

Remark 1. In case there is a tie between different subsets $\mathbf{K}_i$ and $\mathbf{K}_j$ leading to incompatible conclusions (as in the famous Bizet-Verdi case) we are not in conditions to choose between $Cont(A \wedge K_i)$ and $Cont(A \wedge K_j)$ since they are identical. So both conditionals are false. This is what has been called a *Gestalt Effect* in Pizzi [13].

The relativization to $\mathbf{K}$ of the truth of $A > C_i$ might be dropped by existential quantification over $\mathbf{K}$, in other words by saying that there is some $\mathbf{K}$ that has the mentioned properties. The minimal set of conditions b1-b4) is sufficient to define what we could call *rational* inference, but it is plausibile to require a further restriction. Such condition says that whenever any $K_i \wedge A$ ε-implies some C_i all the elements of $\mathbf{K}_i$ must have the property of being *essential* to such a derivation (otherwise the information content of $\mathbf{K}_i$ could turn out to be higher than the content of $\mathbf{K}_j$ simply for the occurrence in it of some irrelevant statement A). This means that, for every P belonging to some $\mathbf{K}_i$, we have to require also the following condition beyond b1–b4:

c) For every P and every $\mathbf{K}_i$, if P belongs to $\mathbf{K}_i$ and $K_i \wedge A$ $\varepsilon-$ *implies* B, the conjunction of statements belonging to $\{\mathbf{K}_i - \{P\}\} \cup \{A\}$ does not ε-imply B.

Note that clause c) solves many cases of irrelevance due to the high probability of the conclusion. If for instance $Pr(C_i) = 1 - \varepsilon$, it follows that, for every consistent $A, Pr(C_i|A) = 1 - \varepsilon$, but this makes irrelevant every element of $\mathbf{K}_i$ since it is drawn simply via the laws of the background logic. In the same vein, notice that if $\mathbf{K}_i$ contains only B and A this does not legitimate $A >_C B$ even if $Pr(B|A \wedge K_i) = 1$ for every A and every B. This feature marks an important difference with Stalnaker-Lewis logics, since they accept the controversial law:

$$(A \wedge B) \supset (A >_C B)$$

Given the preceding conditions, we might distinguish between rational and reasonable inference, by asking that an inference is *reasonable* when and only when it satisfies beyond b1)-b4) the supplementary clause c). So every reasonable inference is also rational, but not vice versa[9].

---

the next raven will be black and also with the conclusion that the next raven will be of some other color. But since we have to choose, in Rescher's view, the "most plausible" subset, we are guided by a rule which says "When the initial evidence exhibits a marked logical pattern, then pattern- concordant statements are – ceteris paribus – to be evaluated as more plausibile than pattern-discordant ones" (p. 226). This criterion introduces a certain arbitrarity, while the Principle of the Uniformity of Nature appears to provide a firmer foundation to inductive reasoning.

[9] The distinction here drawn between rational and reasonable inference could be different if the comparison were made between epistemic utilities and not between information contents. On the notion of epistemic utility see Hintikka and Pietarinen [4].

Up to now we have characterized the properties of factual, afactual and counterfactual conditionals. What about abductive conditionals? Let us introduce the symbol $>_A$ to denote the abductive conditional. For instance $A >_A C_i$ may be "the match lit (A), so it has been scratched (C_i)" or also "if the match lit, this means that it has been scratched". The reliability of the conclusion lies in the fact that it is preferred in the set of other possible incompatible conclusions, as for instance the one embodied in the conditional "the match lit, so it has been put into fire". It is remarkable that abductive reasoning has important features in common with counterfactual reasoning, abstracting from the fact that the antecedent is normally belived to be true and not false. In both cases, in fact, we are faced with a rational choice between incompatible conclusions. Let A be "Smith has been killed". Let **K** be for example a set which includes

1. Smith has been killed in New York by only one person who had the keys of the room
2. No one had the keys except Brown and White
3. White was in Patagonia at the moment of the murder
4. White had a strong interest in killing Smith.

A consequence of **K** is the disjunction:
   D : Brown is the murderer or Smith is the murderer $(C_1 \vee C_2)$
Now there is a subset $\mathbf{K}_i$ of **K** i.e. $\{1, 2, 3\}$ such that jointly with A ε-implies

5. Brown is the murderer (C_1)
   while another subset $\mathbf{K}_j$ $\{1, 2, 4\}$ jointly with A ε -implies
6. White is the murderer (C_2)

Note that 5) and 6) are incompatible if conjoined with K, since the premise 1) in **K** states that only one person was responsible of the murder. The conditional "Smith has been killed, so from the given information [it is reasonable to conclude that] Brown is the murderer" $(A >_A C_1)$ is a synthetic expression of what we call here an abductive conditional. What is the specific difference between the reasoning underlying a C-conditional and an A-conditional? Let us recall the schema of Hempel-Oppenheim's Statistical Inference: as is well known, such an inference requires a rule of high probability and also the essentiality of the items occurring in the antecedent. Any C-conditional $A >_C C$ implies that there is a potential *explanans* involving A conjoined with various true presuppositions K_i and an *explanandum* C. But in the case of abductive conditionals the inference, given a certain stock of true presuppositions, is not from the *explanans* to the *explanandum* but in the reverse direction. If A is an *explanandum*, given a certain amount of information represented by K_i, we have at least two *explanantia* $K_i \wedge C_i$ and $K_j \wedge C_j$. The schema of ε-implication is as before, with two important points of difference. In fact we have not as before b.4 $Cont(A \land K_i) \succ Cont(A \land K_j)$

but

b.4* $Cont(C_i \land K_i) \succ Cont(C_j \land K_j)$

Furthermore, the relation between C_i and A is always definable via ε-implication, but we have in place of b.3 what follows:

b.3* $C_i \land K_i$ ε-implies A and $C_j \land K_j$ ε-implies A.

The interesting point of agreement is that, in the case of the example, this choice is performed on the basis that the supposition C_i (that Brown is the murderer) implies the *explanandum* A thanks to a set of data which save more information than the alternative supposition C_j: so C_i is a component of the best explanation, in Hempel's sense, of A.

The two characterizations which we have given for $>_C$ and $>_A$ suggest that we could define an abstract notion of a reasonable inference. This step can be made in different ways. A possibility is to say that a conditional represents a reasonable inference if it is either a C-conditional $A >_C C$ satisfying the clauses b1)–c) of p. 371 or the converse abductive conditional $C >_A A$. We may define then a connective $>>$ as follows:

$(Def >>)$ $A >> C =_{Df} A >_C C \lor C >_A A$

A rough characterization of $>>$ is in saying that C is the best explanatory consequent of A or A is (part of the) best explanatory antecedent of C. Clearly $A >> C$ is independent from the converse $C >> A$, which is equivalent to $C >_C A \lor A >_A C$, so $>>$ is not a symmetric relation. We expect that the two following properties hold for $>>$:

TB1 $(A >> C) \supset \neg(A >> \neg C)$
TB2 $(A >> C) \supset \neg(\neg A >> C)$[10]

Needless to say, $>_A, >_C, >>$ are all non contrapositive. Furthermore, no one of them satisfies Modus Ponens. This is especially clear for the abductive conditionals. To say that A is the best available explanation of C does not mean that A is true given that C is true. In order to reach this conclusion we need a counterproof of A, or some independent evidence for it.

3 Iterated Conditionals and Causal Reasoning

To complete the theory of rational/reasonable inference an important detail needs to be added. The definition of rational and reasonable inference has been introduced in 2) with the restrictive clause that the antecedent clause A and the consequent C are truth-functional statements. But we have to face

---

[10] For this couple of formulas, often termed "Boethius' Theses", see Pizzi [15].

the possibility that a conditional contains another (negated or non-negated) conditional in the antecedent or in the consequent, giving rise to statements which have been named "embedded", "nested" or "iterated" conditionals. The first question to treat in this connection concerns the fact that it has been sometimes claimed that nesting of conditional antecedents lacks an independent sense. This skepticism is embodied in so called Generalized Stalnaker's Thesis.

(GST) $Pr(B > C/A) = Pr(C/A \wedge B)$ $(Pr(A \wedge B) \neq 0)$

But here we have to consider a famous counterexample suggested by R. Thomason:[11]

(Th) If the glass would break if thrown against the wall, then it would break if dropped on the floor.

As Thomason remarked, the logical form of (Th) cannot be $A > B \models C > D$, but $(A > B) > (C > D)$. In fact (Th) exhibits the failure of weakening – which is typical of $>$, not of $\models$ – since the following conditionals is false:

(Th^*) If the glass would break if thrown against the wall and the floor were covered with foam rubber, then it would break if dropped on the floor.

Stalnaker's Thesis indeed suggests that iterated antecedents might be paraphrased into a conjunction. If this were true we would have the permutation of antecedents as a theorem:

$(Perm)$ $(A > (B > C)) \supset (B > (A > C))$

But it is clear that permutation does not work:

(HAB) If you will have headache tomorrow, taking an aspirin you will feel better.

has a meaning which is different from

(AHB) Taking an aspirin, if you will have headache tomorrow you will feel better.

The phrase "Taking an aspirin" in fact receives a different sense when it is in the scope of the supposition concerning an headache tomorrow and in a context in which such information does not exist. Other examples give evidence that in iterated conditionals a premise could be factual in one position, but not factual in a different position. For instance, the supposition "the lamp is alight" may be factual or afactual in the following conditional

(LSD) If the lamp is alight, then if you switch off the light we will be in the dark.

but not in the permutated variant

---

[11] Quoted in van Fraassen [20].

(SLD) If you switch off the light then, if the lamp is alight we will be in the dark.

(SLD) in fact appears to be meaningless or false while (LSD) appears to be true.

In the present section our aim is to outline an analysis of iterated conditionals, both standard and abductive, in the framework of a consequentialist view of conditionals. A useful step is to introduce a more analytical formal language in which the symbol $>$ (which we now stipulate to stand ambiguously for $>_A$ and $>_C$) is indexed by some variables $a, b, c \ldots$ representing intuitively arbitrary rational inferential agents. We add the assumption that every agent $a, b, c \ldots$ is biunivocally associated to a certain set $\mathbf{K}_a, \mathbf{K}_b, \mathbf{K}_c \ldots$ of presupposed information. We will have then an infinite number of conditional operators $>_a, >_b, >_c \ldots$ The intuitive meaning of $A >_a C$ is that the agent a correctly infers C from A (with respect to a certain set $\mathbf{K}_a$ associated to a). Then $(A >_a C) >_b (R >_c Q)$ means then "b reasonably infers, from the fact that a reasonably infers C from A, that c reasonably infers Q from R". The involved sets of information are $\mathbf{K}_a, \mathbf{K}_b, \mathbf{K}_c$. The move from indexed $>$ to non-indexed $>$ is provided by the existential quantification on the variables for agents. In other words, $A > B$ can be made equivalent to $\exists x (A >_x B)$. The intuitive meaning of $A > B$ is then that there is some agent x who reasonably infers B from A and from the background knowledge at his disposal. According to this definition $A > (B > C)$ means then $\exists x (A >_x \exists y (B >_y C))$: For some x, x reasonably infers from A that (for some y, y infers reasonably C from A). A negated conditional $\neg(A > B)$ amounts then to $\neg \exists x (A >_x B)$, (i.e. must be understood as saying that no subject x infers reasonably B from A) and is obviously different from $A > \neg B$. Some remarks are in order.

i. The notion of degree of a conditional is the usual one adopted in conditional logic[12]. We stipulate that if $A >_d B$ is the conditional having the highest degree in a nested formula, the information set $\mathbf{K}_d$ is the *basic* information set: in other words every other information set considered in the formulas is coincident with $\mathbf{K}_d$ save for revisions introduced by the suppositions occurring in lower-degree conditionals (see point iv).

ii. An assumption should be introduced to calculate the information content of $\exists x (A >_x B)$. If $>_x$ stands for *reasonable* inference, it depends on the laws of nature essentially involved in the inference of B from A, so it is natural to think that the information content of $A > B$ is as least as high as the content of the physical laws which are essential to such inference. Of course $Cont\neg(A > B)$ equals $1 - Cont(A > B)$: if no rational subject can make an inference from A to B this makes $A > B$ something which is epistemically vacuous (so something having content near to 0).

---

[12] The conditional degree of a statement S may be simply calculated by 1) replacing $>$ with strict implication 2) eliminating the symbols for strict implications in favor of $\Box$ and truth functional operators and 3) calculating the modal degree of the resulting wff.

iii. In order to make an inference about other inferences we need the special laws which are the *meta-laws* governing the inferential behavior of rational subjects. Such laws are obviously part of the ordinary stock of background knowledge, i.e. of **CR**, but we will also assume that such laws belong to the T in $\mathbf{K}_x$. Some of such laws describe the already defined behavior of any rational subject in calculating information and drawing the "best" conclusion. Other important laws, however, rule the way in which any agent takes into account what other agents know or do.

iv. A metalaw which here we are willing to endorse – but could be ignored in different approaches – is that revision is cumulative: in other words any new supposition S made by some subject y should be added to an information set modified by the suppositions made by all $x_1 \ldots x_n$ in lower degree conditionals.

Let us for instance consider the following nested formula and suppose for sake of simplicity that $>$ is a C-conditional:

(6) $\quad A >_a ((D >_b C) \wedge \neg(H >_c R))$

Here $>_a$ has degree two, while $>_b$ and $>_c$ have degree one. If **K** is the background information set, $\mathbf{K}_a$ is here the basic information set, in the sense that it contains the part of **K** known by the subject a. Then:

1. The inference of a is performed by adding A to $\mathbf{K}_a(A)$
2. The inference of b is performed by adding D to $\mathbf{K}_a(A)(D)$[13]
3. The inference of c is performed by adding H to $\mathbf{K}_a(A)(H)$

Since it is different to add B to $\mathbf{K}(A)$ and to add A to $\mathbf{K}(B)$ this makes clear why $A > (B > C)$ is different from $B > (A > C)$.

This cumulative character of the suppositions should be made explicit by suitable axioms. In the light of the preceding interpretation, for instance, it should be natural to have at our disposition at least two principles, the first of which is obvious:

Ax1 $(A > (A > B)) \equiv A > B$
Ax2 $(A > (B > C)) \supset (A > (A \wedge B > C))$

We may now go back to the question whether nesting is pleonastic or not in the reconstruction of scientific reasoning. Our claim is that nesting is essential to give a correct understanding of important features of scientific arguments. An important argument in favor of the essentiality of nesting concerns causal redundancy in the frame of a counterfactual theory of causation. If e_1 and e_2 are symbols for token events identified by their instant of occurrence and O is an operator forming propositions from token events, our claim is that there is an unlimited number of causal notions, which may differ at least in two features: 1) the degree of the counterfactual that expresses the relation between

---

[13] For the definition of $\mathbf{K}(A)$, to be obiviously extended to $\mathbf{K}b(A)$, see page 367. It is understood that if $\mathbf{K}(A) = \mathbf{K}'$, $\mathbf{K}(A)(D) = \mathbf{K}'(D)$.

cause(s) and effect 2) the additional qualification expressing the explanatory strength of the causes with respect of the effect[14].

The statement $Oe_1 \wedge Oe_2 \wedge (\neg Oe_1 >_C \neg Oe_2)$ (to be read "e_1 is causally relevant for e_2") defines the minimal notion of causality, in the double sense that the conditional has degree one and no supplementary qualification is transmitted. But following the given theory we may define, among other notions, a two-place notion of *causal concurrency* which is as follows:

(CC) if e_1 had not occurred then, in absence of e_2, e_3 had not occurred.

So two concurring causes for e_3 are e_1 and e_2: $\neg Oe_1 >_C (\neg Oe_2 >_C \neg Oe_3)$ Standard examples of overdetermination are clear expressions of concurrency in the given sense: if the first killer had not fired a shot to Smith, the second would have killed Smith (which means: if also the second had not fired his shot then Smith would have not died). The paraphrase of this iterated conditional in terms of inferences performed by rational subjects is not difficult and will be omitted.

What to say about the kind of asymmetrical redundancy called preemption? We assume that preempting is a case of concurrency, but it is an asymmetrical concurrency. A standard definition of preemption says that a cause prevents the action of some other potential cause which would have reached the same effect. As argued in Pizzi [16], preemption should not be confused with causal anticipation or causal delay. The Sarajevo shots have been a triggering cause for the First World War, but the common opinion is that a macroevent classifiable as the First World War would have anyway taken place soon after, in absence of the shots, due to some other potential causes. So strictly speaking this is not a case of pre-emption because the effect-events involved are different. We remark anyway that a full description of causal anticipation might be realized by using an additional statement which has in any case the form of a second degree conditional:

(7) If the Sarajevo shots had not caused the First World War in t, some other event would have caused the First World War in some instant t' posterior to t.

In other cases the seeming preemption is not *anticipated* causation but *delayed causation*. The famous case of the thirsty traveler may be classified in this category, provided we make the reasonable assumption that poison is quicker than dehydration[15]. However, we can imagine a case in which the

---

[14] For this theory see Pizzi [16].
[15] The story says that a traveler has to cross desert with a can full of water, but two enemies try to kill him – the first by making a hole in the can, the second by poisoning the water. The traveler dies without touching the water. If the traveler had not died thirsty he would have died poisoned. But the poison is normally quicker than thirst , so we can say that he would had died poisoned before than the moment in which he really died. So in a sense the hole delayed his death.This

poison takes exactly the same time as thirst in killing the victim or a case in which we are unable to calculate the time of action, so in this case we conventionally stipulate that the two processes take the same time. In this case the first event preempts the other even they are not overdetermining. There is no doubt, to begin with, that the two causes are symmetrically concurring, since it is true both

(8) $\neg Oe_1 >_C (\neg Oe_2 >_C \neg Oe_3)$

and

(9) $\neg Oe_2 >_C (\neg Oe_1 >_C \neg Oe_3)$

But now we have to add the supplementary qualification that one of the two causes preempts the other. What does it mean to preempt? A naïve idea is that to preempt means to interrupt a causal chain. Now a causal chain is often seen as a transference of some quantity (speed, weight, force, energy, ...) from a three-dimensional object to another (as in so-called transference theory of causation). This idea has surely an appeal for physicists and for Aristotelian philosophers, but is insufficiently general. Negative events such as silence, darkness, fast, etc... may be causes or effects, and they are at the origin of the transference of nothing. In the example of the traveler, to say that he died by thirst (absence of water) is to give the example of a negative event causing something. If we switch off the light of the lamp this implies that the pressing of the button causes an interruption of electric current, and this is not clearly a transference of anything.

The idea that we want to propose here is that what preemption blocks is not a causal flux but a possible inference from the effect to one of the causes[16]. More clearly, the inference which is blocked in the case of preemption is the abductive inference from the effect to the preempted cause. When a preempting cause leaves a track in the effect, this means that there is something in the effect which makes an abduction possible for some or all inferential agents: but this inference becomes impossible from the effect to the preempted cause. From the fact that the can has been perforated some rational x infers that no y can infer from the fact that the victim died in the known conditions (empty can etc.) that the victim was poisoned. The correct formal rendering of this simple idea is not straightforward because we have at least two possibilities of formal rendering:

---

asymmetry suggested to R. Smullyan the idea that the poisoner is really more guilty than the perforator.

[16] This is not the proper place to make a comment about the philosophical controversy over preemption/overdetermination. According to the ideas of Lewis and Bunzl no genuine case of overdetermination exists. In fact either 1) the compared effects are really different or 2) a cause preempts another one, as when an electron on a wire prevents another electron to reach the bulb. The following example of of trumping preemption states a case in which we have premption without the interruption of any transmission of energy or of other entities.

a1) $Oe_1 >_C \neg(Oe_3 >_A Oe_2)$
a2) $(Oe_3 >_A Oe_2) >_C \neg Oe_1$

The two formulas are not equivalent since $>_C$ and $>_A$ are not in general contrapositive. Both a1) and a2) actually look suitable to the case of what Jonathan Schaffer in [18] called *Trumping Preemption*. Let us suppose that a major and a sergeant are in front of a corporal, both shout "Charge!" at the same time, and the corporal soon charges (Oe_3). From the rules of military code we understand that the major's order (Oe_1), not the sergeant's order (Oe_2), caused the corporal's decision to charge (*ubi maior minor cessat*). After examining Lewis' and Ramachandran's theories, Schaffer concludes that no one of these theories is able to treat this kind of pre-emption. Our proposal appears to be free from the mentioned difficulties since it does not postulate any interruption of any causal chain. The reason why we say that the major's order preempts the sergeant's order is that, given the mentioned circumstances and the order of the general Oe_1, such an event disallows a correct abduction from the effect Oe_3 to the order of the sergeant Oe_2. Let us simply recall that abduction is inference to the best explanation, and that the military law by which a soldier obeys the order of the higher-degree military is part of the background theory T.

The asimmetry of preemption is then granted by the falsity of $Oe_2 >_C \neg(Oe_3 >_A Oe_1)$ and of $(Oe_3 >_A Oe_1)$.

4 The Role of Iteration in Abductive Reasoning

So what we showed up to now is: i) that iterated conditionals are essential to reconstructing causal reasoning and ii) that abductive conditionals may occur as subclauses of iterated conditionals which are important ingredients of complex causal statements. However, a statement like (a2), in which the abductive conditional occurs in antecedent position, is an additional statement and strictly speaking is not part of the statements which identify the core of the causal statement. We recall here an important remark contained in Goodman, which states that every counterfactual is equivalent to a factual conditional[17]. Goodman suggests that every counterfactual, as for instance

(10) If the match had been scratched it would have lit ($A >_C B$)

is equivalent to

(11) Since that match was not lit it was not scratched ($\neg B >_A \neg A$)

If this remark were correct, every counterfactual would be equivalent to the contrapositive abductive conditional. Unfortunately the development of the semantics for conditional logics made us familiar with the already mentioned

---

[17] See Goodman [7].

idea that conditionals are not in general contrapositive. This means that not every case of iterated counterfactual may be generally turned into an iterated abductive conditional, even this transformation may be a legitimate one in a class of cases which can be exactly delimitated. It is however remarkable that abductive conditionals may occur in iterated conditionals independently from the equivalence with some iterated counterfactual. Some simple examples are in order.

Suppose we accept that C is the best explanation of B ($B >_A C$). Suppose that we know that B' is similar to B in important features and C' is similar to C in important features, given the same informations which make $B >_A C$ a true conditional. So some inferential agent x could infer from $B >_A C$ that $B' >_A C'$: so it turns out that the second degree conditional $(B >_A C) >_C (B' >_A C')$ is a true conditional. The inferential laws which are involved are analogical laws of inference. Note that $B >_A C$ may be a counterfactual or afactual antecedent of $>_C$, and that the schema of the argument to apply here is similar to Thomason' example.

Examples of the previous schema are not difficult to construct when B and C stand for singular propositions expressing real possibilities. For instance, if the best explanation of my cold in the circumstances of yesterday is that I have been under the rain without umbrella, the best explanation of your cold in the same circumstances may be a token-event of the same kind.

But the same schema holds for any $B > C$ which is a counterfactual or counterlegal conditional. Counterlegal or counterpossible suppositions are legitimate suppositions in scientific reasoning[18]. An instance is the following. We know that every planet has an elliptical rotation round the Sun, so to suppose that some planet, say Venus, has a circular rotation round the Sun is to suppose something which is impossible. This means that we should remove from our background knowledge one of Kepler's laws, i.e. the law that every planet has an elliptical orbit, and replace it with a different law. However, we have no reason to reject the more general law.

(12) All planets have orbits of the same form

which is a generic variant of the rejected Keplero's law. It follows then the following conditional is true: If planets had a circular rotation (A) some agent x would infer from this that some rational agent y supposing that Alpha Centauri were a planet (B) would infer that Alpha Centauri would have a circular rotation (C):

(13) $A >_C (B >_C C)$

But note that also the abductive conditional would be appropriate in this example:

(14) $(B >_C C) >_A A$

---

[18] See Pizzi [17].

In fact, if someone is able to infer correctly from the supposition that Alpha Centauri is a planet (B) that Alpha Centauri has a circular orbit (C), the best explanation of this strange argument would be the belief in the false law that planets have a circular orbit. This use of abduction in iterated conditionals will of course seem far-fetched. However, an use of nested abduction may occur more naturally in other contexts: for instance when we want to test the normality of background conditions. A rational agent in normal conditions infers from the existence of smoke the existence of past or present fire. But smoke may be caused in non- normal conditions by other kinds of phenomena (for instance by frozen carbon dioxide, i.e. dry ice). So we could say

(15) If some x were able to exclude from other informations that fire was present, some y would conclude from this and the presence of smoke that dry ice was present.

The form of (15) is $(A >_C \neg B) >_C (C >_A D)$. With a further step, it is not difficult to find examples of abductive statements construed over lower-degree abductive statements. For instance, if someone can make an inference from smoke to dry ice, we might conclude abductively that in those circumstances an abduction from smoke to fire is impossible: $(C >_A D) >_A \neg(C >_A B)$. And this is a meta-abduction about abductions.

5 Abductive Conditionals and Standard Second Degree Conditionals

In the preceding sections we have treated $>_C$ and $>_A$ as operators belonging to the same family, i.e. as subspecies of the same species. But a famous quotation in which Peirce introduces his notion of abduction makes us reflect deeper on the properties of abductive conditionals. In fact the most quoted definition of abduction introduced by Peirce is the following

(16) "The surprising fact, F, is observed; But if H were true, F would be a matter of course. Hence, There is reason to suspect that H is true" [12, 5.189].

Neglecting the condition that C should describe a "surprising" (i.e. improbable) fact, a *prima facie* formal paraphrase of Peirce's analysis is as follows:

(17) $A >_A C$ is true if and only if $C \wedge (A >_C C) \models A$.

But this rendering could be criticized along the same lines used by Thomason against a similar paraphrase of embedded conditionals. In fact, the consequence relation used here is not monotonic, since the addition in the antecedent of some supplementary information D might destroy the validity of the inference. Thus we are justified in supposing that the correct formal rendering is offered by the equivalence

(18) $(A >_A C)$ if and only if $(C \wedge (A >_C C)) >_C A$.

This rendering, however, could also be questioned since $>_C$ appears not to be a proper formalization of "There is reason to suspect". On the other hand, "there is reason to suspect" cannot be an abductive conditional since this would make the definition a circular one. A possible way out is that a proper rendering of the intended meaning would be to put in place of the second occurrence of A the formula $Pr(A) > \delta$, where δ is some threshold probability value. Since we are willing to treat probability as a particular kind of graded modality (see section 1), this is simply a way to say that A is true in a reasonably great class of (epistemically) possible worlds. We have also to remark that the only way to grasp the idea that A is the best explanatory factor of C (an idea which is not explicit in Peirce's words) is to state that A is more information-preserving than every other rival hypotheses $A'_1 \ldots A'_n$, in the terms which have been formulated in section 3. But this or some other qualification does not modify the second order characterization of abduction which is clearly implicit in Peirce's proposal. As a matter of fact, Peirce's definition opens the road to an inquiry about the relations between $>_C$-conditionals and $>_A$-conditionals which may usefully amplify the limits of the theory which has been outlined in the present paper.

References

1. Carnap, R.: *The Continuum of Inductive Methods*. University of Chicago Press, Chicago (1952)
2. Hintikka, J.: *Logic, Language Games and Information*. Oxford University Press, Oxford (1973)
3. Hintikka, J.: A two-Dimensional Continuum of Inductive Methods. In Hintikka, J., Suppes, J. (eds.): *Aspects of Inductive Logics*. North-Holland, Amsterdam (1966) pp. 113–132
4. Hintikka J., Pietarinen, J.: Semantic Information and Inductive Logic. In Hintikka J., Suppes, P., eds.: *Aspects of Inductive Logics*, North-Holland, Amsterdam (1966) pp. 96–112
5. Fattorosi-Barnaba, M., Amati, G.: Modal Operators with Probabilistic Interpretations, *Studia Logica* 46 (anno) pp. 383–393
6. Gaifman, H.: A theory of higher order probabilities. In Halpern, J.Y., ed.: *Reasoning about knowledge*. Morgan Kaufmann, Inc., Los Altos, CA (1986)
7. Goodman, N.: The problem of Counterfactual Conditonals. *Journal of Philosophy* 44 (1947) pp. 113–128
8. Lewis, D.: *Counterfactuals*. Blackwell, Oxford (1973)
9. Lewis, D.: Causation. *Journal of Philosophy* 70 (1973) pp. 556–67
10. Lewis, D. K.: Postscripts to "Causation". In *Philosophical Papers*: Volume II. Oxford University Press, Oxford (1986)
11. Nute, D.: Conditional Logic. In Gabbay, D., Guenthner, F., eds.: *Handbook of Philosophical Logic*. Vol.II. Reidel, Dordrecht (1984) pp. 387–439

12. Peirce, C. S.: *Collected Papers of Charles Sanders Peirce*. Harvard University Press, Cambridge, MA, vols. 1-6, Hartshorne, C. and Weiss, P., eds.; vols. 7-8, Burks, A. W., ed. (1931-1958)
13. Pizzi C.: Gestalt Effects in Abductive and Counterfactual Inference. *Logic Journal of the IGPL* 14 (2006) pp. 257–270
14. Pizzi, C.: Counterfactuals and the Complexity of Causal Notions. *Topoi* 9 (1990) pp. 147–156
15. Pizzi, C.: Cotenability and the Logic of Consequential Implication. *Logic Journal IGPL* 12 (2004) pp. 561–579
16. Pizzi C.: Iterated Conditionals and Causal Imputation. In McNamara, P., Fraakken, H., eds.: *Norms, Logics and Information Systems* IOS Press, Amsterdam, pages (1999) pp. 147–161
17. Pizzi, C.: Deterministic Models and the "Unimportance of the Inevitable". In Magnani, L., Nersessian, N., Pizzi, C., eds.: *Logical and Computational Aspects of Model-Based Reasoning*, Kluwer, Dordrecht (2002) pp. 331–352
18. Schaffer, J.: Trumping Preemption. *Journal of Philosophy* 9 (2000) pp. 165–81
19. Rescher, N.: *The Coherence Theory of Truth*. Oxford University Press, Oxford, (1973)
20. Van Fraassen, B. C.: Probabilities of Conditionals. In Harper and Hooker (eds.): *Foundations of Probability Theory, Statistical Inference and Statistical Theories of Science*, vol.I, Reidel, Dordrecht (1976) pp. 261–308

Peircean Pragmatic Truth and da Costa's Quasi-Truth

Itala M. Loffredo D'Ottaviano and Carlos Hifume

State University of Campinas, Campinas, Brazil
{itala,carlos}@cle.unicamp.br

Summary. In this paper we present a conception of the Peircean pragmatic truth and a formal definition of pragmatic truth, the *quasi-truth* – this concept, previously introduced by da Costa and collaborators, on trying to capture the meaning of the theories of pragmatist thinkers such as Peirce and James, is considered as the truth conception inherent to empirical theories and a generalization (for partial contexts) of Tarski's correspondence characterization of truth. By defining the mathematical concept of partial structure and by using a special semantical approach, we analyze a suitable logic that can be used as the underlying logic for theories whose truth conception is the quasi-truth. We delineate a Kripke model semantics for this logic and among some fundamental results we show that it is a kind of Jaśkowski discussive logic, a paraconsistent modal logic.

This conception of quasi-truth, the logic and the structures here presented can be useful for the analysis of model-based reasoning in empirical theories.

1 Introduction

It is very difficult to develop any theoretical investigation without using the concept of truth. We cannot argue about a theory of truth without using this concept, because questioning a theory is to question its truthfulness, and accepting a theory is to accept it as true. We cannot leave out the concept of truth, as well as we can do with some other concepts.

According to da Costa (see [1]):

[...] we consider the classical concept of truth as a primitive concept. It is presupposed in all our practical and theoretical activities. Philosophically, truth is a final concept, indefinable through other simpler concepts, if we used the term *definition* as a proposition that characterizes and explains, without *petitio principii*, a concept. The sentence itself expressing, in strict sense, the definition of truth would have to be "true".

In [2], Lynch presents some connections between truth and other concepts: it is deeply connected to belief; it is also linked to knowledge; it is a central subject of logic in general; and it is also related to another mysterious concept, reality – in other words, to talk about truth is to talk of reality as it is.

We could investigate about two central subjects, concerning the property or underlying nature of truth:

1. Does truth even have a nature?
2. If so, what kind of nature?

The theories that try to answer this second question are frequently named *robust theories of truth*. Such theories consider that truth is an important property, that requests a substantial and complex explanation. Their defenders are interested in subjects, such as:

> Either does the absolute truth exist, or is every truth somehow either subjective or relative?

> Which type of relationship, if any, relates true propositions with the world?

> Are all truths verifiable by sensitive experiences?

As all such subjects concern the objectivity of truth, according to [2], a fundamental subject for the robust theories is *realism*.

The *deflationary theories of truth* answer the first question negatively, leading to another debate: the deflationists consider that the so-called problem of truth is in fact a pseudoproblem, truth does not constitute a property shared by all the propositions that we consider true. Therefore, the concept of truth should not be understood as expressing such property, but as playing another role, namely by considering that any explanation is unnecessary.

As in the robust theories, if we consider that truth is a property, then it is necessary to specify which things or what kind of things can present such property – the *truth-bearers*: in other words, which things may be either true or false. For Kirkham (see [3]), even in this case there is a lot of confusion: even if the philosophers agreed in identifying by a name the correct truth-bearer, the problem would hardly end, for they could disagree relatively to the nature of the things nominated by every one of those terms.

Among the robust theories of truth we have the correspondence, coherence and pragmatic theories of truth.

Correspondence truth is based on the idea that "truth is correspondence with reality", that is, a truth-bearer is true when the things in the world are as the truth-bearer says they are; if not, the truth-bearer is false. Besides, according to [2], in general it constitutes a realistic vision – if something is true, this does not depend on what everyone believes, truth depends on the world and not on us. However, saying "truth is correspondence with reality" is nothing but a triviality. In order to establish a theory of correspondence truth, it is necessary to establish three of its aspects: which thing has the property of being true, i.e., what is the truth-bearer; the correspondence, i.e., what is the truth relationship; and the "reality" to which the truth-bearer corresponds.

Generically speaking, in a *coherence theory of truth*, a set of two or more beliefs is considered coherent if they "adjust" or "agree" among themselves (see [4]). Hence, the beliefs of an individual are true either if the set of his beliefs is coherent, or a belief is true if it is coherent with other beliefs in a system; on the contrary, they are false. Therefore, instead of being a correspondence relation between a truth-bearer and reality, in coherence theories the truth is a question of relationship between a truth-bearer and another truth-bearer. On account of that, this conception of truth is usually labeled as "epistemic".

The aim of this paper is to present a special conception and a formal definition of *pragmatic truth*, the *quasi-truth*. This concept, previously introduced by da Costa and collaborators, on trying to capture the meaning of the theories of pragmatist thinkers such as Peirce and James, is considered as the truth conception inherent to empirical theories and a generalization, for partial contexts, of Tarski's correspondence characterization of truth.

It is divided into two parts: in the first one, we briefly present an interpretation of Peirce's pragmatic concept of truth; we consider that, although it corresponds to a kind of partial truth, it is a kind of correspondence truth. Therefore, we need to explain what "reality" is, pointing the truth-bearer and the correspondence relationship: the sign, the correspondence relationship and the reality notion, as in Peirce, can be especially understood by adopting the scholastic realism, although Peirce's realism also contains elements of nominalism and idealism.

In the second part, by considering that models are signs, that mental action is an inferential process that uses Peirce's types of reasoning – *abduction*, *deduction* and *induction* –, and that scientific investigation has these same stages – abductive, deductive and inductive –, we present da Costa's formal definition of pragmatic truth, based on models. By defining the mathematical concept of partial structure and by using a special semantical approach, we analyze an appropriate logic that can be used as the underlying logic for theories whose truth conception is da Costa's quasi-truth. We delineate a Kripke model semantics for this logic and among some fundamental results we show that it is a kind of paraconsistent modal Jaśkowski discussive logic.

2 Peircean Pragmatic Truth

2.1 Pragmatism

Pragmatism, the philosophical movement founded by Peirce, can be synthesized by the following passage, considered the "pragmatic maxim":

Consider what effects, that might conceivably have practical bearings, we conceive the object of our conception to have. Then, our conception of those effects is the whole of our conception of the object. [5, 5.402][1]

In this case, practical consequences are those effects of the conception that have influence in our practice, in our action. The meaning of an idea consists of its practical effects on the human experience. This way, if two ideas have the same practical consequences, then they have the same meaning, and ideas without practical consequences do not have any meaning. According to Peirce:

> Pragmatism, then, is a theory of logical analysis, or true definition; and its merits are greatest in its application to the highest metaphysical conceptions. [5, 6.490]

In general, pragmatism can be understood as a method of either explaining ideas or determining meanings, a method that tries to take the techniques of experimental investigation to philosophical analysis. According to Hegenberg and Mota (see [6]), pragmatism is characterized, in a quite wide way, by

i) An specific way of thinking – that approaches what is defended by the British empiricism, ..., concisely, the concretely observable is indispensable for the apprehension of meanings, as well as for the test of beliefs and ideas;

ii) An interpretation of life in evolutionistic terms, ..., concisely, the continuity and development are basic postulates for pragmatism;

iii) Adhesion to a naturalistic psychology – the spirit acts as specific functions of alive organisms, ...

iv) Acceptance of a scientific perspective in which experimentalism prevails.

2.2 Classification of Sciences

One of Peirce's main purposes is to delineate the underlying fundamental principles of the methods used in science, searching for constant elements in the different scientific methods.

Before studying any science, the philosophical thought should begin by a system of logic, whose first task must be establishing the most formal and universal categories of experience. Peirce concludes that there are only three formal elements or universal categories, omnipresent in every and any phenomenon: quality, relationship and representation, later named *firstness*, *secondness*, and *thirdness*. Peirce then distinguishes:

i) Three species of representations (or signs) – similarity (icon), index and symbol;

ii) A triad of conceivable sciences – formal grammar, logic and fomal rethoric;

---

[1] References to Peirce's *Collected Papers* will be designated by [5], followed by the volume and paragraph numbers.

iii) A general division of symbols, common to all those three sciences – terms, propositions and arguments;
iv) Three types of arguments: deduction (symbol), induction (index) and hypothesis or abduction (similarity or icon).

According to Santaella (see [7]), there are three standpoints from which the categories have to be studied: *qualities*, *objects* and *mind*.

From the point of view of qualities (firstness), or ontological point of view, the categories appear as: quality or firstness – the being of positive qualitative possibility; reaction or secondness – action of the current fact; mediation or thirdness – the being of a law that will govern facts in the future.

From the point of view of objects (secondness), or of the existent, the categories appear as: quales – firstness facts; relationships – secondness facts; representations – signs or thirdness facts.

From the point of view of mind (thirdness), the categories appear as: feeling or immediate conscience – firstness signs; sensation of a fact – action sensation and reaction or secondness signs; conception – learning sense, mediation or thirdness signs.

As examples of those categories we have:

i) Firstness – indetermination, vagueness, possibility, originality, coolness, potentiality, quality, feeling;
ii) Secondness – certainty, final, object, correlative, reagent, being linked to, relationship notions, polarity, denial, matter, brute and blind force, compulsion, action-reaction, effort-resistance, here and now, opposition, effect, occurrence, fact, conflict, surprise, doubt, result;
iii) Thirdness – what is in development, generality, continuity, growth, mediation, infinite, intelligence, law, regularity, learning, habit, sign.

Peirce uses his categories as framework to his logical doctrine, as a basis for his classification of sciences. In decreasing order of abstraction:

1 Sciences of Discovery
   1.1 Mathematics
   1.2 Philosophy
      1.2.1 Phaneroscopy
      1.2.2 Normative sciences
         1.2.2.1 Aesthetics
         1.2.2.2 Ethics
         1.2.2.3 Logic or Semeiotics
            1.2.2.3.1 Pure or speculative grammar
            1.2.2.3.2 Critical logic
            1.2.2.3.3 Methodeutics or speculative rhetoric
      1.2.3 Metaphysics
   1.3 Special sciences
2 Sciences of revision
3 Practical sciences.

For Peirce mathematics is the science of exact conclusions, regarding hypothetical states of things, the only science that does not depend on the other ones. It is in philosophy that subjects regarding human experience are discussed and phaneroscopy has as its first task finding the most universal categories of experience.

Normative sciences are those that work either with the ends or the ideals that guide the feeling, the conduct and the human thought; they study the phenomena so that we can act on them and they on us.

Peirce attributes at least two senses to logic: the science of the necessary conditions to reach truth; and the science of the necessary laws of thought. He also considers logic as general semiotics, for treating the general conditions of signs as signs and the laws of thought evolution.

The aim of pure grammar, the first sub-division of logic, is to study all kinds of signs, their nature and meaning. The sign is mediation and belongs to thirdness.

Critical logic corresponds to what we nowadays know as mathematical-logic or logic and its aim is to investigate the conditions of truth of the logical inferences or arguments. However, in this case, Peirce introduces a new type of argument, the abduction – quasi-reasoning, discovery flash, responsible for the creation of hypotheses. So, critical logic was developed as an unified theory of abduction, induction and deduction.

Methodeutics studies the general conditions of the relationship between symbols (and other signs) and their interpretations. It is also a theoretical study whose aim is to examine the appropriate procedures to any investigation (see [7]).

2.3 Reality and Truth

According to the pragmatist maxim, the meaning of a conception of an object is constituted in the totality of its conceivable practical consequences; and its resultant action contains an element capable of moulding a future thought, accomplishing the rational purpose of the conception and having an intellectual element that permeates the deliberate conduct (see [8]). According to Peirce [5, 7.361], thought is rational only when it refers to a possible future.

For Ibri (see [8]), the core of Peirce's conceivable practical consequences is that a positive conception that a real object is supposed to have "should foresee the future course of the experience." And it is the action, or experience, that will reveal if there is a real conformity with the forecast: "correspondence between the theoretical forecast and the temporary course of the facts, the reinforcement of the conception is established in the form of a belief and, otherwise, as a doubt about its truthfulness."

For Peirce [5, 5.372], the distinction between belief and doubt constitutes a practical difference. Beliefs guide our purposes and model our actions. A belief is an indication that some habits are more or less settled down, in such a way that they will determine our actions; they characterize a satisfactory and stable state that we either do not want to avoid, or do not want to shift to another belief; besides, different beliefs are distinguished by the different actions that they originate. On the other hand, doubt does not produce those effects, it constitutes a difficult and uncomfortable state that we struggle getting rid and passing to a state of belief; this uncomfortable state impels us to searching a stable state, a belief.

From the logical point of view, the investigation, the process of establishing stable beliefs, of establishing meanings, happens through three types of reasoning, according to Peirce:

> These three kinds of reasoning are Abduction, Induction, and Deduction. Deduction is the only necessary reasoning. It is the reasoning of mathematics. It starts from a hypothesis, the truth or falsity of which has nothing to do with the reasoning; and of course its conclusions are equally ideal. The ordinary use of the doctrine of chances is necessary reasoning, although it is reasoning concerning probabilities. Induction is the experimental testing of a theory. The justification of it is that, although the conclusion at any stage of the investigation may be more or less erroneous, yet the further application of the same method must correct the error. The only thing that induction accomplishes is to determine the value of a quantity. It sets out with a theory and it measures the degree of concordance of that theory with fact. It never can originate any idea whatever. No more can deduction. All the ideas of science come to it by the way of Abduction. Abduction consists in studying facts and devising a theory to explain them. Its only justification is that if we are ever to understand things at all, it must be in that way. [5, 5.145]

Induction, from a given theory, looks for facts that prove its truthfulness; on the other hand, abduction, from facts, looks for a theory, that is, from the observed experience it constructs the concepts – abduction constitutes the creative reasoning of the ideas of science, of the hypotheses and in a general way of every creation.

In abduction the consideration of the facts suggests the hypotheses. In induction the study of the hypotheses suggests the experiments that bring to light the very facts to which the hypotheses had pointed. [5, 7.218]

In *Questions Concerning Certain Faculties Claimed for Man* (1868) [5, 5.213-263], *Some Consequences of Four Incapacities* (1868) [5, 5.264-317] and *Grounds of Validity of the Laws of Logic: Further Consequences of Four Incapacities* (1869) [5, 5.318-357], Peirce criticizes *cartesianism*, based mainly on the concept of intuition. One of his conclusions is that mental action is an inferential process and that thought only works through signs. According to Santaella (see [9]), such rejection to the cartesian conception is in the basis of the Peircean theories of mental action, signs, cognition, scientific investigation, methods, human insight and discovery, and of pragmatism.

In that way, we consider that the Peircean conception of truth is strongly related to such concepts. We also consider the Peircean truth as a kind of correspondence truth, in other words, a relation among truth-bearers (the signs) and reality. However, such relationship does not express an absolute truth, but a partial truth. This partiality depends on Peirce's sign definition and reality notion.

In spite of Peirce's realism containing elements of the scholastic realism, nominalism and idealism – called "realicism" by Mayorga (see [10]), – sometimes Peirce identifies himself as a "scholastic realist", mainly (but not totally) furthering Duns Scotus' realism. Peirce retakes one of the great controversies of Medium Age, the problem of the universals – that he prefers to denominate the "generals", in defense of his realism, against nominalism.

For Peirce [5, 8.12], the *real* "is that which is not whatever we happen to think it, but is unaffected by what we may think of it".

The *scholastic* or *moderate realism* is placed between two extremes: the *platonic realism* – the universal do exists –, and *nominalism* – the universal is not real. The scholastic realism recognizes that, although the existence only of the individual, the universal can be real.

Nominalism considers that only the individuals exist, only these are real. The universal is not real, because it does not exist, it is mere words as many concepts, and always dependent on the mind.

Peirce considers that the adoption of nominalism is a mistake, because nominalism considers only one way of being, the being of a thing or an individual fact, the existence. For Peirce, the laws and every type of regularity govern future facts – that do not exist –, and they are real; but nominalism cannot explain what is a law, since it recognizes only the current existence. However, realism claims that the existence of individual things is not the only way of being.

Although the universals being mind dependent concepts, they can be considered real. From the epistemological point of view it seems that we acquire knowledge of the world through a generalization process. But, in that way, what the intellect knows on a sensorial object is not what is individual, but only its general characteristics. However, this does not impede that we know the world as really it is (see [10]).

For Peirce [5, 8.12], the real, the independent thing of what someone thinks of it, is not out of the mind, because the intellect attributes an important component to the concept of thing. The real is independent of my thoughts, as well as yours, but it is not independent of thought in general. Thought and human opinion contain an arbitrary, accidental element that depends on the limitations, circumstances and the individual power. But human opinion, in the long run, tends to a defined form, that is the truth. Any human being that has enough information and thinks sufficiently on any subject, will reach a certain defined conclusion, that will be the same one of any other mind under sufficiently favorable circumstances. The arbitrary will or other individual peculiarities of a sufficiently large number of minds can indefinitely postpone the general agreement on a subject, but they cannot affect the character that such opinion will have when it is reached. And the final opinion is the truth, because this opinion is independent, certainly not independent of thought in general, but of everything that is arbitrary and individual in the thought. Everything that will be thought as existing in the final opinion is real, and nothing else. Therefore the scientific method is the best one to acquiring stable beliefs. For although being fallible and partial, such truths will be the fruit of the agreement of a community of investigators and we can have knowledge of the world because human thought, in general, tends to truth.

3 Da Costa's Quasi-Truth

Tarski, when introduced his formal definition of correspondence truth, the "semantical conception of truth for formalized languages", sought "to capture" the existing intentions in "aristotelian classic conception of truth" (see [11–13]). Similarly, Mikenberg et al. (see [14]), tried to represent the "intentions" of the theories of truth of pragmatists such as Peirce and James (cf. [15]): loosely speaking, they say that a sentence is pragmatically true if, in a certain context, "it saves the appearances", i.e., if it is true in the classical correspondence sense.

Mikenberg, da Costa and Chuaqui observe that formal definitions are, at least in principle, neutral, or at least as neutral as the mathematical formulations in which they are represented. From the formalism of set theory, they introduce a formal version of the notion of pragmatic truth, conveniently adapting Tarski's definition.

According to da Costa and French (see [16]), the intentions of the pragmatist vision of truth represent an emphasis on:

i) The nature of the agreement between "imperfect" or "abstract" description and reality;
ii) The empirical consequences of such descriptions, understood as "agreement" with reality, in the classic correspondence sense;
iii) The "complete" or "absolute" truth, again understood in the classic correspondence sense, as (ideal) end of every investigation.

From the naturalistic change in the philosophy of science, the nature and importance of scientific practice have been reevaluated. However, a problem that appears is that no construction of reasoning can accomodate the vagueness and complexities of such practice. According to da Costa and French, a unitary treatment can be built, that incorporates and focuses two fundamental aspects of the epistemic practice in general, concerning the nature of representation used in the scientific reasoning and the epistemic attitudes adopted relatively to it, and the methodology.

The representations are, basically, conceptually incomplete and unfinished, the adopted general attitude is fallibilist. The representations used in the scientific practice are not seen as true in the correspondence sense, but as partially true, approximately true or as containing some truth element – the development of a formal concept of pragmatic truth can eliminate the deficiencies of the attempts of formally capturing such notions.

The definition of quasi-truth offers a way of accomodating the incompleteness inherent to scientific representations, with the introduction of the notion of partial structure, in the semantic approach of theories through the introduction of an adequate model theory.

Tarski's definition is extended to the quasi-truth definition:

i) The notion of structure is extended, by introducing the notion of "partial structure";
ii) The notion of "quasi-truth" is introduced, being a generalization of Tarski's characterization of truth for partial contexts.

In general, when we investigate a certain domain of knowledge, we don't know everything about it, in other words, our information is incomplete or partial. Therefore, we cannot be sure that a particular theory on that domain is true, but we can say that, as much as our information allow us, such theory can be true, i.e., it is quasi-true. According to Hifume (see [17]):

i) When a certain domain Δ of knowledge is investigated, we submit it to a conceptual scheme, in order to systematize and organize the information about it;
ii) That domain is "acted" by a set D of objects, and is studied *via* the analysis of the relations among its elements.
iii) Given a relation R on D, as it frequently happens in scientific contexts, "we do not know" if all the objects of D are related by R;
iv) Therefore, we say that our information about Δ is "incomplete" or "partial."

The introduction of the notions of partial relation and partial structure makes possible to formally accomodate that incompleteness and to represent the information about the investigation domain.

Definition 1. *Let D be a non-empty set. A n-ary partial relation R on D is a triple $\langle R_1, R_2, R_3 \rangle$, where $R_i \cap R_j = \emptyset$, for $i \neq j$, $i, j \in \{1, 2, 3\}$ and $R_1 \cup R_2 \cup R_3 = D^n$, such that:*

i) R_1 is the set of n-tuples that we know that belong to R;
ii) R_2 is the set of n-tuples that we know that do not belong to R;
iii) R_3 is the set of n-tuples that we don't know whether they belong to R or not.

We observe that if $R_3 = \emptyset$, R is an usual n-ary relation, that can be identified with R_1.

Definition 2. *A partial structure A is an ordered pair $\langle D, R_i \rangle_{i \in I}$, where:*

i) D is a non-empty set;
ii) $(R_i)_{i \in I}$ is a family of i-ary partial relations on D.

As in the notion of pragmatic truth correspondence truth is involved, also in the definition of quasi-truth Tarski's characterization of truth is involved. For Tarski, a sentence of a first-order language L is true or false, only relatively to a certain interpretation in a given structure: similarly, a sentence can be quasi-true or quasi-false, only relatively to an appropriate type of structure. But, as in Tarski's characterization only total structures are used (in which the relations are usual, non-partial), intermediate notions of structures are here defined, in order to establish a relationship between partial and total structures.

Definition 3. *A simple pragmatic structure (sps) for a first-order language L is a structure $A = \langle D, R_k, \wp \rangle_{k \in I}$, where:*

i) D is a non-empty set, the universe of A;
ii) R_k is a k-ary family of partial relations on D, for all $k \in I$ (R_k may be empty, for some k);
iii) $\wp$ is a set of sentences of L.

A simple pragmatic structure is a partial structure with a third component: a set of sentences $\wp$ of L, either accepted as true or that are true according to the correspondence theory; these sentences can express either true statements, empirically decidable, or general sentences expressing either laws or theories accepted as true.

Given a simple pragmatic structure, it can be extended to a total structure.

Definition 4. *Let L be a first-order language, $A = \langle D, R_k, \wp \rangle_{k \in I}$ a sps and S a total structure, where L is interpreted. S is an A-normal structure if the following properties hold:*

i) The universe of S is D;
ii) The (total) relations of S extend the correspondent partial relations of A;
iii) If c is an individual constant of L, then c is interpreted in A and S by the same element;
iv) If $\alpha \in \wp$, then S satisfies α, i.e., every sentence of $\wp$ is valid in the structure S, what is denoted by $S \models \alpha$.

Definition 5. *Let L be a language, A a sps and S an A-normal structure. A sentence α of L is quasi-true in the sps A, relatively to S, if α is true in S according to Tarski's definition of truth. Otherwise, α is quasi-false.*

In other words, if α is quasi-true in A then all the logical consequences of α, or α plus the primary declarations $\wp$, should be compatible with any true primary declaration. Hence, α is such that everything happens in the domain of knowledge under investigation Δ as if α was true.

4 A Logic for Quasi-Truth

In [17], chapter 4, it is analyzed a logical system that can serve as the underlying logic to theories that have the quasi-truth as their truth conception. In general, this logic can be used as a deductive logic of science.

In order to build this logic of pragmatic truth, from a first-order language L and a given sps A that interprets L, we consider its A-normal structures as "worlds" of a Kripke structure for the *first-order with equality modal system* $S5Q^=$ – a $S5Q^=$ model. That is, from the universe of a sps A for L, we have several structures (total) in which L can be interpreted, such that any total structure is accessible to the other.

In L (and in A), the possibility operator $\Diamond$ corresponds to the quasi-truth notion (pragmatic truth) and the necessity operator ($\Box$) to the quasi-validity notion (pragmatic validity). In order to formalize these two notions, we deal with two logical systems, $S5Q^=$ and QT.

The pragmatically valid formulas are the formulas α such that $\Box\alpha$ is a theorem of $S5Q^=$. Among these, there are formulas $\Box\Diamond\alpha$ such that $\Diamond\alpha$ is a theorem of $S5Q^=$. We name the first class of formulas *strict-pragmatically valid*, or simply *strictly valid* formulas (the theorems of $S5Q^=$); the second class is named *pragmatically valid* formulas, that are the theorems of da Costa's *system QT – a paraconsistent modal system* associated to $S5Q^=$, a kind of Jaśkowski's discussive logic, a logic for quasi-truth [1].

The language L of QT is the language of $S5Q^=$. The axioms and inference rules are the following, where $\forall\!\!\forall\alpha$ is the closure of α.

Axiom 1. If α is an instance of a classical propositional tautology, then $\Box\forall\!\!\forall\alpha$ is a QT-axiom.
Axiom 2. $\Box\forall\!\!\forall(\Box(\alpha \to \beta) \to (\Box\alpha \to \Box\beta))$
Axiom 3. $\Box\forall\!\!\forall(\Box\alpha \to \alpha)$
Axiom 4. $\Box\forall\!\!\forall(\Diamond\alpha \to \Box\Diamond\alpha)$
Axiom 5. $\Box\forall\!\!\forall(\forall x\alpha(x) \to \alpha(t))$
Axiom 6. $\Box\forall\!\!\forall(x = x)$
Axiom 7. $\Box\forall\!\!\forall(x = y \to (\alpha(x) \leftrightarrow \alpha(y)))$
Axiom 8. In any formula, empty quantifications can be either introduced or suppressed.

Rule 1. $\vdash \Box\forall\!\!\forall\alpha,\ \vdash \Box\forall\!\!\forall(\alpha \to \beta)/\vdash \Box\forall\!\!\forall\beta$
Rule 2. $\vdash \Box\forall\!\!\forall\alpha/\vdash \alpha$
Rule 3. $\vdash \Box\forall\!\!\forall\alpha/\vdash \Box\forall\!\!\forall\Box\alpha$
Rule 4. $\vdash \Diamond\forall\!\!\forall\alpha/\vdash \alpha$
Rule 5. $\vdash \Box\forall\!\!\forall(\alpha \to \beta(x))/\vdash \Box\forall\!\!\forall(\alpha \to \forall x\beta(x))$

Hifume presents specific definitions and proves fundamental results of QT.

Definition 6. *A QT-model is a $S5Q^=$-model.*

Definition 7. *In QT, a well formed formula (wff) α is a semantical-pragmatic consequence of a set Γ of wff of L, what is denoted by $\Gamma \models^p_{QT} \alpha$, if, and only if, there are formulas $\gamma_1, \gamma_2, \ldots, \gamma_n$ in Γ such that $\models_{QT} \Diamond\gamma_1 \wedge \Diamond\gamma_2 \wedge \ldots \wedge \Diamond\gamma_n \to \Diamond\alpha$.*

Theorem 1. *For every wff α of L, α is a theorem of QT if, and only if, $\Diamond\forall\forall\alpha$ is a theorem of $S5Q^=$:*

$$\vdash_{QT} \alpha \quad \Leftrightarrow \quad \vdash_{S5Q^=} \Diamond\forall\forall\alpha.$$

Theorem 2. *If a wff α is a theorem of $S5Q^=$, then it is a theorem of QT:*

$$\vdash_{S5Q^=} \alpha \quad \Rightarrow \quad \vdash_{QT} \alpha.$$

The converse of Theorem 2 is false, i.e., QT is "stronger" than $S5Q^=$. We observe that the Barcan formula, $\forall x \Box \alpha(x) \to \Box \forall x \alpha(x)$, holds in QT; and *Modus Ponens* does not hold, relatively to material implication.

Definition 8. *In QT, a formula α is a syntactic-pragmatic consequence of a set Γ of wffs, what is denoted by $\Gamma \vdash^p_{QT} \alpha$, if there are $\gamma_1, \gamma_2, \ldots, \gamma_n$ in Γ such that $(\Diamond\gamma_1 \wedge \Diamond\gamma_2 \wedge \ldots \wedge \Diamond\gamma_n) \to \Diamond\alpha$ is a theorem in QT.*

Definition 9. *A pragmatic theory whose underlying logic is QT, is a non-empty set Σ of sentences such that, if $\gamma_1, \gamma_2, \ldots, \gamma_n$ are in Σ and $\{\gamma_1, \gamma_2, \ldots, \gamma_n\} \vdash^p_{QT} \alpha$, then α is also in Σ.*

Theorem 3. *If Σ is a pragmatic theory and α is a theorem in QT, then $\alpha \in \Sigma$.*

Definition 10. *Let E be the set of all sentences of QT and Σ a pragmatic theory. Σ is trivial, or overcomplete, if $\Sigma = E$; otherwise, Σ is non-trivial. The theory Σ is inconsistent (contradictory), if there exists at least a sentence α such that $\alpha \in \Sigma$ and $\neg\alpha \in \Sigma$, where $\neg$ is the negation symbol of QT; otherwise, Σ is consistent (non-contradictory).*

Theorem 4. *There are inconsistent, but non-trivial, pragmatic theories.*

According to [18], a logic is *paraconsistent* if it can be used as the underlying logic for inconsistent but non-trivial theories, named *paraconsistent theories*. In this sense, QT is a paraconsistent logic.

Definition 11. *Let α and β be wffs of L:*

1. *Pragmatic implication $\to_p$: $\alpha \to_p \beta =_{df} \Diamond\alpha \to \beta$*
2. *Pragmatic conjunction $\wedge_p$: $\alpha \wedge_p \beta =_{df} \Diamond\alpha \wedge \beta$.*

Remark 1. Let α, β be wffs of L. In general, the *Pseudo-Scotus* Principle does not hold in QT, relatively to the pragmatic implication:

$$\nvdash_{QT} \alpha \to_p (\neg\alpha \to_p \beta).$$

Hence, QT is paraconsistent lato senso, or non-explosive and so paraconsistent, relatively to the pragmatic implication $\to_p$, according to two other definitions of paraconsitent logic of the literature, as for instance, in [19].

Proposition 1. *For every wff α and β in QT, Modus Ponens Rule holds, relatively to the pragmatic implication.*

Theorem 5 (Pragmatic Deduction Theorem). *For every wff α and β, β is a syntactic-pragmatic consequence of α if, and only if, the pragmatic implication $\alpha \to_p \beta$ is a theorem in QT:*

$$\alpha \vdash^p_{QT} \beta \quad \Leftrightarrow \quad \vdash_{QT} \alpha \to_p \beta.$$

Theorem 6 (Completeness). *The wff α is pragmatically valid if, and only if, α is a theorem in QT:*

$$\models_{QT} \alpha \quad \Leftrightarrow \quad \vdash_{QT} \alpha.$$

Theorem 7 (Pragmatic Completeness). *Let Γ be a set of wff and α a wff of L. α is a semantical-pragmatic consequence of Γ if, and only if, α is a syntactic-pragmatic consequence of Γ in QT:*

$$\Gamma \models^p_{QT} \alpha \quad \Leftrightarrow \quad \Gamma \vdash^p_{QT} \alpha.$$

5 Final Considerations

Axiomatization constitutes a formal method of specifying the content of a theory. Given a formal language, from a set of axioms, rules of inference and definitions, the content of the theory can be deductively derived as its theorems. The theory is then identified to the set of axioms and their deductive consequences.

According to da Costa, the axiomatic method leads to a economy of thought: when we study an abstract axiomatics, we are simultaneously treating several theories – all that are framed in the considered axiomatics. By the axiomatic method we can also investigate problems such as the equivalence of theories or the independence of axioms. It also constitutes adequate tools for mathematical work and research. And, in general, the deductive disciplines are based on the norms of the axiomatic method.

In mathematics, from Bourbaki's structural approach, the axiomatic method reaches a high level of precision and development: axiomatizing a mathematical theory consists in defining a type of structure, based on a set theory; a structure consists of a non-empty set and relationships among its elements, satisfying certain conditions imposed by a set of axioms. But such a formalization is essentially syntactic. A type of structure is seen as simply constituting a formal theory, built as a collection of symbols subjected to certain metamathematical rules. However, although such method has suffered critics, according to da Costa and French (see [20]), the axiomatization continues being considered as an important component of the philosophy of science, for its role in the clarification of the basic concepts of a theory; its aid for the comparison of theories; the way it can allow the use of mathematical techniques; and for its usefulness in solving certain philosophical disputes.

However, in the philosophy of science, an alternative method to the axiomatization of theories – essentially syntactic – is the semantical approach of theories or the model theory. According to James (see [16]), the introduction of model theory by Tarski, by formalizing the notion of correspondence truth in terms of "the sentence α is true in a structure S", suggests that other semantical notions could be defined in a similar way. In this approach, the semantical tally of scientific theories should be seen, not as sets of sentences axiomatized in some appropriate formal language, but as classes of models.

In spite of the great development reached by the formal sciences, their representations are still, essentially, deductive, atomist and they use Tarski's conception of truth. Based on Peirce's semeiotic conception of knowledge, on his sign definition, his conceptions of truth and reality, and on his metaphysics, we claim that a more coherent way of representing knowledge and cognitive processes can be obtained – models and systems are kinds of signs, that work with several other kinds of signs through abduction, deduction and induction.

The notions of partial structure and quasi-truth presented in this paper have other important applications in the theory of science, as for instance in the theoretic unification in science [16], in pragmatic probability [21], in the logic of induction, in inconsistent beliefs, in the realism-empiricism debate.

In future works, we intend to pursue our research by introducing and developing a theory of non-classical models, from some Peircean ideas and da Costa's quasi-truth definition.

References

1. da Costa, N.: *O Conhecimento Científico*. 2nd. edn. Discurso Editorial, São Paulo (1999)
2. Lynch, M.: *The Nature of Truth: Classic and Contemporary Perspectives*. The MIT Press, Cambridge (1992)
3. Kirkham, R.: *Theories of Truth: A Critical Introduction*. The MIT Press, Cambridge (1992)
4. Kirkham, R.: Coherence theory of truth. In: *Routledge Encyclopedia of Philosophy*. Volume 9., London/New York, Routledge (1998) 470–2
5. Peirce, C.S.: *Collected Papers of Charles Sanders Peirce*. Harvard University Press, Cambridge, MA (1931-1958) vols. 1–6, Hartshorne, C. and Weiss, P., eds.; vols. 7–8, Burks, A. W., ed.
6. Hegenberg, L., da Mota, O.: Introdução. In: *Semiótica e filosofia: textos escolhidos de Charles Sanders Peirce*, São Paulo, Cultrix (1972)
7. Santaella, L.: *Matrizes da Linguagem e Pensamento: Sonora, Visual, Verbal*. Iluminuras/FAPESP, São Paulo (2001)

8. Ibri, I.: *Kósmos Noētós: a arquitetura metafísica de Charles S. Peirce*. Perspectiva/Hólon, São Paulo (1992)
9. Santaella, L.: *O Método anticartesiano de C.S. Peirce*. Editora Unesp, São Paulo (2004)
10. Mayorga, R.: *On Universals: the scholastic realism of John Duns Scotus and Charles Peirce*. University of Miami, Miami (2002)
11. Tarski, A.: The concept of truth in formalized languages. In Corcoran, J., ed.: *Logic, Semantics, Metamathematics*. Hackett, Indianopolis (1933) 152–278
12. Tarski, A.: The semantic conception of truth. *Philosophy and Phenomenological Research* **4** (1944) 13–47
13. Tarski, A.: *Logic, Semantics, Metamathematics*. 2nd. edn. Hackett, Indianopolis (1983)
14. Mikenberg, I., da Costa, N., Chuaqui, R.: Pragmatic truth and approximation to truth. *The Journal of Symbolic Logic* **51**(1) (1986) 201–21
15. James, W.: Pragmatismo. In: *Os Pensadores*, São Paulo, Nova Cultural (1989) Translated into Portuguese by J.C. da Silva.
16. da Costa, N., French, S.: *Science and Partial Truth: A Unitary Approach to Models and Scientific Reasoning*. Oxford University Press, Oxford (2003)
17. Hifume, C.: *A Pragmatic Theory of Truth: Newton C.A. da Costa's Quasi-Truth*. State University of Campinas, Campinas (2003)
18. D'Ottaviano, I.: On the development of paraconsistent logic and da Costa's work. *The Journal of Non-Classical Logic* **7**(1/2) (1990) 9–72
19. Carnielli, W., Marcos, J.: A taxonomy of C-systems. In Carnielli, W., Coniglio, M., D'Ottaviano, I., eds.: *Paraconsistency - the Logical Way to the Inconsistent*. Marcel Dekker, New York (2002) 1–94
20. da Costa, N., French, S.: The model-theoretical approach in the philosophy of science. *Philosophy of Science* **57** (1990) 248–65
21. da Costa, N.: Pragmatic probability. *Erkenntnis* **25** (1986) 141–62

Sliding Mode Motion Control Strategies for Rigid Robot Manipulators

Antonella Ferrara and Lorenza Magnani

Department of Computer Science, University of Pavia, Pavia, Italy
{antonella.ferrara,lorenza.magnani}@unipv.it

Summary. The paper presents a new control method which achieves motion control for rigid robot manipulators. It is based on sliding mode control techniques and on the compensated inverse dynamics approach. The main advantages of using sliding mode control are robustness to parameter uncertainty, insensitivity to load disturbance, and fast dynamics response, as well as a remarkable computational simplicity with respect to other robust control approaches. Furthermore the proposed approach avoids the estimation of the time-varying inertia matrix. First order and second order sliding mode control laws are presented and in both cases the problem of chattering, typical of sliding mode control, is suitably circumvented. Some simulations results are reported demonstrating the good tracking properties and performances of the proposed control strategy.

1 Introduction

One of the crucial issues in controlling rigid robot manipulators is the necessity of performing path tracking of a desired trajectory. Different solutions to cope with this issue have been proposed during the past years and many approaches have been followed such as feedback linearization [1, 2], model predictive control [3], sliding mode control [4, 5].

In the robotics context, feedback linearization is known as inverse dynamics control [6–8]. The idea is to exactly compensate all the coupling nonlinearities in the dynamical model of the manipulator in a first stage, transforming the nonlinear system into a linear one, by means of a nonlinear coordinate transformation and nonlinear feedback. Then a second stage compensator may be designed based on the linear and decoupled plant.

One of the major drawbacks in using global feedback linearization is that the full Lagrangian dynamical model must be calculated in real-time because the coordinate transformation is a function of system parameters and, hence, sensitive to uncertainties which arise from imprecise knowledge of the kinematics and dynamics, and from joint and link flexibility, actuator dynamics, friction, sensor noise, unknown loads, and unknown environment dynamics.

The large differences in magnitude among the parameters, e.g., between the joint stiffness and the link inertia, can make the computation of the control ill-conditioned and the performance of the system, in terms of convergence, poor. This imposes to couple inverse dynamics approach with robust control methodologies [1].

The robust control technique adopted in the present paper is the sliding mode control methodology [9, 10], due to its advantages, such that robustness to parameter uncertainty, insensitivity to load disturbance, and fast dynamics response, as well as a remarkable computational simplicity with respect to other robust control approaches. Note that, the combination of sliding mode control with compensated inverse dynamics has already been investigated by the authors in [11]. In [5] a similar control strategy is proposed, but it is characterized by an adaptive compensator, and the generated pseudo-sliding is of the first order.

The discussion starts with an introduction on how to apply basic (first order) sliding mode control in connection with inverse dynamics. Then a second order sliding mode approach is presented [12, 13]. It consists of enforcing a second-order sliding mode on a surface $s[x(t)] = 0$ in the system state space, with $\dot{s}[x(t)]$ identically equal to zero, by using a control signal depending on $s[x(t)]$, but directly acting only on $\ddot{s}[x(t)]$ [12, 13].

The major drawback of sliding mode control is the so-called chattering phenomenon which consists of the high frequency switching of the control signal, due to the discontinuous nature of the control strategy, that may introduce problems to the controlled physical system such as disrupting or damaging actuators. In both cases of first order and second order sliding mode control approaches, the problem of chattering is dealt with. In case of first order sliding mode control law, it can only be circumvented by approximating the *sign* function, but so generating pseudo-sliding modes. In case of second order sliding mode control the chattering effect can be made less critical by adopting a continuous control law.

The paper is organized as follows. The next section is devoted to the problem formulation with the dynamics description of the robot manipulator. Some preliminary issues on the use of compensated inverse dynamics method are introduced in Section 3. The formal description of the proposed first order and second order sliding mode control strategy is reported in Section 4 and Section 5 respectively. Simulations results are reported in each section demonstrating the good performances of the proposed strategies.

2 Problem Formulation

Consider the dynamics of an n-joint robot manipulator described by the following Lagrangian model:

$$B(q)\ddot{q} + C(q,\dot{q})\dot{q} + F_v q + g(q) = u \qquad (1)$$
$$Z = \Psi(q)$$

where $q(t) \in \Re^n$ is the vector of joint displacements, $B(q) \in \Re^{n \times n}$ is the inertia matrix, $C(q, \dot{q}) \in \Re^n$ represents centripetal and Coriolis torques, $F_v \in \Re^{n \times n}$ is the friction matrix, $g(q) \in \Re^n$ is the vector of gravitational torques, u is the vector of control torques, and $Z \in \Re^r$ is the output vector.

The aim of the control strategy is to make the output Z of the rigid robot tracking a desired trajectory $Z_d \in \Re^r$ specified by

$$Z_d = \Psi(q_d) \qquad (2)$$

where q_d represents the vector of desired joint displacements.

The model considered in simulation is a two-link planar robot manipulator as shown in Fig. 1 where a_1 and a_2 are links length and $q = [\theta_1 \ \theta_2]^T$ is the vector of joint variables. Its parameters values are listed on Table 1.

The inertia matrix is

$$B(q) = \begin{bmatrix} b_{11}(\theta_2) & b_{12}(\theta_2) \\ b_{21}(\theta_2) & b_{22} \end{bmatrix}$$

with

$b_{11} = I_{l1} + m_{l1}l_1^2 + k_{r1}^2 I_{m1} + I_{l2} + m_{l2}(a_1^2 + l_2^2 + 2a_1 l_2 \cos\theta_2) + I_{m1} + m_{m2}a_1^2$
$b_{12} = b_{21} = I_{l2} + m_{l2}(l_2^2 + a_1 l_2 \cos\theta_2) + k_{r2}^2 I_{m1}$
$b_{22} = I_{l2} + m_{l2}l_2^2 + k_{r2}^2 I_{m1}$

Centripetal and Coriolis torques are

$$C(q, \dot{q}) = \begin{bmatrix} c_{11}(\theta_2, \dot{\theta}_2) & c_{12}(\theta_2, \dot{\theta}_1, \dot{\theta}_2) \\ c_{21}(\theta_2, \dot{\theta}_1) & c_{22} \end{bmatrix} \quad g(q) = \begin{bmatrix} g_1(\theta_1, \theta_2) \\ g_2(\theta_1, \theta_2) \end{bmatrix}$$

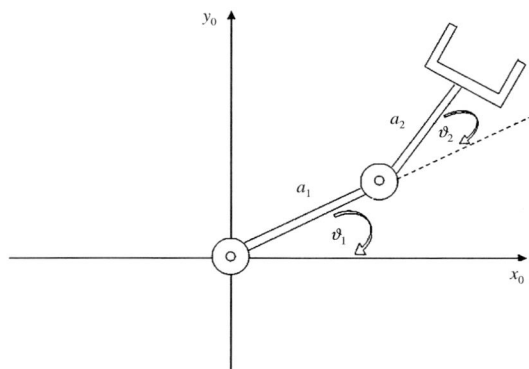

Fig. 1. Two-link rigid robot manipulator.

Table 1. Parameters of the robot manipulator.

Symbol	Parameter	Value
a_i	link length	1 m
l_i	mass center position	0.5 m
m_{l_i}	link mass	50 kg
m_{m_i}	motor mass	5 kg
I_{l_i}	inertia of the link	10 kgm^2
I_{m_i}	inertia of the motor	0.01 kgm^2
k_{r_i}	reduction ratio	100
F_{v_i}	friction coefficient	0.01 Nms/rad

where

$$c_{11} = -m_{l2}a_1 l_2 \sin\theta_2 \dot\theta_2$$
$$c_{12} = -m_{l2}a_1 l_2 \sin\theta_2 (\dot\theta_1 + \dot\theta_2)$$
$$c_{21} = m_{l2}a_1 l_2 \sin\theta_2 \dot\theta_1$$
$$c_{22} = 0$$
$$g_1 = (m_{l1}l_1 + m_{l2}a_1 m_{m2}a_1)g\cos\theta_1 + m_{l2}l_2 g\cos(\theta_1+\theta_2)$$
$$g_2 = m_{l2}l_2 g\cos(\theta_1+\theta_2)$$

and g is the gravitational acceleration, while the friction matrix is

$$F_v = \begin{bmatrix} F_{v1} & 0 \\ 0 & F_{v1} \end{bmatrix}$$

The desired trajectory is given by

$$Z_d = \begin{bmatrix} Z_{d1} \\ Z_{d2} \end{bmatrix} = \begin{bmatrix} 1.6 - 1.6e^{-8t} - 12.8te^{-8t} \\ 3.2 - 3.2e^{-8t} - 25.6te^{-8t} \end{bmatrix} \qquad (3)$$

and the output of the system is

$$Z = \begin{bmatrix} Z_1 \\ Z_2 \end{bmatrix} = \begin{bmatrix} \theta_1 \\ \theta_1 + \theta_2 \end{bmatrix}$$

An unknown load is carried by the robot at the second link assumed to produce a mass increase equal to $\Delta m_2 = 50 kg$. The initial displacements are $(\theta_1, \theta_2) = (\pi/4, -\pi/2)$.

3 Compensated Inverse Dynamics Method

A classical way to design a control law to solve the problem in question is that of inverse dynamics control [6–8]. The basic idea of inverse dynamics control is to transform the nonlinear system (1) into a linear and decoupled plant by using a nonlinear coordinate transformation that exactly compensates all the coupling nonlinearities in the Lagrangian dynamics.

Choosing the following nonlinear feedback control law

$$u = B(q)y + n(q,\dot{q}) \quad n(q,\dot{q}) = C(q,\dot{q})\dot{q} + F_v\dot{q} + g(q) \tag{4}$$

the closed loop system reduces to a decoupled double integrator

$$\ddot{q} = y \tag{5}$$

by designing y as

$$y = -K_P q - K_D \dot{q} + r \quad r = \ddot{q}_d + K_D \dot{q} + K_P q \tag{6}$$

with K_P, K_D selected so that system

$$\ddot{e} + K_D \dot{e} + K_P e = 0$$

obtained by taking the time derivative of (6) and substituting into equation (5), which represents the dynamics of the tracking error $e = Z_d - Z$, is asymptotically stable.

The mayor drawback of inverse dynamics control is that the full Lagrangian dynamic model must be calculated in real-time and the coordinate transformation is a function of system parameters and, hence, sensitive to uncertainties. This imposes the use of a control method which is insensitive to uncertainties such as sliding mode control.

Assuming that Z_d, $\dot{Z}_d$, $\ddot{Z}_d$ are bounded, the time derivative of equation (2), gives

$$\dot{Z} = L_1 \dot{q} \quad \ddot{Z} = L_1 \ddot{q} + L_2 \dot{q} \tag{7}$$

where

$$L_1 = \frac{\partial}{\partial q}\Psi(q) \in \Re^{r \times n} \quad L_2 = \frac{d}{dt}\left[\frac{\partial}{\partial q}\Psi(q)\right] \in \Re^{r \times n} \tag{8}$$

Since system (1) is affected by uncertainties, it is not possible to apply directly the inverse dynamics approach to the system, but a compensated inverse dynamics method is used in analogy with [5], by introducing a matrix $G \in \Re^{r \times n}$

$$G = L_1 B_0^{-1} \tag{9}$$

where B_0 is a nominal form of B such that the on-line estimation of B is not needed.

The compensated inverse dynamics control vector τ is defined as

$$u = G^+(\nu_1 + \nu_2) \tag{10}$$

where $G^+ = G^T(GG^T)^{-1}$ is the generalized inverse of G,

$$\nu_1 = \ddot{Z}_d - K_v \dot{e} - K_p e - K_i \int_0^t e(\tau) d\tau \tag{11}$$

and ν_2 is a compensation signal that will be specified in the sequel, $e = Z - Z_d$ is the tracking error and K_v, K_p, K_i are symmetric and positive defined matrices.

Taking into account equation (7), (9), (11) and letting

$$\nu_1 = L_1 B_0^{-1} \tau - \nu_2 \tag{12}$$

the tracking error dynamics results in

$$\begin{aligned}\ddot{e} + K_v \dot{e} + K_p e + K_i \int_0^t e(\tau)d\tau = \\ = \nu_2 + L_1 B_0^{-1}[(B_0 - B(q))\ddot{q} - C(q,\dot{q})\dot{q} - F_v - g(q)] + L_2 \dot{q}\end{aligned} \tag{13}$$

Now assume that

$$(B_0 - B(q))\ddot{q} - C(q,\dot{q})\dot{q} - F_v - g(q) = M + \Delta M \tag{14}$$

where vector $M \in \Re^n$ is known and $\Delta M \in \Re^n$ is uncertain, but bounded by

$$\|\Delta M\| \leq \Delta_M$$

Equation (13) can now be rewritten in the state space form as

$$\dot{x} = Ax + BN \tag{15}$$

where

$$x = \begin{bmatrix} \int_0^t e(\tau)d\tau \\ e(t) \\ \dot{e}(t) \end{bmatrix} \quad A = \begin{bmatrix} 0 & I & 0 \\ 0 & 0 & I \\ -K_i & -K_p & -K_v \end{bmatrix} \quad B = \begin{bmatrix} 0 \\ 0 \\ I \end{bmatrix}$$

and

$$N = \nu_2 + L_1 B_0^{-1}[M + \Delta M] + L_2 \dot{q}$$

4 First Order Sliding Mode Control

Taking into account the dynamics of the n-joint robot manipulator described in Section 2 and the compensated inverse dynamics approach giving (15), and according to the sliding mode theory [9, 10], a sliding mode manifold can be selected as

$$S = Tx \tag{16}$$

where $T \in \Re^{n \times 3n}$ is a suitable constant matrix.

Define the control law

$$\nu_2 = -L_1 B_0^{-1} M - L_2 \dot{q} - \frac{\rho(TB)^{-1} TAx (S^T TAx)^T}{\|S^T TAx\|} - (TB)^{-1} \Gamma S \tag{17}$$

where α, β are positive constant, $\Gamma > 0$ and

$$\rho \geq \{\|S^T T B L B_0^{-1} \Delta M\| + \alpha e^{-\beta t} + (\gamma - \|\Gamma\| \|S\|) \|S\|\} \frac{\|S^T T A x\| + \alpha e^{-\beta t}}{\|S^T T A x\|^2} + 1 \quad (18)$$

The first and second terms of (17) are needed to cancel the known part of the dynamics whereas parameter ρ is designed to compensate the unknown part such that the reaching condition $S^T \dot{S} < -\gamma \|S\|$ is fulfilled guaranteeing that the state of the system (15) will converge to zero in a finite time and the rigid robot manipulator tracks the desired trajectory Z_d.

In simulations, parameters used for first and second order sliding mode are selected as

$$\alpha = 0.03, \quad \beta = 0.05 \quad (19)$$

$$\Gamma = \begin{bmatrix} 5 & 0 \\ 0 & 10 \end{bmatrix} \quad M = \begin{bmatrix} 100 \\ 100 \end{bmatrix} \quad \rho = 10$$

$$T = \begin{bmatrix} 1 & 0 & 1 & 0 & 1 & 0 \\ 0 & 1 & 0 & 1 & 0 & 1 \end{bmatrix} \quad B_0 = \begin{bmatrix} 200 & 23.5 \\ 23.5 & 122.5 \end{bmatrix}$$

$$L_1 = \begin{bmatrix} 1 & 0 \\ 1 & 1 \end{bmatrix} \quad L_2 = 0$$

and the control gains are selected as

$$K_v = \begin{bmatrix} 7 & 0 \\ 0 & 10 \end{bmatrix} \quad K_p = \begin{bmatrix} 35 & 0 \\ 0 & 70 \end{bmatrix} \quad K_i = \begin{bmatrix} 20 & 0 \\ 0 & 45 \end{bmatrix}$$

Figure 2 shows that, using the inverse dynamics control when the unknown load is carried by the robot at the second link, the performance of the system is poor, i.e. the tracking error does not converge to zero.

In Figure 3 simulations results are reported using the first order sliding mode control law presented in this section, demonstrating that the controlled output tracks the desired reference trajectory although the dynamics of the system is affected by uncertainties.

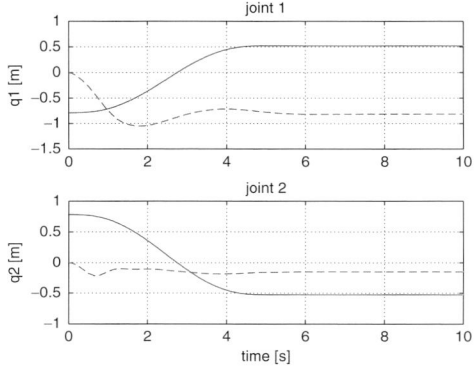

Fig. 2. Trajectory of link 1 and 2 (solid lines) and desired trajectory (dashed lines) using inverse dynamics control.

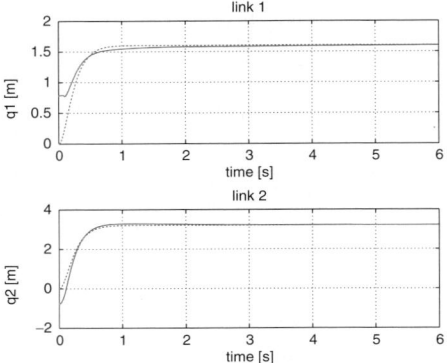

Fig. 3. Trajectory of link 1 and 2 (solid lines) and desired trajectory (dotted lines) using first order sliding mode control.

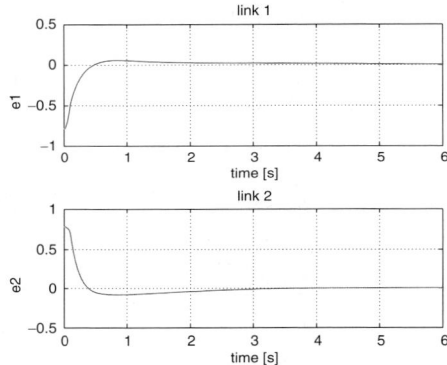

Fig. 4. Tracking error using first order sliding mode control.

The goal of sliding mode control is to force the trajectory error e to zero so that the selected output tracks the desired reference trajectory (3) as shown in Figure 4. Convergence of the system to the sliding manifold is assured as shown in Figure 5. In Figure 6 one can see chattering phenomenon, which always appears in first order sliding mode.

To overcome this problem it is possible to replace the discontinuous term $TAx(S^T TAx)^T / \left\| S^T TAx \right\|$ used in equation (17) with a smooth approximation given by

$$V_{app} = \frac{TAx(S^T TAx)^T}{\|S^T TAx\| + \alpha e^{-\beta t}}$$

so that equation (17) becomes

$$\nu_2 = -L_1 B_0^{-1} M - L_2 \dot{q} - \frac{\rho (TB)^{-1} TAx(S^T TAx)^T}{\|S^T TAx\| + \alpha e^{-\beta t}} - (TB)^{-1} \Gamma S \qquad (20)$$

reducing the chattering phenomenon as shown in Figure 7. The major drawback

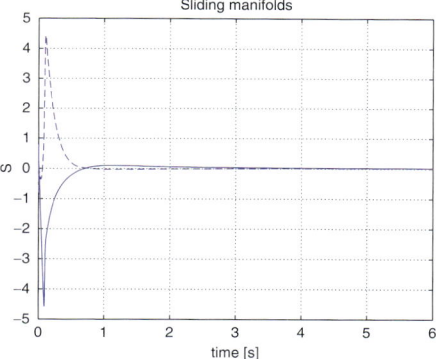

Fig. 5. Sliding manifolds (s_1: solid line, s_2: dashed line).

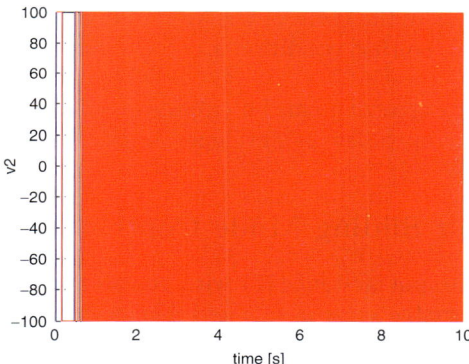

Fig. 6. Chattering phenomenon in first order sliding mode.

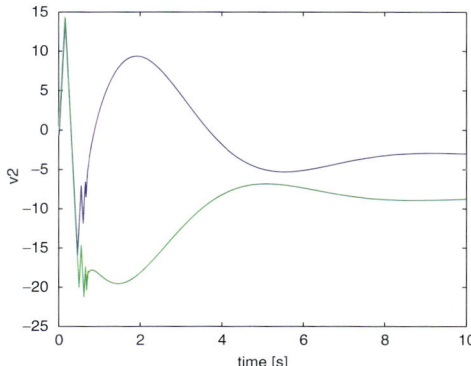

Fig. 7. Control variable without chattering.

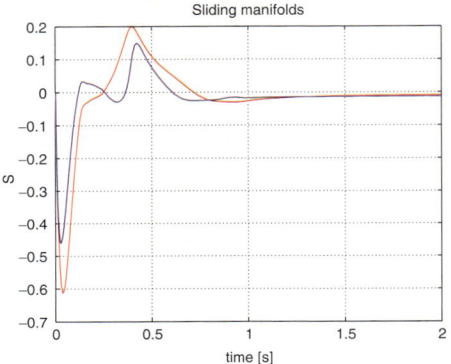

Fig. 8. Condition $s = 0$ is not satisfied using the smooth approximation.

of this approach is that the state of the system does not converge exactly to the sliding manifold, so that condition $s = 0$ can not be assured, as reported on Figure 8.

The problem of chattering can be circumvented by introducing a second order sliding mode control law as developed in the next section.

5 Second Order Sliding Mode Control

Consider the multi-input system (15) with the sliding manifold (16). Differentiate twice the variable s, let $y_1 = s$ and $y_2 = \dot{s}$ and introduce the following auxiliary system

$$\begin{cases} \dot{y}_1 = y_2 \\ \dot{y}_2 = \ddot{s} = T\dddot{x} = T_1\dot{e} + T_2\ddot{e} + T_3\frac{d^3e}{dt^3} \end{cases} \quad (21)$$

where T is chosen as

$$T = [\,T_1\ T_2\ T_3\,] = \begin{bmatrix} 1 & 0 & 1 & 0 & 1 & 0 \\ 0 & 1 & 0 & 1 & 0 & 1 \end{bmatrix}$$

Determining $\ddot{e}$ from equation (13) and taking the time derivative one obtains

$$\ddot{e} = -K_v \dot{e} - K_p e - K_i \int_0^t e(\tau) d\tau + \nu_2 + \quad (22)$$
$$+ L_1 B_0^{-1}\left[(B_0 - B(q))\ddot{q} - C(q,\dot{q})\dot{q} - F_v - g(q)\right] + L_2 \dot{q}$$

$$\frac{d^3 e}{dt^3} = -K_v \ddot{e} - K_p \dot{e} - K_i e + \dot{\nu}_2 + \quad (23)$$
$$+ L_1 B_0^{-1}\left[(B_0 - B(q))\frac{d^3 q}{dt^3} - \dot{B}\ddot{q} - \dot{C}\dot{q} - C\ddot{q} - F_v \ddot{q} - \dot{g}\right] +$$
$$+ \dot{L}_1 B_0^{-1}\left[(B_0 - B(q))\ddot{q} - C\dot{q} - F_v \dot{q} - g\right] + L_2 \ddot{q} + \dot{L}_2 \dot{q}$$

As a consequence $\dot{y}_2$ can be rewritten as

$$\dot{y}_2 = T_1 \dot{e} + T_2 \left[-K_v \dot{e} - K_p e - K_i \int_0^t e(\tau) d\tau + \nu_2 \right] +$$
$$+ T_2 \left\{ L_1 B_0^{-1} [(B_0 - B(q))\ddot{q} - C(q,\dot{q})\dot{q} - F_v - g(q)] + L_2 \dot{q} \right\} +$$
$$+ T_3 \left\{ -K_v \ddot{e} - K_p \dot{e} - K_i e + \dot{\nu}_2 + \right.$$
$$+ L_1 B_0^{-1} \left[(B_0 - B(q)) \frac{d^3 q}{dt^3} - \dot{B}\ddot{q} - \dot{C}\dot{q} - C\ddot{q} - F_v \ddot{q} - \dot{g} \right] \right\} +$$
$$+ T_3 \left\{ \dot{L}_1 B_0^{-1} [(B_0 - B(q))\ddot{q} - C\dot{q} - F_v \dot{q} - g] + L_2 \ddot{q} + \dot{L}_2 \dot{q} \right\}$$

and system (21) become

$$\begin{cases} \dot{y}_1 = y_2 \\ \dot{y}_2 = F(\int e, e, \dot{e}, q, \dot{q}, \ddot{q}, \frac{d^3 q}{dt^3}) + T_3 \dot{\nu}_2 \\ \dot{\nu}_2 = \eta \end{cases} \quad (24)$$

Note that the components F_i of F, even if uncertain, have known bounds $|F_i| < \bar{F}_i$ since the robot manipulator has a limited operational space and the actuators cannot provide unbounded velocities, accelerations and jerks. The control problem can now be viewed as that of steering y_1, y_2 to zero in a finite time in presence of uncertainties and with y_2 not available. Since T_3 is positive definite and diagonal, system (24) in our case can be rewritten component-wise as

$$\begin{cases} \dot{y}_{11} = y_{21} \\ \dot{y}_{12} = y_{22} \\ \dot{y}_{21} = F_1[x, u] + \eta_1 \\ \dot{y}_{22} = F_2[x, u] + \eta_2 \end{cases}$$

i.e. it is composed by two single input single output systems interacting through the term F, but this interaction term is compensated by virtue of the control choice. According to [12], [13], the control signal is chosen as

$$\eta_i = -V_M \alpha \, \text{sign}\left(y_{1i} - \frac{1}{2} y_{1iM}\right)$$

where y_{1iM} is the last singular value of y_{1i} and

$$V_M \geq \left\{ \frac{\bar{F}_i}{\alpha}, \frac{4\bar{F}_i}{3-\alpha} \right\} \quad \alpha =]0, 1]$$

On the basis of the theory of second order sliding mode control [12, 13], it can be claimed that y_1, y_2 converge to zero, so that the relevant output tracks the desired trajectory. Clearly, the discontinuous control η needs to be integrated to obtain signal ν_2 to be included in (12) to determine ν_1. In this way discontinuities are confined to the first derivative of ν_2, and the signal used to control the robot manipulator is continuous. The chattering effect is therefore mitigated, eliminating the drawback which normally limits the applicability of sliding mode control to the robotic context.

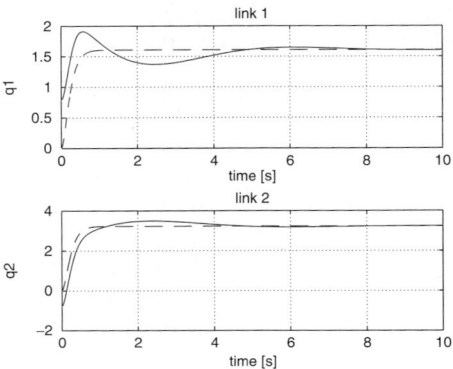

Fig. 9. Trajectory of link 1 and 2 (solid lines) and desired trajectory (dotted lines) using second order sliding mode control.

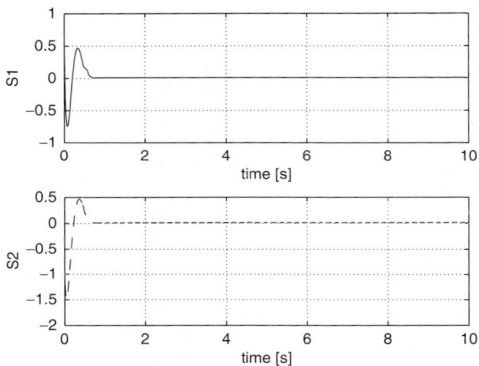

Fig. 10. Tracking error using second order sliding mode control.

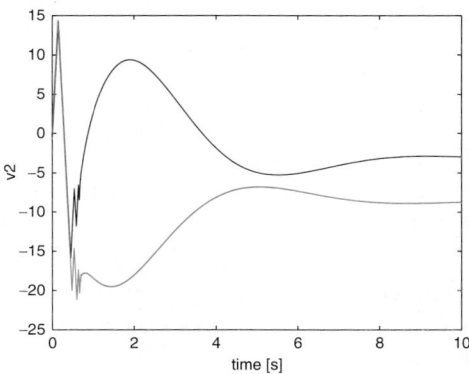

Fig. 11. Second order sliding mode control law.

Figure 9 shows that using second order sliding mode control the controlled output tracks the desired reference trajectory and Figure 10 demonstrates that the tracking error is forced to zero. Moreover, using second order sliding mode the chattering effect is mitigated as shown in Figure 11.

6 Conclusions

In the present paper new sliding mode motion control strategies for robot manipulators are presented. Each first and second order sliding mode control law are introduced and coupled with compensated inverse dynamics method so that the on-line calculation of the time-varying inertia matrix is not needed, providing robustness versus matched uncertainties and disturbances. Moreover, the second order sliding mode control law allows to mitigate chattering problems by confining discontinuities to the derivative of the control law. The proposed control strategies have been tested in simulation with a two link rigid robot manipulator, demonstrating their good performances.

References

1. Abdallah, C., Dawson, D., Dorato, P., Jamshidi, M.: Survey of robust control of rigid robots. *IEEE Contr. Syst. Mag.* **11**(2) (1991) 24–30
2. Kuo, C.Y., Wang, S.: Nonlinear robust industrial robot control. *ASME J. Dyn. Syst. Meas. Control* **11** (1989) 24–30
3. Juang, J.N., Eure, K.W.: *Predictive Feedback and Feedforward Control for Systems with Unknown Disturbance.* NASA/Tm-1998-208744 (1998)
4. Perk, J.S., Han, G.S., Ahn, H.S., Kim, D.H.: Adaptive approaches on the sliding mode control of robot manipulators. *Trans. Control, Autom. and Sys. Eng.* **3**(2) (2001) 15–20
5. Shyu, K.K., Chu., P.H., Shang, L.J.: Control of rigid manipulators via combiantion of adaptive sliding mode control and compensated inverse dynamics approach. *IEE Proc. Control Theory Appl.* **143**(3) (1996) 283–288
6. Adam, M.J.K.: *Basics of robotics: Theory and components of manipulators and robots.* Springer-Verlag, Berlin (1999)
7. Asada, H., Slotine, J.J.E.: *Robot Analysis and Control.* Wiley, New York (1986)
8. Spong, M.W., Lewis, F.L., Abdallah, C.: *Robot Control: Dynamics, Motion Planning, and Analysis.* IEEE Press, New York (1993)
9. Edwards, C., Spurgeon, S.K.: *Sliding Mode Control: Theory and Applications.* Taylor & Francis, U.K. (1998)
10. Utkin, V.I.: *Sliding Modes in Control Optimization.* Springer-Verlag, Berlin (1992)

11. Ferrara, A., Magnani, L.: Motion control of rigid robot manipulators via first and second order sliding modes. *J. Intell. and Robotic Sys.* **48** (2007) 23–26
12. Bartolini, G., Ferrara, A., Usai, E.: Chattering avoidance by second order sliding mode control. *IEEE Trans. Automat. Contr.* **43**(3) (1998) 241–246
13. Bartolini, G., Ferrara, A., Usai, E., Utkin, V.I.: On multi-input chattering-free second-order sliding mode control. *IEEE Trans. Aut. Contr.* **45**(9) (2000) 1711–1717

Model-Based Chemical Compound Formulation

Stefania Bandini, Alessandro Mosca and Matteo Palmonari

Department of Computer Science, Systems and Communication (DISCo),
University of Milano-Bicocca, Milan, Italy
{stefania.bandini,alessandro.mosca,matteo.palmonari}@disco.unimib.it

Summary. Many connections have been established in recent years between Chemistry and Computer Science, and very accurate systems, based on mathematical and physical models, have been suggested for the analysis of chemical substances. However, such a systems suffer from the difficulties of processing large amount of data, and their computational cost grows largely with the chemical and physical complexity of the investigated chemical substances. This prevent such kind of systems from their practical use in many applicative domain, where complex chemical compound are involved. In this paper we proposed a formal model, based on qualitative chemical knowledge, whose aim is to overcome such computational difficulties. The model is aimed at integrating ontological and causal knowledge about chemical compounds and compound transformations. The model allowed the design and the implementation of a system, that is based on the well known Heuristic Search paradigm, devoted to the automatically resolution of chemical formulation problems in the industrial domain of rubber compounds.

1 Introduction

In recent years, many connections between Chemistry and Computer Science have been established in the context of several research areas (e.g. computational representation of atoms and molecules, the storing and searching for data on chemical entities, identification of the relationships between chemical structures and observable behaviors, theoretical elucidation of structures based on the simulation of forces). Researchers in the area of Computational Chemistry sought to develop theoretical and computational methods based on mathematical models for describing and understanding the movement and the function of electrons in molecules, and applied these methods to significant problems of broad chemical interest [1]. Indeed, the term "computational chemistry" is used when a *mathematical* method is sufficiently well developed that it can be automated for implementation on a computer [2–8].

Although such mathematical methods are well-known and there are a number of systems based on them, their computational cost grows largely with the number of electrons [9] and, therefore, with the chemical and physical complexity of the investigated chemical substance. Reasoning on the structural and behavioral change dynamics of chemical compounds (i.e. of chemical substances formed from two or more elements, with a fixed ratio determining the composition) is a hard combinatorial problem, even when a small number of chemical elements are taken into account.

A crucial problem in applied chemistry, in which the physical and chemical complexity of the involved substances can be extremely high is the chemical *Compounding Problem*. The chemical compounding problem consists in the task of generating in an automatic way new complex compound formulations on the basis of a set of desired final behaviors in order to support industrial production processes. The computational limitations of the actual computational chemistry systems suggest that we must rely on different modeling techniques. In other words, as far as the problem of designing and implementing systems that reason and drive transformations of complex chemical substances is concerned, it is a challenging task to overcome the computational intractability of the *quantitative*, mathematical, compound representations.

Now, two questions arise: (1) what does it mean to reason upon chemical compounds taking advantage of a formal model representing *non*-quantitative chemical knowledge (e.g. ontological and causal expert knowledge)? (2) What is the kind of formal representation that allows to automatically transform the formulation of compounds with respect to specific engineering objectives, still preserving their identity as particular chemical compounds (e.g. drug compounds or tyre rubber compounds)?

In the effort to answer these questions, our research led to an epistemological investigation of the qualitative knowledge characterizing chemical compounding problems, and to the definition of a formal representation of that knowledge. A definition of the compound formulation problem (or, compounding problem), stretching some characteristics that have a direct impact on its computational tractability, is given in the next section. The section contains also a brief review of two research areas that are strongly related to the present work. Section 3 provides an introduction to the kinds of knowledge that are involved into the chemical formulation activity, with a focus on the formal ontological axioms defining integrity conditions for the chemical rubber compounds. Section 4 concerns the representation of causal knowledge together with its integration with ontological knowledge into the state space. Concluding remarks end the paper.

2 A Computational Perspective on the Compounding Problem

In industrial domains, the compounding problem, whether for agrochemicals, pharmaceutical, or speciality chemicals areas, deals with the possibility of modifying the formulation of some existing *chemical compound*, in order to gain new compound formulations showing final desired performances. New desired performances for a compound can be originated by specific marketing commitments, design and cost requirements, or by constraints induced by the production process. Desired final performances are always expressed in terms of performance variations with respect to the preexistent compound (e.g. the increase of the *Rolling Resistance* of a tire, together with a reduction of the *Wet Handling*, and the maintenance of all the remaining performances).

Performances are observable behaviors of the chemical compound that are evaluated by means of a specific set of laboratory tests (in which compound behaviors are evaluated in isolation) and environmental tests (in which compound behaviors are evaluated into the final using environment). The system should find a compound structure, whose associated behaviors fit the requirements.

Compounding problem never coincides with a *ex novo* generation of compounds: a problem begins with some form of product specification, together with the specification of new desired performances, and ends with one or more new product specifications that meet the requirements. Therefore, the problem concerns the discovery of a suitable set of transformations that can be applied to the compound formulation in order to obtain a new compound with a specific set of desired performances.

The hard combinatorial nature of the problem essentially depends on the complexity caused by the effects of the application of transformations in compound structures. According to an *holistic* perspective, a transformation in quality/quantity of the elements of a chemical compound (no matter how massive or tiny it is) implies a non-uniform rearrangement of *all* the values of its associated properties, and this makes really problematic to find good a sequence of transformations pointing to the final desired compound. For example, it is usual that the effects of a structural transformation return a compound performing only some of the requirements, and failing on the others; obviously, this situation needs to find further transformations to bridge the gap among the modified object and the desired goal, but there is no guarantee that such transformations exist. This characteristic is common to a number of formulation, design, configuration, or planning problems dealing with entities (e.g. chemical mixtures, blends, and compounds, industrial plant, car engines, rescue or process plans) whose inner structures can be articulated in *parts*.

Generally speaking, transformations on parts produce not uniform transformations of all the wholes' properties, and this makes the search of a sequence of transformations pointing to the solution really problematic (for example, it is usual that the effects of a transformation return a compound performing only some of the requirements, and failing on the others; this situation needs to find further transformations to bring the gap among the modified object and the desired goal, but there is no guarantee that such transformations exist). It is therefore a challenging objective from an AI computational perspective to discover, represent, and exploit domain-specific knowledge with the aim of reducing such an explicit combinatorial complexity.

Since the application of a structural transformation must be evaluated with respect to all the behaviors associated to the compound, a solution to a compounding problem is necessarily a *compromise solution*. In other words, in compounding the *optimum* does not exist. Specific ranges of tolerance have to be defined together with compounding requirements in order to increase the possibilities of bringing a solution. Therefore, the existence of a solution to an instance of the compounding problem is guaranteed only by the accuracy of the compounding expert requirements: if we are looking for an "extravagant" set of performances for chemical compounds devoted to a specific marketplace, there are no guarantees that the compounding will succeed.

2.1 Related Research Areas: Configuration and Planning

There are at least two well known research areas in Artificial Intelligence that are strongly related to the proposed definition of the compounding problem: the area of *Automated Configuration* and that of *Automated Planning*. Configuration and Planning are very close to Compounding, although they have characteristics that does not always perfectly match with our problem.

Configuration can be defined as the design of an individual product by using a set of pre-defined components or component types. Configuration takes into account a set of well-defined restrictions on how the components can be combined together [10]. Planning was emerged as a specific sub-field of Artificial Intelligence with the seminal work of Fikes and Nilsson [11] on the Stanford Research Institute Problem Solver (STRIPS). Newell and Simon's work on GPS [12], Green's QA3 [13, 14] and McCarthy's situation calculus [15] helped to define the classic planning problem and many of their assumptions still influence planning research today. Very briefly, a planning problem is described by a collection of actions, each characterized by their *pre*-conditions (what must be true in order for the action to be executed) and their *post*-conditions (which describe the effect of execution of the action), an initial state of the world and a description of the goals to be achieved. The problem is solved by finding actions that will transform the given initial state into a state satisfying the given goals

Traditional researches in automated configuration and planning have relied on simple and relatively unstructured models of the problem and have placed the emphasis on the development of more efficient algorithms and powerful heuristic control methods. Nevertheless, [16, 17] and others, have recognized that a model typically contains hidden structure that can be exploited by a planner and, under this assumption, several research communities have focused on exploring more articulated modeling choices with the extent of expediting the solution search (see, for example, [18, 19]). Closely related to this perspective are: (i) the logical approach suggested by Kauz and Selman, based on the notion of *Satisfiability* for propositional formulas [20]; (ii) the Lifschitz's approach [21–23], grounded on the *Answer Set Program-*

ming paradigm and related to the researches on *Stable Models* of Subrahmanian and Zaniolo [24]; (iii) the *Model-Checking* approach of Giunchiglia and Traverso [25, 26]; and (iv) the work of Eiter, Faber, Leone, Pfeifer, and Polleres on the DLV^k system [27].

It is beyond the scope of this article to furnish an extensive analysis of the relationships between the compounding problem, on one hand, and configuration and planning problems, on the other (further details can be found in [28]). Nevertheless, we are convinced that most of researches on configuration and planning can be grouped together under the following statements, that actually represent also the background of our work: (i) *the more human knowledge and expertise are embedded in the domain* model, *the less discovery has to be made by the planner*, and (ii) *the correctness of the reasoning system fundamentally depends on the correctness of the* model.

3 Compounding Knowledge

Once a qualitative perspective has been assumed on the compounding problem, two main kinds of knowledge must be considered and integrated: *ontological* and *causal* knowledge. With ontological knowledge we refer to the knowledge that specifies what entities have to be considered as admissible compounds' structures and behaviors (establishing their "integrity conditions" with respect to a domain of interest). This knowledge concerns entities within different perspectives (structural and behavioral), and it guarantees that transformations applied to those entities preserve their ontological integrity. With causal knowledge we refer to knowledge mapping compound transformations at the structural level (i.e. on the compound formulation) to transformations at the behavioral one (i.e. on the tested performances). Causal knowledge in compounding allows to expect the changes on compound behaviors, on the basis of transformations of its chemical formulation.

As for the chemical engineering domain, the automated discovery of ontological and causal knowledge is a problem too hard due to the computational complexity issues (see Section 1). Nevertheless, this knowledge already lives (expressed in *qualitative* terms) in the expert compounding practices and communities, and it can be elicited and formally represented by means of knowledge engineering techniques. Expert knowledge on compounding is often *not* immediately quantifiable, *not* directly math-based, and *not* microscopic; this knowledge has been worked out in chemical industrial context during all the Twentieth Century, producing a number of results that have lead to the success of several Chemical Engineering applications [29, 30].

In this paper, we propose a knowledge model to tackle a compounding problem that is formally based on: (i) a *description logic* (DL) knowledge base (in the language $\mathcal{SHOIN}$), describing ontological representations of compounds' structures and behaviors; (ii) a causal knowledge formal representation, coded into *morphisms* that map structural and behavioral compound representations, at one hand, and structural and behavioral transformations, at the other.

3.1 Mereological Axioms for Compounding

A rubber compound is usually viewed by chemical engineers as a "recipe" or as *a blend of atomic components composed in various proportions*. Atomic components are aggregated into several "systems", in accordance with the *functional role* they have to perform within the compound. Therefore, along this perspective, chemical compounds can be observed as "aggregate objects", a notion for which a wide philosophical literature and different mereological investigations exist (e.g. see [31–35]).

Compounding problems in industry are characterized by the presence of many different formulations for a compound composition; nevertheless, according to the final use of a compound, it is possible to identify a set of necessary boundaries within which all the admissible compound formulations must rely. These boundaries have been represented by means of a formal theory, written in logical terms. Logically speaking, the models of this theory are all those compound formulations that do not cross the chosen ontological boundaries and consequently do not violate ontological integrity constraints.

The model of the ontological knowledge we propose is grounded on a "composed" part-of relation $\prec$, in the sense of Sattler's taxonomy [36] (i.e. a part-of relation that is both *integral* and *functional*). In particular, $\prec$ is a *finite*, *irreflexive*, *asymmetric*, and *intransitive* binary relation. The assumption on functionality (and therefore, on the intransitivity of the part-of relation) is justified by the specific domain we are interested in: all the chemical entities into a compound play a specific functional role with respect to its constitution. DLs are a logic-based formalisms for the representation and reasoning about conceptual knowledge.

In DL, *concepts* are used to describe classes of individuals sharing common properties, and *roles* are used to represent binary relations.

Therefore, let $\prec$ be a primitive role of a DL language standing for "is a functional part of"; it is also useful to introduce the inverse role "has a functional part of" (or $\succ$) as $\succ \doteq \prec^{-1}$. Since functional *part-of* relation is "integral", for an entity to be part of another simply means that the entity must satisfy integrity conditions associated to the relation. Functional part-of is constrained to hold only among entities of a certain predefined kind. Here, the integrity conditions are simply expressed by means of different concept names and value restrictions of the form $\forall R.C$.

As far as compounding is concerned, we know that whatever is the compound of interest, its direct parts must be of type System, and that whatever is the system, its direct parts must be of type GroundElement. This means to assume in our ontological knowledge representation the following General Concept Inclusion (CGI) axioms:

(I1) $\quad\quad\quad\quad\quad\quad\quad\quad$ System $\sqsubseteq \forall \prec$.Compound
(I2) $\quad\quad\quad\quad\quad\quad\quad\quad$ GroundElement $\sqsubseteq \forall \prec$.System

Intuitively, the axioms say that a system may be only a part of a compound and a ground element may be only a part of a system. A so rigid hierarchy is coherent with the abstract representation of compounds as in real problem solving contexts (e.g. [30, 37]).

Ground Elements. Ground elements are the "atomic" entities living in the compounding domain (i.e. they have no parts). It seems reasonable to think that ground elements represent "minimal manageable quantities" of chemical substances: each of them represent a *fixed quantity* of a given substance, characterized by different chemical and physical properties. In concrete domains, ground elements are obviously chosen in accordance with chemical and physical properties.

Attributes. Attributes and properties of a ground element may be represented by introducing in the language specific DL roles, named *functional roles*. A role R is said to be a functional role if and only if $\{(a,b),(a,c)\} \subseteq$ R implies $b = c$. Each concept is characterized by a suitable set of those functional roles. We indicated with NumericalValue a generic filler for functional roles (instances of this generic concept may be integer or real numbers, in accordance with the employed physical measurements). One can formally represent attributes of ground elements by instantiating the following axiom schema:

$$\text{GroundElement} \sqsubseteq f_1.\text{NumericalValue} \sqcap \cdots \sqcap f_n.\text{NumericalValue}$$

where $f_1, \ldots, f_n$ are n generic functional roles.

Exclusive Parts. Close to the axioms (I1) and (I2), it is useful to represent also exclusive relationships among these concepts. In general, a part is said to be "exclusive" if and only if there exists *at most* one whole containing it. Such feature expresses a kind of interdependence among whole and part. In compounding, the introduction of expressions about exclusivity of parts, forces models in having ground elements of certain kind (e.g. NaturalRubber, CarbonBlack) only as constituent parts of specific systems. Exclusive parts are formally represented in DL as particular instances of number restrictions. Number restrictions are concepts of the form $(\geq nr.C)$ (at-least restriction) or $(\leq nr.C)$ (at-most restrictions), where n is a non-negative integer, r is a role, C is a concept. In order to represent exclusive parts of a whole, it is possible to specialize number restrictions by means of the equality symbol $=$, stating that each GroundElement (System) is part-of exactly one System (Compound), as follows:

$$\text{GroundElement} \sqsubseteq (= 1 \prec .\text{System})$$
$$\text{System} \sqsubseteq (= 1 \prec .\text{Compound})$$

Upper and Lower Bounds. Systems may contain ground elements with different quantities. Ontological compounding knowledge provides *upper and lower bounds* of quantity of a ground element a system may contain. To represent the admissible range of quantity of an element is necessary to preserve the integrity of the compound during the formulation activity. The number restriction constructor allows to impose different range of quantities of an element: the $(\geq nr.C)$ and $(\leq nr.C)$ concept constructors can be combined in order to set upper and lower bounds, as follows:

$$\texttt{System} \sqsubseteq (\geq 1 \succ) \sqcap (\leq n \succ)$$

where n is an integer that will be instantiated according to concrete applications. Technically, the quantity of a substance in a system corresponds to the cardinality of the set of *has-part-of*-fillers of this system. From (I2) one can say that these fillers are all from the category of `GroundElement`. A system *must* have *at least* one ground element as its part, that is, a lower bound not inferior to 1. If we consider the relationships between `Compound` and `System`, the same situation arises: compounds may contain a number of systems, but *at least* one system must be contained. Therefore, ground elements must be considered as *essential parts* of systems, and systems as *essential parts* of compounds [38].

Atomicity. Atomicity immediately follows from the introduction of the ground elements. We resort to translate atomicity into the non-existence of fillers of the part-of relation in correspondence to ground elements. On the other hand, we can state that compounds cannot be part of any other entity.

$$\texttt{GroundElement} \sqsubseteq \forall \succ .\bot$$
$$\texttt{Compound} \sqsubseteq \forall \prec .\bot$$

In the rest of the paper, we present "tread tire compounds" as a paradigmatic example of chemical compound. The formulae we will introduce have to be understood as a specialization of the mereological theory introduced so far and, as a consequence, the involved concepts respect the ontological constraints.

3.2 The Case of Rubber Compounds for Tread Tire

The "tread" is the part of the tire in contact with the road. The profile and rubber compound are chosen on the basis of the use of the tire. The following logical formulae are a fragment of our mereology for compounding; in particular, the introduced formulae show some key elements of the ontological theory for tread rubber compound formulation (in what follows we grouped together entities that agree on the same mereological level). The set of axioms guarantees that if a model of these statements exists, then this model describes a compound for the production of tread tire in the industrial field of interest.

Model-Based Chemical Compound Formulation 421

$$\text{TreadCompound} \equiv \text{Compound} \sqcap (= 1 \succ .\text{PolymericMatrix})$$
$$\sqcap (= 1 \succ .\text{Vulcanization})$$
$$\sqcap (= 1 \succ .\text{ProcessAid}) \sqcap (= 1 \succ .\text{Antidegradant})$$
$$\sqcap (= 1 \succ .\text{ReinforcingFiller})$$
$$\sqcap ((\geq 0 \succ .\text{Softener}) \sqcap (\leq 1 \succ .\text{Softener}))$$

It is standard that a rubber compound devoted to tread tire production is made of at least five essential systems [6, 37, 39, 40]: **(1)** the `PolymericMatrix` is the system that contains *polymers* (e.g. Natural rubber, Butadiene rubber, Styrenebutadiene rubber) and it plays a decisive functional role in tread compound. The final tread compound formulation will contain a suitable subset of discrete amounts of those polymers. **(2)** The `Vulcanization` system provides suitable chemicals for the compound vulcanization process. The system is made of "vulcanization chemicals" (e.g. `Sulphur`, `Peroxides`, `Urethane`), "vulcanization accelerators" (e.g. `Guanidines`, `Thiazoles`), "activators" (e.g. `MetalOxides`, `FattyAcids`, `SaltFattyAcids`), and "vulcanization inhibitors" (chemicals based on phthalimide sulfenamides). **(3)** The `ProcessAid` system, whose aim is to enable a rubber compound to be fabricated with less energy, is made of "peptizers" (e.g. `Renacit`) and "plasticizers" (e.g. `oil`); **(4)** the `Antidegradant` system is made of "antioxidants" and "antiozonats" that have been developed to inhibit the action of oxygen and ozone. Finally, **(5)** the `ReinforcingFiller` is defined as the ability of fillers to increase the stiffness of unvulcanized compounds, and the reinforcement effect of a filler shows up especially in its ability to change the viscosity of a compound; reinforcing fillers are `CarbonBlack` and `Silica`. Further systems may be present in a tread tire compound, such as the `Softener`, the `Extenders`, and the `Tackifier` systems, depending on the application context.

$$\text{PolymericMatrix} \equiv \text{System} \sqcap (= 100 \succ .(\text{NaturalRubber} \sqcup \text{ButadieneRubber}))$$
$$\text{Vulcanization} \equiv \text{System} \sqcap ((\geq 1 \succ .\text{Sulphur}) \sqcap (\leq n \succ .\text{Sulphur}))$$
$$\sqcap ((\geq 1 \succ .hasFamilyName.\text{Accellerant})$$
$$\sqcap (\leq m \prec .hasFamilyName.\text{Accellerant}))$$
$$\sqcap ((\geq 2 \succ .\text{ZincOxide}) \sqcap (\leq p \succ .\text{ZincOxide}))$$
$$\sqcap ((\geq 2 \succ .\text{StearicAcid}) \sqcap (\leq p \succ .\text{StearicAcid}))$$
$$\text{ProcessAid} \equiv \text{System} \sqcap ((\geq 1 \succ .hasFamilyName.\text{Peptizer})$$
$$\sqcap (\leq z \succ .hasFamilyName.\text{Peptizer}))$$
$$\sqcap (= z \succ .hasFamilyName.\text{Plasticizer})$$
$$\text{ReinforcingFiller} \equiv \text{System} \sqcap ((\geq 1 \succ .\text{CarbonBlack}) \sqcap (\leq n \succ .\text{CarbonBlack}))$$
$$\sqcap ((\geq 1 \succ .\text{Silica}) \sqcap (\leq \frac{n}{2} \prec .\text{Silica}))$$

The `PolymericMatrix`, the `Vulcanization`, the `ProcessAid`, and the `ReinforcingFiller` are systems. The polymeric matrix has 100 parts as a blend of natural and synthetic rubber or, alternatively, 100 parts of natural or synthetic rubber alone. Parts of the vulcanization system are the Sulphur, the Oxide Zinc and the Stearic Acid in a predefined *quantity*. The possibility of selecting parts by their membership to specific chemical families is exploited in the definition of the vulcanization system. A vulcanization system contains some quantity of a (not further specified) element in the family of the `Accellerant`, while a process aid system takes part from the `Peptizer` and `Plasticizer` family. A reinforcing filler contains Carbon Black and Silica in a predefined *ratio*. The possibility of representing two quantities in a certain ratio is crucial in compounding: the presence of a ground element often asks for the presence of another one (e.g. the "activator-activated" couples of chemicals)[1].

$$\begin{aligned}
\text{CarbonBlack} &\equiv \text{GroundElement} \sqcap (= 1 \prec .\text{ReinforcingFiller}) \\
&\sqcap hasSurfaceArea.\text{NumericalValue} \\
&\sqcap hasPorosity.\text{NumericalValue} \\
&\sqcap hasTortuosity.\text{NumericalValue} \\
\text{Renacit} &\equiv \text{GroundElement} \sqcap (= 1 \prec .\text{ProcessAid}) \\
&\sqcap hasDensity.\text{NumericalValue} \sqcap hasFamilyName.\text{Peptizer} \\
\text{NaturalRubber} &\equiv \text{GroundElement} \sqcap (= 1 \prec .\text{PolymericMatrix}) \sqcap hasStructure.\text{CIS} \\
&\sqcap hasMolecularWeight.\text{NumericalValue} \sqcap hasFamilyName.\text{Polymer} \\
\text{StyrenebutadieneRubber} &\equiv \text{GroundElement} \sqcap (= 1 \prec .\text{PolymericMatrix}) \\
&\sqcap hasMolecularWeight.\text{Value} \sqcap hasFamilyName.\text{Polymer} \\
\text{ButadieneRubber} &\equiv \text{GroundElement} \sqcap (= 1 \prec .\text{PolymericMatrix}) \\
&\sqcap hasStructure.\text{CIS} \sqcap hasFamilyName.\text{Polymer} \\
&\sqcap hasMolecularWeight.\text{Value}
\end{aligned}$$

Carbon Black, Renacit, Natural rubber, Butadiene rubber, and *Styrene butadiene rubber* are ground elements and exclusive parts of the reinforcing filler system, the process aid system, and the polymeric matrix system, respectively. Carbon black is characterized by a specific value of "surface area", and by a specific "microstructure" (represented in term of its porosity and tortuosity). The Renacit is characterized by a specific value of "density", and by its membership to the family of Peptizers, while the Natural, Butadiene and the Styrene butadiene rubber by a "CIS" configuration, a specific molecular density and by their membership to the family of Polymers.

4 Causal and Ontological Knowledge into the State Space

As mentioned in the introduction, causal knowledge provides necessary information to compute (and, forecast) the application of compound structural transformations, on one side, and the effects these structural transformations have in behavioral terms (e.g. it is known that an increase of the amount

---

[1] Since the syntax of the $\mathcal{SHOIN}$ description logic does not allow to express individual variables, the m, n, p, z symbols need to be instantiated with appropriate integers once the axioms are taken to represent ontological knowledge in a specific compounding domain.

of Silica worsens abrasive and resistance behaviors of a tread rubber compound), on the other. Causal knowledge has been formally represented within the search paradigm [41–43], by means of a set of *transitions* of the state space, and a set of *morphisms* linking the different dimensions of which the space is made.

More precisely, the state space has been finally defined as the *product* of three different labeled transition systems, corresponding to three different levels of abstraction. The first one of these systems represents compound formulations: states are logical descriptions of concrete compound formulations, as introduced in Section 3.2, while transitions are transformations of these formulations (i.e. discrete *increases*, discrete *reductions*, and substitutions of ground elements). Transitions of this system can be formally represented as a set of functions from compound formulations to compound formulations, with (i) domain dependent *pre*-conditions, listing prerequisites that must be satisfied by the compound mereological structure, and (ii) *post*-conditions, which specify precisely how the structure must be transformed. For example, in what follows we sketch the definition of a transition representing an instance of the substitution class:

$$f_{cis_+}^{\mathbf{NR}}(r) \to r',$$

The function $f_{cis_+}^{\mathbf{NR}}$ returns a compound r', that is equal to r, except for the natural rubber of r, that has been substituted with an alternative natural rubber with a greater *cis* value (where "*cis*" refers to a basic property of polymers coming from the specific geometrical atoms arrangement). The satisfiability of the pre-conditions and the consistency of the application of the recipe transformations essentially depend on the mereological structure of the involved compound. In particular, a quantity increase of a *part e* cannot be applied to a given compound: (i) if some pre-conditions on its applicability are not satisfied (observe that these integrity constraints, rising from the semantics of the logical formulae introduced in Section 3.2, are imposed in order to discard the computation of usefulness compounds during the formulation activity), and (ii) if the effects of this application produce a new compound containing an amount of e that turn out to be outside its admitted range.

The second and the third labeled system represent the synthesis of two levels of behavioral evaluation of the compound. On one hand, a compound is evaluated by means of a specific set of mechanical laboratory tests that return *quantifiable* properties. On the other one, the final performances evaluation is provided by means of tests studying the interactions of the compound within its application environment and under different conditions (the final performances of a tread rubber compound, as an example, are evaluated under wet and dry road conditions, irregular terrains, extreme temperatures, and so on). The formulation expert knowledge has specific heuristics to trace back the *qualitative* results of these tests to the behaviors of a single compound or of an identified aggregate of compounds. Qualitative information about final performances of a compound can be thus inferred and computationally managed, once a suitable metric has been provided with the help of expert chemical engineers.

The knowledge on how a chemical compound need to be modified in order to obtain a new compound with final desired properties essentially concerns the applicability of transformations at the structural level, and the effects they have on the associated behavioral levels. Once the three labeled transition systems have been defined, the problem solving knowledge can thus be understood and formally represented by means of a couple of *morphisms*, mapping states to states and transitions to transitions of the different systems.

A morphism $\Gamma \to \Gamma'$ between transition systems can be intuitively introduced as a pair (σ, λ), where σ is a function on states, preserving initial states, and λ is a partial function λ on the transition labels. The morphism maps a transition of Γ to a transition of Γ', whenever this makes sense; in other words, if (p, α, q) is a transition in Γ then $(\sigma(p), \lambda(\alpha), \sigma(q))$ is a transition in Γ' provided that $\lambda(\alpha)$ is defined.

The role we assign to morphisms here is strongly connected to the task of relating transformations at one representation level with transformations at the other ones. Morphisms carry expert causal knowledge linking structural transformations on the compound to behavioral ones and, therefore, they allow to forecast behavioral changes of a chemical compound during the searching activity of new formulation (e.g. morphisms represent the fact that an increase of the amount of Silica worsens abrasive and resistance behaviors, by appropriately mapping the structural transformation "Silica Increase" to "Abrasive Decrease" and "Resistance Decrease" behavioral transformations). Observe that the definition of these compounding morphisms depends exactly on the acquired expert problem-solving causal knowledge.

We omit here the formal definition of the product of labeled transition systems [28], but we furnish a diagrammatic representation of it in the figure below. Figure 1 shows the structure we obtained by connecting the three systems we mentioned above, and the resulting state space in which our compounding system operates.

The states $s = \langle c, l, h \rangle$ and $s' = \langle c', l', h' \rangle$, for which the three dimensions are represented in Figure 1, are elements of the state space (c stands for *c*ompound structure, l for *l*ow-level behaviors and h for *h*igh-level behaviors or compound performances). τ, τ' are compounding morphisms such that $\tau(s_c) = s_l$ and $\tau'(s_l) = s_h$; these morphisms represent that a specific compound formulation s_c is associated to compound behaviors s_l and compound performances s_h. Note that the association is plainly given at the beginning of the compounding problem, where a compound structure together with its associated behavioral evaluations is given as input of the problem. On the other hand, the association must be computed during the solution searching process, on the basis of the structural transformations the initial compound is subject step by step. In fact, the existence of morphisms representing causal knowledge in compounding is mandatory for the construction of new states and transitions in the structure of the state space.

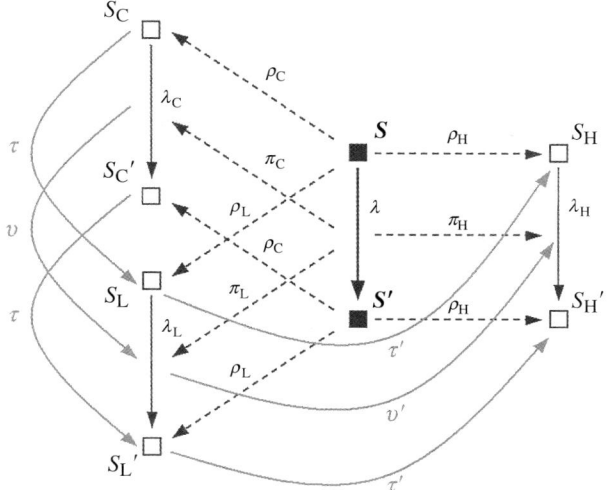

Fig. 1. Diagrammatic representation of the state space.

In Figure 1, an illustration of the compounding morphisms is given, with respect to projection morphisms of the product. The application of the transition λ to s, returning a new state s', is generated first of all by the application of a compound structural modification:

$$\lambda_C \colon S_C \to S_C$$

such that $\lambda_C(s_c) = s'_c$. The application of λ_C leads to a partial state $\langle c', l, h \rangle$, that it is not well defined, because it does not respect the constraints coming from causal knowledge and coded into the inter-dimensional morphisms. Therefore, in order to obtain a new correct state s', representing a feasible solution of the compounding problem, the transformations associated to λ_C have to be applied in the remanent behavioral dimensions. We can say that (s, λ_i, s') is a transition of the state space, written,

$$(s, \lambda, s') \in \to$$

if and only if $s' = \langle c', l', h' \rangle$, and $\tau(s'_c) = s'_l$ and $\tau'(s'_l) = s'_h$. Given by the introduced morphisms, the state components l' and h' are obtained by mapping the transition λ_C to transitions λ_L and λ_H.

5 A System for the Rubber Compound Formulation

Heuristic search algorithms occupy a fundamental place among all the artificial intelligence problem solving methods; these algorithms explore a solution space, in order to find optimal solutions to a given problem [42]. They require a representation of (1) a *state space*, and (2) the choice of a *search algorithm*, possibly relying on good heuristic functions; moreover, their efficiency strongly depends on the involved formal representations.

The model we introduced combines ontological and causal knowledge in order to compute feasible solutions to the compounding problem. In this direction, a *compounding problem* is an instance of the state space introduced in Section 4, together with an initial state and a goal state. In the case of chemical compound formulation, the initial state would be any triple $\langle c, l, h \rangle$, provided that $\tau(c) = l$ and $\tau'(l) = h$. The *goal state* is partially specified in terms of required compound behaviors and performances: no information about the compound formulation performing these behaviors and performances is available as a component of the goal state.

From our computational perspective, if no ontological information on the states had been provided, every state in the state space would be generated and explored by the search algorithm as a feasible solution of the problem. On the contrary, with the support of the ontological representation, a state s' representing a compound formulation that resolves to be ontologically inconsistent (i.e. a formulation that is inconsistent with the ontological axioms of Section 3) is discarded by the system (i.e. pruned from the search tree). Since efficiency constraints do not allow to exploit an automated DL reasoner for checking the consistency of a compound formulation with respect to the axioms, compound formulations have been coded in an object-oriented data structure and the ontological constraints have been coded as pre- and post-conditions of the transition operators.

The proposed knowledge model, not only provides a sound representational framework for the state space, but it also allows to reduce the complexity of the solution space, exploiting the integration of ontological and causal knowledge. In fact, following both the constraints coming from the ontological representation of the chemical compounds, and the mapping between structural and behavioral transformations, the expansion of the search tree is exempted from computing useless ontologically inconsistent chemical compounds. In other words, all the possible expansions for the tree must respect the ontological consistency requirements of the involved compounds (e.g. if the given compounding problem concerns rubber compounds for tyre industry, a chewing gum must not be computed as "feasible solution" in the search tree). This reduces the combinatory of the searching process and minimize the computational effort of the implemented system.

The above computational model has been already exploited in solving the Chemical Formulation Problem in the domain of "rubber compound" production for tire industry [28] (this work has been part of the larger project "P-Truck", made in collaboration with the Business Unit Truck of Pirelli Tire S.p.A. [44]). In this context, a specific experimental campaign has been devised and encouraging experimental results have been obtained from the application of several search algorithms (namely A*, IDA*, Iterative Expansion and Branch and Bound) to a state space defined and implemented according to the present knowledge model [45]. The IDA* algorithm has proved to be suitable for solving this kind of problem and, actually, more efficient and faster than the other experimented techniques. The results obtained by means of this algorithm have actually enabled the construction of new and performing rubber compounds. In particular, an automatic system has been developed and successfully tested on a significant number of prototypical chemical compounding problems (e.g. the problem of increasing the *Tread Tear Resistance* or, in a slightly more elaborate case, the problem of increasing the *Rolling Resistance*, together with a reduction of the *Wet Handling*, and the maintenance of all the remaining performances).

6 Concluding Remarks

Our research has been addressed to exploit a knowledge model in order to design and implement a system based on the Search paradigm. The system is devoted to perform searching in the chemical engineering area, improving its efficiency by suitable knowledge-based heuristics. This means that the integration between ontological and causal knowledge into our model produces immediate effects on the expansion rate of the search tree, with a considerable reduction in the time and space consumption for the system.

Recently, we are also engaged in investigating the use of Genetic Algorithms (GAs) [46, 47] to solve this kind of formulation problem. Compared to the other techniques that have been presented in this paper, GAs have the advantage that they enable to navigate even huge state spaces in an intelligent and efficient way, by means of a set of stochastic operators based on the Darwinian principles of biological evolution applied to a population of potential solutions. Compared to deterministic algorithms such as A*, IDA*, Iterative Expansion and Branch and Bound, GAs usually don't consider a large number of possible solutions, which are cut off the search process by means of a selection strategy which emulates natural selection. Several possibilities exist to apply GAs on the compounding problem, depending on which structures are chosen as the potential solutions to be evolved (or *individuals*, according to the GAs terminology). The use of Genetic Algorithms for the Compounding Problem is also motivated by the fact that classical AI algorithms generally work on decision tree structures and cut off the search process some subtrees, depending on some conditions. The eventuality that one of that subtrees contains one optimal solution is not remote, especially when large search spaces are considered. Working with a population of potential solutions, and being based on stochastic operators, GAs should enable an intelligent exploration of larger regions of the search space. In other words, the advantage of GAs for formulation problems using large quantities of data should not only be a lower computational effort, but also a higher quality of the solutions found.

References

1. Grant, G., Richards, W.: *Computational Chemistry.* Volume 29 of Oxford Chemistry Primers. Oxford University Press (1995)
2. Cohen-Tannoudji, C., Diu, B., Laloe, F.: *Quantum Mechanics* Volume I & II. John Wiley & Sons (1977)
3. MacQuarrie, D.: *Quantum Chemistry.* Prentice Hall (1983)
4. Hammond, B., Lester, W., Reynolds, P.: *Monte Carlo Methods in Ab Initio Quantum Chemistry.* World Scientific (1994)
5. Parr, R., Yang, W.: *Density Functional Theory of Atoms and Molecules.* Oxford University Press (1989)
6. Hoffmann, W.: *Rubber Technology Handbook.* Oxford University Press, New York (1989)
7. Burkert, U., Allinger, N.: *Molecular Mechanics.* American Chemical Society (1982)
8. Schlick, T.: *Molecular Modeling and Simulation.* Springer Verlag (2002)
9. Young, D.: *Computational Chemistry : A Practical Guide for Applying Techniques to Real World Problems.* Wiley-Interscience (2001)
10. Mittal, S., Frayman, F.: Towards a generic model of configuration tasks. In: *Proc. of the 11th IJCAI*, Detroit, MI (1989) 1395–1401
11. Fikes, R., Nilsson, N.J.: Strips: A new approach to the application of theorem proving to problem solving. *Artificial Intelligence* **2** (1971) 189–208
12. Newell, A., Simon, H.A.: Gps, a program that simulates human thought. In Feigenbaum, E.A., Feldman, J., eds.: *Computers and Thought.* McGraw-Hill (1963) 279–293
13. Green, C.C.: Theorem proving by resolution as a basis for question-answering systems. In Meltzer, Michie, eds.: *Machine Intelligence* 4. Edinburgh University Press, Edinburgh (1969)
14. Green, C.C.: Application of theorem proving to problem solving. In: *IJCAI1.* (1969) 219–239
15. McCarthy, J., Hayes, P.J.: Some philosophical problems from the standpoint of artificial intelligence. In Meltzer, B., Michie, D., eds.: *Machine Intelligence* 4. Edinburgh University Press (1969) 463–502
16. Fox, M., Long, D.: The automatic inference of state invariants in tim. *Journal of AI Research* **9** (1998) 367–421
17. Gerevini, A., Schubert, L.: Inferring state constraints for domain-independent planning. In: *AAAI '98/IAAI '98: Proceedings of the Fifteenth National/Tenth Conference on Artificial Intelligence/Innovative Applications of Artificial intelligence*, Menlo Park, CA, USA, American Association for Artificial Intelligence (1998) 905–912
18. Borrett, J.E., Tsang, E.P.K.: A context for constraint satisfaction problem formulation selection. *Constraints* **6** (2001) 299–327
19. Westfold, S., Smith, D.: *Synthesis of efficient constraint satisfaction programs* (1998)
20. Kautz, H., Selman, B.: Planning as satisfiability. In: *ECAI '92: Proceedings of the 10th European conference on Artificial intelligence*, New York, NY, USA, John Wiley & Sons, Inc. (1992) 359–363
21. Lifschitz, V.: Answer set programming and plan generation. *Artificial Intelligence* **138** (2002) 39–54

22. Lifschitz, V.: Answer set planning. In: *ICLP*. (1999) 23–37
23. Lifschitz, V., Turner, H.: Representing transition systems by logic programs. In: *LPNMR*. (1999) 92–106
24. Subrahmanian, V.S., Zaniolo, C.: Relating stable models and ai planning domains. In: *ICLP*. (1995) 233–247
25. Giunchiglia, F., Traverso, P.: Planning as model checking. In: *ECP '99: Proceedings of the 5th European Conference on Planning*, London, UK, Springer-Verlag (2000) 1–20
26. Spalazzi, L., Traverso, P.: A dynamic logic for acting, sensing, and planning. *Journal of Logic Computation* **10** (2000) 787–821
27. Eiter, T., Faber, W., Leone, N., Pfeifer, G., Polleres, A.: A logic programming approach to knowledge-state planning, ii: the dlvk system. *Artificial Intelligence* **144** (2003) 157–211
28. Mosca, A.: A theoretical and computational inquiry into the Compounding Problem. Ph.D. thesis, Department of Computer Science, Systems, and Communication - University of Milano-Bicocca, Italy (2005)
29. Himmelblau, D.M., Riggs, J.B.: *Basic Principles and Calculations in Chemical Engineering*. 7 edn. Prentice Hall Professional Technical Reference (2003)
30. Duncan, T.M., Reimer, J.A.: *Chemical Engineering Design and Analysis, An introduction*. Cambridge Series in Chemical Engineering. Cambridge University Press (1998)
31. Fine, K.: Compounds and aggregates. *Nous* **28** (1992) 137–158
32. Husserl, E.: Logische Untersuchungen. Zweiter Band. Untersuchungen zur Phnomenologie und Theorie der Erkenntnis. Halle: Niemeyer (1900/1901) [2nd ed. 1913; Eng. trans. by J. N. Findlay: *Logical Investigations*, Volume Two, London: Routledge & Kegan Paul (1970)
33. Rescher, N.: Axioms for the part relation. *Philosophical Studies* **6** (1955) 8–11
34. Montague, R.: On the nature of certain philosophical entities. *The Monist* **53** (1969) 159–194
35. Simons, P., Dement, C.: Aspects of the mereology of artifacts. In Poli, R., Simons, P., eds.: *Computers and Thought*. Kluwer Academic Publishers (1996) 255–276
36. Sattler, U.: Description logics for the representation of aggregated objects. In W.Horn, ed.: *Proceedings of the 14th European Conference on Artificial Intelligence*, IOS Press, Amsterdam (2000)
37. Gent, A.E.: *Engineering with rubber, how to design rubber components*. Hanser Publisher, New York (1992)
38. Simons, P.: *Parts: A Study In Ontology*. Clarendon Press, Oxford (1987)
39. White, J.L.: *Rubber processing, Technology - Materials - Principles*. Hanser Publisher, Munich Vienna New York (1995)
40. Roberts, A.D., ed.: *Natural rubber science and technology*. Oxford University Press, Ney York (1988) s. 161.
41. Newell, A., Simon, H.A.: Computer science as empirical inquiry: symbols and search. *Commun. ACM* **19** (1976) 113–126
42. Korf, R.E.: Artificial intelligence search algorithms. In: *Algorithms and Theory of Computation Handbook*, CRC Press, 1999. CRC Press (1999)
43. Korf, R.E.: Search: A survey of recent results for Artificial Intelligence. In Shrobe, H.E., A.A., eds.: *Exploring Artificial Intelligence: Survey Talks from the National Conferences on Artificial Intelligence*, San Mateo, CA, Kaufmann (1988) 197–237

44. Bandini, S., Manzoni, S., Sartori, F.: Knowledge maintenance and sharing in the KM context: the case of P–Truck. In Cappelli, A., Turini, F., eds.: *AI*IA 2003: Advances in Artificial Intelligence, Proceedings of 8th Congress of the Italian Association for Artificial Intelligence*, Pisa (I), September 2003. Volume 2829 of Lecture Notes in Artificial Intelligence, Berlin, Heidelberg, Springer-Verlag (2003) 499–510
45. Bandini, S., Mosca, A., Vanneschi, L.: Towards the use of genetic algorithms for the chemical formulation problem. In Manzoni, S., Palmonari, M., Sartori, F., eds.: *Proceedings of the 9th Congress of the Italian Association for Artificial Intelligence (AI*IA 2005)*, Workshop on Evolutionary Computation (GSICE 2005), Milano, Centro Copie Bicocca (2005) ISBN 88-900910-0-2.
46. Holland, J.H.: *Adaptation in Natural and Artificial Systems*. The University of Michigan Press, Ann Arbor, Michigan (1975)
47. Goldberg, D.E.: *Genetic Algorithms in Search, Optimization and Machine Learning*. Addison-Wesley (1989)

Model-Based Reasoning for Self-Repair of Autonomous Mobile Robots*

Michael Hofbaur[1], Johannes Köb[1], Gerald Steinbauer[2], and Franz Wotawa[2]

[1] Institute for Automation and Control, Graz University of Technology, Graz, Austria
  michael.hofbaur@tugraz.at,johannes.kob@tugraz.at
[2] Institute for Software Technology, Graz University of Technology, Graz, Austria
  gstein@ist.tugraz.at,wotawa@ist.tugraz.at

Summary. Retaining functionality of a mobile robot in the presence of faults is of particular interest in autonomous robotics. From our experiences in robotics we know that hardware is one of the weak points in mobile robots. In this paper we present the foundations of a system that automatically monitors the driving device of a mobile robot. In case of a detected fault, e.g., a broken motor, the system automatically re-configures the robot in order to allow to reach a certain position. The described system is based on a generalized model of the motion hardware. The path-planner has only to change its behavior in case of a serious damage. The high-level control system remains the same. In the paper we present the model and the foundations of the diagnosis and re-configuration system.

1 Introduction

Retaining the functionality of a mobile robot even in the presence of faults in its hardware is of particular interest. This fact becomes even more important in the case of truly autonomous systems, who are carrying out tasks without or at least with limited possibility for interacting with a human operator. Hardware faults like broken or overheated motors, are well known phenomena in the robotics domain. Even in commercial or safety-critical applications the reliability of robotics hardware is limited and tends to fail frequently. See for example [1] for a qualitative and quantitative estimation about the reliability of robotics hardware which justifies these observations.

In general a robot will not be able to successfully finish its task in the case of a fault in its hardware. If the robot should be able to deal with such situations, the robot control system has to be enriched with the capability for reasoning about such faults. Furthermore, the control system should be

---
* This research has been funded in part by the Austrian Science Fund (FWF) under grant *P17963-N04*. Authors are listed in alphabetic order.

able to adapt its behaviors in oder to compensate the faults. If the damage of the hardware is not too bad we might distinguish three different scenarios. In the first scenario, the robot is able to retain its full *physically* functionality. This means that the control system was able to detect and locate the fault in the hardware. Furthermore, the control system was able to repair the fault by taking an appropriate repair action like, e.g., restarting the faulty hardware component. Moreover, an appropriate repair action also can be a reconfiguration of the hardware or its low-level control software. This is only successful if the robot offers a certain level of redundancy. An example for this situation is the omnidirectional drive of robots of our Middle-Size team [2]. This omnidirectional drive comprises four motors with omni-wheels in a cross arrangement. If one of the motors fails, the robot is able to retain its omni-directional motion capability by a reconfiguration of the low-level drive controller. The remaining three motors offers enough actuation for controlling all three DOF in the plane.

In the second scenario the damage of the hardware can be not sufficiently compensated and therefore the functionality of the robot degrades. This means the robot control system is not able to take appropriate actions in order to retain the full functionality of the robot. But the robot is not yet doomed to fail in carrying out its task as long the robot control system is able to detect this scenario. Moreover, if the control system is able to reason about this degradation it can adapt its behavior in order to compensate the limited functionality. Therefore, the robot may still be able to finish its task. An example for such a situation is an omnidirectional drive with three omni-wheels. If one motor fails the drive can be reconfigured to an differential drive. In most cases a robot is still able to carry out its task with a differential drive. But most likely without the same performance.

Finally, if faults in the robots hardware occur and none of the two scenarios above fit, the robot loses its physically capabilities to carry out its task. Moreover, the knowledge about this fact is still valuable because the high level control may set the robot to a safe state and may informs a human operator.

The already mentioned ideas behind retaining functionality which can be classified as self-healing or self-repair capabilities of technical systems are stimulated by nature. For example, animals are capable of fulfill task even in case of severe restrictions to physical capabilities like broken legs. The reason for this behavior is that animals can (at least limited) reason about their state and capabilities and about their environment. This reasoning capabilities depend on the kind of animal. Apes for example perform more sophisticated task than insects. However, both of them can adapt to their internal current state and the environment. This includes self-healing as well as some sort of reconfiguration and adaption processes in cases a complete healing is not possible. One research area with growing importance is the area of self-healing and self-adapted systems. The contribution of this article in this area is to investigate frameworks and concepts that allow machines to reason about themself and to take appropriate actions in case of failure (almost) without human interaction.

The ability to self-repair is closely related to abductive and deductive reasoning. Abduction is used to find a cause for an observation. In our domain, i.e., the domain of mobile robots, an observation is might be a certain position which is in contradiction with the predicted position after executing an action, e.g., moving from one position to another. The cause of such an observation, e.g., a blocking wheel, gives information about how to overcome this problem. For example, the system might deductively decide to turn off the wheel and to re-configure the whole drive in order to retain the ability to move. Hence, the first step in self-repair is to use abduction to find a plausible reason for an observation, and the second step involves reasoning about re-configuration using the identified reason.

In this paper we present a generalized framework for improving the robustness of the motion of mobile robots. The framework is able to recognize and to handle the three scenarios outlined above. The framework comprises three parts. Which are extensions to the well-known hybrid control architecture for mobile robots [3]. The first part is a generalized meta-model about the capabilities of the motion system. It models all possible nominal and faulty operational modes of the motion hardware. The second part is a model-based diagnosis engine. The engine is able to detect and localize faults in the robots hardware by reason about the meta-model and current observations of the system. The last part is a model-based motion controller. Based on a concrete motion model this controller carries out the motion control by mapping immediate control signals provided from a path-planner to low-level commands for the actuators. The concrete motion model is an instantiation of the meta-model where the estimated operational mode of the hardware is used as a parameter. One important issue is that the meta-model remains the same in all scenarios. It is created by an engineer concerning all motion constraints of the hardware. The fault-detection and the adaptation of the controller is autonomously carried out by the framework without interaction with an operator or engineer.

We continue paper with a deeper description of the proposed framework in Section 2. In Section 3 we discuss the model-based diagnosis engine. In the next section, we show how the meta-model is created and used for motion control. In Section 5 we present related research. Finally, we will draw some conclusions.

2 Framework

In order to improve the robustness of mobile robots and to meet the stated requirements we propose a generalized framework for fault-tolerant motion control. A picture of the framework is shown in Figure 1. The framework is a

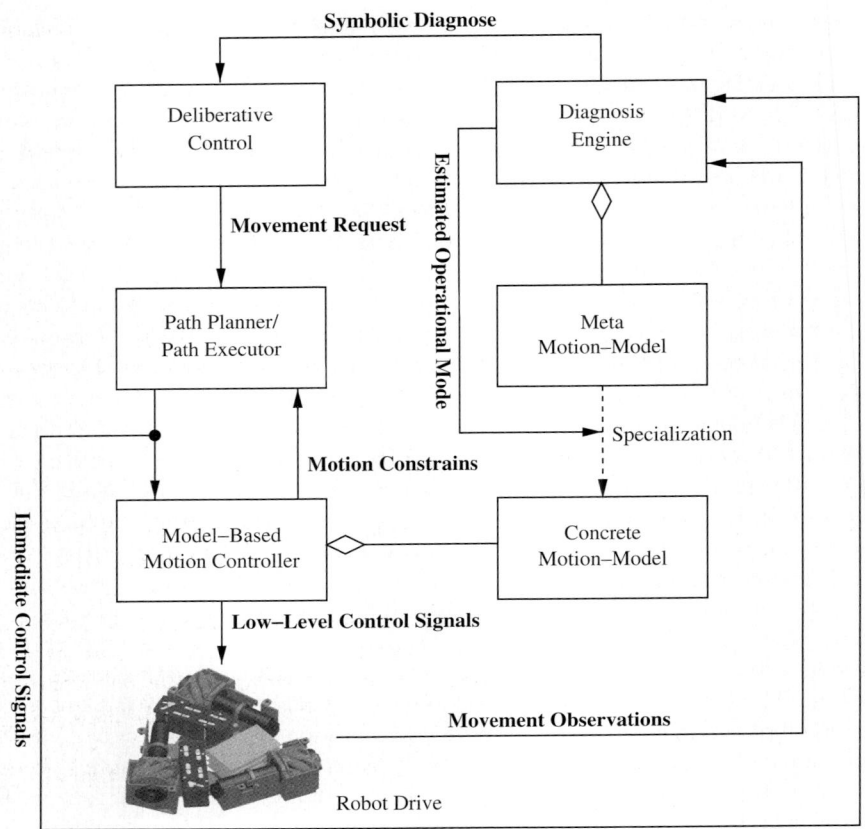

Fig. 1. The generalized framework for fault-tolerant motion control.

general pattern for the intelligent and robust control of a mobile robot. Within this paper, we discuss the framework for the improvement of the reliability of motion control. However, the framework can be easily adapted to other aspects of mobile robots like manipulators or even control software in general.

The left three components in the figure form a classical hybrid robot control architecture. On top of the architecture a deliberative component performs longterm planning and reasoning on an abstract logic-based level. Desired actions like movements are committed to the path-planner. The path-planner tries to find a path to the desired position based on a model about the world and the kinematic capabilities of the robot. If a path is found, the path-executor moves the robot along this path. Usually, this is done by sampling the path and committing immediate control signals like velocity and acceleration in the global or local reference frame to a motion controller. The motion controller is responsible for converting these immediate control signals into appropriate control signal for the actuators. In many applications a reactive obstacle avoidance behavior is also part of this control chain.

In general the motion controller and sometimes even the path-planner contain an engineered model about the kinematics of the robot. Usually, this model represents the nominal behavior of the system. Therefore, in the presence of a fault in the robots hardware the motion controller or even the path-planner tends to fail to appropriately move the robot along the desired path. This happens because the implicit model of the robot kinematics does not adapt to the new situation.

In order to improve the robustness of the motion control and to be able to handle faults in the robots hardware, we propose an enriched framework. The additional parts of this framework are: (1) a meta-model of the kinematics of the robot, (2) a model-based diagnosis engine and (3) a model-based motion controller.

In contrast to an implicit engineered motion-model the meta-model is a generalized motion-model. It is build up by a combination of all motion constraints of the robot. The individual motion constraints are defined by the number, the arrangement and type of all the wheels of the robot. See [4] for an introduction to kinematic constraints. In Section 4 we present an example. The advantage of this meta-model is that it contains the kinematic models for all possible operational modes of the robot. In this context we understand operational modes as the different situations with the presence of no or different combinations of faults in the hardware. It is clear that the hardware shows different behaviors in the different operational modes. The meta-model has to be engineered only once and remains the same as long the robots hardware does not change.

This meta-model is used by the model-based diagnosis engine to detect and identify faults in the hardware. For this purpose the meta-model is used to predict the behavior of the robot in all possible operational modes. It has to be noticed that there is exactly one nominal mode (no faults) and a number of abnormal modes (different faults). The number of the abnormal modes depends on the number of possible combinations of different faults. The diagnosis engine permanently compares the results of the predictions of the behavior in the different modes with the actual observations of the system. If there is a deviation between the prediction of the nominal behavior and the observed behavior, a fault has been detected. Possible observations are, e.g., the actual movement of the robot measured by odometry, the supply current or torque of a motor, optical flow in a camera image. By tracking of the abnormal behaviors the most probable abnormal operational mode is estimated. This estimated mode also contains information about the exact root cause of the fault, e.g., which motor is broken. Model-based diagnosis is introduced in more details in the next section.

So far the framework is able to detect and identify faults. In order to react on such faults the framework has three possibilities. First, the framework creates a concrete instance of the meta-model. This model mimics the behavior of the kinematic in the estimated operational mode. Such a model is used by the model-based motion controller to provide an appropriate mapping of the immediate control signals to the low-level control signals of the actuators. The adaptation of this mapping to the concrete motion model can be carried out autonomously by the controller. Such an adaptation may cause an degradation of the functionality as noticed in the example in the introduction. Furthermore, the concrete motion-model is able to postulate additional motion constraints. Advanced path-planning algorithms [5, 6] are able to take advantage from such constraints in order to find an appropriate path. But it is neither guaranteed that the model-based controller is able to find an appropriate mapping nor that the path-planner is able to find an executable path in the presence of additional motion constraints. However, this information are also valuable. The deliberative control can be informed that because of the serious damage of the hardware the robot is unable to move in a desired way. This information may be used by the deliberative component to stop the robot and to inform an operator about the situation.

The proposed framework is able to detect faults in the hardware and to react to these faults. Either the model-based motion controller and the path-planner adapt itself to the new situation or a dangerous situation can be recognized and appropriate action can be taken.

In the next section we discuss model-based diagnosis in more details.

3 Diagnosis

There are several different definitions of diagnosis. All of them are related to certain tasks which can be subsumed under the term "diagnosis". Diagnosis can be the identification of a malformed behavior. This sub-task of diagnosis is known as fault identification. For example monitoring approaches or verification approaches implement fault identification. After fault identification the focus is on identifying the root cause, i.e., the diagnosis, of the faulty behavior. This task is referred as fault detection. Finally, someone is interested in changing the system in a way such that the systems behaves like expected. This final task is fault repair.

In this section we focus on fault detection and fault localization. In particular we introduce the basic ideas of model-based diagnosis [7] and explain how these ideas apply for the diagnosis part of fault tolerant motion control. In model-based diagnosis the fault detection and localization part is based on a model of the systems. This model captures the intended behavior of the system and allows for computing predictions of the behavior of the system in a particular situation. When comparing these predictions with the observed behavior of the real system, we distinguish two situations. Either the predictions are consistent with the observations, or not. The latter case is obviously an indicator for the manifestation of a fault as the system did not behave in the indented manner. In order to find the root cause that is responsible for the faulty behavior, the idea of model-based diagnosis is to use the same model directly. This can be achieved by making explicit assumptions about the state of the system with the model. For example, if we know that the given behavior of a component is only correct when the component is working correctly, then we store this information as some sort of implication of the form "The component is correct implies the behavior" in the model. This allows us to reason about assumptions about the state of a component (e.g., faulty or not) when computing predictions. We now define diagnosis as task of identifying system assumptions that allow for predicting values which do not contradict the available observations. Formally, we say that a set of assumptions A is a diagnosis for a model M and observations OBS, if $A \cup OBS \cup M$ is a consistent (logical) theory. Note that this definition has the advantage that a faulty behavior of components or parts of the system has not to be known. Only the correct behavior is necessary to allow for testing consistency with given observations.

Reiter [7] introduced this consistency-based diagnosis definition. In Reiter's paper beside the basic definitions and properties an algorithm for computing all diagnosis, i.e., causes for a certain behavior, can be found which was corrected later on by Greiner et al. [8]. Beside the consistency-based diagnosis definition there is an abductive diagnosis definition. In abductive diagnosis the observations have to be explained by a given theory or model. This is definitely stronger than pure consistency-based diagnosis where only consistency between assumptions and observations have to be ensured. Friedrich et al. [9] introduced a definition of abductive diagnosis which is used for therapy. Although, abductive diagnosis is stronger than consistency-based diagnosis Torasso and Console [10] proved that both approaches are to some extent equivalent. The equivalence can be shown when using consistency-based diagnosis where also fault models are used. Fault models capture the faulty behavior of components and thus allows for exactly predicting the observed behavior which is the same when using abduction.

In consistency-based diagnosis there has been a long debate about the use of fault models. Struss and Dressler [11] motivate the use of fault models in order to remove non-intuitive diagnosis candidates which may occur when using consistency-based diagnosis. Unfortunately, the complexity of diagnosis increases when using fault models which makes it hard to use in practice. As an alternative to consistency-based diagnosis with fault models Friedrich et al. [12] developed a method which makes use of model extension in order to remove non-intuitive solutions instead of fault models. There idea is to introduce rules which handles physical impossibilities. Another important argument which prevents from using fault models in all applications is the fact that the faulty behavior of systems is not always known in advance. For example, in software debugging someone cannot state how a statement will fail. This is also true in the hardware domain where the exact behavior in case of a fault is not available. Consider for example a nuclear plant. Nobody wants to crash the safety system of such a plant to get knowledge about all effects on the environment. The approach we are using describing in the paper can be classified as an abductive diagnosis approach. The correct and the faulty behavior is used in order to explain a given observation. In the next paragraph we discuss the modeling issue and afterwards our modeling approach in more detail.

From its beginning model-based diagnosis has been developed in several directions. These developments have been driven by the different characteristics of the application areas and their corresponding definition and use of models. A coarse partitioning of model is to distinguish qualitative and quantitative models. Qualitative models are models that use only finite value domains. Examples for the use of qualitative models for diagnosis can be found in [13]. In this paper we focus on quantitative models. This is because at the sensor/actuator level we have real-value domains and control loops which can be most efficiently represented as differential or difference equations. In order to combine quantitative modeling in terms of difference and algebraic equations with the need for making assumptions about the current state of the system (or at least a part of the system, e.g., a component) explicit we use the well developed modeling paradigm of hybrid automata [14, 15]. A hybrid automata comprises states and connection between these states. A state represents assumptions about the current behavior of the system. These assumptions are represented by discrete modes that are assigned to state variables. States are connected if there is a possible transition of the system from one to another behavior. The behavior of the system itself in a particular state is given by difference and algebraic equations that are assigned to each state. The task of diagnosis using hybrid automata is to identify a state where the corresponding behavior does not contradict the observations.

Figure 2 depicts a simple hybrid automata that models the behavior of a differential drive. The model distinguish four states. Each state represents an operational mode of the drive. Either both motors are working as expected, the left or the right motor is broken or both motors are broken. A broken motor provides no torque anymore.

In order to support the component-oriented modeling paradigm the composition of hybrid automata is of particular interest. In [16] the authors describe how different hybrid automata can be efficiently combined to form an automaton for a system comprising different components. Moreover, the concept of hybrid automata can be extended by assigning probability values to mode transitions. This allows for selecting most probable transitions and thus provides a heuristics to improve searching for explanations, i.e., consistent states.

Figure 3 shows a situation where our model of the differential drive (Fig. 2) is used in order to determine the current system state. The figure shows the real behavior of a robot and its intended behavior. After $t = 2.2s$ the predicted and the observed behavior deviate which causes a transition from state $ok(M_r) \land ok(M_l)$ to $\neg ok(M_r) \land ok(M_l)$. Note that the right motor is not working correctly and, therefore, the robot is going into a circular trajectory

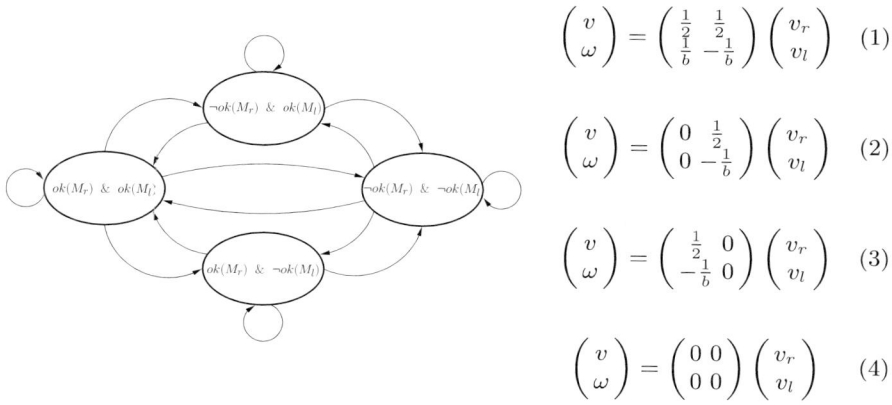

Fig. 2. A simple hybrid automata modeling a differential drive. On the left the automata of the different operational modes are shown. On the right the kinematic equations are shown for the different operational modes: (1) nominal mode, (2) $\neg ok(M_r)$ mode, (3) $\neg ok(M_l)$ mode and (4) $\neg ok(M_r) \wedge \neg ok(M_l)$ mode. v and w denotes the translational and rotational velocity. v_r and v_l denote the velocity of the right and left wheel.

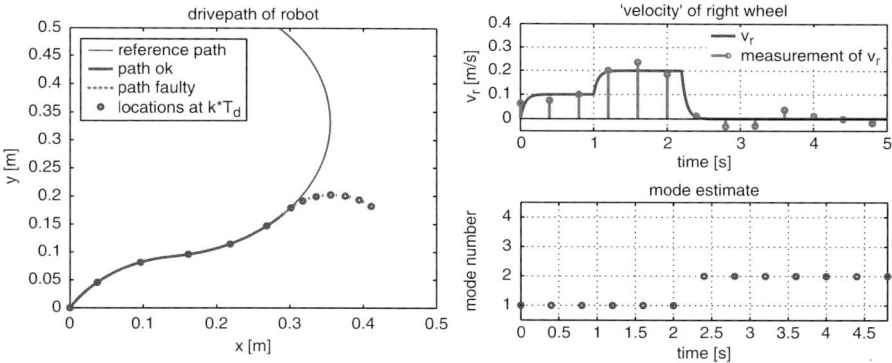

Fig. 3. The left diagram shows the execution of a path by a differential drive. The upper right diagram shows the velocity at the right wheel. After 2.2 s the right motor fails. This fact can be deduced from the decrease of the speed. The lower right diagram shows the estimated mode of the drive. 1 represents the nominal mode. 2 represents the $\neg ok(M_r)$ mode.

instead of a straight line as expected. Because in reality the observed behavior may deviate because of inaccuracies and uncertainties of measurements we use a tolerance interval. If the predicted value is within this interval no deviation and thus inconsistency is said to be detected.

The diagnosis procedure can be stated as follows:

1. **Initialization:** Let H comprise all states. *Note that H comprise all possible states that my explain a certain behavior.*
2. For all elements $s \in H$ do:
   a) Compute the continuously valued state variables and output variables for the next point in time using the given input variables and command variables.
   b) If the computed output values are not consistent with the observed values, then remove s from H. Search for state s' which is connected with s where the computed next values are equivalent to the observed values and add s' to H.
3. Goto 2

The described diagnosis procedure implements a multi-hypothesis tracking procedure where all hypotheses that explain the observed behavior are element of H. This procedure can be improved by storing a limited number of hypotheses, e.g., only the 5 with the highest probability. If using a probability measure for ranking diagnoses, transition probabilities between states can be introduced in the hybrid automaton. More details about multi-hypothesis tracking can be found in [15, 17].

Once the diagnosis engine identifies the most likely diagnosis, a new model is generated which is used in the model-based motion control component of the system. We explain the computation of the adapted model from a general model of the kinematics of the system in more detail in the next section.

4 Model-Based Controller

The task of our model-based motion controller is to convert motion commands from the path planer / path executor into an appropriate actuation of the robot's wheels. In detail, it converts a requested motion

$$\dot{\xi}_R = \begin{bmatrix} \dot{x}_R \\ \dot{y}_R \\ \dot{\Theta} \end{bmatrix} \quad (5)$$

into a desired rotational speed $\dot{\varphi}$ and, in case of a steered wheel, in its steering angle β for each wheel of the robot. Low level control of each wheel will then take care of the appropriate actuation of the electrical drives.

A typical way to do this is to take the robot's kinematics configuration and derive a fixed relationship between the motion command and the wheel actuation. This is usually done by computing the instantaneous center of rotation (ICR) for $\dot{\xi}_R$ and assuming that a higher level path planner takes the kinematic

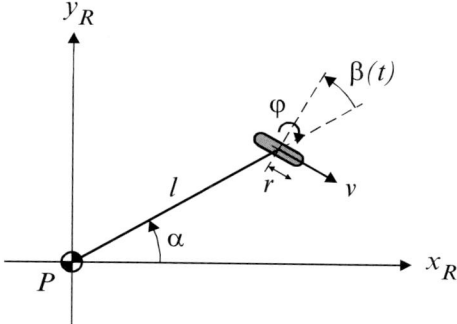

Fig. 4. Parameters for a steered wheel.

constraints of the robot's configuration into account. However, one considers the fault-free configuration at design time only. Our goal is different in that we allow changed configurations (e.g. due to a fault in the steering mechanism of a wheel) and thus have to take care of changed kinematic constraints.

Our meta motion model provides the specification for each wheel in terms of its geometric alignment with respect to the robots reference frame and the operational/fault modes for its actuation. Thus each wheel specifies a rolling and a sliding constraint for the robot's kinematics. Figure 4 shows this specification for a steered wheel that leads to the following rolling and sliding constraints (see [4] for details)

$$\begin{bmatrix} \sin(\alpha + \beta(t)) & -\cos(\alpha + \beta(t)) & -l\,\cos(\beta(t)) \end{bmatrix} \dot{\xi}_R - r\dot{\varphi} = 0 \\ \begin{bmatrix} \cos(\alpha + \beta(t)) & \sin(\alpha + \beta(t)) & l\sin(\beta(t)) \end{bmatrix} \dot{\xi}_R = 0. \tag{6}$$

Combining the constraints of all (n) wheels of the robot, we obtain the vector-form of the constraints

$$\mathbf{J}_1(\beta_s)\dot{\xi}_R - \mathbf{J}_2\dot{\phi} = \mathbf{0} \tag{7}$$

$$\mathbf{C}_1(\beta_s)\dot{\xi}_R = \mathbf{0} \tag{8}$$

where β_s denotes the vector of all steering angles and $\dot{\phi}$ denotes the vector of angular speeds for the wheels.

One can compute the space of the admissible velocities $\mathcal{Z}$ through the *null-space* or *kernel*:

$$\mathcal{Z} = \text{kernel}(\mathbf{C}_1(\beta_s)) \subseteq \mathbb{R}^3. \tag{9}$$

However, one has to make sure, that an admissible velocity $\dot{\xi} \in \mathcal{Z}$ can be actuated (controlled) through the robot's wheels, i.e. it leads to a non-zero $\dot{\phi}$ in (7). Re-writing (7) we obtain

$$\mathbf{J}_2^{-1}\mathbf{J}_1(\beta_s)\dot{\xi}_R = \dot{\phi} \tag{10}$$

and it becomes evident that we can compute the *non-controllable* velocities $\bar{\mathcal{S}}$ by means of

$$\bar{\mathcal{S}} = \text{kernel}(\mathbf{J}_1(\beta_s)) \subseteq \mathbb{R}^3. \tag{11}$$

Whenever the two spaces do intersect, i.e. $\mathcal{Z} \cap \bar{\mathcal{S}} \neq \emptyset$, we have to refine the admissible velocities $\mathcal{Z}$ to exclude those movements that cannot be actuated through the robot's wheels. By computing the complement of $\bar{\mathcal{S}}$

$$\mathcal{S} = \text{kernel}(\bar{\mathbf{S}}^T) \subseteq \mathbb{R}^3, \tag{12}$$

where $\bar{\mathbf{S}}$ denotes the matrix of basis vectors for $\bar{\mathcal{S}}$ we obtain the *controllable* velocities so that the intersection

$$\mathcal{Z} \cap \mathcal{S} =: \dot{\Xi} \tag{13}$$

defines the space of *admissible and controllable* velocities for a given configuration of the robot. Our model-based controller deduces this vector space on-line for the configuration of the robot that is identified through the diagnosis engine [18]. Knowing this space, it checks whether a requested velocity from the path planner / path executor unit can be accepted through

$$\dot{\xi}_R \in \dot{\Xi}. \tag{14}$$

Whenever this condition holds, it proceeds with computing the robot's instantaneous center of rotation (ICR), that, in turn, will lead to the steering angles β_i and angular velocities $\dot{\varphi}_i$ for each individual wheel.

Violating (14) indicates, that the current path of the robot is infeasible. The controller provides $\dot{\Xi}$ to the path planner so that this unit can derive an alternative path that is acceptable for the robot's kinematics in the current (faulty) configuration.

All changes within the kinematics that do not reduce the vector space for the admissible velocities (typically single faults in the robot's drives) are dealt with automatically, since the computation of the ICR and the consecutive deduction of steering angles and angular velocities utilizes a motion model (7-8) that reflects the currently active kinematic constraints of the robot and not just the nominal fault-free case. In this sense, we obtain an automatic *model-based* configuration of the motion controller that directly handles an onset of single faults in the robot's drive, recognizes the controller's limits for re-configuration and communicates the relevant implications of changed kinematic constraints to the higher level control hierarchy.

5 Related Research

Diagnosis and autonomous reconfiguration of autonomous systems has been in the focus of research for decades. There are a wide range of different approaches for diagnosis and reconfiguration. These approaches differ mainly in the used type of models (qualitative, quantitative or hybrid) and their deduction process (probabilistic state estimation, rule-based systems or logic inference).

The Livingstone architecture proposed by Williams and colleagues [19] was used by the space probe *Deep Space One* to detect failures in the its hardware and to recover from them. The fault detection and recovery are based on model-based reasoning. In [20] and [21] particle filter techniques were used to estimate the state of the robot and its environment. These estimations together with a model of the robot were used to detect faults. The most probable state is derived from unreliable measurements. The advantage of this approach is that it is able to handle non-Gaussian uncertainties of the robot's sensing and acting as well as uncertainties in its environment. Other approaches which are based on Kalman-filter are only able to account Gaussian uncertainties.

In [22] the authors present a framework for detection and repair of faults in the control system of autonomous mobile robots. The framework used a model-based approach for fault detection. The information about the state of the system is mainly obtained from the observation of communication between software modules. The repair is done by a systematic restart of the effected software components. In [23] model-based diagnosis was used to establish a functional reconfiguration of a telecommunication system in order to ensure the desired functionality. In this application the reconfiguration was used to allow the system to incorporate new requirements.

Beside model-based reasoning for self-adaptivity and self-repair other techniques like case-based reasoning has been used. Anglano and Montani [24] applied case-based reasoning for self-healing of software systems. They applied their methodology to a large Internet application with promising results. However, a more elaborated case study or some empirical evaluation had been left for future research. Another different approach is the use of genetic programming. In a book [25] the authors provide several examples of how to use genetic programming to really implement evolvable hardware. Such an evolvable hardware has the capabilities of self configuration whenever needed. Although, the introduced example applications are steps in a right direction self adaptivity is still limited to small areas. In contrast to [25] our approach is based on a model of the behavior and, therefore, determinism is given.

Other application areas of self-adaptive system include software engineering and programs where models and architectural descriptions are used to allow programs reasoning about themself. For example, Sztipanovits and Karsai [26] for example presented a model-based approach for supporting self-adaptability of software. Other model-based approaches in this application area [27, 28] mainly deal with fault localization of software during the design phase and not during lifetime. However, some of the techniques especially about modeling can be used for reasoning during runtime and thus help to locate a fault.

6 Conclusion

In this paper we present a framework which integrates model-based fault diagnosis and reconfiguration of the hardware of an autonomous mobile robot. The aim of the framework is that a robot is able to fulfill its task even in the presence of faults. The robustness of a system which uses the framework is achieved by reconfiguration of low-level control. In the case of a fault the system is able to retain the functionality. The presence of a fault is detected and identified using model-based reasoning. The used model is a hybrid model composed of all kinematic constraints of the motion system. Moreover, if the functionality could not be fully retained, at least the high-level controller is informed about this circumstance. The novel idea in this paper is the integration of autonomous diagnosis and reconfiguration in the same framework. Furthermore, the same meta-model about all possible operational modes of the hardware is used for fault diagnosis and reconfiguration which helps to reduce the amount of necessary modeling.

References

1. Carlson, J., Murphy, R.: How UGVs physically fail in the field. *IEEE Transactions on Robotics* **21** (June 2005) 423–437
2. Steinbauer, G., Faschinger, M., Fraser, G., Mühlenfeld, A., Richter, S., Wöber, G., Wolf, J.: Mostly Harmless Team Description. In: *Proceedings of the International RoboCup Symposium.* (2003)
3. Murphy, R.: *Introduction to AI Robotics.* MIT Press (2002)
4. Siegwart, R., Nourbakhsh, I.: *Introduction to Autonomous Mobile Robots.* MIT Press (2004)
5. Latombe, J.C.: *Robot Motion Planning.* Eight reprint edn. Kluwer Academic Publisher (2004)
6. Choset, H., Lynch, K., Hutchison, S., Kantor, G., Burgard, W., Kavarki, L., Thrun, S.: *Principles of Robot Motion. Theory, Algorithms and Implementations.* MIT Press (2004)
7. Reiter, R.: A theory of diagnosis from first principles. *Artificial Intelligence* **32** (1987) 57–95
8. Greiner, R., Smith, B., Wilkerson, R.: A correction to the algorithm in Reiter's theory of diagnosis. *Artificial Intelligence* **41** (1989) 79–88
9. Friedrich, G., Gottlob, G., Nejdl, W.: Hypothesis classification, abductive diagnosis and therapy. In: *Proceedings of the International Workshop on Expert Systems in Engineering*, Vienna, Springer Verlag, Lecture Notes in Artificial Intelligence, Vo. 462 (1990)
10. Console, L., Dupré, D., Torasso, P.: On the relationship between abduction and deduction. *Journal of Logic and Computation* **1** (1991) 661–690
11. Struss, P., Dressler, O.: Physical negation – Integrating fault models into the general diagnostic engine. In: *International Joint Conference on Artificial Intelligence (IJCAI)*, Detroit (1989) 1318–1323

12. Friedrich, G., Gottlob, G., Nejdl, W.: Physical impossibility instead of fault models. In: *National Conference on Artificial Intelligence (AAAI)*, Boston (1990) 331–336. Also appears in Readings in Model-Based Diagnosis (Morgan Kaufmann, 1992).
13. Hamscher, W., Console, L., de Kleer, J., eds.: *Readings in Model-Based Diagnosis*. Morgan Kaufmann Publishers, Inc. (1992)
14. Hofbaur, M., Williams, B.: Mode estimation of probabilistic hybrid systems. In Tomlin, C., Greenstreet, M., eds.: *Hybrid Systems: Computation and Control, HSCC 2002*. Volume 2289 of Lecture Notes in Computer Science. Springer Verlag (2002) 253–266
15. Hofbaur, M.: *Hybrid Estimation of Complex Systems*. Volume 319 of Lecture Notes in Control and Information Sciences. Springer Verlag (2005)
16. Hofbaur, M., Wotawa, F.: A causal analysis method for concurrent hybrid automata. In: *Proceedings of the 21st National Conference on Artificial Intelligence (AAAI-06)*. (2006) 840–846
17. Li, X., Bar-Shalom, Y.: Multiple-model estimation with variable structure. In: *IEEE Trans. Automatic Control*. Volume 41. (1996) 478–494
18. Köb, J.: Modellbasierte Regelung eines Roboterfahrwerkes. Master's thesis, Institute for Automation and Control, Graz University of Technology (2005)
19. Williams et al., B.: Remote agent: To boldly go where no AI system has gone before. *Artificial Intelligence* **103** (1998) 5–48
20. Dearden, R., Clancy, D.: Particle filters for real-time fault detection in planetary rovers. In: *Proceedings of the Thirteenth International Workshop on Principles of Diagnosis*. (2002) 1–6
21. Verma, V., Gordon, G., Simmons, R., Thrun, S.: Real-time fault diagnosis. *IEEE Robotics & Automation Magazine* **11** (2004) 56–66
22. Steinbauer, G., Wotawa, F.: Detecting and locating faults in the control software of autonomous mobile robots. In: *19th International Joint Conference on Artificial Intelligence (IJCAI-05)*, Edinburgh, UK (2005)
23. Stumptner, M., Wotawa, F.: Model-based reconfiguration. In: *Proceedings Artificial Intelligence in Design*, Lisbon, Portugal (1998)
24. Anglano, C., Montani, S.: Achieving self-healing in autonomic software systems: A case-based approach. In Czap, H., Unland, R., Branki, C., Tianfield, H., eds.: *Self-Organization and Autonomic Informatics (I)*. Volume 135 of Frontiers in Artificial Intelligence and Applications. IOS Press (2005) 267–281
25. Higuchi, T., Liu, Y., Yao, X., eds.: *Evolvable Hardware*. Springer (2006)
26. Sztipanovits, J., Karsai, G.: A model-based approach to self-adaptive software. *IEEE Intelligent Systems* **14** (1999) 46–53
27. Friedrich, G., Stumptner, M., Wotawa, F.: Model-based diagnosis of hardware designs. *Artificial Intelligence* **111** (1999) 3–39
28. Peischl, B., Wotawa, F.: Automated source level error localization in hardware designs. *IEEE Design & Test of Computers* **23** (2005) 8–9

Application of Bayesian Inference to Automatic Semantic Annotation of Videos

Fangshi Wang[1,2], De Xu[1], Hongli Xu[1], Wei Lu[2], and Weixin Wu[1]

[1] School of Computer & Information Technology, Beijing Jiaotong University, Beijing China
{fshwang,dxu,hlxu}@bjtu.edu.cn
[2] School of Software, Beijing Jiaotong University, Beijing, China
wlu@bjtu.edu.cn

Summary. It is an important task to automatically extract semantic annotation of a video shot. This high level semantic information can improve the performance of video retrieval. In this paper, we propose a novel approach to annotate a new video shot automatically with a non-fixed number of concepts. The process is carried out by three steps. Firstly, the semantic importance degree (SID) is introduced and a simple method is proposed to extract the semantic candidate set (SCS) under considering SID of several concepts co-occurring in the same shot. Secondly, a semantic network is constructed using an improved K2 algorithm. Finally, the final annotation set is chosen by Bayesian inference. Experimental results show that the performance of automatically annotating a new video shot is significantly improved using our method, compared with classical classifiers such as Naïve Bayesian and K Nearest Neighbor.

1 Introduction

In recent years, many research works focus on Content-Based Video Retrieval (CBVR) using the perceptual features. But the results of CBVR can not make the users satisfied because human beings judge the video similarity mainly according to their understanding of video contents, not only to the similarity in visual features of videos. Users often want to retrieve video at the semantic level. However, a gap remains between the low-level feature descriptions, such as colors, textures, shapes and motions, and the semantic descriptions of objects, events, scenes, people and concepts that are meaningful to users. The low-level feature can be easily extracted from the videos.

In order to bridge the semantic gap, experts or specified users annotate the video shot with semantic concepts manually, which is very costly and time consuming. New technologies are needed for reducing annotation costs. At present, most methods of annotating video with concepts automatically or semiautomatically are based on supervised classification [1–4]. Experts or

Fig. 1. A random selection of key frames from the training set.

users specify several classes, a system can construct one or more classifiers through learning from the training set which is built manually by users and includes a small number of samples. The perpetual features of a new video are extracted automatically and input into the well-trained classifiers. Then the result of the classification is the semantic annotation of the new shot. The classes defined in such methods are mutually exclusive, so each video shot can have only one semantic concept.

One concept is not enough to completely summarize a video shot with rich contents. For example, we can see "mountain", "sky" and "water" in the first picture of Figure 1. Three concepts are needed to describe the shot. It should not only belong to any one of the three classes. Obviously, semantic concepts do not occur independently or are not isolate from each other, and the mutual relationship between them should be taken into account.

Intuitively it is clear that the presence of a certain concept suggests a high possibility of detecting certain other concepts. Similarly some concepts are less likely to occur in the presence of others. The detection of "car" boosts the chances of detecting "road", and reduces the chances of detecting "waterfall". It might also be possible to detect some concepts and infer more complex concepts based on their relation with the detected ones. Naphade proposed the MultiNet as a way to represent higher level probabilistic dependencies between concepts [5]. The classes and structure of the classification frameworks were either decided by experts or specified by users. Moreover, the structure will become very large with the number of the classes increasing. If there are n classes, there will be n variable nodes and $n(n-1)/2$ function nodes and $n(n-1)$ edges in the MultiNet.

MediaNet [6] can automatically select the salient classes from annotated images and discover the relationship between concepts by using external knowledge resources from WordNet. However, the relationships between concepts in MediaNet are too complex. There are not only perceptual relationships such as "equivalent", "specializes", "co-occurs" and "overlaps", but also semantic relationships such as "Synonymy/Antonymy","Hypernymy/Hyponymy", "Meronymy/Holonymy", "Troponymy" and "Entailment", which are summarized into a small subclass of all these relationships by clustering subsequently.

Jeon et al. [7] proposed a cross-media relevance model (CMRM) to learn the joint distribution of blobs and words in order to perform both image annotation and ranked retrieval. Lavrenko et al. [8] proposed a Continuous-space Relevance Model (CRM) to compute a joint probability of image features over different regions in an image using a training set and used this joint probability to annotate and retrieve images. Both CMRM and CRM depend on automatic segmentation of image, and the overall annotation performance is strongly affected by the quality of segmentation. Feng et al. [9] replaced blobs with rectangular blocks and modeled image keywords using a Multiple-Bernoulli Relevance Model (MBRM) to automatically annotate and retrieve images/videos. Words are modeled using a multiple Bernoulli process and images modeled using a kernel density estimate. But all annotation obtained by the three models have a fixed number of words. The length of the annotation is determined by users and has a direct influence on the recall and precision. In addition, it is not reasonable to label every shot with a fixed number of concepts, no matter whether the shot content is rich or not.

In fact, while annotating the video, we only concern about whether concept B is also present in the same frame if concept A is present. Furthermore, none of the previous work considers the different degrees of several co-occurring concepts in one video shot. For example, in the first picture of Figure 1, we first pay attention to the water region which occupies most part of the image, then the mountain region, at last the sky region. The concept corresponding to the large region in the image is more important, otherwise less important.

In the previous works discussed above, most methods annotate a video shot with one concept. Even though some methods can annotate a shot with several concepts, none of them has considered the different importance degrees of these concepts co-occurring in the same shot. In this paper, we have developed a novel approach to annotate a new video shot automatically with a non-fixed number of concepts under the consideration of the semantic importance degree.

The paper is organized as follows. The Semantic Importance Degree (SID) is introduced and a simple but efficient approach is proposed to extract Semantic Candidate Set (SCS) in section 2. This is followed by an improved K2 algorithm to build a semantic network in section 3. Section 4 describes a method to select the reasonable concepts from the SCS by Bayesian Inference and automatically annotate a video shot with a varied number of concepts. The experimental results are given in section 5. Finally, section 6 concludes the paper.

2 Obtaining the Semantic Candidate Set of a New Video

The training set is constructed by manually annotating the key frames of video shots, which is regarded as Ground Truth (GT). There is an annotation with 1-4 concepts for each key frame in the training set.

At first, the semantic importance degree is introduced. If there is only either the background or the foreground, the importance degrees of several concepts are sorted from strong to weak according to their covered areas in the image from large to small. For example, in the first picture of Figure 1, the sorted importance degrees from strong to weak are "water","mountain" and "sky". If there are both the background and the foreground, the foreground always precedes the background. For example, in the second picture of Figure 1, even though "boat" occupies smaller area than "water", "boat" is more important than "water".

Multiple concepts of one frame are sorted from strong to weak according to their importance degrees. If there are several concepts whose importance degrees are about equal, then they are sorted randomly. The semantic classes are automatically obtained from the GT of training set. Each concept is considered a semantic class.

Next, the Semantic Candidate Set (SCS) is obtained under the consideration of SID. SCS is a set of N most probable concepts, which is used to show the semantic characteristics of a frame. It is supposed that all concepts actually annotated for a testing frame are in its SCS, the actual concepts are chosen from the SCS by Bayesian Inference. The procedure of obtaining SCS is as follows.

Step 1. All the subclass centers are calculated for each concept. The frames in the training set are classified according to their semantic concepts. If a frame includes concept S, it belongs to class S regardless of the position of S in the annotation. So a frame might belong to k classes since it includes k concepts. Because concept S might be present in the $1st$, $2nd$, $3rd$ or $4th$ position in the annotation of a frame, there are four subclass centers Sub_1^S, Sub_2^S, Sub_3^S and Sub_4^S at most for one semantic class. Not all the four subclass centers always exist at the same time. If S is not present in the first position in the annotation of any frame in the training set, then there is not the first subclass center Sub_1^S for S.

The ith existing subclass center $Sub_i^S (i = 1, 2, 3, 4)$ for S is obtained as follows.

$$Sub_{i,k}^S = \frac{1}{|T_i^S|} \sum_{j=1}^{|T_i^S|} f_{j,k} \qquad f_j \in T_i^s, \quad k = 0,..., dim - 1 \qquad (1)$$

where dim is the dimension of the visual feature vector, T_i^s is the set of the samples with S in the ith position in their semantic annotation, $|T_i^S|$ is the number of the samples in T_i^s, f_j is the jth frame in T_i^s, $f_{j,k}$ is the kth visual

Table 1. Weights corresponding to subclass centers.

	weights corresponding to subclass centers			
There are 4 subclass centers	0.5	0.3	0.2	0.1
There are 3 subclass centers	0.6	0.3	0.1	
There are 4 subclass centers	0.7	0.3		
There is 1 subclass center	1			

feature element of frame f_j, $Sub^S_{i,k}$ is the kth visual feature element of the ith subclass center of concept S.

Step 2. Formula (2) is used to calculate the global class center C^s of concept S, where w_j is the weight obtained through many experiments, shown in Table 1.

$$C^S = \sum_{j \wedge \exists Sub^S_i} w_j Sub^S_i \quad (2)$$

where $\exists Sub^S_i$ represents that the ith subclass center of S exists. For example, if there are only two subclass centers for S, Sub^S_2 and Sub^S_4, then the global class center of S is calculated as $C^S = 0.7 \times Sub^S_2 + 0.3 \times Sub^S_4$.

Step 3. Formula (3) is used to compute the distances between the key frame F of a new shot and every global class center, which are denoted as $Dist[1], Dist[2], \ldots, Dist[n]$.

$$Dist[S] = \sqrt{\sum_{k=0}^{dim-1}(F_k - C^S_k)^2} \quad (S = 1, ..., n) \quad (3)$$

where F_k is the kth visual feature element of the testing key frame. C^S_k is the kth feature element of C^S. n is the number of the semantic concepts in the training set.

Step 4. $Dist[1, \ldots, n]$ is sorted from small to large. The N most probable concepts, denoted as S1,..., SN corresponding the N smallest distances, consist of SCS. In our experiment, the result is the best for $N = 4$. S1 is regarded as the first concept of the new shot. This process of extracting SCS is named Semantic Importance Degree Method (SID).

3 Constructing a Semantic Network Based on Bayesian Network

Bayesian Network (BN), also known as Belief Network, is a graphical model that efficiently encodes the joint probability distribution over a set of random variables. Bayesian Network is selected to construct the Semantic Network, in which a node represents a semantic concept and an edge represents the dependency relationship between two concepts.

Two reasons prompted the selection of Bayesian Networks for learning statistical dependencies between concepts. First, there are algorithms to learn both the parameters and the topology of a Bayesian Network. If the nodes in a Bayesian Network represent concepts, then the algorithms are actually learning statistical relationships among the concepts. Second, once built, the Bayesian Network can answer arbitrary probabilistic questions about the concepts (e.g., joint probability for the values of any two nodes).

The traditional K2 algorithm [10], denoted as TK2, is a representative of general BN structure learning algorithms. It takes a data set and a node order as input and constructs the BN structure as output. Some of the sequences among concepts are easy to be determined, such as "sky" and "cloud". The presence of "cloud" in a frame makes sure of the presence of "sky", but otherwise not the truth. It means that "cloud" should precede "sky" and the direction of the edge should be from "cloud" to "sky". But as for other concepts, such as "water" and "animal", it is difficult even for a expert to determine their sequence. The TK2 algorithm has a serious drawback. If we choose the variable order carelessly, the built network structure may fail to reveal many conditional independencies among the nodes. Let n be the number of the nodes, i.e. the number of the concepts in our paper. Thus, in the worst case, we have to explore $n!$ node orderings to find the best one.

We propose an improved method which loosens the requirement of complete node ordering by allowing partial node ordering or even dose not need for node ordering.

First, users are allowed to manually input prior structure which they can make sure, e.g. cloud→sky represents the directed edge from cloud to sky. Users maybe input a partial node ordering or even nothing at all.

Second, the system can determine automatically the complete node ordering by an improved topological sorting. An array $batch[1...n]$ is used to record the number of batch in which InDegree of every node becomes zero. For example, in Figure 2(a), node v1 and v2 are the nodes in the first batch with zero InDegree, so $batch[v1] = batch[v2] = 1$. In the process of topological sort, if node v1 and its incident edges are removed before v2, then v3 is the only node in the second batch with zero InDegree and will precede v5 in the node ordering. If node v2 and its incident edges are removed before v1, then v5 is

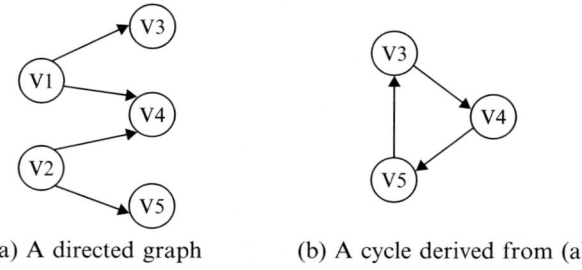

(a) A directed graph   (b) A cycle derived from (a)

Fig. 2. Graph and cycle.

the only node in the second batch with zero InDegree and will precede v3 in
the node ordering. The structure of BN will be influenced by the difference
sequence between v3 and v5. So v3 and v5 should be labeled with the same
batch number, which means that the sequence between the two nodes is not
sure and should be determined through a criterion. After all nodes in the
i-th batch with zero InDegree are deleted from the queue, the current nodes
with zero InDegree are labeled with the batch number $i+1$ and inserted into
the queue successively. In Figure 2(a), v3,v4 and v5 are all labeled with the
same batch number 2 because their InDegrees become zero after removing all
nodes, i.e. node v1 and v2, in the first batch. The improved topological sort
is as follows.

Procedure Improved_topo_sort ()
{Calculate and store the InDegree of every node into array InDegree[1..n];
 Initialize the batch number of each node to 0;
 Initialize queue; $b = 1$;
 Enqueue nodes with 0 InDegree and assign 1 to their batch numbers;
 while(queue is not empty)
   { p=queue[front];   f=$batch[p]$; b++ ;
     while($batch[p]$==f and front < rear)
       { for(k=1;k≤ n; k++)
           if(GM[p][k]!=0) InDegree[k]- - ;   //GM is adjacency Matrix;
         front++;   //Delete the front element from the queue;
         p=queue[front];   //Let p be the current front of the queue;
       }
     set b to the batch number of nodes with 0 InDegree and Enqueue them;
   }
}

Having sorted topologically, the sequence of departing from the queue is
the node ordering, each node having a batch number.

Third, an improved K2 algorithm is used to construct BN. In TK2 algorithm, the parent nodes of a node can only be selected from those preceding
the node itself in the node ordering. In fact, several nodes whose InDegrees
become zero at the same time enter the queue in a random order and the
sequence between any two of them is not sure.

Our improvement is to choose the parents for a node X_i of the kth batch
from those whose batch numbers are not greater than k. A node with the
same batch number with X_i and following X_i maybe becomes X_i's parent.
The improved K2 algorithm (IK2) is as follows.

Procedure IK2 ()
{for ($i = 1$; $i \leq n$; $i++$)
  { $pa[i]=\phi$;   Pold=0;   proceed=true;
    while (proceed==true && $|pa[i]| < u$)

```
    { Let Z be the node in Candidate(i)–pa[i] that maximizes g(i, pa[i]∪{Z});
      Pnew=g(i, pa[i] ∪ {Z}); // g( ) is the scoring function;
      if ( Pnew>Pold) then
        { Pold=Pnew;      pa[i] = pa[i] ∪ {Z}; }
      else proceed=false;
    }
  }
  if( IsCycle()) Eliminate_Cycle;
}
```

where $pa[i]$ is the set of the ith node's parent nodes, u is the maximum number of a node's parents, Candidate(i) is the set of nodes whose batch number is not bigger than that of node i. The scoring function is as follows.

$$g(i, pa[i]) = \prod_{j=1}^{q_i} \frac{1}{(N_{ij}+1)!} \prod_{k=1}^{r_i} (N_{ijk})! \qquad (4)$$

where r_i is the number of the values of X_i, $r_i = 2$ because X_i only has two values {presence, absence}, N_{ijk} is the number of samples in D in which node X_i has the kth value and $pa[i]$ has the jth value, and $N_{ij} = \sum_{k=1}^{r_i} N_{ijk} = N_{ij1} + N_{ij2}$.

However, it maybe generates a cycle among nodes labeled with the same batch number, for example, in Figure 2(b). Function $Iscycle$ is used to judge whether there is a cycle in the built BN structure. If there is a cycle, then return TRUE; otherwise FALSE. $Eliminate\_Cycle$ is used to remove the edge pointing node X whose scoring function value is the smallest among all nodes in the cycle.

So we can make sure that the posterior probability $P(\text{ST}|D)$ of the structure ST is the biggest among those of all structures, where D is the data set. Having constructed the semantic network, the parameters, i.e. the conditional probability of each node, are learned by standard statistic method given the BN structure. If there is the prior knowledge, IK2 algorithm is denoted as IK2_A, otherwise as IK2_B.

4 Inferring the Final annotation Set of a New Video

Having obtained the semantic candidate set (SCS) and the first concept S1 of the new shot, we should have a way to determine which of the others in SCS are also present in the same shot. Bayesian inference is used to calculate the conditional probabilities of the other concepts given S1. Suppose that the initial current evidence set is $Evidence\_Set = \{S1 = 1\}$ (1 means presence, 0 means absence). If $P(S2 = 1|Evidence\_Set) > \sigma$, then S2 will be assigned to the new shot; If $P(S3 = 1|Evidence\_Set) > \sigma$, then S3 will be assigned to the new shot. If $P(S4 = 1|Evidence\_Set) > \sigma$, then S4 will be assigned to the new shot. We obtain the final annotation set of the shot by dynastic Bayesian inference because *Evidence_Set* varies during inferring.

First, the directed graph of Bayesian Network is transformed to a chordal graph through moralization and triangulation. Then a join tree is built from the chordal graph, which consists of cliques node and separator sets (abbreviated as sepset). The belief potentials of each clique and each sepset are initialized according to the conditional probabilities of the nodes in Bayesian Network [11]. Suppose that the initialized join tree is denoted as *JT*. The procedure of obtaining the final annotation set (FAS) is as follows.

Procedure Infer_FAS(*JT*)
{ *New_Evidence* = {S1};   *Evidence_Set* = ϕ;
  $SCS = \{S1, \ldots, SN\}$
  While (NotEmpty (*New_Evidence*)) do
    {Input *New_Evidence* into *JT* and modify the potentials in the *JT*;
     Perform global propagation to make the potentials of *JT*
     locally consistent;
     Evidence_Set=*Evidence_Set* $\bigcup$ *New_Evidence*;
     New_Evidence = ϕ;
     for (each concept *Si* in *SCS*) do
       if ((*Si* not in *Evidence_Set*) and ($P(Si = 1|Evidence\_Set) > \sigma$))
         then *New_Evidence* = *New_Evidence* $\bigcup$ {*Si*};
    }
}

where *Evidence_Set* is used to store all evidence, and it is the final annotation set (FAS) of a shot after the procedure stops. NotEmpty (*New_Evidence*) is used to judge whether the set is empty. If it is empty, then return FALSE, otherwise TRUE.

Huang [11] did not give the stopping condition and it was determined by human. In **Infer_FAS**, a variable *New_Evidence* is introduced to store the newly generated evidence after inferring every time in order to judge when to stop the inference procedure. The inferring procedure will stop when generating no more new evidence, i.e. *New_Evidence* is empty. Initially, *New_Evidence* includes only one concept S1. After *New_Evidence* is input into *JT* and incorporated to *Evidence_Set*, it is set to empty subsequently. After inferring given *Evidence_Set* each time, the concept whose conditional probability is larger than σ becomes a new evidence and is put into *New_Evidence*. Repeat the above process until there is no more new evidence generated after inference.

Also, Huang [11] did not give a way to determine the threshold. In our experiment, the threshold σ is determined as follows.

$$\sigma = \frac{1}{N_T} \sum_{k=1}^{N_T} \sum_{i=2}^{\#C_k} P(S_i^k | S_1^k, \ldots, S_{i-1}^k) \quad (5)$$

where N_T is the number of samples in training set, $\#C_k$ is the number of actual concepts in Ground Truth of the kth sample, S_i^k is the ith actual concept of the kth sample, $P(S_i^k|S_1^k,...,S_{i-1}^k)$ is the conditional probability of S_i^k given $S_1^k,...,S_{i-1}^k$. In a word, σ is the average conditional probability of all concepts of Ground Truth over all samples in the training set. The threshold can be calculated automatically and adaptively for different data sets, avoiding setting the threshold manually.

5 Experiments

We have chosen videos of different kinds including landscape, city and animal from website www.open-video.org to create a database of 7.5 hours of videos. Data from 98 video clips has been used for the experiments.

First, we use the tool VideoAnnEx developed by IBM to partition every video clip into several shots (http://www.research.ibm.com/VideoAnnEx). The key frames of each shot are extracted automatically using the method in [12] to form the samples set. The perceptual features such as HSV accumulated Histogram and Edge Histogram are extracted automatically from the sample set and stored into a video database after being normalized.

Then the training set is automatically constructed from the sample set using the method in [5]. K-means clustering was performed on the sample set and selected one sample closest to the cluster center. All selected samples form the training set in order to improve the chance of obtaining a good classifier from the annotated data alone. Examples of selected key frames with the complete and sorted annotation are shown in Figure 1. 14 different concepts are extracted automatically from the ground truth annotation of the training set built manually, and their presence frequency is shown in Table 2. So our system can also work if the data set varies and more concepts are added.

In our experiments, there are 2764 key frames for training and 1353 key frames for testing. It was implemented by VC++ on a PC machine with AMD Athlon 2500+ CPU, 256M memory and Windows XP environment.

5.1 Evaluation Metric of Annotation

Suppose that there are n concepts. The evaluation metric used in other related works are the average precision and the average recall over n concepts, which are calculated as follows.

Table 2. 14 concepts with their presence frequency (%) in training set.

concept	car	road	bridge	building	waterfall	water	boat
percentage	3.08	6.84	2.66	19.39	1.14	42.97	6.08
concept	cloud	sky	snow	mountain	greenery	land	animal
percentage	13.31	44.87	6.84	14.45	32.7	25.48	25.48

$$AP = \frac{1}{n}\sum_{i=1}^{n} precision[c_i] \tag{6}$$

$$AR = \frac{1}{n}\sum_{i=1}^{n} recall[c_i] \tag{7}$$

where c_i ($i=1,\ldots,n$) is the ith concept,

$$precision[c_i] = \frac{N_{correct}}{N_{label}} \tag{8}$$

$$recall[c_i] = \frac{N_{correct}}{N_{ground\_truth}} \tag{9}$$

where $N_{correct}$ is the number of samples correctly annotated with a given concept c_i, N_{label} is the number of samples automatically annotated with c_i, N_{ground_truth} is the number of samples having c_i in GT.

For example, suppose that there are 4 concepts and the testing set is $\{f1, f2, f3\}$. The true annotation (i.e. ground truth) is as follows.

$$f1\{c1, c3\}, f2\{c1, c2\}, f3\{c4\}$$

If System I automatically annotates the three testing frames as follows

$$f1\{c1, c3\}, f2\{c1, c3\}, f3\{c2\}$$

then

$precision[c1] = 2/2 = 100\%,\qquad recall[c1] = 2/2 = 100\%$

$precision[c2] = 0/1 = 0,\qquad recall[c2] = 0/1 = 0$

$precision[c3] = 1/2 = 50\%,\qquad recall[c3] = 1/1 = 100\%$

[3]$precision[c4] = 0/0 = 1,\qquad recall[c4] = 0/1 = 0$

$$AP = \frac{1}{n}\sum_{i=1}^{n} precision[c_i] = (1 + 0 + 0.5 + 1)/4 = 62.5\%$$

$$AR = \frac{1}{n}\sum_{i=1}^{n} recall[c_i] = (1 + 0 + 1 + 0)/4 = 50\%$$

---

[3] When the denominator is zero, the quotient is set to 1. In fact, it is all right as long as it is destined as the same value under such a case.

AP and AR condition each other for their values. One is higher while the other is lower and vice versa. If the annotation result of another system are AP=60% and AR=52.5%, it is difficult to decide which of the two systems is better.

Suppose that System II automatically annotates the three testing frames as follows.

$$f1\{c1, c2, c3\}, f2\{c1, c3\}, f3\{c2\}$$

then

$precision[c1]$ = 2/2 =100%, $\quad\quad recall[c1]$ = 2/2= 100%

$precision[c2] = 0/2 = 0,$ $\quad\quad recall[c2] = 0/1 = 0$

$precision[c3] = 1/2$ =50%, $\quad\quad recall[c3] = 1/1$= 100%

$precision[c4] = 0/0$ =1, $\quad\quad recall[c4] = 0/1$= 0

$$AP = \frac{1}{n}\sum_{i=1}^{n} precision[c_i] = (1+0+0.5+1)/4 = 62.5\%$$

$$AR = \frac{1}{n}\sum_{i=1}^{n} recall[c_i] = (1+0+1+0)/4 = 50\%$$

System II annotates frame $f1$ with one more wrong concept c3 than System I does. It is obvious that System I is better than System II. But the conclusion is that both systems have the same performance using Formula (6) and (7) as evaluation metric because the length of semantic annotation (i.e. the number of concepts in annotation) is not considered. If AP and AR are the same respectively, the more the difference between the length of the annotation given by system and that of the ground truth annotation is, the worse the performance of the system is.

A new evaluation metric is proposed as follows.

$$SCORE = w_1*(AP+AR) + w_2*\left(1 - \frac{\text{abs}(AL - |AGT|)}{|SCS| - |AGT|}\right) \quad\quad (10)$$

where AL is the average length of annotation given for all the testing frames by the system, for example, AL in System I is 5/3 and AL in System II is 2. $|AGT|$ is the average length in the Ground Truth annotation of all the testing frames, in the above example, $|AGT|$=5/3, $|SCS|$ is the average length in $|SCS|$, and equals 4 in our experiments. If FAS is the same with SCS, then $AL = 4$ and the second item of Formula (10) becomes 0. Weight w1 and w2 are set to 0.9 and 0.1 respectively. The value of the second item is between 0 and 0.1 and that of $SCORE$ is between 0 and 1.9. The proportion of annotation length in $SCORE$ is only about 0.1/1.9=5.3%. AP and AR are still the main evaluation factor.

5.2 Experiments of Extracting SCS

Benitez [6] concluded that different classifiers were evaluated including K-nearest neighbors, one-layer neural network, and mixture of experts, of which K-nearest neighbor was shown to outperform the rest. So we compare our method of extracting the semantic candidate set described in section 2 with K Nearest Neighbor (KNN) and Naïve Bayesian (NB) classifier. Four most probable concepts are also chosen for KNN and NB to build their SCSs. Four standards are used to measure the performance of the three methods, three of them are the same with formula (6) (7) and (10) respectively, the fourth one is as follows.

$$AF = \frac{1}{n}\sum_{i=1}^{n} C\_first[c_i] \quad (11)$$

where

$$C\_first[c_i] = \frac{N_{correct\_first}}{N_{first}} \quad (12)$$

where N_{first} is the number of samples with concept c_i in the first position in Ground truth, $N_{correst\_first}$ is the number of the samples whose first concept in SCS is the same as that in Ground truth, i.e. c_i, $N_{correct}$ in formula (8) and (9) is the number of samples having a given concept c_i in its SCS correctly, N_{label} in formula (8) is the number of samples having c_i in SCS. In our experiments, $|AGT| = 2.4487$, $|SCS| = |AL| = 4$.

Table 3 shows the average $C\_first(AF)$, the average precision (AP) and the average recall (AR) over all concepts in SCS, and $SCORE$. It indicates KNN consistently outperforms NB, which conforms to the conclusion drawn by Benitez [6]. It also indicates that all metrics of SID are the biggest among the three methods, especially AR and AF of SID are much bigger than those of NB and KNN methods. So we have two reasons to expect that the annotation performance obtained by SID could be the best among the three methods. First, the larger the recall is, the more the SCS covers the correct concepts. Second, the first concept in SCS is the first evidence during Bayesian inference and the accuracy of the evidence is very crucial to the annotating results.

Table 3. The average precision (AP), average recall (AR) and average $C\_first(AF)$ of semantic candidate set before inference.

Method	AP	AR	AF	SCORE	# concepts with recall>0	concepts with recall=0
NB	0.274	0.444	0.137	0.6462	13	waterfall
KNN	0.378	0.490	0.182	0.7808	12	waterfall, bridge
SID	0.378	0.719	0.364	0.9872	14	—

5.3 Experiments of Constructing Semantic Network

We designed 21 node orderings and input them into TK2 algorithm [11], IK2_A algorithm respectively. Three classes of BN structures are built. The structures learned by TK2 algorithm are denoted as Class ST^1, the structures learned by IK2_A as Class ST^2, and a structure learned by IK2_B algorithm as Class ST^3. There are 21 structures in Class ST^1 and ST^2 respectively, and only one in Class ST^3.

Figure 3 shows one sample network built by the three algorithms respectively.

5.4 Selection of Inferring Threshold σ

Inferring threshold has a direct influence on the annotation performance. Formula (5) is used to calculate the inference threshold, which is regarded as the default threshold. In order to describe briefly, $A + B$ is used to denote the inference using A algorithm to extract SCS and using B algorithm to build the semantic network.

Figure 4 shows the curves of annotation score of SID+ IK2_B along with different thresholds. The $SCORE$ corresponding to the default threshold is called default $SCORE$. It indicates that the default $SCORE$ of SID +IK2_B is just the maximum. Limited to space, Figure 5 only shows the

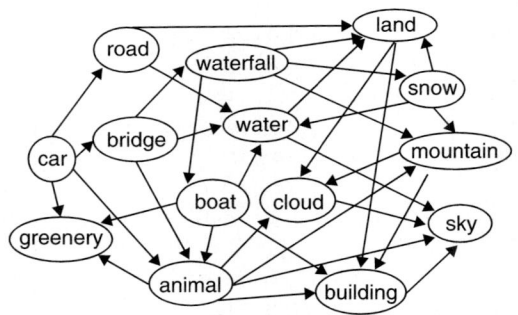

(a) A BN built by TK2 algorithm.

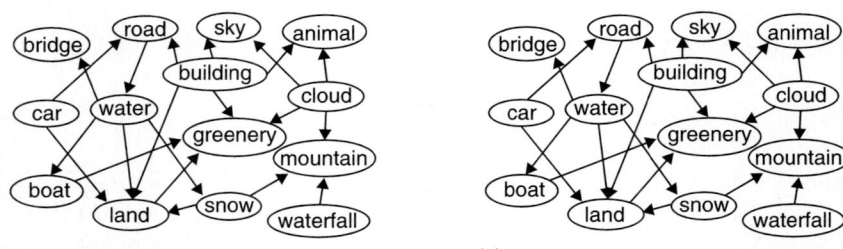

(b) A BN built by IK2_A algorithm. (c) A BN built by IK2_B algorithm.

Fig. 3. The semantic networks built by TK2, IK2_A and IK2_B algorithm.

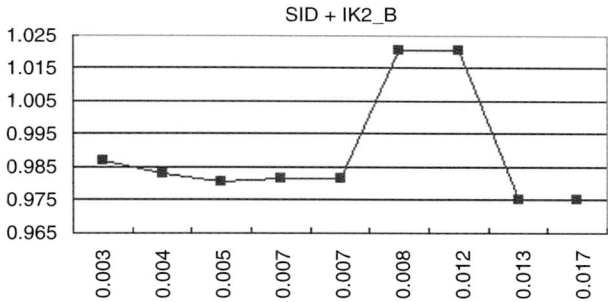

Fig. 4. The curves of annotation score along with different thresholds. The default threshold is 0.00816 and the default $SCORE$ is just the maximum 1.0208.

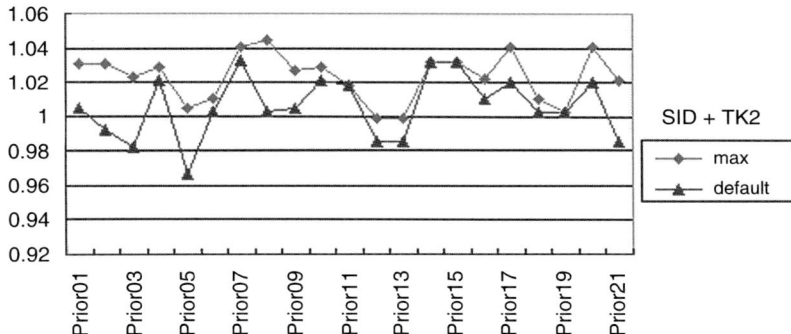

(a) The default $SCORE$ is less by 0.0162 (1.58%) than the average maximum.

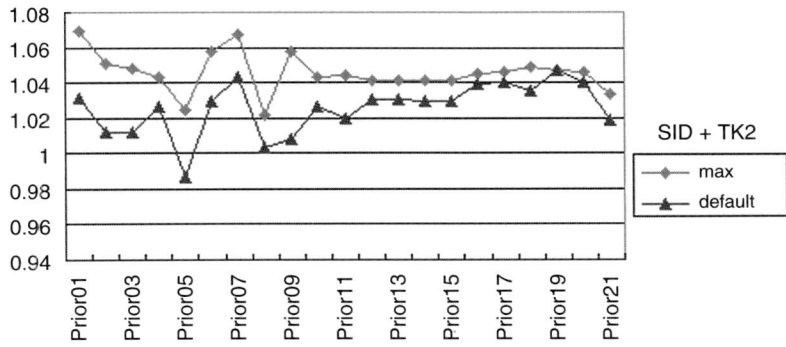

(b) The default $SCORE$ is less by 0.0199 (1.91%) than the average maximum.

Fig. 5. The comparison between the average default $SCORE$ and max $SCORE$.

comparing curves of the default $SCORE$ and the maximum $SCORE$ of SID +TK2 and that of SID +IK2_A. The horizontal coordinate axis is the different node orders and the vertical coordinate axis is $SCORE$. Sequence max is the maximum $SCORE$ under 21 node orders. Sequence $default$ is the $SCORE$ corresponding to the default threshold calculated by Formula

Table 4. The mean performance of annotating under all 21 node orders. # con is the number of concepts with recall > 0. Irtc is the increasing rate of the annotation performance of FAS higher than that of SCS. 'Drtm' is the decreasing rate of the default $SCORE$ lower than the maximum $SCORE$.

Method		MAL	MAP	MAR	MS	#con	MSC	Irtc	Drtm
NB(The $SCORE$ of its SCS is 0.6462)									
TK2	ST^1	3.1271	0.313	0.3721	0.6851	12	0.6729	4.13%	1.89%
IK2_A	ST^2	2.5707	0.3496	0.3237	0.6733	13	0.6959	7.69%	1.02%
IK2_B	ST^3	2.7034	0.3356	0.3377	0.6733	13	0.6896	6.72%	0
KNN(The $SCORE$ of its SCS is 0.7808)									
TK2	ST^1	3.5664	0.3186	0.3994	0.7181	10	0.6742	-16.1%	16.1%
IK2_A	ST^2	3.6263	0.3387	0.4380	0.7767	10	0.7232	-7.4%	7.4%
IK2_B	ST^3	2.5361	0.3844	0.3105	0.6949	9	0.7198	-7.8%	7.8
SID(The $SCORE$ of its SCS is 0.9872)									
TK2	ST^1	3.229	0.4508	0.6100	1.0609	14	1.0045	1.75%	1.58%
IK2_A	ST^2	2.951	0.48	0.5853	1.0652	14	1.0256	3.89%	1.91%
IK2_B	ST^3	2.411	0.4905	0.5353	1.0259	14	1.0208	3.4%	0

(5) under all 21 node orders. Figure 5(a) indicates that the average default $SCORE$ of SID+TK2 is only lower about 0.0162(1.58%) than the average maximum $SCORE$ over all 21 node orders. Figure 5(b) shows that the average default $SCORE$ of SID+IK2_A is only lower about 0.0199(1.91%)than the average maximum $SCORE$ over all 21 node orders.

In Table 4, column "Drtm" is the decreasing rate of the default $SCORE$ lower than the maximum $SCORE$ in all the annotation obtained by combining different method of extracting SCS and that of building semantic network. It indicates that the average default $SCORE$ is a little lower than the corresponding average maximum $SCORE$ expect for KNN algorithm.

Although the default $SCORE$ is not equal to the maximum $SCORE$, its advantage is no need for users to determine the inferring threshold and relieves users' burden. The function of manually setting the inferring threshold is also provided in our system. The following $SCORE$s are all default $SCORE$s.

5.5 Experimental Results of Annotation

FAS is selected from SCS by Bayesian Inference on the three classes of structures respectively. The performance of annotating is also measured using formula (10). The results are shown in Table 4.

Figure 6 shows the performance curves of all the annotation obtained by combining SID with three algorithms of building semantic network.

First, it indicates that IK2_A has the same performance with TK2 or outperforms TK2 under the same node order.

Second, it also shows that the fluctuating range of IK2_A under 21 node orders is much smaller than that of TK2, which means IK2_A is influenced by the correctness of the node order much smaller than TK2.

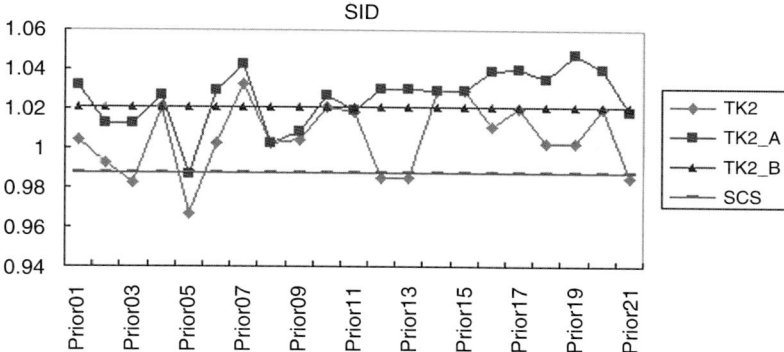

Fig. 6. The $SCORE$ curves of annotating results along with different node orders Sequence SCS corresponds to the $SCORE$ of SCS obtained by SID method. The default $SCORE$ curves of annotating results obtained by combining SID with several structure learning algorithm.

Third, it indicates that IK2_A outperforms IK2_B under some node orders, such as prior01 and prior07 and so on, and the result is reversed under some other node orders, such as prior02 and prior05 and so on. This is because IK2_A uses the node orders to orient the edges, and IK2_B completely depends on the training set to orient the edges. If the node order is correct, IK2_A can correctly determine the direction of edges. But IK2_B may produce the bias due to ignoring the correct node orders. So IK2_A outperforms IK2_B. If the wrong, IK2_A is influenced by it and gives the wrong direction of edges. But IK2_B can avoid the influence of wrong node order. So IK2_B outperforms IK2_A.

Forth, Table 4 shows the mean performance of annotation obtained by combining different methods of extracting SCS with different methods of building semantic network over all concepts under all 21 node orders. MAL, MAP and MAR are the mean values of AL, AP and AR respectively under all 21 node orders. MS and MSC are the mean value of sum ($sum = AP + AR$) and $SCORE$ respectively under all 21 node orders. It indicates that the inference performance on SCS obtained using SID is significantly higher than that using NB and KNN algorithm, no matter which algorithm of learning structure is used. It conforms to our expectation. The recall values of all concepts in SID are larger than 0, and that of some concepts with low presence frequency in NB and KNN are zero. This is because KNN and NB algorithm are sensitive to the distribution of each concept in the training set. SID does not suffer from such a problem because it is not sensitive to the presence frequency of each concept. Even a concept with low presence frequency can be assigned to a video shot if its semantic class centre is close to the shot based on visual features. It indicates that SID is more robust than NB algorithm and KNN algorithm.

Fifth, the column "Irtc" in Table 4 is the increasing rate of the annotation performance of FAS higher than that of SCS. It indicates that he $SCORE$ of FAS are higher than that of SCS after inferring except for KNN algorithm. The first reason is that KNN is sensitive to the presence frequency of concepts. Two concepts with low presence frequency, e.g. waterfall and bridge, are missed during the process of extracting SCS and they can not appear in FAS certainly. Moreover, the concepts having the high correlation with the two concepts are perhaps missed during inferring process. The second reason is that the concepts are sorted according to their presence frequency during the process of extracting the SCS using KNN algorithm. Suppose that the first concept in SCS is S1 and there is a concept S2 which should be assigned to the testing frame and has a very low presence frequency. Its conditional probability given S1 is calculated as follows.

$$P(S2|S1) = P(S2,S1)/P(S1) \qquad (13)$$

It is known that P(S1) is the highest and P(S2) is low in SCS. The joint probability P(S2, S1) is even lower than P(S2), so P(S2|S1) could not be high. If the inferring threshold σ is set higher, then S2 will be discarded. Table 4 shows that the recall values of several concepts indeed become zero after inferring.

The annotation performance using KNN algorithm increases along with the threshold σ decreasing. When σ is closed to zero, FAS is just SCS and the $SCORE$ reach the maximum. The average length of annotation obtained using KNN algorithm is also bigger than that of annotation obtained using other methods. Table 3 shows that KNN outperforms significantly NB in extracting SCS, but Table 4 shows that the advantage of KNN over NB is not obvious after inferring.

The concepts in SCS obtained using SID are sorted according to the similarity of the visual features. As long as the visual features of a semantic class centre is close to that of the testing frame, even though the concept has a very low presence frequency, it can be put in the first position in SCS and kept in FAS.

Figure 7 shows SCSs extracted by three methods and the automatic annotation of several testing samples using three methods to obtain SCS and three methods to construct BN. We can see that there is the most important concept in each frame. For example, the most important concept in Figure 7(a) is "waterfall", that in Figure 7(b) is "boat" and that in Figure 7(c) is "animal". SID can correctly annotate the most important concepts and put it in the first position of the annotation in most cases.

In general, SID is simple but efficient, and the performance is much better than KNN and NB algorithm in automatically annotating a shot. Although the IK2 outperform TK2 only a little, its outstanding advantage over TK2 is no need for users to give the complete node ordering.

We do not compare the results of our method with that of other papers because it is unfair to make a direct quantitative comparison with their

 GT: waterfall mountain	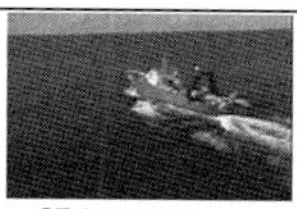 GT: boat water	 GT: animal water
NB	NB	NB
SCS: snow mountain building greenery	SCS: mountain road bridge building	SCS: animal water road greenery
NB+TK2 :	NB+TK2 :	NB+TK2 :
snow mountain building greenery	mountain building	animal water greenery
NB+IK2_A :	NB+IK2_A :	NB+IK2_A :
snow greenery mountain building	mountain building	animal greenery
NB+IK2_B :	NB+IK2_B :	NB+IK2_B :
snow greenery mountain building	mountain road bridge	animal road greenery
KNN	KNN	KNN
SCS: water mountain greenery sky	SCS: water animal greenery sky	SCS: water animal building greenery
KNN+TK2 :	KNN+TK2 :	KNN+TK2 :
water mountain greenery sky	water animal greenery sky	water animal building greenery
KNN+IK2_A :	KNN+IK2_A :	KNN+IK2_A :
water greenery sky mountain	water sky	water building
KNN+IK2_B :	KNN+IK2_B :	KNN+IK2_B :
water greenery sky mountain	water animal greenery	water animal greenery
SID	SID	SID
SCS: water mountain greenery snow	SCS: boat water animal cloud	SCS: animal boat water cloud
SID+TK2 : waterfall greenery	SID+TK2 : boat water animal	SID+TK2 : animal water
SID+IK2_A : water mountain	SID+IK2_A : boat water	SID+IK2_A : animal water
SID+IK2_B :	SID+IK2_B : boat animal cloud	SID+IK2_B : animal cloud
water mountain greenery		

Fig. 7. Automatic annotation results comparison.

method using different data set. We do not obtain the standard video data set in TRECVID to measure the algorithm performance now.

6 Conclusions

This paper presents a novel and efficient method for annotating video shots based on Bayesian inference. There are two main contribution of this work. The first is to introduce the semantic importance degrees of several concepts co-occurring in the same shot and propose a novel method to obtain SCS considering SID.

The second is to propose an improved K2 algorithm to discover the relationship of co-occurring concepts.

The experiment results have shown that SID method has outperformed significantly KNN and NB algorithm in annotating a new shot, and the performance of annotation with non-fixed length after inference is higher than that with fixed length before inference. Our future work focuses on improving accuracy by combining audio features and automatic determination of the weight value for subclass centers.

References

1. Szummer, M., Picard, R.: Indoor-outdoor image classification. In Syeda-Mahmood, T., ed.: *IEEE International Workshop in Content-Based Access to Image and Video Databases*, Bombay, India, IEEE Computer Society Press, U.S. (1998) 42–51

2. Vailaya, A., Jain, A., Zhang, H.: On image classification: city vs. landscape. In Li, C.S., ed.: *IEEE Workshop on Content-Based Access of Image and Video Libraries*, Santa Barbara, CA, USA., IEEE Computer Society Press, U.S. (1998) 3–8
3. Barnard, K., P. Duygulu, D., Forsyth, N., de Freitas, Blei, D., Jordan, M.: Matching words and pictures. *Journal of Machine Learning Research* **3** (2003) 1107–1135
4. Tseng, B.T., Lin, C.Y., Naphade, M.R., Natsev, A., Smith, J.: Normalized classifier fusion for semantic visual concept detection. In Torres, L., Garcia, N., eds.: *International Conference on Image Processing*, Barcelona, Spain, I.E.E.E. Press (2003) 535–538
5. Naphade, M.R.: A Probabilistic Framework For Mapping Audio-visual Features to High-Level Semantics in Terms of Concepts and Context. PhD thesis, University of Illinois at Urbana-Champaign (2001)
6. Jiménez, A.B.B.: Multimedia Knowledge:Discovery, Classification, Browsing, and Retrieval. PhD thesis, Columbia University (2005)
7. Jeon, J., Lavrenko, V., Manmatha, R.: Automatic image annotation and retrieval using cross-media relevance models. In Clarke, C., ed.: *Proceedings of the 26th International ACM SIGIR Conference on Research and Development in Information Retrieval*, Toronto, Canada, ACM Press (2003) 119–126
8. Lavrenko, V., Manmatha, R., Jeon, J.: A model for learning the semantics of pictures. In Saul, L.K., Weiss, Y., Bottou, L., eds.: *Proceedings of the Seventeenth Annual Conference on Neural Information Processing Systems*, Vancouver, British Columbia, Canada, MIT Press (2004) 553–560
9. Feng, S.L., Manmatha, R., Lavrenko, V.: Multiple bernoulli relevance models for image and video annotation. In Davis, L., ed.: *IEEE Conference on Computer Vision and Pattern Recognition*, Washington DC, USA, IEEE Computer Society (2004) 1002–1009
10. Cooper, G., Herskovits, E.: A Bayesian method for the induction of probabilistic networks from data. *Machine Learning* **9(4)** (1992) 309–347
11. Huang, C.: Inference in belief networks:a procedural guide. *International Journal of Approximate Reasoning* **11** (1994) 1–158
12. W, F., De, X., Weixin, W.: A cluster algorithm of automatic key frame extraction based on adaptive threshold. *Journal of Computer Research and Development* **42(10)** (2005) 1752–1757

An Algebraic Approach to Model-Based Diagnosis

Shangmin Luan[1], Lorenzo Magnani[2], and Guozhong Dai[3]

[1] Institute of Software, Chinese Academy of Sciences and School of Software, Beijing Institute of Technology, Beijing, P.R. China
shangmin@iscas.cn
[2] Department of Philosophy and Computational Philosophy Laboratory, University of Pavia, Pavia, Italy and Sun Yat-sen University, Guangzhou, P.R. China
lmagnani@unipv.it
[3] Institute of Software, Chinese Academy of Sciences, Beijing, P.R. China
guozhong@admin.iscas.ac.cn

Summary. Traditional approaches to computing minimal conflicts and diagnoses use search technique. It is well known that search technique may cause combination explosion. Algebraic approach may be a way to solve the problem. In this paper we present an algebraic approach to model-based diagnosis. A system with an observation can be represented by a special Petri net PN, checking whether there is a conflict between the correct system behavior and the observation corresponds to checking whether there exists a marking $M \in \mathcal{R}(M_0)$ such that $M(p_1)$ and $M(p_2)$ are not zero, where p_1 and p_2 are labeled with the output of the system and its negation respectively. Furthermore, we show that $M = M_0 + \mathcal{C}X$ is such a marking, where M_0 is the initial marking, $\mathcal{C}$ is the incidence matrix of PN, and X is the maximal vector in $\{V|V$ is a $\{0,1\}$-vector and for each transition t, if $V(t) = 1$, then there is a firing sequence $t_1, t_2, \ldots, t_m, t\}$. Then, we present an algorithm to compute the maximal vector X in $VSE(PN)$ in polynomial time. Once the maximal vector in $VSE(PN)$ is generated, we can check whether there is conflicts between the correct system behavior and the observation. We also present algorithms for computing minimal conflicts and diagnoses by using the above algorithm. Compared with related works, our algorithm terminates in polynomial time if the inputs of the each component in the system are not more than a given constant.

1 Introduction

Model-based approach to diagnosis is an important method for system diagnosis, and its main problem is to compute minimal conflicts and diagnoses. The approach ws first introduced by Reiter [1] and later extended by de Kleer [2]. The diagnosis algorithm begins with a description of a system and an observed system behavior (called an observation). If the observation conflicts with the correct system behavior, the system needs to be diagnosed. A system is described by a set of first-order sentences, but it is well known that the consistency check of a set of first-order sentences is undecidable. So, a system is described by a set of propositional sentences later on [3–5]. The diagnosis process involves two steps. First, all minimal conflicts are generated, then diagnoses can be obtained from minimal conflicts. Haenni [6] presents an algorithm for generating diagnoses from minimal conflicts. Reiter [1] computes diagnosis by HS-$tree$. However there are two problems with HS-$tree$. One problem with a complete HS-$tree$ is that the size of the tree grows exponentially with the size of the incoming collection of conflicts. Although diagnoses are generated more efficiently by pruned HS-$tree$, the construction of pruned HS-$tree$ is difficult to organize such that no unnecessary results are generated, because unnecessary subtrees are detected only after the entire subtree has been generated. Another problem is that the order of the incoming conflict is not specified, the process of the construction of the tree is not predicted and the desirable results may be not generated.

Many algorithms for computing diagnoses and consequences are based on the resolution proof procedure. Darwiche [3, 4] presented an approach computing diagnoses by using structured system description. Marquis [7] and Haenni [8] introduced algorithms for computing consequences. These algorithms have low efficiency for the reason that they generate all resolvents in the process of computing minimal conflicts. In fact, only part of all consequences are related to minimal conflicts. Darwiche [5] introduced an algorithm for diagnosing discrete-event systems. Val [9–11] presented an approach able to find consequences for some restricted propositional language. Algorithms for diagnosing special structured systems are also presented in several works, such as polynomial algorithms for tree-structured systems [12, 13], and the algorithm for diagnosing dynamic systems [14]. Fast algorithms for computing k-th diagnosis are given in [15, 16]. Luan, Magnani and Dai [17] give an algorithm to compute minimal conflicts by using structural information. The algorithm terminates in polynomial time for a system with special structure, such as the tree-structured system. Magnani [18] also discusses this topic from the philosophical point of view.

In this paper, we first give rules to transform a system with an observation into a Petri net, and show that the correct system behavior conflicts with an observation if and only if, for places p_1 and p_2 labeled with the output of the system and its negation in its corresponding Petri net $\langle S, T, F, W, M_0 \rangle$, $M(p_1)$ and $M(p_2)$ are not zero, where $VSE(N, M_0)$ is the set $\{V | V$ is a $\{0, 1\}$-vector and for each transition t, if $V(t) = 1$, then there is a firing sequence $t_1, t_2, \ldots, t_m, t\}$. Furthermore, algorithms for computing minimal conflicts and diagnoses are introduced.

The rest of the paper is organized as follows. Section 2 shows concepts and notations for model-based diagnosis and Petri net. In section 3, we introduce rules to transform a system with an observation into a Petri net PN. Furthermore, we prove that the correct system behavior conflicts with the observation if and only if, for place p_1 and p_2 labeled with the output of the system and its negation, $M(p_1) \neq 0$ and $M(p_2) \neq 0$, where $M = M_0 + \mathcal{C}X$, X is the maximal vector in $\{V|V$ is a $\{0,1\}$-vector and for each transition t, if $V(t) = 1$, then there is a firing sequence $t_1, t_2, \ldots, t_m, t\}$. In section 4, we present an algorithm for computing the maximal vector in $\{V|V$ is a $\{0,1\}$-vector and for each transition t, if $V(t) = 1$, then there is a firing sequence $t_1, t_2, \ldots, t_m, t\}$. Its correctness and time complexity are also shown. In section 5, the algorithms computing minimal conflicts and diagnoses are introduced. And our approach introduced in this paper is compared with related works in section 6. Section 7 concludes this paper.

2 Preliminary

This section introduces concepts and definitions related to model-based diagnosis and Petri net.

Definition 1. *A system description is a 4-tuple $\langle \mathcal{P}, \mathcal{A}, \mathcal{C}o, \Delta \rangle$, where $\mathcal{P}$ is a set of non-assumable atomic propositions; $\mathcal{A}$ is a set of assumable atomic propositions; $\mathcal{C}o$ is a set of components in a system; and Δ is a set of propositional formulas constructed from atoms in $\mathcal{P}$ and $\mathcal{A}$. This is illustrated as follows.*

Example 1. The system description of the system in figure 1 is as follows.
$\mathcal{P} = \{A, B, C, D, E\}$, $\mathcal{A} = \{Ok(x), Ok(y), Ok(z)\}$, $\mathcal{C}o = \{x, y, z\}$
$\Delta = \{Ok(x) \to (A \equiv \neg C), Ok(y) \to (A \wedge B \equiv D), Ok(z) \to (C \vee D \equiv E)\}$

The function of a component in a system can be described by an equivalent propositional formula which is true under the assumption that the component is in a normal state. Hence, each formula in Δ is an implication. The left of an implication is $Ok(x)$ which means that component x is in a normal state, and the right of an implication is an equivalent formula which describes the relationship between an output and inputs of a component.

Definition 2. *An observation Φ of a system is a set of literals constructed from the literals in $\mathcal{P}$. $\langle \mathcal{P}, \mathcal{A}, \mathcal{C}o, \Delta, \Phi \rangle$ means that Φ is an observation of $\langle \mathcal{P}, \mathcal{A}, \mathcal{C}o, \Delta \rangle$.*

Example 2. $\{\neg A, B, \neg E\}$ is an observation for the system in example 1.

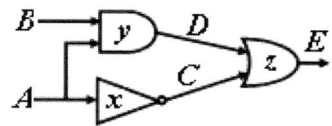

Fig. 1. An Example of a System.

Definition 3. *A conflict of $\langle \mathcal{P}, \mathcal{A}, Co, \Delta \rangle$ is a subset Co' of Co such that $\Delta \cup \Phi \cup \{Ok(x)|x \in Co'\}$ is inconsistent. A conflict Co' is minimal if and only if no proper subset of Co' is a conflict.*

Definition 4. *A diagnosis for $\langle \mathcal{P}, \mathcal{A}, Co, \Delta, \Phi \rangle$ is a minimal subset Co' of Co such that $\Delta \cup \Phi \cup \{\neg Ok(x)|x \in Co'\} \cup \{Ok(x)|x \in Co - Co'\}$ is consistent.*

Theorem 1. *$Co' \subseteq Co$ is a conflict for $\langle \mathcal{P}, \mathcal{A}, Co, \Delta, \Phi \rangle$ if and only if $\{Ok(x)| x \in Co'\}$ is minimally inconsistent with a subset of $\Delta \cup \Phi$ [1].*

By theorem 1, $\{x|Ok(x) \in \Gamma\}$ is a minimal conflict if Γ is a minimally inconsistent subset of $\Delta \cup \Phi \cup \mathcal{A}$. So minimal conflicts can be constructed from minimally inconsistent subsets.

Definition 5. *A Petri net is a 5-tuple $PN = (S, T, F, W, M_0)$ [21], where*

1. *$S = \{p_1, p_2, \ldots, p_m\}$ is a finite set of places.*
2. *$T = \{t_1, t_2, \ldots, t_n\}$ is a finite set of transitions.*
3. *$F \subseteq (P \times T) \cup (T \times P)$ is a set of arcs(flow relation).*
4. *$W : F \to \{1, 2, 3, \ldots\}$ is a weight function.*
5. *$M_0 : P \to \{0, 1, 2, 3, \ldots\}$ is the initial marking.*
6. *$S \cup T \neq \emptyset$ and $S \cap T = \emptyset$.*

$N = (S, T, F, W)$ is called a Petri net structure and denoted by N. A Petri net with the initial marking M_0 is denoted by $PN = (N, M_0)$.

Definition 6. *(domain, codomain, preset, postset, marking)*

1. *$dom(F) \cup cod(F) = S \cup T$, where*
   $dom(F) = \{x \in S \cup T| \exists y \in S \cup T : (x, y) \in F\}$
   $cod(F) = \{x \in S \cup T| \exists y \in S \cup T : (y, x) \in F\}$
2. *$\forall x \in S \cup T, \cdot x = \{y|(y \in S \cup T) \wedge ((y, x) \in F)\}$ is called preset of x, and $x^\cdot = \{y|(y \in S \cup T) \wedge ((x, y) \in F)\}$ is called postset of x.*
3. *A marking is a function $M : P \to \{0, 1, 2, 3, \ldots\}$.*

Definition 7.

1. *A transition $t \in T$ is called enabled at M if and only if $\forall p \in \cdot t : M(p) \geq w(p, t)$. $M[t\rangle$ means that t is enabled at M.*
2. *Suppose M is a marking and t is enabled at M, the firing of t generates a new marking M':*

$$M'(p) = \begin{cases} M(p) + w(t, p) & \text{If } p \in t^\cdot - \cdot t \\ M(p) - w(p, t) & \text{If } p \in \cdot t - t^\cdot \\ M(p) - w(p, t) + w(t, p) & \text{If } p \in \cdot t \cap t^\cdot \\ M(p) & \text{Otherwise} \end{cases}$$

This is denoted by $M[t\rangle M'$.

3. *A firing sequence is a sequence* $\sigma = M_0, t_1, M_1, t_2, \ldots, t_k M_k$ *or simple* $t_1, t_2, \ldots, t_k$ *such that* $\forall 1 \leq i \leq k$, $\exists t_i \in T : M_{i-1}[t_i\rangle M_i$. M_k *is reachable from* M_0, *and denoted by* $M_0[\sigma\rangle M_k$.
4. *For a marking* M, *the set of all reachable markings from* M *in PN is denoted by* $\mathcal{R}(N, M)$ *or simply* $\mathcal{R}(M)$.
5. *If* $\sigma \in T^*$, $\exists M \in R(M_0)$: $M_0[\sigma\rangle M$, $\#(\sigma/t)$ *denotes the occurrence times of* t *in* σ.
6. X *is called a firing count vector if* $X(i) = \#(\sigma/t_i), i \in \{1, 2, \ldots, n\}$.
7. *For each* $t \in T$, *if* t *appears at most one time in* $\sigma = t_1, t_2, \ldots t_k$, *i.e.*, $\#(\sigma/t) \leq 1$, *then this is denoted by* $M_0[\sigma\rangle_1 M_k$. $\mathcal{R}_1(M_0)$ *is used to denote the set of such kind of reachable markings.*

Definition 8. *1. For a Petri net PN, matrix* $C : S \times T \to Z$ *is called its incidence matrix, where* $C(p, t) = a(t, p) - a(p, t)$, Z *is the integer set.*

$$a(t,p) = \begin{cases} W(t,p), & (t,p) \in F; \\ 0, & \text{Otherwise.} \end{cases}$$

$$a(p,t) = \begin{cases} W(p,t), & (p,t) \in F; \\ 0, & \text{Otherwise.} \end{cases}$$

2. If $M[\sigma\rangle M'$, $M, M' \in \mathcal{R}(M_0)$, $\sigma \in T^*$, *then* $M' = M + CX$ *is called the state equation for* $PN = (N, M_0)$, *where* $X(t) = \#(\sigma, t)$ *is the firing count vector.*

Definition 9. *A Petri net is acyclic if there is no circles in it.*

Theorem 2. *Suppose* $PN = (S, T, F, W, M_0)$ *is a Petri net, where* M_0 *is the initial marking,* C *is the incidence matrix of PN. If* $M \in \mathcal{R}(M_0)$, *then there exists a n-vector* X *such that* $M = M_0 + CX$ *[21].*

The condition in theorem 2 is necessary, but not sufficient, i.e., if $M \in \mathcal{R}(M_0)$, then there must exist a nonnegative integer n-vector X such that $M = M_0 + CX$; but M may not be in $\mathcal{R}(M_0)$ if there exists a nonnegative integer solution for equation $M = M_0 + CX$. For an acyclic Petri net PN, the condition is sufficient and necessary [21].

Theorem 3. *For an acyclic Petri net* $PN = (S, T, F, W, M_0)$, C *is its incidence matrix.* $M \in \mathcal{R}_1(M_0)$ *if and only if there exists a* $\{0, 1\}$-*vector* X *such that* $M = M_C + CX$ *[21].*

3 Transforming a System with an Observation into a Petri Net

In this section, we present rules to transform a system with an observation into a Petri net PN, and prove that the correct system conflicts with the observation if and only if, for places p_1 and p_2 labeled with the output of the system and its negation respectively, $M(p_1) \neq 0$ and $M(p_2) \neq 0$, where $M = M_0 + CX$, X is the maximal vector in $VSE(PN)$.

Let $outputs(p)$ denote the number of outputs of p. The following rules are used to transform a system with an observation into a Petri net (N, M_0).

1. For each literal L in Φ, a place labeled with L is constructed.
2. For each component x in the system, there is a formula r, $Ok(x) \to (F \equiv C)$, to describe its function, where C is the output of x, F is constructed from the the inputs of x. The formula is firstly transformed into its conjunction normal form $Conj(r)$. For each clause $L_1 \vee L_2 \ldots L_n \vee L$ in $conj(r)$, where L is C or $\neg C$, it is equivalent to the rule $\neg L_1, \neg L_2, \ldots, \neg L_n \to L$. Suppose $Rule(r)$ denotes the set of the rules corresponding to the clauses in $conj(r)$.
3. For each rule $r : L_1, L_2, \ldots \to L_n$ in $Rule(r)$, a place is constructed for each literal L in r if there is no place labelled with L; a transition t corresponding to r is constructed as follows: the preset of t is the set of places labelled with $L_1, L_2, \ldots$ and L_n, and the postset of t is the set of place labelled with L.
4. For $t \in T$ and $p \in P$, if $(t,p) \in F$, then $W(t,p)$ is $outputs(p) + 1$; If $(p,t) \in F$, then $W(p,t)$ is 1.
5. For a place p, if p is the place labelled with L in the observation, $M_0(p) = outputs(p) + 1$; if p is the place labelled with $Ok(x)$, where x is a component, $M_0(p) = outputs(p) + 1$; otherwise, $M_0(p) = 0$. This is summarized as follows:

$$M_0(p) = \begin{cases} outputs(p) + 1, & p \text{ corresponds to a literal in the observation;} \\ outputs(p) + 1, & p \text{ corresponds to } Ok(x); \\ 0, & \text{Otherwise.} \end{cases}$$

By the above rules, a system with an observation can transformed into a Petri net. This is illustrated by the following example.

Example 3. The system in example 1 with the observation in example 2 is transformed into its corresponding Petri net $PN = (S, T, F, W, M_0)$ as follows.

For component x, it is a negation gate. $Ok(x) \to (A \equiv \neg C)$ is transformed into conjunction form: $\neg Ok(x) \vee \neg A \vee C$ and $\neg Ok(x) \vee A \vee \neg C$. These two clauses correspond to the following two rules, respectively: $Ok(x), \neg A \to C$ and $Ok(x), A \to \neg C$, i.e., $Rule(Ok(x) \to (A \equiv \neg C)) = \{Ok(x), \neg A \to C; Ok(x), A \to \neg C\}$. For the two rules, we construct places p_1, p_2, p_3, p_4, p_5 labeled with $Ok(x), A, \neg A, C$ and $\neg C$, respectively. Corresponding to the two rules, two transitions t_1 and t_2 are constructed.

For component y, it is an "and" gate. $Ok(y) \to ((A \wedge B) \equiv D)$ is transformed into conjunction normal form: $\neg Ok(y) \vee \neg A \vee \neg B \vee D$, $\neg Ok(y) \vee A \vee \neg D$, and $\neg Ok(y) \vee B \vee \neg D$. The three clauses are transformed into the following three rules, respectively: $Ok(y), A, B \to D$, $Ok(y), \neg A \to \neg D$ and $Ok(y), \neg B \to \neg D$, i.e., $Rule(Ok(y) \to (A \wedge B \equiv D)) = \{Ok(y), A, B \to D;$

$Ok(y), \neg A \to \neg D$; $Ok(y), \neg B \to \neg D$}. For the three rules, we construct places p_6, p_7, p_8, p_9 and p_{10} labeled with $Ok(y)$, B, $\neg B$, D and $\neg D$, respectively. The places corresponding to A and $\neg A$ have been constructed before, so we need not construct them again. Three transitions t_3, t_4 and t_5 are constructed to Correspond to the three rules.

For component z, it is an "or" gate. $Ok(z) \to ((C \vee D) \equiv E)$ is transformed into conjunction normal form: $\neg Ok(z) \vee C \vee D \vee \neg E$, $\neg Ok(z) \vee \neg C \vee E$ and $\neg Ok(z) \vee \neg D \vee E$. These two clauses are transformed into the following three rules, respectively: $Ok(z), \neg C, \neg D \to \neg E$, $Ok(z), C \to E$ and $Ok(z), D \to E$, i.e., $Rule(Ok(z) \to ((C \vee D) \equiv E)) = \{Ok(z), \neg C, \neg D \to \neg E; Ok(z), C \to E; Ok(z), D \to E\}$. For the three rules, we construct places p_{11}, p_{12} and p_{13} labeled with $Ok(z)$, E and $\neg E$, respectively. The places corresponding to C, $\neg C$, D and $\neg D$ have been constructed before, so we need not construct them here. Corresponding to the three rules, three transitions t_6, t_7 and t_8 are constructed.

So, a Petri net $PN = (S, T, F, W, M_0)$ corresponding to the system in example 1 with the observation in example 2 is defined as follows.

$S = \{p_1, p_2, p_3, p_4, p_5, p_6, p_7, p_8, p_9, p_{10}, p_{11}, p_{12}, p_{13}\}$.

$T = \{t_1, t_2, t_3, t_4, t_5, t_6, t_7, t_8\}$.

$F = \{(p_1, t_1), (p_1, t_2), (p_2, t_2), (p_3, t_1), (p_3, t_4), (p_6, t_3), (p_6, t_4), (p_6, t_5), (p_2, t_3),$
$(p_7, t_3), (p_8, t_5), (p_4, t_7), (p_5, t_6), (p_9, t_8), (p_{10}, t_6), (p_{11}, t_6), (p_{11}, t_7),$
$(p_{11}, t_8), (t_1, p_4), (t_2, p_5), (t_3, p_9), (t_5, p_{10}), (t_4, p_{10}), (t_7, p_{12}), (t_6, p_{13}),$
$(t_8, p_{12})\}$.

The weights of arcs are given below.

$W(p_1, t_1) = W(p_1, t_2) = W(p_2, t_2) = W(p_3, t_1) = W(p_3, t_4) = W(p_6, t_3) = W(p_6, t_4) = W(p_6, t_5) = W(p_2, t_3) = W(p_7, t_3) = W(p_8, t_5) = W(p_4, t_7) = W(p_5, t_6) = W(p_9, t_8) = W(p_{10}, t_6) = W(p_{11}, t_6) = W(p_{11}, t_7) = W(p_{11}, t_8) = W(t_7, p_{12}) = W(t_6, p_{13}) = W(t_8, p_{12}) = 1$, $W(t_1, p_4) = W(t_2, p_5) = W(t_3, p_9) = W(t_5, p_{10}) = W(t_4, p_{10}) = 2$.

Initial marking M_0 is defined as follows. p_1, p_6 and p_{11} correspond to $Ok(x)$, $Ok(y)$ and $Ok(z)$, so $M_0(p_1) = 3$, $M_0(p_6) = 4$, $M_0(p_{11}) = 4$. The observation given in example 2 is $\{\neg A, B, \neg E\}$, p_3, p_7 and p_{13} correspond to $\neg A$, B and $\neg E$, respectively. Hence, $M_0(p_3) = 3$, $M_0(p_7) = 2$, $M_0(p_{13}) = 1$. $M_0(p_2) = M_0(p_4) = M_0(p_5) = M_0(p_8) = M_0(p_9) = M_0(p_{10}) = M_0(p_{12}) = 0$. The petri net is shown in figure 2.

Proposition 1. *For a system with n components, each component in it has one output and at most m inputs, its corresponding Petri net has at most $2 * n * (1 + m)$ places and $n * 2^{m+1}$ transitions.*

Proof. For a component, since it has one output and at most m inputs, so there are at most $2 * (1 + m)$ places and 2^{m+1} transitions needed to be constructed

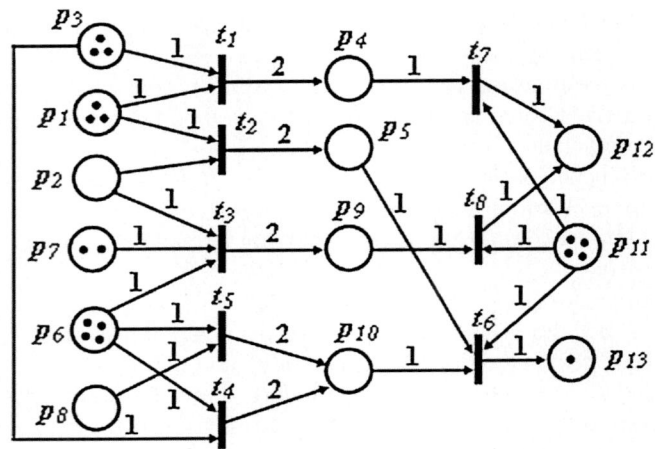

Fig. 2. Petri Net Representation of a System with an Observation.

in the Petri net corresponding to the system. Since there are n components in the system, there are at most $2*n*(1+m)$ places and $n*2^{m+1}$ transitions in the Petri net corresponding to the system.

Meseguer [20] also gives rules to transform a rule base into a Petri net. A marking is said to represent conflict if there are at least two places p_1 and p_2 such that $M(p_1)$ and $M(p_2)$ are not zero, where p_1 and p_2 correspond to L and $\neg L$ respectively. M_{CTR} is used to denote the set of markings represent conflicts, and M_{ini} is used to denote the set of initial markings. The approach is based on the following theorem.

Theorem 4. *For a rule base KB, $PN = (S, T; F, M_0)$ is its corresponding Petri net by rules in reference [20]. KB is inconsistent if and only if there are markings $M \in M_{ini}$, $M' \in M_{CTR}$ and a $\{0,1\}$-vector X such that $M' = M + CX$ [20].*

By theorem 4, many linear equations are needed to decide whether they have solutions, since there may be more than one markings in M_{ini} and M_{CTR}. Furthermore, by the definition of M_{CTR} in [20], it may be an infinite set. His approach is only true for an acyclic Petri net, but not true for cyclic Petri net. In fact, for a rule base, its corresponding Petri net may not be acyclic. In this paper, we will introduce a new approach to solve these problems. By our rules, the initial marking M_0 is generated in the process of construction of the Petri net PN corresponding to a system with an observation. And our algorithm is also true for cyclic Petri net.

Definition 10. *Suppose X is a n-vector.*

1. $|X| = \sum_{i=1}^{n} X(i)$.

2. X is nonnegative if its components are not negative. $X \geq 0$ represents that X is nonnegative.
3. X is positive if its components are positive. $X > 0$ represents that X is positive.

Definition 11. *For a Petri net $PN = (N, M_0)$, $\mathcal{C}$ is its incidence matrix.*

1. $VS(PN) = \{X | X \text{ is a } \{0,1\}\text{-vector such that } M = M_0 + \mathcal{C}X \geq 0\}$.
2. $VSE(PN) = \{X | X \in VS(PN) \text{ and for each transition } t, \text{ there is a firing sequence } t_1, t_2, \ldots, t_m, t \text{ in } PN \text{ if } X(t) = 1\}$.
3. *For a vector X in $VSE(PN)$, X is called the maximal vector in $VSE(PN)$ if $|X_1| < |X|$ for each $X_1 \in VSE(PN)$ and $X \neq X_1$*

Proposition 2. *For a Petri net PN, $VS(PN) = VSE(PN)$ if PN is acyclic.*

Proof. For each $X \in VS(PN)$, $M = M_0 + \mathcal{C}X$. If PN is acyclic, by theorem 3, $M \in \mathcal{R}_1(M_0)$, and there is a fire sequence σ, such that $M_0[\sigma\rangle M$ and, for each $t \in T$, $t \in \sigma$ if $X(t) = 1$. Hence, for each $t \in T$, there is a firing sequence $t_1, t_2, \ldots, t_m, t$ if $X(t) = 1$. So the theorem is true.

Lemma 1. *For a system with an observation, its corresponding Petri net is $FN = (S, T, F, W, M_0)$. For $X_1, X_2 \in VSE(PN)$, $t \in T$, a new vector X is constructed as follows.*

$$X(t) = \begin{cases} 1, & X_1(t) \neq 0 \text{ or } X_2(t) \neq 0; \\ 0, & X_1(t) \text{ and } X_2(t) \text{ are both zero.} \end{cases}$$

Then $X \in VSE(PN)$.

Proof. We first show that $M = M_0 + \mathcal{C}X \geq 0$. Suppose $M_1 = M_0 + \mathcal{C}X_1$, $M_2 = M_0 + \mathcal{C}X_2$, $\mathcal{C}(p)$ is the row of $\mathcal{C}$ labelled by p, then $M(p) = M_0(p) + \mathcal{C}(p) * X$.

If $M_1(p)$ or $M_2(p)$ are more than zero, without loss of generality, it is assumed that $M_1(p) > 0$. Then by the process of construction of the Petri net corresponding to the system with the observation, $M_0(p) = |\{(p,t)|(p,t) \in F\}| + 1$, or there is a positive integer $|\{(p,t)|(p,t) \in F\}| + 1$ in $\mathcal{C}(p)$, and $X_1(t) = 1$, so, $X(t) = 1$. Suppose that n is the sum of the negative numbers in $\mathcal{C}(p)$, we also know that $\mathcal{C}(p,t) + n = 1$. So, $M(p) = M_0(p) + \mathcal{C}(p) * X > 0$.

If both $M_1(p)$ and $M_2(p)$ are zero, then $X_1(t) = 0$ and $X_2(t) = 0$ for $\mathcal{C}(p,t) \neq 0$, and $M_0(p) = 0$. So, $X(t) = 0$ for $\mathcal{C}(p,t) \neq 0$. $M(p) = M_0(p) + \mathcal{C}(p) * X = 0$. Hence, $M \geq 0$ if $M_1 \geq 0$ and $M_2 \geq 0$.

For each $t \in T$, if $X(t) = 1$, then $X_1(t) = 1$ or $X_2(t) = 1$. Since X_1 and X_2 is in $VSE(PN)$, there is a firing sequence σ such that $t \in \sigma$. Hence, $X \in VSE(PN)$.

Theorem 5. *For a system with an observation, its corresponding Petri net is $PN = (S, T, F, W, M_0)$. If PN is acyclic, then there is maximal vector X in $VSE(PN)$.*

Proof. The vectors in $VSE(PN)$ can be ordered by $\geq$. Since $VSE(PN)$ is finite, hence, there must exist a vector X such that $|X| \geq |X_1|$ for each $X_1 \in VSE(PN)$. Suppose there exist two such vectors X_1 and X_2, $X_1 \neq X_2$, such that $|X'| \leq |X_1|$ and $|X'| \leq |X_2|$ for $X' \in VSE(PN)$. By lemma 1, a new vector X is constructed, $|X| > |X_1|$ and $|X| > |X_2|$. This conflicts with the hypothesis $|X| \leq |X_1|$ and $|X| \leq |X_2|$. Hence, the lemma is true.

Theorem 6. *For a system with an observation, its corresponding Petri net is $PN = (S, T, F, W, M_0)$, $\mathcal{C}$ is the incidence matrix of PN. There are conflicts between the correct system behavior and the observation if and only if, for the places p_1 and p_2 corresponding to the output of the system and its negation, both $M(p_1)$ and $M(p_2)$ are not zero, where $M = M_0 + \mathcal{C}X$ and X is the maximal vector in $VSE(PN)$.*

Proof. A transition corresponds to a rule, and a firing of a transition corresponds to an application of the rule. So, if both $M(p_1)$ and $M(p_2)$ are not 0, a conflict can be deduced from $\mathcal{A} \cup \varPhi \cup \{r | r$ corresponds to a transition t and $X(t) = 1\}$, i.e., there are conflicts between the correct system behavior and the observation.

Conversely, if there are conflicts between the correct system behavior and the observation, then a contradiction can be deduced from some formulae in $\mathcal{A} \cup \varPhi \cup \varDelta$. By the rules for constructing a Petri net corresponding to a system with an observation, an application of a formula corresponds to firings of transitions. So, these applications of the formulae correspond to a firing sequence σ_2. And the numbers of tokens in p_1 and p_2 are not 0 after σ is fired, i.e., $M_2(p_1) > 0$ and $M_2(p_2) > 0$. Suppose X_2 is the firing count vector of σ, $M_2 = M_0 + \mathcal{C}X_2 \geq 0$. Hence, $M = M_0 + \mathcal{C}X \geq M_2$, i.e., $M(p_1) > 0$ and $M(p_2) > 0$, where X is the maximal vector in $VSE(PN)$. ◇

Theorem 6 may not be true for the maximal vector $X \in VS(PN)$. This is illustrated by the following example.

Example 4. For the Petri net PN in figure 3, $\mathcal{C}$ is the incidence matrix of the petri net. The initial marking is $M_0 = (2, 0, 0)^T$. $X = (1, 1)^T$ is the maximal vector in $VS(PN)$, $M = M_0 + \mathcal{C}X = (1, 1, 1)^T \geq 0$. Suppose PN corresponds to a system with an observation, p_2 and p_3 correspond to the output of the system and its negation, respectively. Both $M(p_2)$ and $M(p_3)$ are 1, but there is no conflicts between the correct system behavior and the observation. X is in $VS(PN)$, but not in $VSE(PN)$ since each transition in PN is not enabled at M_0.

By theorem 6, once we have the maximal vector X in $VSE(PN)$, we can check whether there is conflict between the correct system behavior and an observation. In next section, we will introduce an algorithm to compute the maximal vector X in $VSE(PN)$.

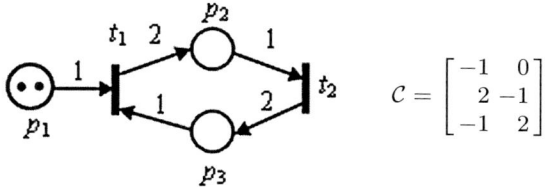

Fig. 3. A Cyclic Petri Net PN and its Incidence Matrix.

4 An Algorithm to Compute the Maximal Vector

By theorem 6, for a Petri net, if we have the maximal vector in $VSE(PN)$, we can check whether there is conflict between the correct system behavior and the observation. In this section, we present an algorithm for computing the maximal vector in $VSE(PN)$, and show the correctness and time complexity of the algorithm.

We first give an algorithm to compute the maximal vector in $VSE(PN)$ for an acyclic Petri net PN. Let X be the n-vector $(1, 1, 1, \ldots, 1)$, i.e., it is assumed that each transition appears in a firing sequence. Then, we check which transitions are not enabled. $\mathcal{C}(k)$ denotes the k-th row of $\mathcal{C}$. The process treats incidence matrix $\mathcal{C}$ row by row and begins with the first row of $\mathcal{C}$. Suppose that the k-th row is under consideration now.

1. If $M_0(p_k) + \mathcal{C}(p_k) * X \geq 0$, we can not decide which transitions are not enabled, so, X is not modified.
   a) If k is the number of columns of $\mathcal{C}$, then the process terminates and X is the maximal vector in $VSE(PN)$;
   b) otherwise, $k + 1$ row of $\mathcal{C}$ is treated.
2. If $M_0(p_k) + \mathcal{C}(p_k) * X < 0$, then $X(t) = 0$ for t that $\mathcal{C}(k, t) < 0$. The reason is as follows:
   a) If $M_0(p_k) \neq 0$, by the rules used to construct the Petri net, $M(p_k) = outputs(p_k) + 1$, so $M_0(p_k) + \mathcal{C}(p_k) * X > 0$; this conflicts with $M_0(p_k) + \mathcal{C}(p_k) * X < 0$. Hence, $M(p_k) = 0$.
   b) For $i (1 \leq i \leq$ the number of columns of $\mathcal{C})$ such that $\mathcal{C}(p_k, t_i)$ is positive, then $X(t_i) = 0$; otherwise, i.e., $X(i) = 1$, by the process of construction of the Petri net, $\mathcal{C}(p_k, t_i) = outputs(p_k) + 1$, so $M_0(p_k) + \mathcal{C}(p_k) * X > 0$.

   So, the transition t does not appear in any firing sequence if $\mathcal{C}(p_k, t) < 0$, Hence, $X(t) = 0$.

   This is further illustrated by example 5.

Example 5. The incidence matrix $\mathcal{C}$ for the Petri net in figure 2 is as follows.

$$C = \begin{array}{c} p_1 \\ p_2 \\ p_3 \\ p_4 \\ p_5 \\ p_6 \\ p_7 \\ p_8 \\ p_9 \\ p_{10} \\ p_{11} \\ p_{12} \\ p_{13} \end{array} \begin{bmatrix} -1 & -1 & 0 & 0 & 0 & 0 & 0 & 0 \\ 0 & -1 & -1 & 0 & 0 & 0 & 0 & 0 \\ -1 & 0 & 0 & -1 & 0 & 0 & 0 & 0 \\ 2 & 0 & 0 & 0 & 0 & 0 & -1 & 0 \\ 0 & 2 & 0 & 0 & 0 & -1 & 0 & 0 \\ 0 & 0 & -1 & -1 & -1 & 0 & 0 & 0 \\ 0 & 0 & -1 & 0 & 0 & 0 & 0 & 0 \\ 0 & 0 & 0 & 0 & -1 & 0 & 0 & 0 \\ 0 & 0 & 2 & 0 & 0 & 0 & 0 & -1 \\ 0 & 0 & 0 & 2 & 2 & -1 & 0 & 0 \\ 0 & 0 & 0 & 0 & 0 & -1 & -1 & -1 \\ 0 & 0 & 0 & 0 & 0 & 0 & 1 & 1 \\ 0 & 0 & 0 & 0 & 0 & 1 & 0 & 0 \end{bmatrix}$$
$$ \; t_1 \; t_2 \; t_3 \; t_4 \; t_5 \; t_6 \; t_7 \; t_8$$

The initial marking is $M_0 = (3, 0, 3, 0, 0, 4, 2, 0, 0, 0, 4, 0, 0)^T$. Suppose $X = (1, 1, 1, 1, 1, 1, 1, 1)^T$, since $M_0(p_1) + C(p_1)X \geq 0$, X need not be changed. Then the second row is under consideration further. $M_0(p_2) + C(p_2)X = -2 < 0$, $C(p_2, t_2) < 0$ and $C(p_2, p_3) < 0$, so, $X(t_2) = X(t_3) = 0$. X has been changed now. The process is repeated again from first row of C. For $X = (1, 0, 0, 1, 1, 1, 1, 1)^T$, $M_0(p_i) + C(p_i)X \geq 0$, $i = 1, \ldots, 4$, so, X need not be changed. For $i = 5$, since $M_0(p_i) + C(p_i)X < 0$, and $C(p_5, t_6) < 0$, hence $X(t_6) = 0$, i.e., $X = (1, 0, 0, 1, 1, 0, 1, 1)^T$. Because X has been changed, the process should be repeated again from first row of C. For $i = 1, \ldots, 7$, $M_0(p_i) + C(p_i)X \geq 0$, X need not be changed. For $i = 8$, $M_0(p_i) + C(p_i)X = -1 < 0$, and $C(p_8, t_5) < 0$, so $X(t_5) = 0$, i.e., $X = (1, 0, 0, 1, 0, 0, 1, 1)^T$. Because X have been changed, the process should be repeated from first row of C. For $i = 1, \ldots, 8$, $M_0(p_i) + C(p_i)X \geq 0$. For $i = 9$, $M_0(p_i) + C(p_i)X = -1 < 0$, and $C(p_9, t_8) < 0$, $X(t_8) = 0$, i.e., $X = (1, 0, 0, 1, 0, 0, 1, 0)^T$. The process begins from the first row of C again. For $i = 1, \ldots, 13$, $M_0(p_i) + C(p_i)X \geq 0$, the process terminates and the generated vector $X = (1, 0, 0, 1, 0, 0, 1, 0)^T$ is the maximal vector in $VSE(PN)$. Since the Petri net is acyclic, for each Petri net is acyclic, for each transition t that $X(t) = 1$, there $t_2, \ldots, t_k$ such that t is in the firing sequence. In fact, there is a firing sequence that contains all the transitions that $X(t) = 1$.

The algorithm to compute the maximal vector in $VSE(PN)$ for an acyclic Petri net is shown as follows. C is the incidence matrix of PN, and X is a vector with 1, i.e., $X = (1, 1, \ldots, 1)^T$.
Algorithm AcyclicMaximalVector(C, X)
  m = the number of lines in matrix C;
  n = the number of columns in matrix C;
  i = 1; %The process begins from first row of the matrix.
  While $((i \leq m)$ and $(M_0(p_i) + C(p_i)X \geq 0))$ do
    i = i+1;
  If $(M_0(p_i) + C(p_i)X < 0)$ then

```
    {
      For(j = 1 to n) do
        If (C(p_i, t_j) < 0) then X(t_j) = 0;
      X_1 = AcyclicMaximalVector(C, X);
      Return X_1
    }
  Else Return X;
End(Algorithm).
```

Theorem 7. *If PN is acyclic, then AcyclicMaximalVector(C, X) terminates in finite steps, and the returned vector X is the maximal vector in $VSE(PN)$.*

Proof. We first show AcyclicMaximalVector(C, X) terminates in finite steps. If $M_0(p_i) + C(p_i)X \geq 0$ for each $i(1 \leq i \leq m)$, then the procedure terminates in n steps(m is the number of rows of C, i.e., the number of the places in PN). If AcyclicMaximalVector(C, X) is recursively called, the procedure is called at most n times, where n is the number of the transitions in PN. Since the procedure is called when $M_0(p_i) + C(p_i)X < 0$, for this case, at least one component of X is changed from 1 to 0, so, X become a vector with 0 after the procedure is called n times. If X is a vector with 0, $M_0 + CX \geq 0$, the procedure terminates after it check each row of C. Hence, the procedure is called at most n times and terminates. This show that the procedure is an algorithm.

By the process of constructing X, we know that $M = M_0 + \mathcal{X}X \geq 0$. In order to prove $X \in VSE(PN)$, we should show that for each t, there exists a firing sequence $t_1 t_2 \ldots t_k$ containing t if $X(t) = 1$. For a transition t, there is at least one p_i such that $C(p_i, t) < 0$. If $X(t) = 1$ then $C(p_i, t) * X(t) < 0$. Since $M_0 + CX \geq 0$, for each $p \in {}^\cdot t$, $M_0(p) \neq 0$ or there is at least one transition t' such that $X(t') = 1$ and $C(p_j, t')$ is positive; since $X(t') = 1$, t' is treated as t, and this process is repeated and terminates since PN is acyclic. The transition sequence obtained by the above process is reversed and the resulted sequence is an firing sequence and t is last one in the sequence. So, $X \in VSE(PN)$.

We now show X is the maximal vector in $VSE(PN)$. Suppose that X' is the maximal vector in $VSE(PN)$, then $X(t) \leq X'(t)$ for each $t \in T$. For each $t_1 \in T$, $X(t_1) \leq X'(t_1)$, and $X(t_1) \neq X'(t_1)$ only if $X(t_1) = 0$ and $X'(t_1) = 1$. $X(t_1) = 0$ shows that $X(t_1)$ is changed from 1 to 0 in the process of computing. A component of X is changed from 1 to 0 if there exists a positive integer k such that $M_0(p_k) + C(p_k)X < 0$. So, there exists a positive integer k such that $M_0(p_k) + C(p_k)X' < 0$ since $X'(t_1)$ is changed from 1 to 0. But, by the definition of $VSE(PN)$ and the assumption that X' is the maximal vector in $VSE(PN)$, $M_0(p_k) + C(p_k)X' \geq 0$ for each k. So, it is impossible that $X(t_1) \neq X'(t_1)$, i.e., $X(t_1) = X'(t_1)$ is true for each $t \in T$. Hence, X is the maximal vector in $VSE(PN)$.

By theorem 6, for a system with an observation, if its corresponding Petri net is acyclic, then an algorithm for checking whether there are conflicts

between the correct system behavior and the observation is given as follows.
Algorithm Con-Check-Acyclic(PN)
   Construct the incidence $\mathcal{C}$ of PN;
   Construct $|T|$-vector $X = (1, \ldots, 1)^T$;
   Call procedure $X = $ AcyclicMaximalVector($\mathcal{C}, X$);
   $M = M_0 + \mathcal{C}X$;
   Check whether both $M(p_1)$ and $M(p_2)$ are not zero, where places p_1 and p_2 correspond to the output of the system and its negation. If there are such places, then there are conflicts.
End

Definition 12. *For a Petri net $PN = (S, T, F, W, M_0)$, arcs in $T \times S$ are called input arcs.*

The algorithm AcyclicMaximalVector($\mathcal{C}, X$) returns the maximal vector for an acyclic Petri net, but may not return correct result for a cyclic Petri net. The rest of this section will show the way to treat a Petri net with circles.

Example 6. For the Petri net in (a) of figure 4, the initial marking is $M_0 = (2, 0, 2)$.

Since p_3 is in the circle, and $M_0(p_3) = 2 \neq 0$, so the arc (t_2, P_3) in the circle is removed from PN, and the new Petri net PN_1 is shown in (b) of figure 4. Maximal-Vector($\mathcal{C}, X$) is called to compute the maximal vector in $VSE(PN_1)$, and the returned vector is $X = (1, 1)^T$. In fact, X is also the maximal vector in $VS(PN)$. If there are two places corresponding to the output of the system and its negation respectively, then the correct system behavior conflicts with the observation.

Theorem 8. *Suppose PN is the corresponding Petri net to a system with an observation. Cir is a circle in PN, and there is a place p in Cir such that there are tokens in p, then the edge $(t, p) \in Cir$ is removed from PN, and the resulted Petri Net is denoted as PN_1. The maximal vector in $VSE(PN_1)$ is the maximal vector in $VSE(PN)$.*

Proof. It is assumed that X is the maximal vector in $VSE(PN)$, X' is the maximal vector in $VSE(PN_1)$, then $X \geq X'$.

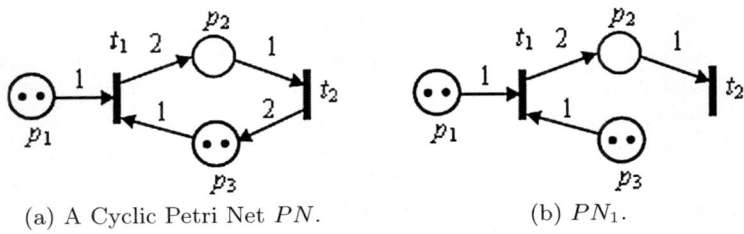

(a) A Cyclic Petri Net PN. (b) PN_1.

Fig. 4. A Cyclic Petri Net is transformed into Acyclic One.

Suppose σ is a firing sequence of PN. since the remove of edge (t,p) just change the state of p, but there are tokens in p, so the remove of (t,p), does not influence other transitions. Hence, σ is also a firing sequence of PN_1 So, $X \leq X'$. This is concluded by $X = X'$.

Theorem 9. *Suppose PN is the corresponding Petri Net to a system with an observation. It is also assumed that a circle Cir is in PN. The edges in Cir from a transition to a place are removed from PN, and the resulted Petri net is denoted as PN_1. $M_1 = M_0 + C_1 X_1$, where X_1 is the maximal vector in $VSE(PN_1)$, and C_1 is the incidence matrix of PN_1.*

1. *If $M_1(p) = 0$ for each p in the circle, then X_1 is the maximal vector in $VSE(PN)$.*
2. *If there is a place p in the circle such that $M_1(p) \neq 0$, then the arc (t,p) in the circle is removed from PN, the resulted Petri net PN_2 is acyclic. The maximal vector in $VSE(PN_2)$ is the maximal vector in $VSE(PN)$.*

Proof. Let X be the maximal vector in $VSE(PN)$.

1. Since $M_1(p) = 0$ for each p in the circle, for each $t \in \cdot p$, t does not occur in any firing sequence of PN_1. Hence, for a firing sequence σ in PN_1 and PN, σ does not contain transitions in the circle. So, σ is a firing sequence of PN_1 if and only if σ is a firing sequence of PN, i.e., $X(t) = X_1(t)$ for each $t \in T$. Hence, $X = X_1$.
2. Let X_2 is maximal vector in $VSE(PN_2)$. We show that $X_2 = X$. For a transition $t \in T$, if $X(t) \neq 1$, by the definition of $VSE(PN)$, there is a firing sequence $\sigma = t_1 t_2 \ldots t_k t$ of PN.

   a) If transitions in σ are not in the circle, then σ is also a firing sequence of PN_2.

   b) Otherwise, suppose that t_j in σ is the first transition that becomes disenabled in PN_2, but for t_k $(1 < k < j)$, t_k becomes enabled after t_{k-1} is fired in PN_2. The reason is that there is a place p in the circle and a transition $t' \in t_1, t_2, \ldots, t_{j-1}$ such that the edge (t',p) is in circle, but it is removed from PN. So, the firing of t' cause the number of tokens in PN, but not change the number of tokens in PN_2 since the edge (t',p) does not exist in PN_2. By the assumption of the theorem, a edge (t',p) in the circle is removed if $M_1(p) \neq 0$. Hence, there is a firing sequence σ_1 whose transitions are not in the circle, and the number of the tokens in p is changed after the transitions in σ_1 are fired sequentially. If some transitions in σ_1 appear in σ, then all such transitions are removed from σ_1 and the resulted sequence is denoted as σ_2, then we extend σ as follows: $t_1 t_2 \ldots t_{j-1} \sigma_2 t_j \ldots t$. For the transitions like t_j are treated as t_j, and the resulted sequence is σ', then σ' is a firing sequence of PN_2 and the last transition in σ' is t.

   Hence, $X_2(t) = X(t) = 1$.

If $X_2(t) = 1$, then there is a firing sequence σ of PN_2 such that t is in σ. σ is also a firing sequence of PN, so, $X(t) = 1$.

Hence, for $t \in T$, $X(t) = 1$ if and only if $X_2(t) = 1$. So, $X = X_2$, i.e., the maximal vector in $VSE(PN_2)$ is the maximal vector in $VSE(PN)$.

Example 7. For the Petri net PN in Example 4, there is no token in each place in the circle, and there is no firing sequence to change the number of each place in the circle. So, the inputs arcs in the circle are removed from the Petri net PN, and the resulted Petri net PN_1 is as follows. The maximal vectors of the Petri nets in Figure 3 and in Figure 5 are the same vector $(0,0)^T$.

Example 8. For the Petri net PN shown in figure 6, there is no place p in the circle such that $M_0(p) \neq 0$. So, all input arcs in the circle are deleted from PN, and the resulted Petri net PN_1 is acyclic. Maximal-Vector($\mathcal{C}$, X) is called to compute the maximal vector in $VSE(PN_1)$, and the returned vector is $(0, 1, 1)^T$. $M = M_0 + \mathcal{C}_1 X = (2, 1, 0, 1)$, where $\mathcal{C}_1$ is the incidence of PN_1. For p_2 in the circle, $M(p_2) = 1 \neq 0$, So, the arc (t_1, p_2) in the circle is deleted from PN, and the resulted Petri net PN_2 is shown in (b) of figure Maximal-Vector($\mathcal{C}$, X) is called to compute the maximal vector in $VSE(PN_2)$, the returned vector $X = (1, 1, 1)^T$ is the maximal vector in $VSE(PN)$.

Example 9. Suppose PN' is the Petri net resulted by removing all input arcs in the circle, X is the maximal vector in $VSE(PN_1)$, and $M' = M_0 + \mathcal{C}_1 X$. In a circle, there may be several places whose corresponding values in M_0 or M' are not zero. PN' is obtained by removing input arcs in the circle. In fact, only part of these arcs need to be deleted in order to remove the circle, and this does not affect the inconsistency check.

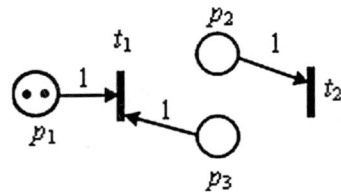

Fig. 5. PN_1.

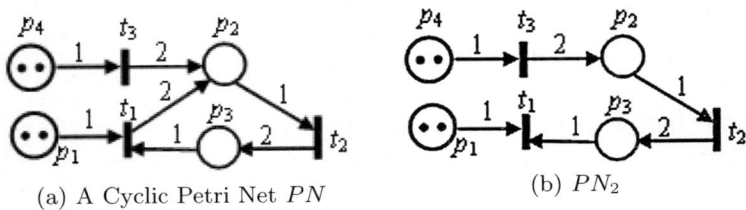

(a) A Cyclic Petri Net PN (b) PN_2

Fig. 6. A Cyclic Petri Net is transformed into Acyclic One.

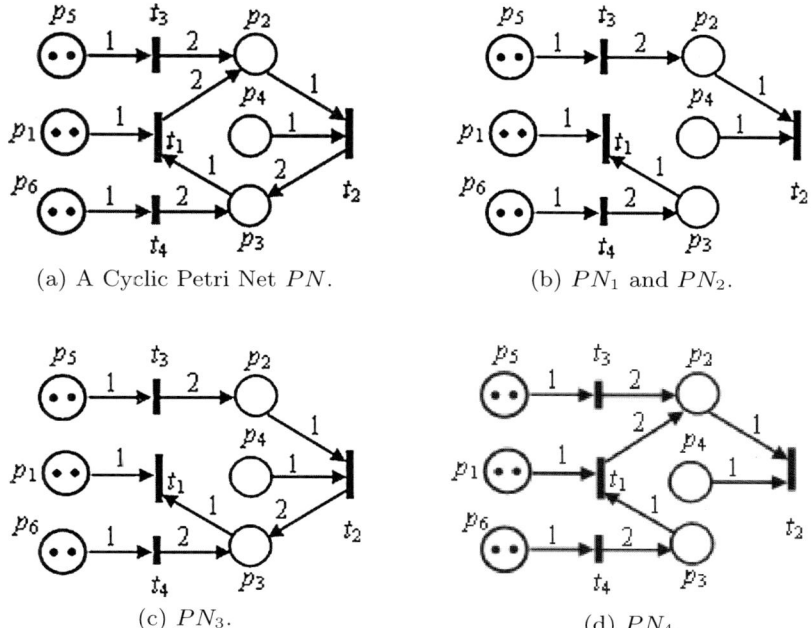

Fig. 7. A Cyclic Petri Net is transformed into Acyclic One.

For example, the Petri net in (a) of figure 7 has a circle, and for each place p in the circle, $M_0(p) \neq 0$, so, all the input arcs in the circle are removed from PN, and the resulted Petri net PN_1 is shown in (b) of figure 7, Maximal-Vector($\mathcal{C}$, X) is called to compute the maximal vector in $VSE(PN_1)$, the returned vector is $X = (1, 0, 1, 1)$. $M = M_0 + \mathcal{C}_1 X = (1, 2, 2, 0, 1, 1)$, where $\mathcal{C}_1$ is the incidence of PN_1. Since $M(p_2) = M(p_3) = 2 \neq 0$ for the places p_2 and p_3 in the circle, the arcs (t_1, p_2) and (t_2, p_3) are deleted from PN, and the resulted Petri net PN_2 shown in (b) of figure 7 is acyclic. Maximal-Vector($\mathcal{C}$, X) is called to compute the maximal vector in $VSE(PN_2)$, the returned vector is $X = (1, 0, 1, 1)$. X is also the maximal vector in $VSE(PN)$.

One of the two input arcs (t_1, p_2) and (t_2, p_3) is removed from PN, and the resulted Petri nets PN_3 and PN_4 shown in (c) and (d) of figure 7 are also acyclic. And the same maximal vector is generated.

If there is more than one circle in a Petri Net PN, by theorem 9, the maximal vector in $VSE(PN)$ can be computed recursively. The following example shows the way.

Example 10. A Petri net PN is shown in (a) of figure 8. There are two circles in the Petri net. One is $t_1 p_2 t_2 p_3 t_1$ denoted by Cir_1, another is $t_6 p_6 t_5 p_8 t_6$ denoted by Cir_2. For the Petri net in figure 8, the maximal vector in $VSE(PN)$ is $X = (1, 1, 1, 1, 1, 1)^T$, which can be generated by following steps.

1. First, the circle Cir_1 is selected to be treated. All the input arcs in Cir_1 are removed from PN, and the resulted Petri Net is denoted as PN_1 shown in (b) of figure 8, By theorem 9, the maximal vector in PN_1 should be computed, but there is also a circle in PN_1.

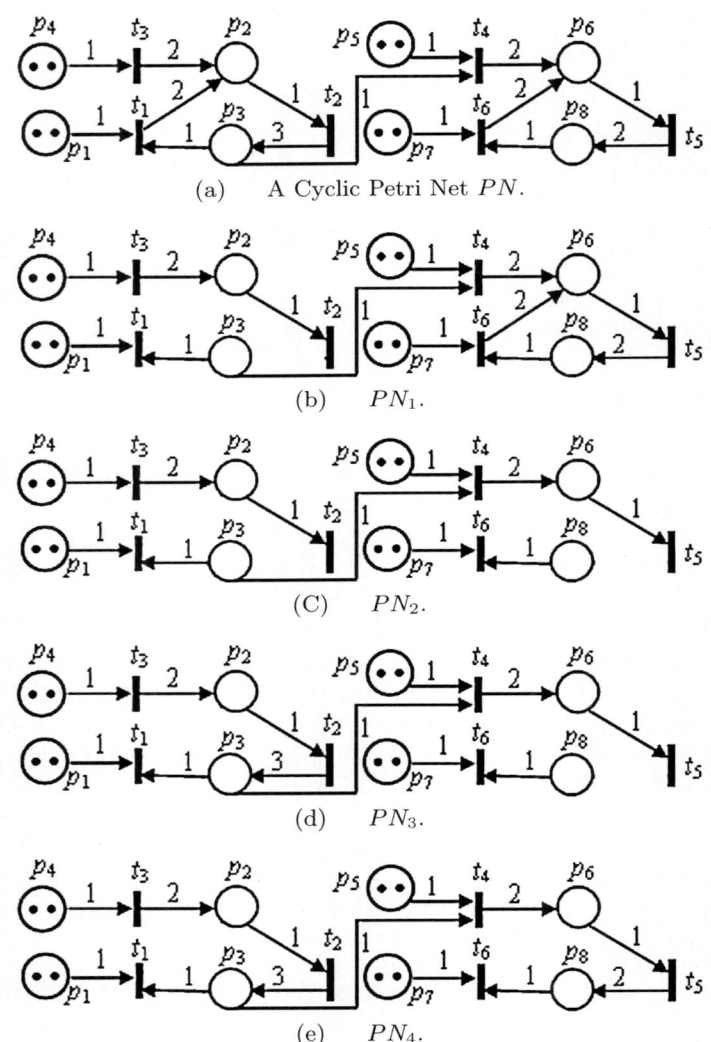

Fig. 8. A Cyclic Petri Net with two Circles.

2. There is also a circle in PN_1. So, the input arcs in Cir_2 are also removed from PN_1, and the resulted Petri Net is denoted as PN_2 shown in (c) of figure 8. Since there are no circles in PN_2, so Maximal-Vector(PN_2) is called to computed the maximal vector in $VSE(PN_2)$, and $X = (0,1,1,0,0,0)^T$ is the returned vector, $M_2 = M_0 + C_2 X = (2,1,0,1,2,0,2,0)^T$, where C_2 is the incidence of PN_2.
3. Since $M_2(p_2) = 1 \neq 0$ and $M_0(p_2) = 0$, the arc (t_1, p_2) is removed from PN, and the circle Cir_1 is removed. So, all the other edges in Cir_1 except (t_1, p_2) are are added to PN_2, and the resulted Petri net is denoted by shown in (d) of figure 8.
4. Since there are no circles in PN_3, Maximal-Vector(C_3, X) is called to compute the maximal vector in $VSE(PN_3)$, the returned vector is $X = (1, 1, 1, 1, 1, 0)^T$, $M_3 = M_0 + C_3 X = (1, 1, 1, 1, 1, 1, 2, 0)^T$, where C_3 is the incidence of PN_3. Since $M_3(p_6) = 1 \neq 0$ and $M_0(p_6) = 0$, So, all the other edges in Cir_2 except (t_6, p_6) are added to PN_3, and the resulted Petri net is denoted by resulted Petri net is denoted by PN_4 shown in (e) of figure 8.
5. C_4 is the incidence of PN_4. Since there are no circles in PN_4, Maximal-Vector(C_4, X) is called to compute the maximal vector in $VSE(PN_4)$, the returned vector is $X = (1, 1, 1, 1, 1, 1)^T$, which is also the maximal vector in $VSE(PN)$.

By theorem 8, in order to compute the maximal vector in $VSE(PN)$, if there are places in circles whose tokens are not zero, then the input arc to such places in the circles are removed from PN, and the resulted Petri net is denoted PN_1. By theorem 9, we can compute the maximal vector in $VSE(PN_1)$, and this process is denoted by Maximal-Vector-General(PN_1, $Circle$) which will be given later, $Circle$ is the set of circles in PN_1. So, the process of compute maximal vector in $VSE(PN)$ is given as follows:
Maximal-Vector-Generality(PN)
   $Circle$ is the set of circles in PN;
   $PN_1 = PN$;
   If ($Circle$ is empty) then
     { C is the incidence matrix of PN;
       $X = (1, 1, \ldots, 1)^T$;
       Return X = Maximal-Vector(C, X);
       Return X;
     }
   Else
     { For (each $Cir \in circle$) do
       If (there is a place $p \in Cir$ such that $M_0(p) \neq 0$) then
         { Remove Input Arcs (t, p) in Cir from PN_1;
           The resulted Petri Net is also denoted as PN_1;
           $Circle = Circle - \{Cir\}$;
         }

           $\langle X, PN_2 \rangle$ =Maximal-Vector-General(PN_1, $Circle$);
           Return X;
      }
End(Procedure).
   By theorem 9, Maximal-Vector-General(PN, $Circle$) is described as follows.
Procedure Maximal-Vector-General(PN, $Circle$)
   If (there are circles in PN) then
      { Cir is a circle in PN;
        Remove Input Arcs in Cir from PN and PN_1 is the resulted Petri Net;
        $\langle X_2, PN_2 \rangle$ =Maximal-Vector-General(PN, $Circle$);
        C_2 is the incidence matrix of PN_2;
        M_0 is the initial marking of PN_2;
        $M_2 = M_0 + C_2 X_2$;
        If (there is a place p in $Cir \in Circle$ such that $M_2(p) \neq 0$) then
           { For (each p in $Cir \in Circle$ such that $M_2(p) \neq 0$) do
             Add the arcs in Cir to PN_2 except $(t, p) \in Cir$
             The resulted Petri net is denoted as PN_3;
             C_3 is the incidence matrix of PN_3;
             $X = (1, 1, \ldots, 1)^T$;
             $X' =$ Maximal-Vector(C_3, X);
             Return $\langle X', PN_3 \rangle$;
           }
        Else Return $\langle X_1, PN_2 \rangle$;
      }
   Else
      { C is the incidence matrix of PN;
        $X = (1, 1, \ldots, 1)^T$;
        $X =$ Maximal-Vector(C, X);
        Return $\langle X, PN \rangle$;
      }
End(Procedure).
   The following theorem shows the correctness of the procedure Maximal-Vector-General(PN, $Circle$).

Theorem 10. *Maximal-Vector-General(PN, Circle) returns the maximal vector in $VSE(PN)$ if there are no circles in PN.*

Proof. Maximal-Vector-General(PN, $Circle$) is a recursive description of the theorem 9, so, the theorem is true.

   Theorem 11 shows the correctness of Maximal-Vector-Generality(PN).

Theorem 11. *Maximal-Vector-Generality(PN) returns the maximal vector in $VSE(PN)$.*

Proof. Maximal-Vector-Generality(PN) combines the steps in theorem 8 and 9. So, by theorem 8 and 10, the theorem is true.

Theorem 12. *Suppose $|T| = m$ and $|S| = n$. The time complexity of Maximal-Vector($\mathcal{C}, X$) is $O(n * m^2)$.*

Proof. The number of call to Maximal-Vector($\mathcal{C}, X$) is at most m times. And Maximal-Vector($\mathcal{C}, X$) terminates at most $n*m$ steps. So, the time complexity of the procedure is $O(n * m^2)$.

Theorem 13. *Suppose $|T| = m$ and $|S| = n$. The time complexity of Maximal-Vector-General(PN, $Circle$) is $O(n * m^3 + n^2 * m^2)$.*

Proof. The number of call to Maximal-Vector-General($PN, Circle$) is at most $n+m$. And Maximal-Vector-General($PN, Circle$) terminates at most $K_1 * n * m^2 + K_2 * n * m$ steps, where K_1 and K_2 are constants. So, the time complexity of the procedure is $O((n + m) * (n * m^2))$, i.e., $O(n * m^3 + n^2 * m^2)$.

Theorem 14. *Suppose $|T| = m$ and $|S| = n$. The time complexity of Maximal-Vector-Generality(PN) is $O(n * m^3 + n^2 * m^2)$.*

Proof. Maximal-Vector($\mathcal{C}, X$) terminates at most $max\{K_1 * n^2 * m, (K_2 * n * m^3 + n^2 * m^2 + K_3 * n * m)\}$ steps, where K_1, K_2 and K_3 are constants, where K_1 and K_2 are constants. So, the time complexity of the procedure is $O(n * m^3 + n^2 * m^2)$.

5 Algorithms for Diagnosing a System

Computing minimal conflicts and diagnoses are two related topics for model-based diagnoses. In this section, we show a way to compute minimal conflicts and diagnoses by using the consistency check algorithm given in section 4.

5.1 An Algebraic Approach to Computing Minimal Conflicts

In this section, algorithms for computing minimal conflicts is introduced.

Definition 13. *Suppose σ is a firing sequence $t_1 t_2 \ldots t_m$, for each $t_i (i = 1, 2, \ldots, m)$, $\sigma_1 = t_1 \ldots t_{i-1} t_{i+1} \ldots t_m$ is not a firing sequence, then σ is called a minimal firing sequence.*

Lemma 2. *Suppose PN is the Petri Net corresponding to a system with an observation. It is also assumed that σ is a firing sequence, $M_0[\sigma\rangle M$. p_1 and p_2 correspond to the output of the system and its negation, respectively. $S_1 = \bigcup_{t \in \sigma} \cdot t$, $Con = \{Ok(x) | p \in S_1$ and p corresponds to $Ok(x)\} \cap \mathcal{A}$ is a minimal conflict if σ is a minimal firing sequence such that $M(p_1) \neq 0$ and $M(p_2) \neq 0$.*

Proof. Suppose X is a vector constructed as follows: $X(t) = 1$ if $t \in \sigma$. Then, $M = M_0 = CX$. If $M(p_1) \neq 0$ and $M(p_2) \neq 0$, by theorem 6, there is a conflict between the correct system behavior and the observation. Furthermore, Con is a conflict. Suppose that Con is not minimal, then it is deduced that σ is not a minimal too. This is conflicts with the assumption. Hence, the lemma is true.

Definition 14. *Suppose* $PN = (S, T, F, W, M_0)$ *is a Petri net, and* $S_1 \subseteq T \cup S$. $\cdot S_1 = \bigcup_{e \in S_1} \cdot e$.

Suppose that PN is the Petri net corresponding to a system $\langle \mathcal{P}, \mathcal{A}, \mathcal{C}o, \Delta \rangle$ with an observation Φ, by lemma 2, if minimal firing sequences is generated, minimal conflicts can be constructed from the generated minimal firing sequences. For each $t \in \sigma$ and $p \in \cdot t$, there is one transition $t' \in \cdot p$, $X(t) = 1$, and $t' \in \sigma$. So, the process of constructing minimal firing sequences is as follows: For p_1 and p_2 corresponding to the output of the system and its negation, $t_1 \in \cdot p_1$, $t_2 \in \cdot p_2$, $\{t_1, t_2\}$ is constructed if $X(t_1) = X(t_2) = 1$; $\{t_1\}$ is constructed if $X(t_1) = 1$ and $X(t) = 0$ for each $t \in \cdot p_2$; $\{t_2\}$ is constructed if $X(t_2) = 1$ and $X(t) = 0$ for each $t \in \cdot p_1$. Furthermore, if $\cdot(\cdot t_1) \neq \emptyset$, then a minimal firing sequence should contain one transition t' in $\cdot(\cdot t_1)$ and $X(t') = 1$. For t_2, it is similar to t_1, i.e., if $\cdot(\cdot t_2) \neq \emptyset$, then a minimal firing sequence should contain one transition $t'' \in \cdot(\cdot t_2)$ with $X(t'') = 1$. For t' and t'', the above process is repeated until there $\cdot(\cdot t')$ and $\cdot(\cdot t'')$ is empty. The above process is summarized by the following procedure.

Algorithm Minimal-Conflict($\mathcal{P}, \mathcal{A}, \mathcal{C}o, \Delta, \Phi$)
  Construct a Petri Net PN corresponding to $\langle \mathcal{P}, \mathcal{A}, \mathcal{C}o, \Delta \rangle$ with Φ;
  X = Maximal-Vector-Generality(PN);
  $M = M_0 + CX$; $MC = \emptyset$; $TMC = \emptyset$;
  p_1 and p_2 correspond to the output of the system and its negation;
  If ($M(p_1) = 0$ or $M(p_2) \neq 0$) then { return $\emptyset$; exit()};
  If (for each $t \in \cdot p_1$, $X(t) = 0$) then $MC = MC \cup \{\{t_2^0\}|t_2 \in \cdot p_2, X(t_2) = 1\}$;
  Else If (for each $t \in \cdot p_2$, $X(t) = 0$) then
      $MC = MC \cup \{\{t_1^0\}|t_1 \in \cdot p_1, X(t_1) = 1\}$;
  Else $MC = MC \cup \{\{t_1^0, t_2^0\}|t_1 \in \cdot p_1, t_2 \in \cdot p_2, X(t_1) = X(t_2) = 1\}$;
  While ($nonempty(MC)$) do
    { $T_1 = delete(MC)$;
      If (there no $t_j^0 \in T_1$) then $\{TMC = TMC \cup \{T_1\}\}$;
      Else
        { For ($t_j^0 \in T_1$) do
          { $T_1 = T_1 - \{t_j^0\} \cup \{t_j^1\}$; $TT = \{T_1\}$;
            $S_1 = \cdot t_j \cap \{p|\cdot p \neq \emptyset\}$;
            While ($nonempty(S_1)$) do
              { $p = delete(S_1)$;
                $TT_1 = \cdot p$; $TT_2 = \emptyset$;
                If ($nonempty(TT_1)$) then
                  { For (each $t \in TT_1$ and $X(t) = 1$) do

$$TT_2 = \{T' \cup \{t^0\}|T' \in TT, t^0 \notin T' \text{ and } t^1 \notin T'\};$$
$$TT = TT_2;$$
}
}
$MC = MC \cup TT;$
}
}
}
$Con = \emptyset;$
For $(T' \in TMC)$ do
  $Con = Con \cup \{\{L|L \text{ is the literal corresponding to } t \in T'\}\};$
$Conflicts = \{\mathcal{F} \cap \mathcal{A}|\mathcal{F} \in Con\};$
For $(Co_1, Co_2 \in Conflicts \text{ and } Co_1 \subset Co_2)$ do
  $Conflicts = Conflicts - \{Co_2\};$
$MinConflicts = \emptyset;$
For (each $\mathcal{A}' \in Conflits$) do
  $MinConflicts = MinConflicts \cup \{\{x|Ok(x) \in \mathcal{A}'\}\};$
Return $MinConflicts;$
End(Algorithm)

The process is illustrated by the following example.

Example 11. For the system in example 1 with the observation in example 2, its corresponding Petri net is shown in example 3, the maximal vector $X = (1, 0, 0, 1, 0, 0, 1, 0)^T$ is given in example 5. Since $M = M_0 + \mathcal{C}X = (2, 0, 1, 1, 0, 3, 2, 0, 0, 2, 3, 1, 1)$. Since, p_{12} and p_{13} correspond to E and $\neg E$, respectively, and $M(p_{12}) = M(p_{13}) = 1$, a conflict occurs between the correct system behavior and the observation. In row 13 of the incidence matrix $\mathcal{C}$, only $\mathcal{C}(p_{13}, t_6)$ is nonzero, but $X(t_6) = 0$, this shows that $\neg E$ is not deduced from the rule corresponds to t_6. Since $M_0(p_{13}) = 1$, so, $\neg E$ is only a fact. In row 12 of $\mathcal{C}$, $\mathcal{C}(p_{12}, t_7)$ and $\mathcal{C}(p_{12}, t_8)$ are nonzero. $X(t_8) = 0$ shows that E is not deduced from the rule corresponds to t_8. $X(t_7) = 1$ shows that E can be deduced from the rule corresponds to transition t_7. Column 7 of $\mathcal{C}$ has two negative number $\mathcal{C}(p_4, t_7) = \mathcal{C}(p_{11}, t_7) = -1$; this means p_4 and p_{11} are the inputs of t_7. p_{11} corresponds to $Ok(z)$, and p_{11} has no inputs since there are no positive number in row 11 of $\mathcal{C}$. In row 4 of $\mathcal{C}$, $\mathcal{C}(p_4, t_1) = 2$ shows that t_1 is the input of p_4. In column 1 of $\mathcal{C}$, $\mathcal{C}(1, 1) = -1$ and $\mathcal{C}(3, 1) = -1$ are negative; this shows that p_1 and p_3 are inputs of t_1, t_1 is the inputs of p_4. Since $X(1) = 1 > 0$, so t_1 is enabled. In row 1 and 3, there are no positive numbers; this show that p_1 and p_3 have no inputs. So, the rules corresponding to t_1 and t_7 are used in the process of deduction of E. Hence, it is deduced that $\{x, z\}$ is the minimal conflict.

Theorem 15. *Minimal-Conflicts($\mathcal{P}$, $\mathcal{A}$, Co, Δ, Φ) returns the set of minimal conflicts for a system with an observation.*

Proof. The procedure enumerates possible firing sequences, and by lemma 2, the theorem is true.

By the algorithm in [6], diagnoses can be computed from minimal conflicts. In next section, we will introduce an algorithm to compute diagnoses directly by using the consistency check algorithm given in section 4.

5.2 An Algebraic Approach to Computing Diagnoses

Definition 15. *Suppose PN is the Petri net corresponding to a system with an observation. p_1 and p_2 correspond to the output of the system and its negation, M_0 is the initial marking. Both $M(p_1)$ and $M(p_2)$ are not zero, where $M = M_0 + CX$, X is the maximal vector in $VSE(PN)$.*

1. *$T' \subseteq T$ is a called a cut if and only if, for PN_1 which is the Petri net by removing the transitions in T' from PN, C_1 is the incidence matrix of PN_1, X_1 is the maximal vector in $VSE(PN_1)$, $M_1 = M_0 + C_1 X_1$, $M_1(p_1)$ or $M_1(p_2)$ are 0.*
2. *$T' \subseteq T$ is a called a minimal cut if and only if T' is a cut, but for a proper subset T'' of T' is not a cut.*

Lemma 3. *Suppose PN is the Petri Net corresponding to a system with an observation. It is also assumed that σ is a firing sequence, $M_0[\sigma\rangle M$, $M(p_1)$ and $M(p_2)$ are not zero, where p_1 and p_2 correspond to the output of the system and its negation, respectively. T' is a subset of T. $S_1 = \bigcup_{t \in T'} \cdot t$, $Con = \{Ok(x) | p \in S_1 \text{ and } p \text{ corresponds to } Ok(x)\} \cap \mathcal{A}$ is a minimal cut if T' is a minimal cut.*

Proof. The proof of the lemma is similar to the lemma 2.

Suppose that PN is the Petri net corresponding to a system $\langle \mathcal{P}, \mathcal{A}, \mathcal{C}o, \Delta \rangle$ with an observation Φ, by lemma 3, if we have minimal cuts, we can construct diagnoses. Suppose X is the result returned by Maximal-Vector-Generality(PN), p_1 and p_2 correspond to the output of the system and its negation. $M = M_0 + \mathcal{C}X$, where $\mathcal{C}$ is the incidence matrix of PN. If both $M(p_1)$ and $M(p_2)$ are not 0, then there is a conflict between the correct system behavior and the observation. So, if $\{t^0 | t \in \cdot p_1, X(t) = 1\}$ is not empty, then it is a minimal cut. Similarly, if $\{t^0 | t \in \cdot p_2, X(t) = 1\}$ is not empty, then it is a minimal cut. Suppose T' is a minimal cut, for $t \in T'$, $S' = \cdot t$. For $p \in S'$, if $\{t | t \in \cdot p \text{ and } X(t) = 1\} \neq \emptyset$, then $T' - \{t\} \cup \{t | t \in \cdot p \text{ and } X(t) = 1\}$ is also a minimal cut. So, we give the following procedure to compute diagnoses.
Algorithm Diagnoses($\mathcal{P}, \mathcal{A}, \mathcal{C}o, \Delta, \Phi$)
  Construct a Petri Net PN corresponding to $\langle \mathcal{P}, \mathcal{A}, \mathcal{C}o, \Delta \rangle$ with Φ;
  $X = $ Maximal-Vector-Generality(PN);
  $M = M_0 + \mathcal{C}X$; $TMC = \emptyset$;
  p_1 and p_2 corresponding to the output of the system and its negation;
  If $(M(p_1) \neq 0$ and $M(p_2) \neq 0)$ then
    $MC = \{\{t^0 | t \in \cdot p_1, X(t) = 1\}\} \cup \{\{t^0 | t \in \cdot p_2, X(t) = 1\}\}$;
  Else $MC = \emptyset$;%The correct system behavior conforms with the observation

While $(nonempty(MC))$ do
  { $T_1 = delete(MC)$;
    If (there no $t_j^0 \in T_1$) then $\{TMC = TMC \cup \{T_1\}\}$;
    Else
      { For $(t_j^0 \in T_1)$ do
        { $T_1 = T_1 - \{t_j^0\}$; $MC = MC \cup \{T_1 \cup \{t_j^1\}\}$;
          $S_1 = \cdot t_j \cap \{p| \cdot p \neq \emptyset\}$;
          For (each $(p \in S_1)$) do
            { $S_2 = \{t^0 | t \in \cdot p, t^0 \notin T_1, t^1 \notin T_1, \text{ and } X(t) = 1\}$;
              If $(nonempty(S_2))$ then
                $MC = MC \cup \{T_1 \cup S_2\}$;
            }
        }
      }
  }
$Con = \emptyset$;
For $(T' \in TMC)$ do
  $Con = Con \cup \{\{L | L \text{ is the literal corresponding to } t \in T'\}\}$;
$Dia = \{\mathcal{F} \cap \mathcal{A} | \mathcal{F} \in Con\}$;
For $(Co_1, Co_2 \in Conflicts \text{ and } Co_1 \subset Co_2)$ do
  $Dia = Dia - \{Co_2\}$;
$MDia = \emptyset$;
For (each $\mathcal{A}' \in Dia$) do
  $MDia = Dia \cup \{\{x | Ok(x) \in \mathcal{A}'\}\}$;
Return $MDia$;
End(Algorithm)

The process is illustrated by the following example.

Example 12. For the system in example 1 with the observation in example 2, its corresponding Petri net is shown in example 3, the maximal vector $X = (1, 0, 0, 1, 0, 0, 1, 0)^T$ is given in example 5. Since $M = M_0 + \mathcal{C}X = (2, 0, 1, 1, 0, 3, 2, 0, 0, 2, 3, 1, 1)$. Since, p_{12} and p_{13} correspond to E and $\neg E$, respectively, and $M(p_{12}) = M(p_{13}) = 1$, a conflict occurs between the correct system behavior and the observation. $\cdot p_{13} = \{t_6\}$, $X(t_6) = 0$, so, there is no minimal cut generated. $\cdot p_{12} = \{t_7, t_8\}$, $X(t_7) = 1$ and $X(t_8) = 0$, so a minimal cut, $\{t_7\}$ is constructed. For the minimal cut $\{t_7\}$, $\cdot t_7 = \{p_4\}$, and $\cdot p_4 = \{t_1\}$, and $X(t_1) = 1$, so another minimal cut $\{t_1\} = (\{t_7\} - \{t_7\}) \cup \{t_1\}$ is generated. So, all the minimal cuts are generated, and the corresponding minimal conflicts is $\{x\}$ and $\{z\}$.

Theorem 16. *Diagnoses($\mathcal{P}, \mathcal{A}, Co, \Delta, \Phi$) returns the set of diagnoses for a system with an observation.*

Proof. The procedure enumerates possible cuts, and by lemma 3, the theorem is true.

6 Related Works

Reiter [1] also introduces an approach to computing the diagnoses the *HS-tree*. But the problem with a complete *HS-tree* is that the size of the tree grows exponentially with the size of the incoming collection of conflict sets. Although diagnoses are generated more efficiently by pruned *HS-tree*, but the construction of pruned *HS-tree* is difficult to organize such that no unnecessary results are generated, because unnecessary subtrees are detected only after the entire subtree has been generated. Another problem is that he does not specify the order of the incoming conflict sets, the process of the construction of the tree is not predicted and the desirable results may be not generated. Our approaches can mechanically generate all the minimal conflicts.

Another approach related to computing minimal conflicts and diagnoses is to use resolution operator to generate consequences, such as the works in [3, 4, 7, 8]. Although these approaches can be applied to computer minimal conflicts and diagnoses, they have low efficiency since they generate all consequences. In fact, only part of consequences are related to minimal conflicts. The most recent work using resolution operator to compute consequences is given by Haenni [8]. He uses possibility function to decide resolution operator is applied to which clauses, but he does not show a way to construct such a function. So, his approach is difficult to apply to a practical problem. He also claims that LB his algorithm returned is sound, but not complete; and UB that his algorithm returned is not complete and sound. The following example shows that LB is not only non-sound, but also non-complete, and the same for UB.

Suppose $P = \{A, B, C, D\}$, $A = \{Ok(a), Ok(c), Ok(d), Ok(e)\}$, $\Gamma = \{Ok(a) \lor A \lor B \lor C, Ok(c) \lor A \lor C, Ok(c) \lor C \lor D, Ok(d) \lor D, Ok(e) \lor B\}$, $h = AB$By the algorithm Haenni [8] introduces, $H = h = A$, $\Gamma_H = \mu(\{Ok(a) \lor A \lor B \lor C, Ok(c) \lor A \lor C, Ok(c) \lor C \lor D, Ok(d) \lor D, Ok(e) \lor B, A\}) = \{Ok(a) \lor A \lor B \lor C, Ok(c) \lor A \lor C, Ok(c) \lor C \lor D, Ok(d) \lor D, Ok(e) \lor B, A\}$; $\Gamma = Cons_A(\Gamma_H) \cap D_A = \Gamma_H$; $\Gamma_0 = Cons_A(\Gamma_H) \cap D_A = \emptyset$; $\Gamma_A = \Gamma_B = \Gamma_C = \Gamma_D = \emptyset$.

Suppose the order of clauses which resolution operator is applied to is as follows.

1. Suppose the selected clause from Γ is A, then $R = \emptyset$ since Γ_A is empty; $\Gamma = \{Ok(a) \lor A \lor B \lor C, Ok(c) \lor A \lor C, Ok(c) \lor C \lor D, Ok(d) \lor D, Ok(e) \lor B\}$; Γ_0 keeps unchanged; $\Gamma_A = \Gamma_A \cup \{\neg A\} = \{\neg A\}$; $S = \mu(\Gamma \cup \Gamma_A \cup \Gamma_B \cup \Gamma_C \cup \Gamma_D) = \{Ok(a) \lor A \lor B \lor C, Ok(c) \lor A \lor C, Ok(c) \lor C \lor D, Ok(d) \lor D, Ok(e) \lor B, \neg A\}$; $\Gamma = \Gamma \cap S = \{Ok(a) \lor A \lor B \lor C, Ok(c) \lor A \lor C, Ok(c) \lor C \lor D, Ok(d) \lor D, Ok(e) \lor B\}$; $\Gamma_A = \Gamma_A \cap S = \{\neg A\}$; $\Gamma_B = \Gamma_B \cap S = \emptyset$; $\Gamma_C = \Gamma_C \cap S = \emptyset$; $\Gamma_D = \Gamma_D \cap S = \emptyset$.
2. Suppose the selected clause from Γ is Γ is $Ok(a) \lor A \lor \neg B \lor C$ and the selected literal is A, then $R = \{Ok(a) \lor \neg B \lor C\}$; $\Gamma = \{Ok(c) \lor A \lor C, Ok(c) \lor \neg C \lor \neg D, Ok(d) \lor D, Ok(e) \lor B, Ok(a) \lor \neg B \lor C\}$; Γ_0 keeps unchanged; $\Gamma_A = \Gamma_A \cup \{Ok(a) \lor A \lor \neg B \lor C\} = \{\neg A, Ok(a) \lor A \lor \neg B \lor C\}$; $S = \mu(\Gamma \cup \Gamma_A \cup \Gamma_B \cup \Gamma_C \cup \Gamma_D) = \{Ok(a) \lor \neg B \lor C, Ok(c) \lor A \lor C,$

An Algebraic Approach to Model-Based Diagnosis   493

$Ok(c) \vee \neg C \vee \neg D$, $Ok(d) \vee D$, $Ok(e) \vee B$, $\neg A\}$, $\Gamma = \Gamma \cap S$ $\{Ok(a) \vee \neg B \vee C$, $Ok(c) \vee A \vee C$, $Ok(c) \vee \neg C \vee \neg D$, $Ok(d) \vee D$, $Ok(e) \vee B\}$; $\Gamma_A = \Gamma_A \cap S = \{\neg A, Ok(a) \vee A \vee \neg B \vee C\}$; $\Gamma_B = \Gamma_B \cap S = \emptyset$; $\Gamma_C = \Gamma_C \cap S = \emptyset$; $\Gamma_D = \Gamma_D \cap S = \emptyset$.

3. Suppose the selected clause from Γ is $Ok(e) \vee B$ and the selected literal is B, then $R = \emptyset$; $\Gamma = \{Ok(a) \vee \neg B \vee C Ok(c) \vee A \vee C Ok(c) \vee \neg C \vee \neg D Ok(d) \vee D\}$; Γ_0 keeps unchanged; $\Gamma_B = \Gamma_B \cup \{Ok(e) \vee B\} = \{Ok(e) \vee B\}$; $S = \mu(\Gamma \cup \Gamma_A \cup \Gamma_B \cup \Gamma_C \cup \Gamma_D) = \{Ok(a) \vee \neg B \vee C, Ok(e) \vee B, Ok(c) \vee A \vee C, Ok(c) \vee \neg C \vee \neg D, Ok(d) \vee D, \neg A\}$; $\Gamma = \Gamma \cap S = \{Ok(a) \vee \neg B \vee C, Ok(c) \vee A \vee C, Ok(c) \vee \neg C \vee \neg D, Ok(d) \vee D\}$; $\Gamma_A = \Gamma_A \cap S = \{\neg A, Ok(a) \vee A \vee \neg B \vee C\}$; $\Gamma_B = \Gamma_B \cap S = \{Ok(e) \vee B\}$; $\Gamma_C = \Gamma_C \cap S = \emptyset$; $\Gamma_D = \Gamma_D \cap S = \emptyset$.

4. Suppose the selected clause from Γ is $Ok(a) \vee \neg B \vee C$ and the selected literal is B, then $R = \{Ok(a) \vee Ok(e) \vee C\}$; $\Gamma = \{Ok(c) \vee A \vee C, Ok(c) \vee \neg C \vee \neg D, Ok(d) \vee D, Ok(a) \vee Ok(e) \vee C\}$; Γ_0 keeps unchanged; $\Gamma_B = \Gamma_B \cup \{Ok(e) \vee B\} = \{Ok(e) \vee B, Ok(a) \vee \neg B \vee C\}$; $S = \mu(\Gamma \cup \Gamma_A \cup \Gamma_B \cup \Gamma_C \cup \Gamma_D) = \{Ok(a) \vee \neg B \vee C, Ok(e) \vee B, Ok(c) \vee A \vee C, Ok(c) \vee \neg C \vee \neg D, Ok(d) \vee D, Ok(a) \vee Ok(e) \vee C, \neg A\}$; $\Gamma = \Gamma \cap S = \{Ok(c) \vee A \vee C, Ok(c) \vee \neg C \vee \neg D, Ok(d) \vee D, Ok(a) \vee Ok(e) \vee C\}$; $\Gamma_A = \Gamma_A \cap S = \{\neg A, Ok(a) \vee A \vee \neg B \vee C\}$; $\Gamma_B = \Gamma_B \cap S = \{Ok(e) \vee B, Ok(a) \vee \neg B \vee C\}$; $\Gamma_C = \Gamma_C \cap S = \emptyset$; $\Gamma_D = \Gamma_D \cap S = \emptyset$.

5. Suppose the selected clause from Γ is $Ok(a) \vee Ok(e) \vee C$ and the selected literal is C, then $R = \emptyset$; $\Gamma = \{Ok(a) \vee \neg B \vee C, Ok(c) \vee A \vee C, Ok(c) \vee \neg C \vee \neg D, Ok(d) \vee D\}$; Γ_0 keeps unchanged; $\Gamma_C = \Gamma_C \cup \{Ok(a) \vee Ok(e) \vee C\} = \{Ok(a) \vee Ok(e) \vee C\}$; $S = \mu(\Gamma \cup \Gamma_A \cup \Gamma_B \cup \Gamma_C \cup \Gamma_D) = \{Ok(a) \vee \neg B \vee C, Ok(e) \vee B, Ok(c) \vee A \vee C, Ok(c) \vee \neg C \vee \neg D, Ok(d) \vee D, \neg A\}$; $\Gamma = \Gamma \cap S = \{Ok(a) \vee \neg B \vee C, Ok(c) \vee A \vee C, Ok(c) \vee \neg C \vee \neg D, Ok(d) \vee D\}$; $\Gamma_A = \Gamma_A \cap S = \{\neg A, Ok(a) \vee A \vee \neg B \vee C\}$; $\Gamma_B = \Gamma_B \cap S = \{Ok(e) \vee B, Ok(a) \vee \neg B \vee C\}$; $\Gamma_C = \Gamma_C \cap S = \{Ok(a) \vee Ok(e) \vee C\}$; $\Gamma_D = \Gamma_D \cap S = \emptyset$.

6. Suppose the selected clause from $Ok(c) \vee \neg C \vee \neg D$ and the selected literal is C, then $R = \{Ok(a) \vee Ok(e) \vee Ok(c) \vee \neg D\}$; $\Gamma = \{Ok(c) \vee A \vee C, Ok(a) \vee Ok(e) \vee Ok(c) \vee \neg D, Ok(d) \vee D\}$; Γ_0 keeps unchanged; $\Gamma_C = \Gamma_C \cup \{Ok(c) \vee \neg C \vee \neg D\} = \{Ok(c) \vee \neg C \vee \neg D, Ok(a) \vee Ok(e) \vee C\}$; $S = \mu(\Gamma \cup \Gamma_A \cup \Gamma_B \cup \Gamma_C \cup \Gamma_D) = \{Ok(a) \vee \neg B \vee C, Ok(e) \vee B, Ok(c) \vee A \vee C, Ok(c) \vee \neg C \vee \neg D, Ok(d) \vee D, Ok(a) \vee Ok(e) \vee C, \neg A\}$; $\Gamma = \Gamma \cap S\{Ok(c) \vee A \vee C, Ok(a) \vee Ok(e) \vee Ok(c) \vee \neg D, Ok(d) \vee D\}$; $\Gamma_A = \Gamma_A \cap S = \{\neg A, Ok(a) \vee A \vee \neg B \vee C\}$; $\Gamma_B = \Gamma_B \cap S = \{Ok(e) \vee B\}$; $\Gamma_C = \Gamma_C \cap S = \{Ok(c) \vee \neg C \vee \neg D, Ok(a) \vee Ok(e) \vee C\}$; $\Gamma_D = \Gamma_D \cap S = \emptyset$.

7. Suppose the selected clause from $Ok(d) \vee D$ and the selected literal is D, then $R = \emptyset$; $\Gamma = \{Ok(c) \vee A \vee C, Ok(a) \vee Ok(e) \vee Ok(c) \vee \neg D\}$; Γ_0 keeps unchanged; $\Gamma_D = \Gamma_D \cup \{Ok(d) \vee D\} = \{Ok(d) \vee D\}$; $S = \mu(\Gamma \cup \Gamma_A \cup \Gamma_B \cup \Gamma_C \cup \Gamma_D) = \{Ok(a) \vee \neg B \vee C, Ok(e) \vee B, Ok(c) \vee A \vee C, Ok(c) \vee \neg C \vee \neg D, Ok(d) \vee D\}$; $\Gamma = \Gamma \cap S = \{Ok(c) \vee A \vee C, Ok(a) \vee Ok(e) \vee Ok(c) \vee \neg D\}$;

$\Gamma_A = \Gamma_A \cap S = \{\neg A, Ok(a) \vee A \vee \neg B \vee C\}$; $\Gamma_B = \Gamma_B \cap S = \{Ok(e) \vee B\}$; $\Gamma_C = \Gamma_C \cap S = \{Ok(c) \vee \neg C \vee \neg D, Ok(a) \vee Ok(e) \vee C\}$; $\Gamma_D = \Gamma_D \cap S = \{Ok(d) \vee D\}$.

8. Suppose the selected clause from Γ is $Ok(a) \vee Ok(e) \vee Ok(c) \vee \neg D$ and the selected literal is D, then $R = \{Ok(a) \vee Ok(e) \vee Ok(c) \vee Ok(d)\}$; $\Gamma = \{Ok(c) \vee A \vee C\}$; $\Gamma_0 = \{Ok(a) \vee Ok(e) \vee Ok(c) \vee Ok(d)\}$ $\Gamma_D = \Gamma_D \cup \{Ok(a) \vee Ok(e) \vee Ok(c) \vee \neg D\} = \{Ok(d) \vee D, Ok(a) \vee Ok(e) \vee Ok(c) \vee \neg D\}$; $S = \mu(\Gamma \cup \Gamma_A \cup \Gamma_B \cup \Gamma_C \cup \Gamma_D) = \{Ok(a) \vee \neg B \vee C, Ok(e) \vee B, Ok(c) \vee A \vee C, Ok(c) \vee \neg C \vee \neg D, Ok(d) \vee D, Ok(a) \vee Ok(e) \vee C\}$; $\Gamma = \Gamma \cap S = \{Ok(c) \vee A \vee C\}$; $\Gamma_A = \Gamma_A \cap S = \{\neg A, Ok(a) \vee A \vee \neg B \vee C\}$; $\Gamma_B = \Gamma_B \cap S = \{Ok(e) \vee B\}$; $\Gamma_C = \Gamma_C \cap S = \emptyset$; $\Gamma_D = \Gamma_D \cap S = \{Ok(d) \vee D, Ok(a) \vee Ok(e) \vee Ok(c) \vee \neg D\}$.

It is assumed that the resource is used out at this time, and the procedure terminates and returns the sets $LB = \{Ok(a) \vee Ok(e) \vee Ok(c) \vee Ok(d)\}$ and $UB = \{Ok(c), Ok(a) \vee Ok(e) \vee Ok(c) \vee Ok(d)\}$. $Ok(a) \vee Ok(e) \vee Ok(c) \vee Ok(d)$ is not a prime implicate since $Ok(e) \vee Ok(c) \vee Ok(d)$ is a implicate of Γ. This shows that LB is sound for the consequence set, but not sound for the prime implicate set. UB does not contains all the prime implicates and there exist sentences in UB which are not consequences of Γ, so UB is not sound and complete for the consequence set and the prime implicate set.

Many fast algorithms are introduced for some special systems, such as the algorithms introduced by Darwiche [5], Val [9–11], Fattah and Dechter [12], Stumptner and Wotawa [13]. And the algorithms terminate in polynomial time some special system, such as the algorithm introduced in [13] terminates for the tree-structured system if the inputs of each component in the system are not more than two. The algorithms presented in section 5 also terminates in polynomial time for some special structured systems. Furthermore, our approach may provide a way to introduce approximate algorithm for model-based diagnosis.

7 Conclusion

In this paper, we first present a way to transform a system with an observation into a Petri net, and show that a contradiction occurs between the correct system behavior and the observation if and only if, for places p_1 and p_2 labeled with the output of the system and its negation respectively in its corresponding Petri net $PN = \langle P, T, W, M_0 \rangle$, $M(p_1)$ and $M(p_2)$ are not zero, where $M = M_0 + CX$, X is the maximal $\{0, 1\}$-vector in $VSE(PN)$. Then we introduce an algorithm to compute the maximal vector X in $VSE(PN)$. Furthermore, algorithms for computing minimal conflicts and diagnoses are introduced. Compared with other related works, our approach terminates in polynomial time for some special structured systems.

As we know that computing minimal conflicts and diagnoses is NP-hard, so, the efficient way to solve the problem is to introduce approximate algorithm to compute minimal conflicts and diagnoses. Our approach may be a way to introduce approximate algorithm to model-based diagnosis. We will focus on this topic in future.

Acknowledgement. This work is supported by National Grand Fundamental Research 973 Program of China under Grant No. 2002CB312103.

References

1. R. Reiter, A theory of diagnosis from first principles, *Artificial Intelligence*, vol. 32, pp. 57–95, 1987.
2. J. d. Kleer, A. K. Mackworth, and R. Reiter, Charactering diagnoses and systems, *Artificial Intelligence*, vol. 56, pp. 197–222, 1992.
3. A. Darwiche, Model-based Diagnosis using causal networks, *Proc. Int. Joint Conf. on Artificial Intelligence*, pp. 211–217, Montreal, Canada, August 1995.
4. A. Darwiche, Model-based diagnosis using structured system descriptions, *Journal of Artificial Intelligence Research*, vol.8, pp. 165–222, 1998.
5. A. Darwiche, and G. Provan, The Effect of observation on the complexity of model-based diagnosis, *Proc. American National Conf. on Artificial Intelligence*, pp. 94–99, Providence, Rhode Island, USA, July 1997.
6. R. Haenni, Generating diagnoses from conflict sets, *Proc. Florida Artificial Intelligence Research Symposium*, pp. 120–124, Florida, USA, May 1998.
7. P. Marquis, Consequence finding algorithms, in: *(eds. S. Kholas, J. Moral) Handbook on Deafeasible Reasoning and Uncertainty Management Systems*, pp. 41–145. Kluwer Academic, Boston, 2000.
8. R. Haenni, A query-driven anytime algorithm For argumentative and abductive reasoning, *Proc. First Int. Conf. on SoftWare*, pp. 114–127, Belfast, Northern Ireland, April 2002.
9. A. del Val, The complexity of restricted consequence finding and abduction, *Proc. 17th American National Conf. on Artificial Intelligence*, pp. 337–342, Texas, USA, July 2000.
10. L. Simon and A. del Val, Efficient consequence finding, *Proc. Int. Joint Conf. on Artificial Intelligence*, pp. 359–365, Seattle, Washington, USA, August 2001.
11. A. del Val, A new method for consequence finding and compilation in restricted languages, *Proc. Sixteenth American National Conf. on Artificial Intelligence*, pp. 259–264, Florida, USA, July 1999.
12. Y. E. Fattah, and R. Dechter, Diagnosing tree-decomposable circuits, *Proc. Int. Joint Conf. on Artificial Intelligence*, pp. 572–578, Montreal, Canada, August 1995.
13. M. Stumptner, and F. Wotawa, Diagnosing tree-structured systems, *Artificial Intelligence*, vol. 127, pp. 1–29, 2001.
14. Bartlomiej Gorny, Antoni Ligeza, Model-based diagnosis of dynamic Systems: systematic conflict generation, in: L. Magnani, N. J. Nersessian and C. Pizzi (eds.) *Logical and Computational Aspects of Model-based Reasoning*, pp. 273–291, Kluwer Academic Publisher, 2002.
15. I. Mozetic, A Polynomial-time algorithm for model-based diagnosis, *Proc. 10th European Conf. on Artificial Intelligence*, pp. 729–733, Vienna, Austria, August 1992.

16. R. L. Childress, and M. Valtorta, Polynomial-time model-based diagnosis with the critical set algorithm, *Proc. Fourth Int. Fourth Int. Workshop on Principles of Diagnosis*, pp. 166–177, Aberystwyth, Wales, UK, 1993.
17. S. Luan, L. Magnani, G. Dai, Algorithms for Computing Minimal Conflicts, *Logic Journal of IGPL*, Vol.14 No.2, 391–406, June 2006.
18. L. Magnani, *Abduction, Reason, and Science: Processes of Discovery and Explanation*, New York, Kluwer Academic/Plenum Publishers, 2001.
19. N. J. Nilsson, *Artificial Intelligence: A New Synthesis*, Morgan Kaufmann, San Fransisco, 1999.
20. P. Meseguer. A new method to checking rule bases for inconsistency: a Petri net approach. In *Proceedings of the 9th European Conference on Artificial Intelligence (ECAI-90)*, Stockholm, August 1990, pp. 437–442.
21. Tadao Murata, Petri Nets: Properties, analysis and applications, *Proceedings of the IEEE*, Vol. 77, No 4, April, 1989, 541–580.

CYBERNARD: A Computational Reconstruction of Claude Bernard's Scientific Discoveries

Jean-Gabriel Ganascia[1] and Claude Debru[2]

[1] LIP6 - University Pierre and Marie Curie (Paris VI), Paris, France,
jean-gabriel.ganascia@lip6.fr
[2] Department of Philosophy, Ecole Normale Supérieure, Paris, France,
claude.debru@ens.fr

Summary. With epistemological insight and artificial intelligence techniques, our aim is to reconstruct Claude Bernard's empirical investigations with a computational model. We suppose that Claude Bernard had in mind what we call "kernel models" that contain the basic physiological concepts upon which Claude Bernard builds his general physiological theory. The "kernel models" provide a simplified view of physiology, where the internal environment – the so-called "milieu intérieur" –, mainly the blood, plays an essential role. According to this perspective, we assume that the "kernel models" allow Claude Bernard to make some hypotheses and to draw out their logical consequences. More precisely, the role of the "kernel models" is twofold: on the one hand, they help to generate and manage working hypotheses, for instance to enumerate the probable effects of a toxic substance, on the other hand, they derive, by simulation, the most plausible consequences of each of those hypotheses. We shall show how those "kernel models" can be specified using both description logics and multi-agent systems. Then, the paper will explain how it is possible to build, on these "kernel models", a virtual experiment laboratory, which lets us construct and conduct virtual experiments that play a role similar to the role of thought experiments. More generally, the paper constitutes an attempt to correlate Claude Bernard's experiments, achieved to corroborate or refute some of his working hypotheses, to virtual experiments emulated on "kernel models".

1 The CYBERNARD Project

Claude Bernard (1813–1878) was one of the most eminent 19th century physiologists. He was a pioneer in many respects. He introduced the concept of internal environment (the "Milieu intérieur") [1], which corresponds to today's principle of "homeostasis". He investigated and enlightened many physiological mechanisms, e.g. the glycogenic liver function [2], effects of carbon monoxide, [3] and [4], effects of curare [5] and [3], etc. But, Claude Bernard was not only a great physiologist; he was also a theoretician who generalized his experimental method in his famous book, "Experimental Medicine" [6], which is nowadays a classic that all young students in medicine are supposed to have read.

However, there is debate in the epistemology community about the importance of the book. Some think that Claude Bernard revolutionized the physiology while others consider that he is only a great physician who successfully tried to vulgarize his scientific works. In a way, the structure of the book makes this debate possible since the first part exposes abstract principles on which relies a general experimental method, while the second exemplifies the application of the method on discoveries that are mainly derived from Claude Bernard's own work. Therefore, it could be possible to interpret the experimental method as an introduction to the description of Claude Bernard's personal scientific contribution. On the other hand, some philosophers think that the "Experimental Medicine" [6] played the same role for the 19th and 20th century physiology that the Descartes "Discourse on Method" for the 17th century physical sciences. In modern terms, it originated a "change of paradigm" in experimental medicine. Even if the knowledge of physiological mechanisms is far more detailed today than it was at the Claude Bernard's time and if the statistical techniques make the analysis of experimental data more rigorous, the principles on which relies the methodology of clinical experimentations are based on the same theoretical foundations. It is the argument of those who promote the "Experimental Medicine" as a key contribution for the modern medicine.

The CYBERNARD project aims at contributing to this debate by the achievement of a computer model and by a computer assisted diachronic analysis of Claude Bernard's texts. More precisely, the goal of the CYBERNARD project is twofold. The first is to clarify and to generalize the experimental method by formalizing it with artificial intelligence techniques and by simulating it on computers. It will then be possible to understand in what respect this method is general and can be applied to contemporaneous clinical medicine. Once this first goal will be achieved, we shall attempt synchronous reconstructions of some of the Claude Bernard's scientific discoveries, i.e. reconstructions of the discoveries that he has described at the end of his life, in his large audience papers. The second goal is then to confront the original Claude Bernard's scientific texts – i.e. his personal notes, scientific papers, etc. – to the reconstruction of his own work he made when he wrote the "Experimental Medicine" [6]. Our aim is to understand the effective status of the method described in the "Experimental Medicine": does it correspond to the actual method that Claude Bernard used or to an ideal reconstruction of what it should have been This confrontation can be called a diachronic reconstruction, since it is to compare the own Claude Bernard's latest reconstruction of his work to its effective ideas as they were expressed in his papers and published articles at the time of discovery. Three teams participate to the CYBERNARD project, which is highly interdisciplinary: an artificial intelligence group headed by Jean-Gabriel

Ganascia, the ACASA team, belonging to the LIP6 computer science laboratory, the epistemology department of the École Normale Supérieure directed by Claude Debru and the linguistic team of the ITEM laboratory that is specialized in genetic criticism.

This paper relates a joint work of the ACASA team and the epistemology department of École Normale Supérieure that is part of the CYBERNARD project. Within this work, our aim is to reconstruct Claude Bernard's empirical investigation with a computational model that simulates his experimental method. We are mainly interested in his investigations of carbon monoxide and curare effects. To start, we shall refer to two of Claude Bernard's texts, [5] and [3], where he rationalizes his own discoveries. In parallel, with the help of philologists, we shall confront Claude Bernard's rational reconstruction of his own previous discoveries with his former reasoning as it appeared in his writings. However, this paper focuses only on the first point.

The first part recalls the Claude Bernard's experimental method. The second is dedicated to the description of a two level model build to simulate the experimental method. The third formalizes the Bernard's medical ontology. The fourth describes the notion of "kernel model"; the fifth, the virtual laboratory on the top of which virtual experiments may be done. A sixth section presents the hypothesis generation module. The final and last part envisages possible generalizations of the experimental method and of its simulation to multi-scale "kernel models".

2 The Experimental Method

According to Claude Bernard's views, scientific investigation cannot be reduced to the sole observation of facts nor to the construction of theories that have not been previously confirmed by empirical evidence. In other words, Claude Bernard is neither an inductivist who reduces the scientific activity to the pure induction of general rules from particulars, nor an idealist – or a neo-Platonist – who thinks that ideal, pure and perfect theories are given before any experimentation. The experimental method he promotes begins with an initial theory, which is usually built from passive observations or preconceived ideas. When the phenomenon is unknown, some experiments "to see" are done.

For instance, when Claude Bernard investigated the effects of the curare, he began with some general experiments in order to see what happened and to provide a first idea. Claude Bernard does not detail the way the first idea or the initial theory is built. It corresponds to an intuition or to what he called a feeling that has to be validated and refined or adjusted according to empirical results generated by relevant experiments. The experimental method starts there.

In other words, once an initial theory is given, scientists must design an experimental apparatus able to test (corroborate or refute) the given theory. The experiments are viewed as "provoked" observations generated by an adequate device; those observations are compared with the expectations derived from the given theory. Their cautious analysis helps to revise, correct, refine or validate the current theory. The inferences that are involved in such an analysis clearly correspond to abduction, since it is to try to explain observations by modifying theories. However, Claude Bernard's trail of thought cannot be simply reduced to abduction. The experimental method, iterated until the theory predicts all current experimental results, makes use of abduction, deduction, analogical reasoning and induction.

More precisely, the experimental method described by Claude Bernard is an iterative procedure of theory refinement that proceeds in three steps, each step involving a specific scientific function:

Experimentation: an hypothesis that has to be validated is given. It is called an idea or a theory. For the sake of clarity, we shall refer to it as the *current theory*. The first step is to design an experimental apparatus able to generate observations that can be compared to expectations derived from the current theory. In other words, the experimentation is designed to test the hypothesis under investigation, i.e. the current theory.

Observation: the second step consists in collecting observations from the designed experiments. It is not only a receptive step, since the experimenter may interpret observations and note unexpected details.

Analysis: this third step is the most crucial an original. It is to confront the current theory predictions to the observations and to generate plausible hypotheses that may transform the current theory when its predictions are not in accordance with the experimental observation.

The key question concerns the analysis and, consequently, the hypothesis generation: how, from a set of observations that invalidates a set of theories, would it be possible to generate new theories that will then be evaluated and refined until experiments will fully validate them? That step plays a crucial role in the experimental method. One has to clarify and to generalize it if we want to model and to simulate the method. In other words, designing an experiment to validate or invalidate a theory is a very complicated task that requires intuition, skill and imagination. It is out of the scope of our project to automatize such a design.

On the other hand, the observation is mainly a matter of patience. Nowadays, it may appear that censors and computers could both help looking out and gathering data. Therefore, it is not central to the experimental method that mainly has to analyze observational data and then to generate new theories. Our point is to automate the analysis of experimental results and the hypothesis generation process that corresponds to the most crucial step. We focus on it in this paper. We assume that abduction plays an important role here, since it is to explain experimental results by modifying the current theory.

Abductive reasoning makes generally use of background knowledge on the top of which hypothesis are formulated. Considering all Claude Bernard's hypotheses and revisions, it appears that they had some resemblance; they were formulated using the same words; they seemed to be generated from the same "ontology". In the late reconstruction of his discoveries, Claude Bernard elicited the "ontology" he had in mind. The next section describes it.

3 The Claude Bernard's Ontology

To have a clear understanding of the Claude Bernard's ontology and of its originality, one has first to cast a glance at previous medical conceptions. Let us first recall that the old theory of fluids introduced by Galen (131–201), during the 2nd century, and very much developed by Santorio Sanctorius (1561–1636) in the early 1600's was prevalent in the 17th and 18th century European medical schools. According to this theory, the body is made of solid tissues and fluids, which naturally tend to become corrupted without excretions and perspiration. As a consequence, most of the diseases and of the body dysfunctions are due to fluid corruption. At the end of the 18th century, inspired by the physics and the chemistry, François–Xavier Bichat (1771–1802) and François Magendie (1783–1855), who was the Claude Bernard's professor, studied the body anatomy and the organs. The physiology was then viewed as a physical interaction between organs. As a consequence, the causes of body dysfunctions and diseases were attributed to organ damages. Post-mortem dissection could then help to diagnose the organs responsible of the diseases. Claude Bernard opposed to this reduction of organs to physical bodies; he thought that organs are not only inert solid tissues, but that each of them has its autonomy and its own functions, which have to be investigated. More precisely, in his writings (cf. [3] and [4]), Claude Bernard presumes that organisms are composed of organs, themselves analogous to organisms since each of them has its own aliments, poisons, excitations, actions etc. Organs are categorized into three classes – skeleton, tissues (e.g. epithelium, glandular tissue or mucous membrane) and fibers (i.e. muscles and nerves) – that are recursively subcategorized into subclasses, sub-subclasses etc. Each class and subclass has its own characteristics, which can easily be formulated, according to Claude Bernard's explanations.

The internal environment – i.e. the "milieu intérieur" –, mainly the blood, carries organ poisons and aliments, while the organ actions may have different effects on other organs and, consequently, on the whole organism. More precisely, for Claude Bernard, the life is synonymous of exchanges. The organisms exchange through the external medium that is the air for outside animals or the water for fish. The external medium may also carry aliments, poisons etc. Similarly, organs can be viewed as some sorts of organisms living in the body and participating to its life. Their life is also governed by exchanges; but the medium that supports exchanges is not air or water; it is the so-called "milieu intérieur", which mainly corresponds to blood. The Claude Bernard's ontology may easily be derived from these considerations. It is then easy to formulate it in an ontology description language.

For instance, below are some of the previous assertions expressed with description logics.

The organs belong to the class Organ and are all parts of the organism:

$$Organ \sqsubseteq \exists PART.Organism \qquad (1)$$

The organs are tissues, skeleton or fibers:

$$Organ \equiv Tissue \sqcup Skeleton \sqcup Fiber \qquad (2)$$

$$Tissue \sqcap Fiber = \bot \qquad (3)$$

$$Tissue \sqcap Skeleton = \bot \qquad (4)$$

$$Fiber \sqcap Skeleton = \bot \qquad (5)$$

Fibers may be nerves or muscles:

$$Fiber \equiv Nerve \sqcup Muscle \qquad (6)$$

Nerves may be sensitive or motor:

$$Nerve \equiv Sensitive\_Nerve \sqcup Motor\_Nerve \qquad (7)$$

Epithelium, glandular tissue, mucous membrane etc. are tissues:

$$Tissue \sqsupseteq Epithelium \sqcup Glandular\_Tissue \sqcup Mucous\_Membrane \sqcup etc. \qquad (8)$$

Each organ can be viewed as some sort of organism that has its own nutriments, its own poisons, its own actions, etc.

$$Organ \sqsubseteq \exists Aliment \qquad (9)$$

$$Organ \sqsubseteq \exists Poison \qquad (10)$$

$$Organ \sqsubseteq \exists Action \qquad (11)$$

$$etc. \qquad (12)$$

The physiological ontology plays a crucial role in the way Claude Bernard erected new hypotheses. It can be considered as a clue for the discovery process. All scientific hypotheses obviously depend on the concepts with which they may be expressed. On the one hand, when a concept is lacking, one may miss some efficient hypotheses; on the other hand, the presence of some useless concepts leads to formulate misleading and confusing explanations. For instance, the old fluid theory precluded the observation of correlations between the evolution of the scurvy disease and the presence of fruit and vegetable in nutriments. Claude Bernard himself was unable to precisely locate the effects of curare, despite his relentless work during more than twenty years; one explanation could be that the concept of motor nerve ending did not belong to his ontology.

The question is how the ontologies are originated? What is their relevancy? And how do they evolve? Up to now, we don't yet feel able to provide fully convincing answers; but our goal within this work is to contribute to get a better understanding of those ontology evolution processes. In the case of Claude Bernard, the ontology here described corresponds to the one he gave at the end of his scientific life, in his large audience papers (e.g. [3] or [5]) and books [6]. There is no doubt that it appears naive and wrong with respect to the modern medical knowledge. Nevertheless, the main question for us does not concern its today relevance, but its evolution during Claude Bernard's scientific career.

This paper is focused on the rational reconstruction of Claude Bernard's own discoveries that he achieved by himself when he was famous. Our ultimate goal is to go further and to confront this late and personal reconstruction of Claude Bernard's scientific discoveries to the actual Claude Bernard's discovery process as it appears through informal notes, laboratory books, scientific papers etc.

4 Two-Level Model

As previously stated, abduction played a crucial role in Claude Bernard's investigations. More precisely, he always considered an initial hypothesis, which he called an idea or a theory. He then tried to test it by designing in vivo experiments. According to the observational results of his experiments, he changed his hypotheses, until he reached a satisfying theoretical explanation of empirical phenomena.

4.1 "Kernel models"

To design a computational model that simulates the intellectual pathway leading Claude Bernard to his discovery, we have supposed that he had in mind what we call "kernel models" that contain basic physiological concepts – such as internal environment, organ names etc. – upon which he builds his theories. More precisely, theories correspond to hypothetical organ functions that Claude Bernard want to elucidate, while "kernel models" describe the physical architecture of the organism.

The "kernel models" enable Claude Bernard to hypothesize tentative assumptions and draw out their logical consequences. These "kernel models" constitute the core on which the reasoning process is based; they correspond to putative architectures of the organisms. Depending on the question under investigation, they may be more or less simplified. For instance, if one want to investigate the hart function, it is not necessary to detail the precise role of all muscles. Our aim is to build and to simulate those "Kernel models" using multi-agent architectures. Such simulations have to show, on a simplified view, both the normal behavior of the organism and the consequences of an organ dysfunction.

4.2 Working Hypotheses Management

The second level of the considered model manages hypotheses relative to the function of different organs. Each working hypothesis is evaluated through empirical experiments. Claude Bernard assumes that one can use toxic substances as tools of investigation – he evokes the idea of "chemical scalpel" – to dissociate and identify the functions of different organs. He presupposes, as an underlying principle, that each toxic substance neutralizes one organ first. When a toxic substance affects an organ, the anatomy of death shows how the organism behaves without the poisoned organ. Nevertheless, even when laying down such a presupposition, the investigation puzzles lot physiologists, because it is a double entry enigma: they have to elucidate both the organs corrupted by toxic substances and the function of affected organs.

Two questions need to be solved when we want to rationally reconstruct the discovery process: how are working hypotheses generated and how are validating experiments designed? In order to answer these questions, we add to the "kernel model" a working hypothesis management module that has both to guide working hypothesis generation and to design experiments. Once an hypothesis is made, virtual experiments have to simulate, on the top of the "kernel model", the probable observable consequences of this hypothesis, which helps designing real experiments. Such virtual experiments play a role analogous to thought experiments in traditional physics: they are required as a preliminary step to any empirical experiment. For the sake of clarity, let us recall that thought experiments are experiences that scientists do not conduct in the outside world, but only in their head. One may attempt to describe some of those thought experiments with computer models that can be simulated on computers.

In case of Claude Bernard, we have found in his writings personal notes describing ideas of experiments. Some of them correspond to experiments that are achieved, while most of them remain imaginary. Our aim is to simulate those ideas of experiments with "kernel models" and to understand the place of those experiments in the discovery process with the hypothesis management module.

5 "Kernel Model" Simulation

The "kernel models" contain organs and connections between organs through the internal environment, mainly the blood, and direct connections. Both organs – e.g. muscles, hart, lung, nerves etc. – and connections between organs are represented using automata, i.e. entities characterized by their inputs, their outputs and their internal state. A "kernel model" may then be viewed as a network of automata. Each organ corresponds to an automaton with an internal environment plus external or internal excitations as inputs, organ actions and modified internal environment as outputs and a symbol characterizing the state. It is possible, for the internal environment, to lose or gain some substance, for instance oxygen, and some pressure when passing by an organ. In the usual case, e.g. for muscles, the input internal environment corresponds to arterial blood while the output corresponds to venous blood. Most of the connections correspond just to transmitters that associate the outputs of some organs to the inputs of others. Nevertheless, connections may also act as crossing points, for instance, as an artery splitting or as a vein join that divide or concentrate the flow.

From a computational point of view, each organ is viewed as an agent [7] that communicates with other organs and evolves in the "milieu intérieur" viewed as the internal environment. As a consequence, the organism is modeled as a synchronous multi-agent system, where each agent has its own inputs, transfer function and states. The organ activation cycle follows the blood circulation. The time is supposed to be discrete and after each period of time, the states of the different automata belonging to the "kernel model" and their outputs are modified.

A first implementation was programmed in JAVA using object oriented programming techniques. It helped both to simulate the "kernel model" evolutions and to conduct virtual experimentations (see section 6) on those "kernel models", which fully validates our first ideas concerning the viability of the notion of "kernel model". Within this implementation, organs and connections between organs are associated to objects. The instantiation and inheritance mechanisms facilitated the programming. However, since our ultimate goal is to simulate the hypothesis generation and especially the abuctive reasoning on which relies the discovery process (see section 7), we are currently rebuilding "kernel models" using logic programming techniques on which it is easy to simulate logical inferences, whatever they are, either deductive or abductive.

The logic programming implementation of the "kernel model" is programmed in SWI Prolog[3]. It makes use of modules to emulate object oriented programming techniques, i.e. mainly the instantiation, inheritance and message sending mechanisms. The resulting program looks like a collection of modules similar to the one given in Figure 1. Each of those modules describes a class of organs, e.g. muscles. Finally, on the top of the inheritance hierarchy of modules, there is a conjunction of literals corresponding to a virtual organism expressed as a network of connected organs. Once an initial condition and some ulterior events are given, it is possible to make the organism evolve by itself and to print states characterizing this evolution.

---

[3] See http://www.swi-prolog.org/ for more details.

```
:- module(organ, []).
inherit(organ, automata).

%%%%%%%%%%% Output %%%%%%%%%%%%%%%%%%%%%%%%%%%%
output(O, E, dead, S):- invoke(O, transmit, [dead, E, S]).
output(O, E, fresh, S):- invoke(O, transmit, [fresh, E, S]).
output(O, E, weary, S):- invoke(O, transmit, [weary, E, S]).

%%%%%%%%%%%% Transitions %%%%%%%%%%%%%%%%%%%%%%%%%%%
transition(_, _, dead, dead).
transition(O, E, fresh, fresh) :- invoke(O, keep_fresh, [E]),!.
transition(_, _, fresh, weary).
transition(O, E, weary, fresh) :- invoke(O, recovery, [E]),!.
transition(O, E, weary, weary) :- invoke(O, subsistence, [E]),!.
transition(_, _, weary, dead).

%%%%%%%%%%%%%%%% transmission %%%%%%%%%%%%%%%%%%%%%%%
shift_pressure(O, E, S) :- val_al(E, blood, B), val_al(B, pressure, P),
            invoke(O, reduction, [pressure, R]), NP is P*R,
            add_al(B, [pressure, NP], NB), add_al(E, [blood, NB], S).

transmit(O, State, Input, S) :- invoke(O, shift_pressure, [Input, NE]),
            invoke(O, blood_components, [L]), val_al(NE, blood, B),
            invoke(O, transmit_blood, [State, L, B, NB]),
            add_al(NE, [blood, NB], S).

transmit_blood(_, _, [], B, B) :- !.
transmit_blood(O, State, [Comp | L], B, SB) :-
            invoke(O, consumption, [Comp, State, C]),
            val_al(B, Comp, VC), (VC > C -> NVC is VC - C; NVC is 0),
            add_al(B, [Comp, NVC], NB),!,
            invoke(O, transmit_blood, [State, L, NB, SB]).
%%%%%%%%%%%%%%%%% procedures %%%%%%%%%%%%%%%%%%%%%%

keep_fresh(O, E) :- invoke(O, blood_components, [L]), val_al(E, blood, B),
            forall( member(Comp, [pressure|L]),
                    ( val_al(B, Comp, V),
                      invoke(O, threshold_min, [Comp, Th]), V >= Th)).
            forall( member(Comp, [pressure|L]),
                    ( val_al(B, Comp, V),
                      invoke(O, threshold_recovery, [Comp, Th]), V >= Th)).
subsistence(O, E) :- invoke(O, blood_components, [L]), val_al(E, blood, B),
            forall( member(Comp, [pressure|L]),
                    ( val_al(B, Comp, V),
                      invoke(O, threshold_subsistence, [Comp, Th]), V >= Th)).

%%%%%%%%%%%%%%%%% constants %%%%%%%%%%%%%%%%%%%%%%%
blood_components(_, [oxygene, glucide, lipid]).
consumption(_, oxygene, fresh, 3).
consumption(_, oxygene, weary, 6).
...

reduction(_, pressure, 0.95).
...
```

Fig. 1. Here is the SWI Prolog code for a simplified virtual organ.

6 Virtual "Thought Experiments"

Once the "kernel model" is built, it is not only possible to simulate normal organism behavior, but also to introduce pathologies (i.e. organ deficiencies) in the multi-agent system that models the organism and then emulate its evolution. In a way, these abnormal behavior simulations can be viewed as virtual experiments, or as "thought experiments": they help to draw consequences of virtual situations under a working hypothesis, i.e. a supposition concerning both the effect of a substance on some organs and the function of the implied organs. In order to complete the range of virtual experiments, we introduce, according to Claude Bernard's practices, some virtual experimental operators, such as injection and ingestion of substances, application of tourniquet on members, excitations, etc.

For instance, if one wants to understand the effects of a substance A, one can hypothesize that its concentration in the blood may affect such or such organ subclass that has such or such function in the organism. Under these hypotheses, it is possible with the "kernel model" simulation to predict the consequences of a direct injection of A combined with any combination of experimental operations (applying a tourniquet on a member and/or exciting another part of the organism before or after injecting the substance A etc.). In other words, it is possible to specify virtual experiments and to anticipate the subsequent model behavior under a working hypothesis.

For the sake of clarity, let us consider the experimental device described by Claude Bernard in [3] with the help of Figure 2. In this experiment, Claude Bernard mentioned that curare has been introduced on I while a tourniquet was applied on N. Let us now suppose that one lay down, as a tentative hypothesis, that curare only affects the muscles – that corresponds to one of

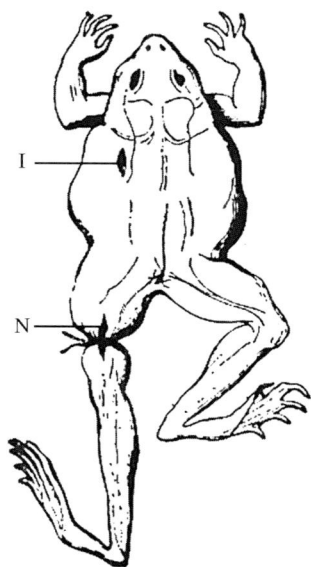

Fig. 2. This schema was published by Claude Bernard in [3]. It summarizes an experiment in which an incision has to be done on I to introduce curare, while a tourniquet is applied to the left thigh, on N.

the Claude Bernard hypotheses – but neither the sensitive nerves, nor the motor nerves, then the frog perceives excitations while the muscles belonging to all the organism are unable to move, except those on the left leg, because the tourniquet protects them from the curare effect. Let us now imagine that we excite the right leg on a "kernel model" built to model this experiment. It has to provoke a reaction on the left leg of the "kernel model", while other virtual limbs are not able to move because of the curare effect. This can be deduced from the current hypothesis. The role of the virtual experiment is to automatically generate such evolutions from an adequate "kernel model". One can also envisage to browse all the hypotheses, i.e. all the organ dysfunctions, which could generate the same behaviors. The virtual experiment may then prove the viability of the experiment.

7 Abduction

The previous section presented the virtual experiment laboratory built over the "kernel model". However, as suggested, the virtual experiments are achieved under working hypotheses that assume, for instance, that a substance A affects such or such a function of such or such an organ class. Being given a toxic substance, one has to explore all the possible hypotheses and, suggest, for each, experiments that could corroborate or refute them by showing observable consequences. It is the role of the working hypothesis management module to investigate all these hypotheses. Nevertheless, the goal is neither to achieve, nor to generate experiments, as would be the case with a robot scientist (see for instance [8]); it is just to reconstruct the scientific steps of Claude Bernard by simulating hypothesis exploration and by providing, for each hypothesis, the key experiments carried out by Claude Bernard.

More precisely, the computer reconstruction of "kernel models" shows that tentative explanations are built on three levels. The first corresponds to the ontological level. As previously said, it is out of the scope of the present study to automatically create new concepts. In a way, the ontology transformation may be assimilated to some kind of paradigm shift. In the future, it may be a very exciting challenge to tackle this problem, but up to now it appears to be premature.

The second level covers hypothetical function of organs. The aim of scientific discovery would undoubtedly be to elucidate the organ function. The study of toxic substance effects may be viewed as a mean to investigate those organ functions. However, today it seems too difficult to automate the generation of those functions. Therefore, we do not focus our study on this point.

Our present goal is more modest: being given a physiological ontology and explicit theories about organ functions, it is to find out the effect of toxic substances. That corresponds to the third level of investigation. More precisely, the computer has to browse all the possible effects of a toxic substance, i.e. all the organs that may be affected by the substance of which we investigate the effects. Under each of the plausible hypotheses, experiments are formulated with "kernel models" that may be simulated on a computer and then confronted to empirical observations. It is then possible either to invalidate or to confirm each of the plausible explanations. Both explorations of all the tentative explanations and attempts to confirm or disconfirm plausible explanations belong to abductive inference processes. Let us note that one can test the consistency of our model, i.e. that one can check that it is in accordance with the empirical evidences as they are mentioned by scientists. Moreover, anotations containing original experiments and observations are associated to each of the plausible hypotheses. It may help epistemologists and historians of science to understand the way research were conducted.

8 Conclusion

A first version of both the "kernel model" and the virtual laboratory are programmed in Java. They allowed us to build virtual experiments associated with different working hypotheses about the toxic effects of carbon monoxide and curare. It was then possible to correlate those virtual experiments to actual experiments done by Claude Bernard, and then to corroborate or refute working hypotheses according to the observations. As a consequence, we are able to computationally reconstruct part of Claude Bernard's intellectual pathway. A second implementation using logic programming techniques is now under construction. The reason is that it seems easier to model abductive reasoning using logic programming than traditional object oriented programming languages. We hope to reproduce the different steps of the Claude Bernard's toxic substance investigations, mainly carbon monoxide and curare.

However, this work relies on a fixed ontology, which biases the investigation and may prevent discovery. For instance, Claude Bernard's study of curare toxic effect was precluded by the absence of the motor nerve ending concept. Our further research will concern the way the "kernel models" evolve in Claude Bernard's research, especially the way both the Claude Bernard's ontology and the hypotheses concerning the different organ functions were transformed during Claude Bernard's scientific life. The detailed study of Claude Bernard's personal writings and scientific papers with genetic criticism techniques will help us in such an investigation.

We also investigate the possibility to build multi-scale "kernel models" in which physiological behaviors can be studied at different scales – organ, cell, molecule etc. –. It should open new perspectives to modern clinical medicine. As a matter of fact, principles on which lay down Claude Bernard empirical method are always valid, even if the "kernel models" considerably changed with time. Today, the effect of new substances is usually studied at the cell or molecule scale, while the organ scale was dominant at Claude Bernard's epoch. A model that could help to simulate effects of physiological dysfunctions at different levels would be of great help to determine the effects of new substances by recording different experiments and by ensuring that all the plausible hypotheses have already been explored.

References

1. Grmek, M.: *Le legs de Claude Bernard.* Fayard (1997)
2. Prochiantz, A.: *Claude Bernard: la révolution physiologique.* Presses Universitaires de France, Paris (1990)
3. Bernard, C.: Études physiologiques sur quelques poisons américains. *Revue des deux mondes* **53** (1864) 164–190
4. Grmek, M.: *Raisonnement expérimental et recherches toxicologiques chez Claude Bernard.* Droz, Genève (1973)
5. Bernard, C.: *Leçon sur les effets des substances toxiques et médicamenteuses.* Cours de médecine du collège de France. J.-B. Baillière et Fils, Paris (1857)
6. Bernard, C.: *An Introduction to the Study of Experimental Medicine.* Macmillan & Co., Ltd. (1927) First English translation by Henry Copley Greene.
7. Russel, S., Norvig, P.: *Artificial Intelligence a Modern Approach.* Series in Artificial Intelligence. Prentice Hall (1995)
8. King, R., Whelan, K., Jones, F., Reiser, P., Bryant, C., Muggleton, S.: Functional genomic hypothesis generation and experimentation by a robot scientist. *Nature* **427** (2005) 247–252

Do Computational Models of Reading Need a Bit of Semantics?

Remo Job[1] and Claudio Mulatti[2]

[1] Università degli Studi di Trento, Trento, Italy
   remo.job@unitn.it
[2] Università degli Studi di Padova, Padova, Italy
   claudio.mulatti@unipd.it

Summary. Coltheart, Rastle, Perry, Langdon, and Ziegler [1] claim that "the psychology of reading has been revolutionized by the development of computational models of visual word recognition and reading aloud". They attribute this to the fact that a computational model is a computer program – an algorithm – "that is capable of performing the cognitive task of interest and does so by using exactly the same information-processing procedures as are specified in a theory of how people carry out this cognitive activity" [1, p. 204]. According to this view, the computational model is the theory, not a simple instantiation of a theory. In this paper we argue that computational models of reading have indeed helped in dealing with such a complex system, in interpreting the phenomena underlying it, and in making sense of the experimental data. However, we also argue that it is crucial for a model of reading to implement a computational semantic system that is as yet a missing component of all computational models. We provide two reasons for such a move. First, this would allow explaining some phenomena arising from the interaction of semantics and lexical variables. We will review the following empirical findings: faster response times to polysemic words [2] and slower response times to synonyms [3]; the leotard [4] and turple effects [5]; and the asymmetry of the neighbourhood density effect in free and conditional reading [6]. Second, such an "enriched" model would be able to account for a richer set of tasks than current computational models do. Specifically, it would simulate tasks that require access to semantic representation to be performed, such as semantic categorization and semantically-based conditional naming. We will present a computational instantiation of a semantic module that accounts for all the described phenomena, and that has helped in generating predictions that guides on-going experimental activity.

1 What a Computational Model Is (for Us)

Without theories (models[3]), our ability to understand the great deal of data generated by experimental observation would be sporadic and limited to isolated facts or cases. Models of cognitive processes, then, constitute frameworks

---

[3] The terms *theory* and *model* are treated as interchangeable.

which help scientists in dealing with such complex systems, in interpreting the phenomena and in making sense of the experimental data. A model of a cognitive process describes and explains that cognitive process. For example, a model in the visual word recognition field – under the assumption that it is an appropriate model – describes and explains the processes underling reading.

Ontogenetically, models are first expressed verbally[4]. In the visual word recognition domain, a verbal model describes and explains processes trough the utilization of natural language (sentences), graphical supports (flowcharts – the so called boxes and arrows models), or both. A verbal model, then, is a qualitative one. Shortcomings of purely verbal theories are vagueness, ambiguity and imprecision [7][5], reluctance to falsification, and confusability, in the sense that a qualitative description is not easily distinguishable from other qualitative descriptions [8]. Last but not least, verbal model are too easily adaptable, extendible, to new data, even if inconsistent (provided the inconsistencies are "comfortable" [9]). However, to regard verbal models as always inadequate would be an error, since they do present with positive aspects: "[they] attract the expression of creative ideas, when the database is still too sparse to reasonably constrain more formal models. [they also] attract the organization of results coming from a broad variety of tasks" [9, p. 1312].

When a model is no longer a sketch of a cognitive process but, rather, it describes and explains the procedures involved in that cognitive function in a greater detail, the model is often (that is, when possible) translate into a *computational*[6] *model*. A computational model simulates a mental function by

---

[4] Generally, this statement is false. As Jacobs and Grainger [9] pointed out in the field of word recognition, "two distinct approaches to model construction emerge from the literature [...]. The first, which may be coined the *gardeners approach* (or 'the model is not the theory'), can be caricaturized as consisting in 'growing' a model or network that mimics in some respect a human cognitive function, without necessarily having an explicit theory of that function [...]. The second strategy could be coined the *architects approach* (or 'the model is the theory'). In line with the central dogma of cognitive science [10], some continue to argue that it is the right approach to start with a fully specified theory (based on general principles) and then (if one wishes) to implement it as an algorithmic model". The gardeners first develop a computational model that works and then they develop a theory compatible with the model. The architects first develop a theory compatible with the data and then they develop a computational model to test the theory. If the focus is restricted only to the architects approach, the dead end is overcome since the statement gets true.

[5] Broadbent [7] was able to demonstrate how a single algorithm could explain four patterns of results which were previously explained by four different verbal models, so giving a clear example of the explanatory inadequacy of verbal models.

[6] Marr [12] reserved the term *computational* for highest level description, whereas he called *algorithmic* the simulation models. Here, no distinction between the terms algorithmic and computational are made, so that the terms computational and algorithmic are interchangeable.

implementing a theory. More precisely, a computational model is a computer program -an algorithm- "that is capable of performing the cognitive task of interest and does so by using exactly the same information-processing procedures as are specified in a theory of how people carry out this cognitive activity" [1, p. 204]. Therefore, the computational model is the theory, not a simple instantiation of a theory (but see Norris [11]).

Coltheart et al. [1] go on to say that "the psychology of reading has been revolutionized by the development of computational models of visual word recognition and reading aloud", which is agreeable since computational modeling has obvious benefits. Firstly, in order to be implemented in the form of a computational simulation a model needs to be represented in an explicit form which imply a level of specification that typically eludes verbal theories. Secondly, the modeler needs to solve issue at a local level he/she may not even be aware of. Thus, the attempt to develop a computational model may interact with the theory itself, giving raise to a reciprocal improvement: the translation itself can be a productive process since this operation can uncover gaps or inconsistencies. Furthermore, since computational models are built on mathematical laws, the principles of their operations are explicit; this rigorous declaration of a theory can enable more accurate communication of ideas and reduce the scope for misinterpretation.

Jacobs and Grainger [9, p. 1312] listed a few "possible drawbacks of algorithmic models [which] are the dangers that they fossilize thinking and restrict creativity more than verbal models; that they focus the model builders attention too much on [...] implementation details that are irrelevant and thus obscure the discovery of general principles; that, in absence of a computational theory, they are not more than mimicry [12]; or that they cannot explain much if they still have to be explained themselves" [13].

The computational model itself can become the subject of investigation. A computational model simulates cognitive function. Such simulation not only enables theories to be put on the test, but it provides tools for investigating the theory itself and making likens between alternative theories. A computational model can be used to explore a theory in ways that would otherwise be either beyond the scope and the possibility of experimental investigation – as, for example, the quantitative estimation of the relative weight of competitive procedures in determining the output of a function – or too complex to be faced purely from the behavioral data.

Moreover, computational models help in generating predictions that can guide future experimental activity. The practice of modeling, generating predictions and testing these predictions, leads to improvements in the model itself, and doing so improves knowledge about the cognitive function mimed by the model.

In the next sections, we will briefly address some issues arising in the study of reading aloud single words. We will then describe a computational model of reading aloud and visual word recognition. After that section, we will show how computational modeling should be used *in practice* by deriving some predictions from a model and testing them. Finally, we will discuss how the model accounts for the body of empirical data.

2 Reading Aloud: Some Issues

Word reading is a complex cognitive operation requiring, at least, a mechanism that maps print into semantics and (or) into phonology. Not surprisingly, theories of word recognition usually posit some sort of lexical path (print to sound) and some sort of semantic path (print to meaning), although that distinction might not be so explicit and many researchers [14, 15] believe that meaning can only be accessed once the phonological representation has been retrieved (print to sound to meaning). Even if different theories might not – and often do not – converge on what the minimal set of assumptions needed is, what the nature of the computations and of the representations is, or what the structures of the processes involved look like, there is an aspect that is common to all of them: whereas the process of deriving sound from print is described in details, the process of accessing the meaning is usually under-specified in terms of both the representations involved and the procedures operating on those representations (e.g. [1, 16]). The causes of this deficiency have to be tracked back to the nature of the semantic representations itself, which is fleeting and hard to capture, to entrap into a describable format, despite the considerable amount of empirical evidence and notable recent theoretical contributions (e.g. [17–19]).

This vagueness in the verbal descriptions of the semantic system directly reflects into the computational models that from those theories are derived. Although currently available computational models of reading and visual word recognition do a great job in explaining/simulating (various portions of) the set of data reported in the literature (e.g. written frequency, letter length, orthographic neighborhood size, orthographic neighborhood frequency, regularity, position of irregularity, body-rime consistency) they do so without implementing any semantic system. Noteworthy, all the effects that they explain/simulate are phenomena that can be ascribed to the operations of the lexical system: since they do not implement any semantic system, they cannot account for phenomena arising within the semantic system itself or due to the interchange of information between the semantic system and other systems, e.g. the orthographic input system.

To avoid the problems arising while trying to model the semantic system from a purely theoretical starting point, in the work here reported we choose a different approach. We select (on the base of both personal preferences and explanatory power) a suitable computational model of word recognition, and look at the implications ensuing from adding a minimal semantic system. This approach has some immediate benefits. It allows testing the plausibility of the assumption underlying the model in contexts different from those the model was originally developed for. It also allows testing the minimal computational apparatus needed to simulate (at least some) semantic effects. Moreover, it allows a better understanding of the dynamic of processing of the already implemented components, such as the orthographic and phonological systems.

We will now briefly describe the computational model we selected, the *Dual Route Cascaded* model (DRC, [1]).

3 The DRC Model

Architecture of the model

The general architecture of the DRC is outlined in Figure 1. It can be split into two parts: a. parallel search within a dictionary, the *lexical* routine; b. serial conversion of graphemes into phonemes, the *non-lexical* routine.

The Feature and Letter Identification levels, as well as the Phonemic Buffer, are shared by the two routines. Each unit in the Feature level represents one of a letter's features, each unit in the Letter level represents one letter of the alphabet, and each unit in the Phonemic Buffer represents one phoneme of the target language.

The *lexical* routine's specific components are (a) the Orthographic Input Lexicon and (b) the Phonological Output Lexicon. The lexicons consists of lexical entries which are localist nodes that represent each word known to the model in terms of its spelling (in the orthographic input lexicon) and sound (in the phonological output lexicon). The lexical routine works in parallel.

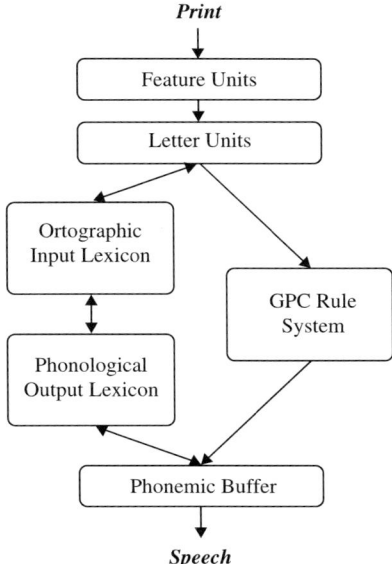

Fig. 1. Outline of the DRC model.

The model assumes: interactive activation between levels – with compatible units activating each other and incompatible units inhibiting each other, with the exception of the connections between the units in the orthographic and phonological lexicons, which are only excitatory; inhibition within levels – units belonging to the same level inhibit one another through inhibitory lateral connections.

Presenting a word to the model causes the activation of the visual feature units involved. Subsequent to the Feature level, activation is cascaded across all remaining levels. The features activate Letter level representations, which activate units in the Orthographic Input Lexicon which in turn activate units in the Phonological Output Lexicon. Activation then spreads to the Phonemic Buffer. The word is said named when activation of the rightmost phoneme of the words phonological representation in the Phonemic Buffer reaches a prespecified criterion.

Non-words can be named by virtue of the *non-lexical* routine, a mechanism that converts graphemes into phonemes through a procedure that apply grapheme to phoneme conversion (GPC) rules of correspondence. There are three kinds of rules: Single Letter rules, which apply when a single letter maps into a single phoneme; Multi-Letters rules, which apply when a group of letter maps into a single phoneme; Context-Sensitive rules, which apply when preceding or following letters consistently determine the pronunciation of a given grapheme. This mechanism operates serially, left to right, on the output from the Letter level and activates phonemes in the Phonemic Buffer. Letter information becomes progressively available to the non-lexical route. At the first cycle, no letters are available. After a constant number of cycles (10) the first letter is assembled into a phoneme. After this, every 17 cycles another letter becomes available to the routine, until all the letters have been processed or the criterion has been reached.

The amount of activation and inhibition sent between and within levels as well as the relative weigh of the two routes in assembling the stimulus phonology is controlled by a set of 32 parameters. Moreover, using a single parameter set the DRC simulates 18 effects singled out in reading English. Processing fashion and architecture of the model are described in greater detail in Coltheart et al. [1].

Spread of activation

Among the components of the lexical route, activation spreads in a cascaded fashion (*cascaded processing* [20]). In models that operate by thresholded processing, as for example the logogen [21], the processing going on in any module does not begin to affect subsequent modules at an early point in processing; activation in only passed on to the later modules after a threshold is reached in the earlier module.

In models that operate by cascaded processing, as the DRC, there are no thresholds between modules; as soon as there is even the slightest activation in an early module this flows on to later modules. This way of spreading activation is, within the DRC framework, crucial to simulate a few effects such as the effect of orthographic neighborhood size in word reading aloud [22–24].

4 An Emergent Phenomenon: the Orthographic Neighborhood Size Effect

The orthographic neighborhood of a given word is the set of words that can be created by replacing one letter a time of that word. Different words can have neighborhood of different sizes (e.g. CART: calf, calm, card, care, carp, cars, cast, cert, coat, curt, dart, hart, mart, part, tart, wart; FROG: flog, from, grog). Given a word, the size of its orthographic neighborhood influences the time required to read it, indeed, as the number of neighbors increases, reading times decreases [22–24]. Within the DRC framework, this effect naturally emerges from its architecture and processing fashion, indeed "[...] cascaded processing in the model allows [words] to activate orthographically similar words in the orthographic lexicon, and this activation then feeds down to the phonological lexicon and finally to the phoneme system. Because generally the [neighbors'] units that became activated [in the lexicon] share phonemes with the stimulus, phonemic activation generated from the lexical route [...] should facilitate stimulus [reading]" [1].

5 Two Effects

Because of the cascaded processing, a word presented to the model activates all the orthographically similar words. Although the models does not implement any semantic module, cascaded processing allows us to make a rather straightforward prediction: since a word activates all the orthographically similar words in the orthographic lexicon, it activates their semantic representations as well. Two studies seem relevant here, one conducted by Rodd [4], one by Sears, Hino and Lupker [25].

Rodd [4] presented her participants with words, one at a time. They had to perform a *semantic decision*, that is they had to decide whether the words were the name of an animal or the name of something else by pressing one of two buttons. Among the stimuli she used there were name of non-animal things (*leotard*) that had the name of an animal as neighbor (*leopard*). She showed that participants took longer to reject words with an animal name as an orthographic neighbor with respect to words without that sort of neighbors. It must be concluded that *leotard* activated the semantics of *leopard* enough to interfere with the semantic decision process.

Sears, Hino and Lupker ([25], experiment 3; see also [26, 27]) had participants performing a animal/non-animal semantic decision on words varying for orthographic neighbor. Specifically, the stimuli belonging to the non-animal "category" – that is the stimuli requiring a No response – could have a dense or a sparse orthographic neighborhood. The authors observed a facilitatory effect of neighborhood size, that is words with a dense neighborhood were classified as non-animal faster than words with a sparse neighborhood.

A words, then, requires more time to be rejected if it has a neighbors belonging to the target category, less time if it has a dense orthographic neighborhood. This is consistent with what we said earlier in this paragraph: if a word activates its neighbors in the orthographic input lexicon, it also activates their semantic representations, and this influences the performance in semantic tasks. To provide an explanation for those effects we need a semantic module whose architecture is explicitly described, as explicitly described has to be the relations between the semantic module and the orthographic lexicon. Such a module will allow us to explain the above results, and to make new predictions.

6 A Semantic Module

The semantic module consists of a set of units. Each unit is connected with one unit in the orthographic lexicon (multiple mappings will be discussed in section 8). We assume that each single unit represents the meaning of the word it is connected with in the lexicon. Connections between the semantic units and the orthographic units are bidirectional and excitatory. Semantic units are organized by category: all the units representing meanings of words belonging to the same category (e.g. Biological Objects) are connected with a unit representing that category. Semantic units and category units are linked by bidirectional excitatory connections. Connections among category units are inhibitory. The system includes a decisional mechanism that monitors the activity of the category units: a words is recognized as belonging to a given category when the activation in the corresponding category unit passes that of the alternative category units by a given amount (criterion).

7 Two Explanations and a Prediction

Let us first consider how the model incorporating the semantic module can explain the Leotard effect and the orthographic neighbor size effect.

Leotard. The word *leotard* activates, along with its own lexical representation, the lexical representations of its orthographic neighbors. The lexical representation of *leopard*, then, receives activation. *Leotard* and, although to a smaller extent, *leopard* send activation to the semantic units they are connected with, which send activation to the category units they are connected with. Since both the animal category unit and the non-animal category unit receive activation, the competitions between those category units increases thus delaying the response.

Orthographic neighborhood size. The density of the neighborhood was manipulated only for the stimuli not belonging to the category of animals. If also the orthographic neighbors of the stimuli did not belong to the category of animals, then the explanation of the effect would easily follow: when the orthographic neighborhood is dense, more activity would be sent to the non-animal category unit because more lexical units are activated, causing its activity to grow faster with respect to when the words has only few orthographic neighbors. Therefore, the criterion would be reached faster, and the response made earlier.

Intuitively, it is *unlikely* that stimuli not belonging to the category of animals have neighbors belonging to the category of animals. For example, of the twenty randomly selected Italian words not belonging to the category of animals (*dosso, nastro, monte, mondo, miele, letto, lente, polo, fune, rischio, raggio, pianto, laccio, grotta, freno, alba, anta, vaso, vite, borsa*) with an average neighborhood size of eight words, only one (*laccio*) turned out having a neighbors (*luccio*) that is the name of an animal; the remaining 159 neighbors were not names of animals. Noteworthy, the situation is reversed if we shift our attention to the animal names. Indeed, it is *likely* that the orthographic neighbor of an animal name do not belong to the category of animals. Therefore, as the number of neighbors of an animal name increase, the number of orthographic neighbors of that name not belonging to the category of animal increases as well.

A prediction. If as the number of orthographic neighbors of an animal name increases, the number of neighbor not belonging to the category of animals increases, the model predicts an effect of orthographic neighborhood size opposite to that found by Sears et al., for the response "animal" being slower to stimuli with a dense neighborhood with respect to stimuli with a sparse neighborhood. The activation sent to the category of animals competes with that sent to the category of non-animals. Thus, as the number of neighbors not belonging to the category of animals increases, the competition increases, and the time taken to reach the criterion increases as well, delaying the response.

8 Some Data

To address this issue we designed an experiment (see Mulatti and Job [6] for details) where we compared the performance of two groups of participants in two tasks, a free reading and a conditional reading. In the free reading task the participants read all the words they are presented with. As already mentioned, in a reading aloud task the orthographic neighborhood size exerts a facilitation, that is words with many neighbors are read faster than words with few neighbors. In order to have responses comparable with those of the free reading task, rather than using a semantic decision (that requires a manual response) we decided to use a conditional reading task [28]. In such task, the participants have to read only the word belonging to a pre-specified category (e.g. animals) and to withhold the response otherwise. Thus, the conditional naming task involves a covert semantic decision, since it is only after having performed a semantic classification that the stimulus can be read, if it belongs to the pre-specified category, or the response withheld, if the stimulus does not belong to the pre-specified category. In our experiment, the participants performing the conditional reading task had to read only the words belonging to the category of Natural Objects.

The predictions are the following: a) in the free reading task, words with many neighbors are read faster than words with few neighbors; b) in the conditional reading task, words with many neighbors are read slower than words with few neighbors.

The material used in the experiment consisted of seventy-two low frequency words (note: one of the item was removed from the analyses as nearly half of the participants di not recognized it as a word). Half of them were names of things belonging to the category of Natural Objects (NO), half were names of things belonging to the category of Artefacts (A). In a preliminary test, a pool of participant that did not participate to the main experiment score the typicality of each item as a member of the assigned category. No differences between categories resulted in the analysis of the scores distribution. The experimental items were those of the Natural Objects category. Eighteen experimental items had a dense neighborhood (mean: 13.4), eighteen experimental items had a sparse neighborhood (mean: 3.5). Stimuli in the dense and sparse conditions were balanced in terms of typicality, written frequency, and letter length. Noteworthy, the ratio computed comparing the number of neighbors not belonging to the category of Natural Objects with the number of neighbors belonging to the category of Natural Objects was of 5.6 for the dense neighborhood stimuli, and of 1.8 for the sparse neighborhood stimuli, $t(33) = 4.2, p < .001$.

Fourteen participants performed the free reading task. Each participant was asked to read all the words he was presented with as quickly and accurately as possible. The words appeared in the center of a computer screen and stayed on until participant responded. The order of presentation of the stimuli was randomized for each participant. The durations of the intervals between the appearance of the stimuli and the onset of the verbal responses constituted the dependent variable (Reaction Times, RTs).

Twelve participants performed the conditional reading task. They were told to read, as quickly and accurately as possible, only the words denoting objects belonging to the category of Natural Objects and to remain silent otherwise. The order of presentation of stimuli was randomized for each participant. The stimuli appeared in the center of the screen, and stayed on until participant responded. RTs were measured.

Statistical analyses performed on the RTs of correct responses showed that the free reading task was significantly faster than the conditional reading task, and that the main effect of Orthographic Size did not prove significant. Consistently with the prediction, the interaction between the two factors was significant: Whereas in the free reading task words with a dense neighborhood were read faster than words with a sparse neighborhood, in the conditional reading task words with a dense neighborhood were read more slowly than words with a sparse neighborhood.

The results can be summarized as follow. In the free reading, a task that does not (explicitly) require semantic information to be performed, as the number of orthographic neighbors increases, the time to produce a response decreases: the orthographic neighborhood size exerts a facilitatory effect. When the task requires a cover semantic classification, as in the conditional naming task, the characteristics of the semantic representations of the neighbors came into play. Sears et al. [25] showed that as the number of neighbors belonging to the same category as the target word increases, the process of classifying the word is facilitated. We took this as an evidence for our semantic module: the neighbors send activation to the same category unit as the target word; because of this, the activation in that unit reaches the Criterion for the response faster when the target word has many neighbors. On the other hand, we showed that the increase of the number of neighbors that belong to a category different from that of the target word hinders the semantic classification process. We explained this phenomenon within our framework by postulating that the semantically inconsistent neighbors send activation to a category unit that compete with that of the target word, thus slowing the decision process.

9 Multiple Mapping from Orthography to Semantics, and Vice Versa

The semantic model we described in section 6. posits that each orthographic unit maps into one single semantic unit, and that each single semantic unit maps into one single orthographic unit. However, Italian (as many other languages including English) counts both ambiguous[7] words, i.e. words that have more than one meaning, and synonyms, i.e. words that have (roughly) the same meaning. An example of the first class of words would be bank (the rising ground bordering a lake or a river; an establishment for the custody of money); examples of the latter would be couch and sofa, which refer to the same thing. These two classes of words pose a problem for the semantic model

---

[7] Psycholinguistics identified two groups of ambiguous words, homonyms words – different meanings – and polysemous words – the meanings that correspond to a polysemous word share a common core meaning. However, Klein and Murphy [29] showed that even to polysemous words correspond different semantic representations. Because of this, we will include under the same label "ambiguous words" both homonyms and polysemous.

we have proposed, because such words influence behavior in idiosyncratic ways and need to be treated as a specific class of words. Specifically, in lexical decision tasks, where participants have to classify strings of letters as word or non-word, ambiguous word are recognized faster than unambiguous words, whereas synonyms are recognized more slowly than non-synonyms [30]. Thus, the so-called ambiguity effect results in a facilitation, the synonymy effect in an interference.

To accommodate for such effects, the way in which the orthographic units are connected with the corresponding semantic units in the model needs to be changed as follows. The orthographic representation of an unambiguous non-synonymic word maps into one single semantic representation. The orthographic representation of an ambiguous word maps into as many semantic representations as the number of meanings that word has. The orthographic representation of a synonyms maps into one single semantic representation, however, this semantic representation maps into as many orthographic representations as the number of synonyms of that word. Such a modified model accounts for both effects.

The ambiguity effect arises because the orthographic representation of an ambiguous word sends activation to more than one semantic units which feed back activation to the orthographic unit. Therefore, since the orthographic unit of an ambiguous word receives activation from many semantic units, its activation grows faster compared to a unit representing an unambiguous non-synonymic word, facilitating its recognition.

The synonymic effect arises because the semantic representation activated by the synonyms sends activation back to the target synonyms but also to the synonyms of the target. Since there is lateral inhibition among units in the orthographic lexicon, the orthographic unit of the non-target synonyms sends inhibition to the orthographic unit of the target synonyms, slowing the raise of its activation, thus hindering its recognition.

10 Concluding Remarks

Orthographic relationship among words affect semantic processing of those words. This is a sufficient reason to implement a semantic component in a model of reading, as not doing so would prevent the model to account for the available empirical data. However, in addition to this obvious outcome in term of explanatory adequacy, there are several nice side-effects that stems from the enterprise. In fact, implementing a semantic module in the DRC model has allowed us to reach several goals:

a. to test for the reliability of the model in contexts different from those the model was originally built for;
b. to define the minimum computational apparatus needed to simulate semantic effects;

c. to evaluate possible interactions among the already implemented components – namely the orthographic system – and the semantic module;
d. to test the plausibility of the explanations provided to account for the semantic effects obtained in behavioral experiments.

The latter point (d) is worth stressing. The relationship between orthography and semantics was the focus of our original work [6], and it could be considered an instance of the goal (a) since it showed that the DRC model, when incorporating the semantic module we developed, could account for effects of semantic tasks, tasks the model was not developed to account for. The extension of the model to the processing of ambiguous words and synonyms, however, is a further step. In this case, using the same functional architecture and processing assumptions but postulating specific ways of linking the units in the orthographic system and the units in the semantic system, we were able to account for purely semantic phenomena, i.e. ambiguity and synonymity.

References

1. Coltheart, M., Rastle, K., Perry, C., Langdon, R., Ziegler, J.C.: DRC: A dual route cascaded model of visual word recognition and reading aloud. *Psychological Review* **108** (2001) 204–256
2. Hino, Y., Lupker, S.J.: Effects of polysemy in lexical decision and naming: an alternative to lexical access accounts. *Journal of Experimental Psychology: Human Perception and Performance* **22** (1996) 1331–1336
3. Pecher, D.: Perception is a two-way junction: Feedback semantics in word recognition. *Psychonomic Bulletin and Review* **8** (2001) 545–551
4. Rodd, J.: When do leotards get their spots? Semantic activation of lexical neighbors in visual word recognition. *Psychonomic Bulletin and Review* **11** (2004) 434–439
5. Forster, K.I., Hector, J.: Cascaded versus noncascaded models of lexical and semantic processing: The turple effect. *Memory and Cognition* **30** (2002) 1106–1116
6. Mulatti, C., Job, R.: Considerazioni preliminari sull'implementazione di un modulo semantico in un modello computazionale della lettura [Preliminary considerations on the implementation of a semantic module in a computational model of reading]. In P. Giarretta, P. Cherubini, M. Marraffa, eds.: *Cognizione e computazione.* CLEUP, Padova.
7. Broadbent, D.E.: Simple models for experimentable situations. In P.Morris, ed.: *Modelling cognition.* Wiley, New York (1987) 169–185
8. Massaro, D.W.: Understanding mental processes through modeling: Possibilities and limitations. In M. Besson, P. Courrieu, C. Frenck-Mestre, A.M. Jacobs, J. Pynte, eds.: *Language perception and comprehension: Multidisciplinary approaches.* Marseille, France: Centre National de la Recherche Scientifique, Laboratoire de Neurosciences Cognitives (1992) 17–18
9. Jacobs, A.M., Grainger, J.: Models of visual word recognition – Sampling the state of the art. *Journal of Experimental Psychology* **20** (1994) 1311–1334

10. Chomsky, N.: *Aspects of the Theory of Syntax.* The MIT Press, Cambridge, MA (1965)
11. Norris, D.: (2005) How do computational models help us build better theories? In A. Cutler, ed: *Twenty-First Century Psycholinguistics: Four Cornerstones* (2005)
12. Marr, D.: *Vision.* Freeman, San Francisco (1982)
13. Olson, A., Caramazza, A.: The role of cognitive theory in neuropsychological research. In F. Boller, G. Gratman, eds.: *Handbook of neuropsychology.* Amsterdam, Elsevier (1991) 287–309
14. Lukatela, G., Turvey, M.T.: Visual lexical access is initially phonological: 1. evidence from associative priming by words, homophones, and pseudohomophones. *Journal of Experimental Psychology: General* **123** (1994) 107–128
15. Lukatela, G., Turvey, M.T.: Visual lexical access is initially phonological: 2. evidence from phonological priming by homophones and pseudohomophones. *Journal of Experimental Psychology: General* **123** (1994) 331–353
16. Plaut, D.C., McClelland, J.L., Seidenberg, M.S., Patterson, K.E.: Understanding normal and impaired word reading: Computational principles in quasi-regular domains. *Psychological Review* **103** (1999) 56–115
17. Cree, G.S., McRae, K.: Analyzing the factors underlying the structure and computation of the meaning of chipmunk, cherry, chisel, cheese and cello (and many other such concrete nouns). *Journal of Experimental Psychology: General* **132** (2003) 163–201
18. Sartori, G., Lombardi, L.: Semantic relevance and semantic disorders. *Journal of Cognitive Neuroscience* **16** (2004) 439–452
19. Vigliocco, G., Vinson, D.P., Lewis, W.: Representing the meanings of objects and actions words: the featural and unitari semantic space hypotesis. *Cognitive Psychology* **48** (2004) 422–488
20. McClelland, J.L.: On the time relation of mental processes: An examination of systems of processes in cascade. *Psychological Review* **86** (1979) 287–330
21. Morton, J.: Reading, Context and the Perception of Words. Unpublished PhD thesis, University of Reading, Reading, England (1961)
22. Andrews, S.: Frequency and neighborhood effects on lexical access: Activation or search? *Journal of Experimental Psychology: Learning, Memory and Cognition* **15** (1989) 802–814
23. Andrews, S.: Frequency and neighborhood effects on lexical access: Lexical similarity or orthographic redundancy? *Journal of Experimental Psychology: Learning, Memory, and Cognition* **18** (1992) 234–254
24. Sears, C.R., Hino, Y., Lupker, S.J.: Neighborhood size and neighborhood frequency effects in word recognition. *Journal of Experimental Psychology: Human Perception and Performance* **21** (1995) 876–900
25. Sears, C.R., Lupker, S., Hino, Y.: Orthographic neighborhood effects in perceptual identification and semantic categorization tasks: A test of the multiple read-out model. *Perception and Psychophysics* **61** (1999) 1537–1554
26. Forster, K.I., Shen, D.: No enemies in the neighborhood: Absence of inhibitory neighborhood effects in lexical decision and semantic categorization. *Journal of Experimental Psychology: Learning, Memory, and Cognition* **22** (1996) 696–713
27. Carreiras, M., Perea, M., Grainger, J.: Effects of the orthographic neighborhood in visual word recognition: Cross-task comparisons. *Journal of Experimental Psychology: Learning, Memory, and Cognition* **23** (1997) 857–871

28. Job, R., Tenconi, E.: Naming pictures at no cost: Asymmetries in picture and word conditional naming. *Psychonomic Bulletin and Review* **9** (2002) 790–794
29. Klein, D.E., Murphy, G.L.: The representation of polysemous words. *Journal of Memory and Language* **45** (2001) 259–282
30. Hino, Y., Lupker, S.J., Pexman, P.M.: Ambiguity and synonymy effects in lexical decision, naming and semantic catgorizationt tasks: Interactions between otrhography, phonology, and semantics. *Journal of Experimental Psychology: Learning, Memory, and Cognition* **28** (2002) 686–713

Printing: Krips bv, Meppel
Binding: Stürtz, Würzburg